Series Associated with the Zeta and Related Functions

Series Associated with the Zeta and Related Functions

by

H.M. Srivastava

University of Victoria,
Department of Mathematics & Statistics,
Victoria, British Columbia, Canada

and

Junesang Choi

Dongguk University,
Department of Mathematics,
Kyongju, Republic of Korea

Springer-Science+Business Media, B.V.

A C.I.P. Catalogue record for this book is available from the Library of Congress.

Printed on acid-free paper

ISBN 978-90-481-5728-0 ISBN 978-94-015-9672-5 (eBook)
DOI 10.1007/978-94-015-9672-5

Originally published by Kluwer Academic Publishers in 2001.
Softcover reprint of the hardcover 1st edition 2001

Table of Contents

Preface

In recent years there has been an increasing interest in problems involving closed-form evaluations of (and representations of the Riemann Zeta function at positive integer arguments as) various families of series associated with the Riemann Zeta function $\zeta(s)$, the Hurwitz Zeta function $\zeta(s,a)$, and their such extensions and generalizations as (for example) Lerch's transcendent (or the Hurwitz–Lerch Zeta function) $\Phi(z,s,a)$. Some of these developments have apparently stemmed from an over two-century-old theorem of Christian Goldbach (1690–1764), which was stated in a letter dated 1729 from Goldbach to Daniel Bernoulli (1700–1782), from recent rediscoveries of a fairly rapidly convergent series representation for $\zeta(3)$, which is actually contained in a 1772 paper by Leonhard Euler (1707–1783), and from another known series representation for $\zeta(3)$, which was used by Roger Apéry (1916–1994) in 1978 in his celebrated proof of the irrationality of $\zeta(3)$.

This book is motivated essentially by the fact that the theories and applications of the various methods and techniques used in dealing with many different families of series associated with the Riemann Zeta function and its aforementioned relatives are to be found so far only in widely scattered journal articles. Thus our systematic (and unified) presentation of these results on the evaluation and representation of the Zeta and related functions is expected to fill a conspicuous gap in the existing books dealing exclusively with these Zeta functions.

The main objective of this book is to provide a systematic collection of various families of series associated with the Riemann and Hurwitz Zeta functions as well as with many other higher transcendental functions which are closely related to these functions. It aims at presenting a state-of-the-art account of the theory and applications of many different methods (which are available in the rather scattered literature on this subject) for the derivation of the types of results considered here.

In our attempt to make this book as self-contained as possible within the obvious constraints, we include in Chapter 1 (Introduction and Preliminaries) a reasonably detailed account of such useful functions as the Gamma and Beta functions, the Polygamma and related functions, multiple Gamma functions, the Gauss hypergeometric function and its familiar generalization, the Stirling numbers of the first and second kind, and the Bernoulli and Euler numbers and polynomials. In Chapter 2 (The Zeta and Related Functions), on the other hand, we present the definitions and various potentially useful properties (and characteristics) of the Riemann, Hurwitz, and Hurwitz-Lerch Zeta functions, the Polylogarithm and related functions, and the multiple Zeta functions.

In Chapter 3 we begin by providing a brief historical introduction to the main subject of this book. We then describe and illustrate some of the most effective methods of evaluating series associated with the Zeta and related functions. Further developments on the evaluations and (rapidly convergent) series representations of $\zeta(s)$ when $s \in \mathbb{N} \setminus \{1\}$ are presented in Chapter 4 which also deals with various computational results on this subject.

Chapter 5 considers the problem involving computations of the determinants of the Laplacians for the n-dimensional sphere $\mathbf{S}^n$ $(n \in \mathbb{N})$. It is here in this chapter that we show how fruitfully some of the series evaluations (which are presented in the earlier chapters) can be applied in the solution of the aforementioned problem.

The last chapter (Chapter 6) contains a wide variety of miscellaneous results dealing with (for example) the evaluations of the Bernoulli and Euler polynomials at rational arguments, the closed-form summation of several classes of trigonometric series, and the evaluation of certain integrals by means of the Euler-Maclaurin summation formula.

Each chapter in this book begins with a brief outline summarizing the material presented in the chapter, and is then divided into a number of sections. Equations in every section are numbered separately. While referring to an equation in another section of the book, we use numbers like 3.2(18) to represent Equation (18) in Section 3.2 (that is, the second section of Chapter 3).

At the close of each chapter, we have provided a set of carefully-selected problems which are based essentially upon the material presented in the chapter. Many of these problems are generally taken from recent research publications, and (in all such instances) we have chosen to include the precise references for further investigation (if necessary). Another valuable feature of this book is the extensive and up-to-date bibliography on the subject dealt with in the book.

This book is written primarily as a reference work for various seemingly diverse groups of research workers and other users of series associated with the Zeta and related functions. In particular, teachers, researchers, and postgraduate students in the fields of mathematical and applied sciences will find this book to be especially useful, not only for its detailed and systematic presentations of the theory and applications of the various methods and techniques used in dealing with many different classes of series associated with the Zeta and related functions, or for its stimulating historical accounts of a large number of problems considered here, but also for its well-classified tables of series (and integrals) and its well-motivated presentation of many sets of closely related problems with their precise bibliographical references (if any).

Acknowledgements

Many persons have contributed rather significantly to this book, both directly *and* indirectly. Contribution of subject matter is duly acknowledged throughout the text *and* in the bibliography. Indeed we are greatly indebted to the various authors whose works we have freely consulted and who occasionally provided invaluable references and advice serving for the enrichment of the matter presented in this book. The first-named author wishes to express his deep sense of gratitude to his wife *and* colleague, Professor Rekha Srivastava, for her co-operation and support throughout the writing of this book.

The collaboration of the authors on this project was conceptualized as long ago as August 1995, and the preparation of a preliminary outline was initiated in December 1997, during the first-named author's visits to Dongguk University at Kyongju. The first drafts of some of the chapters in this book were written during several subsequent visits of the first-named author to Dongguk University at Kyongju. The final drafts of most of the chapters in this book were prepared during the second-named author's visit to the University of Victoria from August 1999 to August 2000 while he was on Study Leave from Dongguk University at Kyongju. Our sincere thanks are due to the appropriate authorities of each of these universities, to the Korea Research Foundation (Support for Faculty Research Abroad under its Research Fund Program), and to the Natural Sciences and Engineering Research Council of Canada, for providing financial support and other facilities for the completion of this project.

We take this opportunity to express our thanks to the editorial (and technical) staff of Kluwer Academic Publishers (especially the Publishing Editor, Dr. E. A. Liesbeth Mol) for their continued interest in this book and for their proficient (and impeccable) handling of its publication. We also greatly appreciate the generous help and assistance provided by Ms. Ariana Clapton (of the University of Victoria) in the preparation of the *camera-ready* typescript of this book.

Finally, we should like to record our indebtedness to the members of our respective families for their understanding, co-operation, and support throughout this project. The second-named author and his family would, especially, like to express their appreciation for the first-named author and his family's hospitality and every prudent consideration during their stay in Victoria for over one year.

H. M. Srivastava
University of Victoria
Canada

Junesang Choi
Dongguk University
Korea

November 2000

CHAPTER 1

INTRODUCTION AND PRELIMINARIES

In this introductory chapter, we present the definitions and notations (and some of the important properties and characteristics) of the various special functions, polynomials, and numbers, which are potentially useful in the remainder of the book. The special functions considered here include (for example) the Gamma, Beta, and related functions, the Polygamma functions, the multiple Gamma functions, the Gaussian hypergeometric function, and the generalized hypergeometric function. We also consider the Stirling numbers of the first and second kind, the Bernoulli and Euler polynomials and numbers, and the generalized Bernoulli and Euler polynomials and numbers. Relevant connections of some of these functions with other special functions and polynomials, which are not listed above, are also presented here.

1.1 Gamma and Beta Functions

The Gamma function

The origin of the *Gamma function* can be traced back to two letters from Leonhard Euler (1707–1783) to Christian Goldbach (1690–1764), just as a simple desire to extend factorials to values between the integers. The first letter (dated October 13, 1729) dealt with the interpolation problem, while the second letter (dated January 8, 1730) dealt with integration and tied the two together.

The Gamma function $\Gamma(z)$ developed by Euler is usually defined by

$$\Gamma(z) := \int_0^\infty e^{-t}\, t^{z-1}\, dt \quad (\Re(z) > 0). \tag{1}$$

We also present here several equivalent forms of the Gamma function $\Gamma(z)$, one by Weierstrass:

$$\Gamma(z) = \frac{e^{-\gamma z}}{z} \prod_{k=1}^{\infty} \left\{ \left(1 + \frac{z}{k}\right)^{-1} e^{z/k} \right\} \tag{2}$$
$$(z \in \mathbb{C} \setminus \mathbb{Z}_0^-;\ \mathbb{Z}_0^- := \{0,\ -1,\ -2,\ \ldots\}),$$

where γ denotes the Euler-Mascheroni constant defined by

$$\gamma := \lim_{n\to\infty} \left(\sum_{k=1}^{n} \frac{1}{k} - \log n \right) \cong 0.57721\,56649\,01532\,8606\,06512\ldots, \tag{3}$$

and the other by Gauss:

$$\Gamma(z) = \lim_{n\to\infty}\left\{\frac{(n-1)!\,n^z}{z(z+1)\cdots(z+n-1)}\right\}$$
(4)
$$= \lim_{n\to\infty}\left\{\frac{n!\,(n+1)^z}{z(z+1)\cdots(z+n)}\right\}$$
$$= \lim_{n\to\infty}\left\{\frac{n!\,n^z}{z(z+1)\cdots(z+n)}\right\}$$
$$(z \in \mathbb{C}\setminus\mathbb{Z}_0^-),$$

since

$$\lim_{n\to\infty}\frac{n}{z+n} = 1 = \lim_{n\to\infty}\frac{n^z}{(n+1)^z}.$$

In terms of the Pochhammer symbol $(\lambda)_n$ defined (for $\lambda \in \mathbb{C}$) by

(5) $$(\lambda)_n := \begin{cases} 1 & (n=0) \\ \lambda(\lambda+1)\ldots(\lambda+n-1) & (n \in \mathbb{N} := \{1, 2, 3, \ldots\}), \end{cases}$$

the definition (4) can easily be written in an equivalent form:

(6) $$\Gamma(z) = \lim_{n\to\infty}\frac{(n-1)!\,n^z}{(z)_n} \quad (z \in \mathbb{C}\setminus\mathbb{Z}_0^-).$$

By taking the reciprocal of (2) and applying the definition (3), we have

$$\begin{aligned}
\frac{1}{\Gamma(z)} &= z\left[\lim_{n\to\infty}\exp\left\{\left(1+\frac{1}{2}+\cdots+\frac{1}{n}-\log n\right)z\right\}\right]\left[\lim_{n\to\infty}\prod_{k=1}^{n}\left\{\left(1+\frac{z}{k}\right)e^{-z/k}\right\}\right] \\
&= z\lim_{n\to\infty}\left[\exp\left\{\left(1+\frac{1}{2}+\cdots+\frac{1}{n}-\log n\right)z\right\}\cdot\prod_{k=1}^{n}\left\{\left(1+\frac{z}{k}\right)e^{-z/k}\right\}\right] \\
&= z\lim_{n\to\infty}\left\{n^{-z}\prod_{k=1}^{n}\left(1+\frac{z}{k}\right)\right\} \\
&= z\lim_{n\to\infty}\left[\left\{\prod_{k=1}^{n-1}\left(1+\frac{1}{k}\right)^{-z}\right\}\left\{\prod_{k=1}^{n}\left(1+\frac{z}{k}\right)\right\}\right] \\
&= z\prod_{k=1}^{\infty}\left\{\left(1+\frac{z}{k}\right)\left(1+\frac{1}{k}\right)^{-z}\right\},
\end{aligned}$$

which yields Euler's product form of the Gamma function:

(7) $$\Gamma(z) = \frac{1}{z}\prod_{k=1}^{\infty}\left\{\left(1+\frac{1}{k}\right)^{z}\left(1+\frac{z}{k}\right)^{-1}\right\}.$$

When t in (1) is replaced by $-\log t$, (1) is also written in an equivalent form:

(8) $$\Gamma(z) = \int_0^1\left(\log\frac{1}{t}\right)^{z-1}dt \quad (\Re(z) > 0).$$

This representation of the Gamma function as well as the symbol Γ are attributed to Legendre.

Integration of (1) by parts easily yields the functional relation:

$$\Gamma(z+1) = z\,\Gamma(z), \tag{9}$$

so that, obviously,

$$\Gamma(z) = \frac{\Gamma(z+n)}{z(z+1)\cdots(z+n-1)} \qquad (n \in \mathbb{N}_0 := \mathbb{N} \cup \{0\}) \tag{10}$$

which enables us to define $\Gamma(z)$ for $\Re(z) > -n$ $(n \in \mathbb{N}_0)$ as an analytic function except for $z = 0, -1, -2, \ldots, -n+1$. Thus $\Gamma(z)$ can be continued analytically to the whole complex z–plane except for simple poles at $z \in \mathbb{Z}_0^-$.

The representation (2) in conjunction with the well-known product formula:

$$\sin \pi z = \pi z \prod_{n=1}^{\infty} \left(1 - \frac{z^2}{n^2}\right) \tag{11}$$

also yields the following useful relationship between the Gamma and circular functions:

$$\Gamma(z)\,\Gamma(1-z) = \frac{\pi}{\sin \pi z} \qquad (z \notin \mathbb{Z} := \{0, \pm 1, \pm 2, \ldots\}), \tag{12}$$

which incidentally provides an immediate analytic continuation of $\Gamma(z)$ from right to the left half of the complex z-plane.

Several special values of $\Gamma(x)$, when x is real, are worthy of note. Indeed, from (1) we note that

$$\Gamma(1) = \int_0^\infty e^{-t}\,dt = 1, \tag{13}$$

and that $\Gamma(x) > 0$ for all x in the open interval $(0, \infty)$. Thus (12) with $z = \frac{1}{2}$ immediately yields

$$\Gamma\left(\frac{1}{2}\right) = \sqrt{\pi}, \tag{14}$$

which, in view of (1), implies

$$\int_0^\infty \frac{e^{-t}}{\sqrt{t}}\,dt = \sqrt{\pi} \tag{15}$$

or, equivalently,

$$\int_0^\infty \exp\left(-t^2\right)\,dt = \frac{\sqrt{\pi}}{2}. \tag{16}$$

By making use of the relation (10), we obtain

$$\tag{17} \begin{gathered} \Gamma(n+1) = n!; \quad \Gamma\left(n+\frac{1}{2}\right) = \frac{(2n)!\,\sqrt{\pi}}{2^{2n}\,n!}; \\ \Gamma\left(-n+\frac{1}{2}\right) = (-1)^n\,\sqrt{\pi}\,\frac{2^{2n}\,n!}{(2n)!}; \quad (n \in \mathbb{N}_0). \end{gathered}$$

The last two results in (17) also require the use of (14). The formulas listed under (17) enable us to compute $\Gamma(x)$ when x is a positive integer and when x is half an odd integer, positive or negative.

From (2) or (4), it also follows that $\Gamma(z)$ is a meromorphic function on the whole complex z–plane with simple poles at $z = -n$ $(n \in \mathbb{N}_0)$ with their respective residues given by

$$\tag{18} \operatorname*{Res}_{z=-n} \Gamma(z) = \frac{(-1)^n}{n!} \quad (n \in \mathbb{N}_0).$$

Since the function $\Gamma(1/w)$ has simple poles at $w = -1, -\frac{1}{2}, -\frac{1}{3}, \ldots$, which implies that $w = 0$ is an accumulation point of the poles of $\Gamma(1/w)$, the Gamma function $\Gamma(z)$ has an essential singularity at infinity. Furthermore, it follows immediately from (2) that $1/\Gamma(z)$ has no poles, and therefore $\Gamma(z)$ is never zero.

Euler-Mascheroni constant γ

The observation that the series:

$$1 + \frac{1}{2} + \frac{1}{3} + \cdots$$

is divergent is attributed by Jakob Bernoulli (1654–1705) to his brother, Johann Bernoulli (1667–1748) [see Glaisher (1871)], but the connection between the sum: $1 + \frac{1}{2} + \frac{1}{3} + \cdots + \frac{1}{x}$ and $\log x$ was first established by Euler who gave the formula:

$$\tag{19} 1 + \frac{1}{2} + \frac{1}{3} + \cdots + \frac{1}{x} = \gamma + \log x + \frac{1}{2x} - \frac{B_1}{2x^2} + \frac{B_2}{4x^4} - \frac{B_3}{6x^6} + \cdots,$$

where B_1, B_2, B_3, ... are the Bernoulli numbers (see Section 1.6). By setting $x = 10$ in (19), Euler computed the numerical value of γ given by (3).

The true nature of the Euler-Mascheroni constant γ (whether it is an algebraic or a transcendental number) is not known. Indeed it can be expressed as many interesting definite integrals some of which are listed here:

$$\tag{20} \gamma = \int_0^1 \frac{1-e^{-t}}{t}\,dt - \int_1^\infty \frac{e^{-t}}{t}\,dt;$$

$$\tag{21} \gamma = \int_0^\infty e^{-t}\left(\frac{1}{1-e^{-t}} - \frac{1}{t}\right) dt,$$

which is due to Gauss;

(22) $$\gamma = \int_0^1 \left(\frac{1}{\log t} + \frac{1}{1-t} \right) dt;$$

(23) $$\gamma = \int_0^1 \left(1 - e^{-t} - e^{-1/t} \right) \frac{dt}{t};$$

(24) $$\gamma = \int_0^\infty \left(\frac{1}{1+t} - e^{-t} \right) \frac{dt}{t},$$

which is due to Dirichlet;

(25) $$\gamma = \frac{1}{2} + 2 \int_0^\infty \frac{t}{1+t^2} \frac{dt}{e^{2\pi t} - 1},$$

which is called *Poisson's expression* for γ;

(26) $$\gamma = \int_0^1 \left(t - \frac{1}{1 - \log t} \right) \frac{dt}{t \log t};$$

(27) $$\gamma = -\int_0^\infty e^{-t} \log t \, dt = -\int_0^1 \log \left(\log \frac{1}{t} \right) dt.$$

The following integral representations for γ contain circular functions in their integrands:

(28) $$\gamma = -\int_0^\infty \left(\cos t - \frac{1}{1+t^2} \right) \frac{dt}{t};$$

(29) $$\gamma = 1 - \int_0^\infty \left(\frac{\sin t}{t} - \frac{1}{1+t} \right) \frac{dt}{t};$$

(30) $$\gamma = -2 \int_0^\infty \left(\cos t - e^{-t^2} \right) \frac{dt}{t};$$

(31) $$\gamma = \log 2 - \pi \int_0^1 \int_0^{\frac{1}{2}} \tan \left(\frac{\pi t}{2} \right) \left(\frac{\sin \pi t u}{\sin \pi u} - t \right) du dt,$$

which is due to Mikolás [see Campbell (1966, p. 216)];

(32) $$\gamma = \frac{1}{2} + 2 \int_0^\infty \frac{\sin(\tan^{-1} t)}{(e^{2\pi t} - 1) \sqrt{1+t^2}} dt;$$

(33) $$\gamma = \log 2 - 2 \int_0^\infty \frac{\sin(\tan^{-1} t)}{(e^{2\pi t} + 1) \sqrt{1+t^2}} dt,$$

which is due to Jensen [see Erdélyi *et al.* (1953, p. 33, Eq. (13))];

Darling (1997) gave the following integral representation for γ :

$$\gamma = 1 + \int_0^\infty \left(\cos t - \frac{\sin t}{t}\right) \frac{\log t}{t}\, dt, \tag{34}$$

which was proposed earlier by Anglesio (1996a). Anglesio (1996b) also proposed two more integral representations for γ :

$$\gamma = \int_0^\infty \left(\frac{1}{2} - \cos t\right) \frac{dt}{t} \tag{35}$$

and

$$\gamma = \int_0^\infty \left(\frac{1}{1+t} - \cos t\right) \frac{dt}{t}, \tag{36}$$

which is the corrected version of his second proposed formula [see Choi and Seo (1998)].

Knopp (1951) obtained the following expression for γ :

$$\gamma = \frac{1}{2} + \frac{B_2}{2} + \frac{B_4}{4} + \cdots + \frac{B_{2n}}{2n} - (2n+1)! \int_1^\infty \frac{Q_{2n+1}(x)}{x^{2n+2}} dx, \tag{37}$$

where the functions $Q_n(x)$ are defined by

$$Q_n(x) := \begin{cases} x - \dfrac{1}{2} & (n = 1;\ 0 < x < 1), \\ \dfrac{1}{n!} B_n(x) & (n \in \mathbb{N} \setminus \{1\};\ 0 \leq x \leq 1), \end{cases} \tag{38}$$

$B_n := B_n(0)$ and $B_n(x)$ being the Bernoulli numbers and polynomials, respectively (see Section 1.6). As observed by Knopp (1951) by an explicit example with $n = 3$ in (37), the approximate value of γ can easily be calculated with much greater accuracy than before (and, theoretically, to any degree of accuracy whatever) by means of the formula (37).

Pochhammer's symbol and the factorial function

Since $(1)_n = n!$, the Pochhammer symbol $(\lambda)_n$ defined by (5) may be looked upon as a generalization of the elementary factorial; hence the symbol $(\lambda)_n$ is also referred to as the *shifted* factorial.

In terms of the Gamma function, we have [*cf.* Definition (5)]

$$(\lambda)_n = \frac{\Gamma(\lambda + n)}{\Gamma(\lambda)} \quad (\lambda \in \mathbb{C} \setminus \mathbb{Z}_0^-), \tag{39}$$

which can easily be verified. Furthermore, the binomial coefficient may now be expressed as

$$\binom{\lambda}{n} = \frac{\lambda(\lambda-1)\cdots(\lambda-n+1)}{n!} = \frac{(-1)^n (-\lambda)_n}{n!} \tag{40}$$

or, equivalently, as

$$\binom{\lambda}{n} = \frac{\Gamma(\lambda+1)}{n!\,\Gamma(\lambda-n+1)}. \tag{41}$$

It follows from (40) and (41) that

$$\frac{\Gamma(\lambda+1)}{\Gamma(\lambda-n+1)} = (-1)^n\,(-\lambda)_n,$$

which, for $\lambda = \alpha - 1$, yields

$$\frac{\Gamma(\alpha-n)}{\Gamma(\alpha)} = \frac{(-1)^n}{(1-\alpha)_n} \quad (\alpha \notin \mathbb{Z}). \tag{42}$$

Equations (39) and (42) suggest the definition:

$$(\lambda)_{-n} = \frac{(-1)^n}{(1-\lambda)_n} \quad (n \in \mathbb{N};\ \lambda \notin \mathbb{Z}). \tag{43}$$

Equation (39) also yields

$$(\lambda)_{m+n} = (\lambda)_m(\lambda+m)_n, \tag{44}$$

which, in conjunction with (43), gives

$$(\lambda)_{n-k} = \frac{(-1)^k\,(\lambda)_n}{(1-\lambda-n)_k} \quad (0 \leq k \leq n). \tag{45}$$

For $\lambda = 1$, we have

$$(n-k)! = \frac{(-1)^k\,n!}{(-n)_k} \quad (0 \leq k \leq n), \tag{46}$$

which may alternatively be written in the form:

$$(-n)_k = \begin{cases} \dfrac{(-1)^k\,n!}{(n-k)!} & (0 \leq k \leq n), \\ 0 & (k > n). \end{cases} \tag{47}$$

Multiplication formulas of Legendre and Gauss

In view of the definition (5), it is not difficult to show that

$$(\lambda)_{2n} = 2^{2n}\left(\frac{1}{2}\lambda\right)_n\left(\frac{1}{2}\lambda+\frac{1}{2}\right)_n \quad (n \in \mathbb{N}_0), \tag{48}$$

which follows also from Legendre's duplication formula for the Gamma function, *viz*

$$\sqrt{\pi}\,\Gamma(2z) = 2^{2z-1}\,\Gamma(z)\,\Gamma\left(z+\frac{1}{2}\right) \quad \left(z \neq 0, -\frac{1}{2}, -1, -\frac{3}{2}, \ldots\right). \tag{49}$$

For every positive integer m, we have

$$(\lambda)_{mn} = m^{mn} \prod_{j=1}^{m} \left(\frac{\lambda + j - 1}{m}\right)_n \qquad (m \in \mathbb{N};\ n \in \mathbb{N}_0), \tag{50}$$

which reduces to (48) when $m = 2$. Starting from (50) with $\lambda = mz$, it can be proved that

$$\begin{gathered} \Gamma(mz) = (2\pi)^{\frac{1}{2}(1-m)}\, m^{mz-\frac{1}{2}} \prod_{j=1}^{m} \Gamma\left(z + \frac{j-1}{m}\right) \\ \left(z \neq 0, -\frac{1}{m}, -\frac{2}{m}, \ldots;\ m \in \mathbb{N}\right), \end{gathered} \tag{51}$$

which is known in the literature as Gauss's multiplication theorem for the Gamma function.

Stirling's formula for $n!$ and its generalizations

For a large positive integer n, it naturally becomes tedious to compute $n!$. An easy way of computing an approximate value of $n!$ for large positive integer n was initiated by Stirling in 1730 and modified subsequently by De Moivre, who showed that

$$n! \sim \left(\frac{n}{e}\right)^n \sqrt{2\pi n} \quad (n \to \infty), \tag{52}$$

or, more generally, that

$$\Gamma(x+1) \sim \left(\frac{x}{e}\right)^x \sqrt{2\pi x} \quad (x \to \infty;\ x \in \mathbb{R}), \tag{53}$$

where e is the base of the natural logarithm.

For a complex number z, we have the following asymptotic expansion:

$$\begin{gathered} \log\Gamma(z) = \left(z - \frac{1}{2}\right)\log z - z + \frac{1}{2}\log(2\pi) + \sum_{k=1}^{n} \frac{(-1)^{k-1}\, B_k}{2k(2k-1)\, z^{2k-1}} + O\left(z^{-2n-1}\right) \\ (|z| \to \infty;\ |\arg(z)| \leq \pi - \epsilon\,(0 < \epsilon < \pi);\ n \in \mathbb{N}_0), \end{gathered} \tag{54}$$

which, upon taking exponentials, yields an asymptotic formula for the Gamma function:

$$\begin{aligned} \Gamma(z) =& z^z\, e^{-z} \sqrt{\frac{2\pi}{z}} \left[1 + \frac{1}{12\,z} + \frac{1}{288\,z^2} - \frac{139}{51840\,z^3} - \frac{571}{2488320\,z^4}\right. \\ &\left. + \frac{163879}{209018880\,z^5} + \frac{50043869}{75246796800\,z^6} + O\left(z^{-7}\right)\right] \\ &(|z| \to \infty;\ |\arg(z)| \leq \pi - \epsilon\,(0 < \epsilon < \pi)). \end{aligned} \tag{55}$$

The asymptotic formula (55), in conjunction with the recurrence relation (9), is useful in computing the numerical values of $\Gamma(z)$ for large real values of z.

Some useful consequences of (54) or (55) include the asymptotic expansions:

(56)
$$\log\Gamma(z+\alpha) = \left(z+\alpha-\frac{1}{2}\right)\log z - z + \frac{1}{2}\log(2\pi) + O\left(z^{-1}\right)$$
$$(|z|\to\infty;\ |\arg(z)|\leq\pi-\epsilon;\ |\arg(z+\alpha)|\leq\pi-\epsilon;\ 0<\epsilon<\pi),$$

and

(57)
$$\frac{\Gamma(z+\alpha)}{\Gamma(z+\beta)} = z^{\alpha-\beta}\left[1+\frac{(\alpha-\beta)(\alpha+\beta-1)}{2z}+O\left(z^{-2}\right)\right]$$
$$(|z|\to\infty;\ |\arg(z)|\leq\pi-\epsilon;\ |\arg(z+\alpha)|\leq\pi-\epsilon;\ 0<\epsilon<\pi),$$

where α and β are bounded complex numbers.

Yet another interesting consequence of (55) is the following asymptotic expansion of $|\Gamma(x+iy)|$:

(58)
$$|\Gamma(x+iy)| \sim \sqrt{2\pi}\,|y|^{x-\frac{1}{2}}\,e^{-\frac{1}{2}\pi|y|} \quad (|x|<\infty;\ |y|\to\infty),$$

where x and y take on real values.

The Beta function

The *Beta function* $B(\alpha,\,\beta)$ is a function of two complex variables α and β, defined by

(59)
$$B(\alpha,\,\beta) := \int_0^1 t^{\alpha-1}(1-t)^{\beta-1}\,dt = B(\beta,\,\alpha) \quad (\Re(\alpha)>0;\ \Re(\beta)>0)$$

or, equivalently, by

(60)
$$B(\alpha,\,\beta) = 2\int_0^{\pi/2}(\sin\theta)^{2\alpha-1}(\cos\theta)^{2\beta-1}\,d\theta \quad (\Re(\alpha)>0;\ \Re(\beta)>0),$$

which follows from (59) on setting $t=\sin^2\theta$.

The integrals in (59) and (1) are known as the Eulerian integrals of the first and second kind, respectively.

Putting $t=u/(1+u)$ in (59), we obtain the following representation of $B(\alpha,\,\beta)$ as an infinite integral:

(61)
$$B(\alpha,\,\beta) = \int_0^\infty \frac{u^{\alpha-1}}{(1+u)^{\alpha+\beta}}\,du \quad (\Re(\alpha)>0;\ \Re(\beta)>0).$$

The Beta function is closely related to the Gamma function; in fact, we have

(62)
$$B(\alpha,\,\beta) = \frac{\Gamma(\alpha)\,\Gamma(\beta)}{\Gamma(\alpha+\beta)} \quad (\alpha,\,\beta\notin\mathbb{Z}_0^-),$$

which not only confirms the symmetry property in (59), but also continues the Beta function analytically for all complex values of α and β, except when $\alpha,\ \beta \in \mathbb{Z}_0^-$. Thus we may write

$$B(\alpha,\ \beta) = \begin{cases} \displaystyle\int_0^1 t^{\alpha-1}(1-t)^{\beta-1}\,dt & (\Re(\alpha) > 0;\ \Re(\beta) > 0), \\ \dfrac{\Gamma(\alpha)\,\Gamma(\beta)}{\Gamma(\alpha+\beta)} & (\Re(\alpha) < 0;\ \Re(\beta) < 0;\ \alpha,\ \beta \notin \mathbb{Z}_0^-). \end{cases} \tag{63}$$

Next we combine the relationship (62) with (60) and (61), and we obtain the following useful integral formulas:

$$\int_0^{\pi/2} \sin^\mu\theta\,\cos^\nu\theta\,d\theta = \frac{\Gamma\left(\frac{1}{2}\mu+\frac{1}{2}\right)\Gamma\left(\frac{1}{2}\nu+\frac{1}{2}\right)}{2\,\Gamma\left(\frac{1}{2}\mu+\frac{1}{2}\nu+1\right)} \quad (\Re(\mu) > -1;\ \Re(\nu) > -1) \tag{64}$$

and

$$\int_0^\infty \frac{u^{\lambda-1}}{(1+u)^\mu}\,du = \frac{\Gamma(\lambda)\,\Gamma(\mu-\lambda)}{\Gamma(\mu)} \quad (0 < \Re(\lambda) < \Re(\mu)). \tag{65}$$

It may be remarked in passing that the integral (64) provides a generalization of Wallis's formula of elementary calculus, and that (65) with $\mu = 1$ yields the familiar infinite integral:

$$\int_0^\infty \frac{u^{\lambda-1}}{1+u}\,du = \Gamma(\lambda)\,\Gamma(1-\lambda) = \frac{\pi}{\sin\pi\lambda} \quad (0 < \Re(\lambda) < 1), \tag{66}$$

which is usually evaluated in the literature by contour integration [see, for example, Copson (1935, p. 139, Example 1)].

In addition to (61) and (65), by means of suitable substitutions, a number of definite integrals are expressible in terms of the Beta function:

$$\int_0^1 t^{\alpha-1}(1-t)^{\beta-1}\,(1+at)^{-\alpha-\beta}\,dt = (1+a)^{-\alpha}\,B(\alpha,\ \beta) \\ (a > -1;\ \Re(\alpha) > 0;\ \Re(\beta) > 0); \tag{67}$$

$$\int_0^\infty t^{\beta-1}\,(1+at)^{-\alpha-\beta}\,dt = a^{-\beta}\,B(\alpha,\ \beta) \\ (a > 0;\ \Re(\alpha) > 0;\ \Re(\beta) > 0); \tag{68}$$

$$\int_b^a (t-b)^{\alpha-1}\,(a-t)^{\beta-1}\,dt = (a-b)^{\alpha+\beta-1}\,B(\alpha,\ \beta) \\ (a > b;\ \Re(\alpha) > 0;\ \Re(\beta) > 0); \tag{69}$$

$$\int_b^a \frac{(t-b)^{\alpha-1}\,(a-t)^{\beta-1}}{(t-c)^{\alpha+\beta}}\,dt = \frac{(a-b)^{\alpha+\beta-1}}{(a-c)^\alpha\,(b-c)^\beta}\,B(\alpha,\ \beta) \\ (a > b > c;\ \Re(\alpha) > 0;\ \Re(\beta) > 0). \tag{70}$$

The following functional equations for the Beta function can be deduced easily from (59) and (62):

(71) $$B(\alpha,\,\beta+1)=\frac{\beta}{\alpha}B(\alpha+1,\,\beta)=\frac{\beta}{\alpha+\beta}B(\alpha,\,\beta);$$

(72) $$B(\alpha,\,\beta)\,B(\alpha+\beta,\,\gamma)=B(\beta,\,\gamma)\,B(\beta+\gamma,\,\alpha)=B(\gamma,\,\alpha)\,B(\alpha+\gamma,\,\beta);$$

(73) $$B(\alpha,\,\beta)\,B(\alpha+\beta,\,\gamma)\,B(\alpha+\beta+\gamma,\,\delta)=\frac{\Gamma(\alpha)\,\Gamma(\beta)\,\Gamma(\gamma)\,\Gamma(\delta)}{\Gamma(\alpha+\beta+\gamma+\delta)},$$

or, more generally,

(74) $$\prod_{k=1}^{n} B\left(\sum_{j=1}^{k}\alpha_j,\,\alpha_{k+1}\right)=\frac{\Gamma(\alpha_1)\cdots\Gamma(\alpha_{n+1})}{\Gamma\left(\alpha_1+\cdots+\alpha_{n+1}\right)}\quad(n\in\mathbb{N});$$

(75) $$\frac{1}{B(n,\,m)}=m\binom{n+m-1}{n-1}=n\binom{n+m-1}{m-1}\quad(n,\,m\in\mathbb{N}).$$

The incomplete Gamma functions

The *incomplete Gamma function* $\gamma(z,\,\alpha)$ and its *complement* $\Gamma(z,\,\alpha)$ (also known as *Prym's function*) are defined by

(76) $$\gamma(z,\,\alpha):=\int_0^{\alpha} t^{z-1}\,e^{-t}\,dt\quad(\Re(z)>0;\ |\arg(\alpha)|<\pi),$$

(77) $$\Gamma(z,\,\alpha):=\int_{\alpha}^{\infty} t^{z-1}\,e^{-t}\,dt\quad(|\arg(\alpha)|<\pi),$$

so that

(78) $$\gamma(z,\,\alpha)+\Gamma(z,\,\alpha)=\Gamma(z).$$

For fixed α, $\Gamma(z,\,\alpha)$ is an entire (integral) function of z, while $\gamma(z,\,\alpha)$ is a meromorphic function of z, with simple poles at the points $z\in\mathbb{Z}_0^-$.

The following recursion formulas are worthy of note:

(79) $$\gamma(z+1,\,\alpha)=z\,\gamma(z,\,\alpha)-\alpha^z\,e^{-\alpha},$$

(80) $$\Gamma(z+1,\,\alpha)=z\,\Gamma(z,\,\alpha)+\alpha^z\,e^{-\alpha}.$$

The incomplete Beta functions

The *incomplete Beta function* $B_x(\alpha, \beta)$ is defined by

(81) $$B_x(\alpha, \beta) := \int_0^x t^{\alpha-1}(1-t)^{\beta-1}\, dt \quad (\Re(\alpha) > 0).$$

For the associated function:

(82) $$I_x(\alpha, \beta) = \frac{B_x(\alpha, \beta)}{B(\alpha, \beta)},$$

we note here the following properties which are easily verifiable:

(83) $$I_x(\alpha, \beta) = 1 - I_{1-x}(\beta, \alpha),$$

(84) $$I_x(k, n-k+1) = \sum_{j=k}^{n} \binom{n}{j} x^j (1-x)^{n-j} \quad (1 \leq k \leq n),$$

(85) $$I_x(\alpha, \beta) = x\, I_x(\alpha-1, \beta) + (1-x)\, I_x(\alpha, \beta-1),$$

(86) $$(\alpha+\beta-\alpha x)\, I_x(\alpha, \beta) = \alpha(1-x)\, I_x(\alpha+1, \beta-1) + \beta\, I_x(\alpha, \beta+1),$$

(87) $$(\alpha+\beta)\, I_x(\alpha, \beta) = \alpha\, I_x(\alpha+1, \beta) + \beta\, I_x(\alpha, \beta+1).$$

The Error functions

The *error function* $\operatorname{erf}(z)$, also known as the *probability integral* $\Phi(z)$, is defined for any complex z by

(88) $$\operatorname{erf}(z) := \frac{2}{\sqrt{\pi}} \int_0^z \exp\left(-t^2\right) dt = \Phi(z),$$

and its *complement* by

(89) $$\operatorname{erfc}(z) := 1 - \operatorname{erf}(z) = \frac{2}{\sqrt{\pi}} \int_z^\infty \exp\left(-t^2\right) dt.$$

Clearly,

(90) $$\operatorname{erf}(0) = 0 \quad \text{and} \quad \operatorname{erfc}(0) = 1,$$

and, in view of the well-known result (16), we also have

(91) $$\operatorname{erf}(\infty) = 1 \quad \text{and} \quad \operatorname{erfc}(\infty) = 0.$$

The alternative notations:

$$\text{Erf}(z) = \frac{\sqrt{\pi}}{2}\,\text{erf}(z) \quad \text{and} \quad \text{Erfc}(z) = \frac{\sqrt{\pi}}{2}\,\text{erfc}(z) \tag{92}$$

are sometimes used for the error functions. Many authors use the notations $\text{Erf}(z)$ and $\hat{\text{Erf}}(z)$ for the error functions $\text{erf}(z)$ and $\text{Erf}(z)$, respectively, defined by (88) and (92), and $\text{Erfc}(z)$ and $\hat{\text{Erfc}}(z)$ for their complements.

In terms of the incomplete Gamma functions, it is easily verified that

$$\text{erf}(z) = \frac{1}{\sqrt{\pi}}\,\gamma\left(\frac{1}{2}, z^2\right) \quad \text{and} \quad \text{erfc}(z) = \frac{1}{\sqrt{\pi}}\,\Gamma\left(\frac{1}{2}, z^2\right). \tag{93}$$

The Bohr-Mollerup theorem

We have already observed that Euler's definition (1) and its such consequences as (9) and (13) enable us to compute all the real values of the Gamma function from the knowledge merely of its values in the interval $(0, 1)$, as noted in conjunction with (55). Since the solution to the interpolation problem is not determined uniquely, it makes sense to add some more conditions to the problem. After various trials to find those conditions to guarantee the uniqueness of the Gamma function, in 1922, Bohr and Mollerup were able to show the remarkable fact that the Gamma function is the *only* function which satisfies the recurrence relationship and is logarithmically convex. The original proof was simplified, several years later, by Emil Artin, and the theorem together with Artin's method of proof now constitute the *Bohr-Mollerup-Artin theorem*:

Theorem 1.1. *Let $f : \mathbb{R}^+ \to \mathbb{R}^+$ satisfy each of the following properties:*
(a) $\log f(x)$ *is a convex function;*
(b) $f(x+1) = x\,f(x)$ *for all* $x \in \mathbb{R}^+$;
(c) $f(1) = 1$.
Then $f(x) = \Gamma(x)$ for all $x \in \mathbb{R}^+$.

Instead of giving here the proof of Theorem 1.1 [see Conway (1978, p. 179) and Artin (1964, p. 14)], we simply state the necessary and sufficient condition for the logarithmic convexity of a given function.

Theorem 1.2. *Let $f : [a, b] \to \mathbb{R}$ and suppose that $f(x) > 0$ for all $x \in [a, b]$ and that f has a continuous second derivative $f''(x)$ for $x \in [a, b]$. Then f is logarithmically convex if and only if*

$$f''(x)\,f(x) - [f'(x)]^2 \geqq 0 \quad (x \in [a, b]).$$

1.2 Polygamma Functions

The Psi (or Digamma) function

The *Psi* (or *Digamma*) *function* $\psi(z)$ defined by

$$\psi(z) := \frac{d}{dz}\log\Gamma(z) = \frac{\Gamma'(z)}{\Gamma(z)} \quad \text{or} \quad \log\Gamma(z) = \int_1^z \psi(t)\,dt \tag{1}$$

possesses the following properties:

$$\psi(z) = \lim_{n\to\infty}\left(\log n - \sum_{k=0}^{n}\frac{1}{z+k}\right); \tag{2}$$

$$\begin{aligned}\psi(z) &= -\gamma - \frac{1}{z} + \sum_{n=1}^{\infty}\frac{z}{n(z+n)} \\ &= -\gamma + (z-1)\sum_{n=0}^{\infty}\frac{1}{(n+1)(z+n)},\end{aligned} \tag{3}$$

where γ is the Euler-Mascheroni constant defined by 1.1(3).

These results clearly imply that $\psi(z)$ is meromorphic (that is, analytic everywhere in the bounded complex z-plane except for poles) with simple poles at $z = -n$ $(n \in \mathbb{N}_0)$ with its residue -1. Also we have

$$\psi(1) = -\gamma, \tag{4}$$

which follows at once from (3).

The following additional properties of $\psi(z)$ can be deduced from known results for $\Gamma(z)$:

$$\log\frac{\Gamma(z+1)}{\Gamma(z)} = \log z = -\gamma + \sum_{n=0}^{\infty}\left[\frac{1}{n+1} - \log\left(1+\frac{1}{n+z}\right)\right]; \tag{5}$$

$$\psi(z) = \log z - \sum_{n=0}^{\infty}\left[\frac{1}{n+z} - \log\left(1+\frac{1}{n+z}\right)\right], \tag{6}$$

which follows from (3) and (5);

$$\psi(z+n) = \psi(z) + \sum_{k=1}^{n}\frac{1}{z+k-1} \quad (n\in\mathbb{N}); \tag{7}$$

$$\psi(z) - \psi(-z) = -\pi\cot\pi z - \frac{1}{z}; \tag{8}$$

$$\psi(1+z) - \psi(1-z) = \frac{1}{z} - \pi\cot\pi z; \tag{9}$$

$$\psi(z) - \psi(1-z) = -\pi\cot\pi z; \tag{10}$$

$$\psi\left(\frac{1}{2}+z\right) - \psi\left(\frac{1}{2}-z\right) = \pi\tan\pi z; \tag{11}$$

$$\psi(mz) = \log m + \frac{1}{m} \sum_{k=0}^{m-1} \psi\left(z + \frac{k}{m}\right) \quad (m \in \mathbb{N}). \tag{12}$$

Integral representations for $\psi(z)$

Expanding $(1-t)^{-1}$ into a series, integrating term by term, and using (3), we get

$$\psi(z) = -\gamma + \int_0^1 \left(1 - t^{z-1}\right)(1-t)^{-1}\, dt \quad (\Re(z) > 0), \tag{13}$$

which, upon replacing t by e^{-t}, yields

$$\psi(z) = -\gamma + \int_0^\infty \left(e^{-t} - e^{-tz}\right)\left(1 - e^{-t}\right)^{-1} dt \quad (\Re(z) > 0). \tag{14}$$

Making use of (10), it follows from (13) and (14) that

$$\psi(z) = -\gamma - \pi \cot \pi z + \int_0^1 \left(1 - t^{-z}\right)(1-t)^{-1}\, dt \quad (\Re(z) < 1) \tag{15}$$

and

$$\psi(z) = -\gamma - \pi \cot \pi z + \int_0^\infty \left(1 - e^{tz}\right)\left(e^t - 1\right)^{-1} dt \quad (\Re(z) < 1). \tag{16}$$

It is not difficult to derive each of the following integral representations for $\psi(z)$.

$$\psi(z) = \int_0^\infty \left[t^{-1}e^{-t} - \left(1 - e^{-t}\right)^{-1} e^{-tz}\right] dt \quad (\Re(z) > 0), \tag{17}$$

which is due to Gauss and reduces, when $z = 1$, to the integral expression 1.1(21) for the Euler-Mascheroni constant γ;

$$\psi(z) = \int_0^\infty \left[e^{-t} - (1+t)^{-z}\right] t^{-1}\, dt \quad (\Re(z) > 0), \tag{18}$$

which is due to Dirichlet and reduces, when $z = 1$, to 1.1(23);

$$\psi(z) = -\gamma + \int_0^\infty \left[(1+t)^{-1} - (1+t)^{-z}\right] t^{-1}\, dt \quad (\Re(z) > 0); \tag{19}$$

$$\psi(z) = \log z + \int_0^z \left[t^{-1} - \left(1 - e^{-t}\right)^{-1}\right] e^{-tz}\, dt \quad (\Re(z) > 0), \tag{20}$$

$$\psi(z) = \log z - \frac{1}{2z} - \int_0^\infty \left[\left(1 - e^{-t}\right)^{-1} - t^{-1} - \frac{1}{2}\right] e^{-tz} \quad (\Re(z) > 0), \tag{21}$$

$$\psi(z) = \log z + \int_0^\infty \left[\left(1 - e^{-t}\right)^{-1} + t^{-1} - 1 \right] e^{-tz}\, dt \quad (\Re(z) > 0), \tag{22}$$

and

$$\psi(z) = \log z - \frac{1}{2z} - \int_0^\infty \left[\left(e^t - 1\right)^{-1} - t^{-1} + \frac{1}{2} \right] e^{-tz}\, dt \quad (\Re(z) > 0), \tag{23}$$

all four being attributed to Binet;

$$\psi(z) = \log z - \frac{1}{2z} - 2 \int_0^\infty \left(t^2 + z^2\right)^{-1} \left(e^{2\pi t} - 1\right)^{-1} t\, dt \quad (\Re(z) > 0), \tag{24}$$

which, in the special case when $z = 1$, yields 1.1(25).

Gauss's integral representation (17) also yields Malmstén's formula:

$$\begin{aligned} \log \Gamma(z) &= \int_1^z \psi(t)\, dt \\ &= \int_0^\infty \left[(z-1) - \frac{1 - e^{-(z-1)t}}{1 - e^{-t}} \right] \frac{e^{-t}}{t}\, dt \quad (\Re(z) > 0). \end{aligned} \tag{25}$$

From (23) we can deduce that

$$\begin{aligned} \log \Gamma(z) = {} & \left(z - \frac{1}{2}\right) \log z - z + 1 + \int_0^\infty \left(\frac{1}{2} - \frac{1}{t} + \frac{1}{e^t - 1}\right) \frac{e^{-tz}}{t} dt \\ & - \int_0^\infty \left(\frac{1}{2} - \frac{1}{t} + \frac{1}{e^t - 1}\right) \frac{e^{-t}}{t}\, dt \quad (\Re(z) > 0). \end{aligned} \tag{26}$$

The second integral in (26) can readily be evaluated:

$$\int_0^\infty \left(\frac{1}{2} - \frac{1}{t} + \frac{1}{e^t - 1}\right) \frac{e^{-t}}{t}\, dt = 1 - \frac{1}{2} \log(2\pi) \tag{27}$$

by using an elementary (yet technical) separation and recalling a formula for $\log z$:

$$\log z = \int_0^\infty \left(e^{-t} - e^{-tz}\right) \frac{dt}{t} \quad (\Re(z) > 0). \tag{28}$$

Now putting (27) into (26) yields *Binet's first expression for* $\log \Gamma(z)$:

$$\begin{aligned} \log \Gamma(z) = {} & \left(z - \frac{1}{2}\right) \log z - z + \frac{1}{2} \log(2\pi) \\ & + \int_0^\infty \left(\frac{1}{2} - \frac{1}{t} + \frac{1}{e^t - 1}\right) \frac{e^{-tz}}{t} dt \quad (\Re(z) > 0), \end{aligned} \tag{29}$$

which may also give an approximate expression for $\log \Gamma(z)$ just as the special case of 1.1(54) when $n = 0$.

Binet's second expression for $\log \Gamma(z)$ may be written as follows:

$$(30)\qquad \begin{aligned}\log \Gamma(z) = &\left(z - \frac{1}{2}\right) \log z - z + \frac{1}{2} \log (2\pi) \\ &+ 2 \int_0^{\infty} \frac{\arctan (t/z)}{e^{2\pi t} - 1}\, dt \quad (\Re(z) > 0).\end{aligned}$$

From 1.1(12) and (25) we obtain *Kummer's expression for* $\log \Gamma(z)$:

$$(31)\qquad \begin{aligned}\log \Gamma(z) =&\frac{1}{2} \log \pi - \frac{1}{2} \log (\sin \pi z) \\ &+ \frac{1}{2} \int_0^{\infty} \left(\frac{\sinh \left[\left(\frac{1}{2} - z\right) t\right]}{\sinh \left(\frac{1}{2} t\right)} - (1 - 2z)\, e^{-t} \right) \frac{dt}{t} \\ &(0 < \Re(z) < 1).\end{aligned}$$

By applying (31), Kummer derived the following Fourier series for $\log \Gamma(x)$:

$$(32)\qquad \log \Gamma(x) = \frac{1}{2} \log \pi - \frac{1}{2} \log (\sin \pi x) + 2 \sum_{n=1}^{\infty} a_n \sin (2n\pi x) \quad (0 < x < 1),$$

where

(33)

$$\begin{aligned}a_n =& \int_0^{\infty} \left(\frac{2n\pi}{t^2 + 4n^2\pi^2} - \frac{e^{-t}}{2n\pi} \right) \frac{dt}{t} \\ =&\frac{1}{2n\pi} \left[\int_0^{\infty} \left(\frac{1}{1 + t^2} - \cos t \right) \frac{dt}{t} + \int_0^{\infty} \frac{e^{-t} - e^{-2\pi n t}}{t} dt + \int_0^{\infty} \left(\cos t - e^{-t}\right) \frac{dt}{t} \right] \\ =&\frac{1}{2n\pi} (\gamma + \log 2\pi + \log n).\end{aligned}$$

Since

$$(34)\qquad \log (\sin \pi x) = -\log 2 - \sum_{n=1}^{\infty} \frac{1}{n} \cos (2\pi n x) \quad (0 < x < 1),$$

by rewriting (32) in the form:

$$(35)\qquad \begin{aligned}\log \Gamma (x) = &\left(\frac{1}{2} - x\right) (\gamma + \log 2) + (1 - x) \log \pi - \frac{1}{2} \log (\sin \pi x) \\ &+ \sum_{n=1}^{\infty} \frac{\log n}{\pi n} \sin (2\pi n x) \quad (0 < x < 1),\end{aligned}$$

we finally obtain

$$(36)\qquad \begin{aligned}&\log \Gamma(x) = \frac{1}{2} \log (2\pi) \\ &+ \sum_{n=1}^{\infty} \left(\frac{1}{2n} \cos (2\pi n x) + \frac{\gamma + \log (2\pi n)}{\pi n} \sin (2\pi n x) \right) \quad (0 < x < 1).\end{aligned}$$

The Fourier series (36) readily implies each of the integral formulas:

$$\int_0^1 \sin(2\pi n x)\,\log\Gamma(x)\,dx = \frac{\gamma + \log(2\pi n)}{2\pi n} \quad (n \in \mathbb{N}), \tag{37}$$

$$\int_0^1 \cos(2\pi n x)\,\log\Gamma(x)\,dx = \frac{1}{4n} \quad (n \in \mathbb{N}), \tag{38}$$

and

$$\int_0^1 \log\Gamma(x)\,dx = \frac{1}{2}\log(2\pi). \tag{39}$$

Considering 1.1(51) and 1.1(54) when $n = 0$, we readily obtain a more general integral formula than (39) above:

$$\int_x^{x+1} \log\Gamma(t)\,dt = x\log x - x + \frac{1}{2}\log(2\pi) \quad (x \geqq 0), \tag{40}$$

by observing the following relation:

$$\lim_{m\to\infty} \sum_{k=0}^{m-1} \frac{1}{m}\log\Gamma\left(x + \frac{k}{m}\right) = \int_0^1 \log\Gamma(x+u)\,du = \int_x^{x+1} \log\Gamma(t)\,dt.$$

The formula (40) can be extended, without difficulty, to the case when $x \in \mathbb{C} \setminus \mathbb{Z}^-$ ($\mathbb{Z}^- := \{-1, -2, \ldots\}$) and the principal values of the involved logarithms are taken, and can also be further generalized (by iteration) as follows:

$$\begin{aligned}\int_x^{x+n} \log\Gamma(t)\,dt = &\sum_{k=0}^{n-1} (x+k)\log(x+k) - nx - \frac{1}{2}n(n-1)\\ &+ \frac{1}{2}n\log(2\pi) \quad (n \in \mathbb{N}).\end{aligned} \tag{41}$$

Gauss's formulas for $\psi\left(\frac{p}{q}\right)$

Taking $z = \frac{p}{q}$ $(0 < p < q;\ p,\, q \in \mathbb{N})$ in (13) and $t = x^q$, we obtain

$$\gamma + \psi\left(\frac{p}{q}\right) = q\int_0^1 \frac{x^{q-1} - x^{p-1}}{1 - x^q}\,dx, \tag{42}$$

which, upon noting that

$$q\,\frac{x^{p-1} - x^{q-1}}{x^q - 1} = \sum_{k=1}^{q-1} \frac{\omega_k^p - 1}{x - \omega_k} \quad \left(\omega_k := \exp\left(\frac{2\pi i k}{q}\right)\right), \tag{43}$$

readily yields

(44)

$$\gamma + \psi\left(\frac{p}{q}\right) = \frac{\pi}{2q}\sum_{k=1}^{q-1} (2k - q)\sin\frac{2kp\pi}{q} - \sum_{k=1}^{q-1}\left(1 - \cos\frac{2kp\pi}{q}\right)\log\left(2\sin\frac{k\pi}{q}\right).$$

By applying the known trigonometric identities:

$$\sum_{k=0}^{q-1} \cos 2k\alpha = \frac{1}{2} + \frac{\sin(2q-1)\alpha}{2\sin\alpha} \tag{45}$$

and

$$\sum_{k=1}^{q-1} \sin 2k\alpha = \frac{\cos\alpha - \cos(2q-1)\alpha}{2\sin\alpha} \tag{46}$$

in (44), we obtain *Gauss's formula:*

$$\psi\left(\frac{p}{q}\right) = -\gamma - \frac{\pi}{2}\cot\frac{p\pi}{q} - \log q + \sum_{k=1}^{q-1}\cos\frac{2kp\pi}{q}\log\left(2\sin\frac{k\pi}{q}\right) \tag{47}$$
$$(0 < p < q;\ p,\ q \in \mathbb{N}),$$

which implies that, for a positive proper fraction z, the value of $\psi(z)$ can be expressed as a finite combination of elementary functions, and yet, by means of 1.1(10), may be extended to every rational value of z.

By reversing the order of summation, Gauss's formula (47) can be rewritten in the following form:

$$\begin{aligned}\psi\left(\frac{p}{q}\right) =& \gamma - \frac{\pi}{2}\cot\frac{p\pi}{q} - \log q \\ &+ 2\sum_{k=1}^{[(q-1)/2]}\cos\left(\frac{2kp\pi}{q}\right)\log\left(2\sin\frac{k\pi}{q}\right) + r_q(p),\end{aligned} \tag{48}$$

where $[x]$ denotes the greatest integer $\leq x$, and

$$r_q(p) = \begin{cases} (-1)^p \log 2 & \text{if } q \text{ is even,} \\ 0 & \text{if } q \text{ is odd.} \end{cases}$$

If we rewrite (42) as

$$\gamma - q\int_0^1 \frac{x^{q-1}}{1-x^q}dx = -\psi\left(\frac{p}{q}\right) - q\int_0^1 \frac{x^{p-1}}{1-x^q}dx,$$

where the left-hand side is independent of p and only the right-hand side is dependent on p, we can derive *Gauss's second formula:*

$$\sum_{p=1}^{q}\psi\left(\frac{p}{q}\right)\exp\left(\frac{2pk\pi i}{q}\right) = q\log\left[1 - \exp\left(\frac{2k\pi i}{q}\right)\right] \quad (q \in \mathbb{N};\ k \in \mathbb{Z}). \tag{49}$$

Special values of $\psi(z)$

Setting $z = 1$ and $z = \frac{1}{2}$ in (7), we obtain

$$\psi(n) = -\gamma + \sum_{k=1}^{n-1} \frac{1}{k} \quad (n \in \mathbb{N}) \tag{50}$$

and

$$\psi\left(n + \frac{1}{2}\right) = -\gamma - 2 \log 2 + 2 \sum_{k=0}^{n-1} \frac{1}{2k+1} \quad (n \in \mathbb{N}_0), \tag{51}$$

it being understood (here as well as throughout this book) that an empty sum is nil.

By suitably applying the various formulas for the ψ–function, we can derive the following special values of $\psi(z)$:

$$\psi\left(\frac{1}{2}\right) = -\gamma - 2\log 2;$$

$$\psi\left(\frac{1}{3}\right) = -\gamma - \frac{1}{6}\pi\sqrt{3} - \frac{3}{2}\log 3;$$

$$\psi\left(\frac{2}{3}\right) = -\gamma + \frac{1}{6}\pi\sqrt{3} - \frac{3}{2}\log 3;$$

$$\psi\left(\frac{1}{4}\right) = -\gamma - \frac{1}{2}\pi - 3\log 2;$$

$$\psi\left(\frac{3}{4}\right) = -\gamma + \frac{1}{2}\pi - 3\log 2;$$

$$\begin{aligned}\psi\left(\frac{1}{5}\right) = &-\gamma - \frac{1}{2}\pi\sqrt{1 + \frac{2}{5}\sqrt{5}} \\ &- \frac{5}{4}\log 5 - \frac{1}{4}\sqrt{5}\log\frac{1}{2}(3 + \sqrt{5});\end{aligned}$$

$$\begin{aligned}\psi\left(\frac{2}{5}\right) = &-\gamma - \frac{1}{2}\pi\sqrt{1 - \frac{2}{5}\sqrt{5}} \\ &- \frac{5}{4}\log 5 + \frac{1}{4}\sqrt{5}\log\frac{1}{2}(3 + \sqrt{5});\end{aligned}$$

$$\begin{aligned}\psi\left(\frac{3}{5}\right) = &-\gamma + \frac{1}{2}\pi\sqrt{1 - \frac{2}{5}\sqrt{5}} \\ &- \frac{5}{4}\log 5 + \frac{1}{4}\sqrt{5}\log\frac{1}{2}(3 + 2\sqrt{5});\end{aligned}$$

$$\begin{aligned}\psi\left(\frac{4}{5}\right) = &-\gamma + \frac{1}{2}\pi\sqrt{1 + \frac{2}{5}\sqrt{5}} \\ &- \frac{5}{4}\log 5 - \frac{1}{4}\sqrt{5}\log\frac{1}{2}(3 + 2\sqrt{5});\end{aligned}$$

$$\psi\left(\frac{1}{6}\right) = -\gamma - \frac{1}{2}\pi\sqrt{3} - \frac{3}{2}\log 3 - 2\log 2;$$

$$\psi\left(\frac{5}{6}\right) = -\gamma + \frac{1}{2}\pi\sqrt{3} - \frac{3}{2}\log 3 - 2\log 2;$$

$$\psi\left(\frac{1}{8}\right) = -\gamma - \frac{1}{2}\pi(\sqrt{2}+1) - 4\log 2 + \frac{\sqrt{2}}{2}\log(3-2\sqrt{2});$$

$$\psi\left(\frac{3}{8}\right) = -\gamma - \frac{1}{2}\pi(\sqrt{2}-1) - 4\log 2 - \frac{\sqrt{2}}{2}\log(3-2\sqrt{2});$$

$$\psi\left(\frac{5}{8}\right) = -\gamma + \frac{1}{2}\pi(\sqrt{2}-1) - 4\log 2 - \frac{\sqrt{2}}{2}\log(3-2\sqrt{2});$$

$$\psi\left(\frac{7}{8}\right) = -\gamma + \frac{1}{2}\pi(\sqrt{2}+1) - 4\log 2 + \frac{\sqrt{2}}{2}\log(3-2\sqrt{2});$$

$$\psi\left(\frac{1}{10}\right) = -\gamma - \frac{\pi}{2}\left(2\sqrt{1+\frac{2}{5}\sqrt{5}} + \sqrt{1-\frac{2}{5}\sqrt{5}}\right) - \frac{5}{4}\log 5$$
$$- \frac{1}{4}\sqrt{5}\,\log\frac{1}{8}(102+46\sqrt{5}) - 2\log 2;$$

$$\psi\left(\frac{3}{10}\right) = -\gamma + \frac{\pi}{2}\left(2\sqrt{1-\frac{2}{5}\sqrt{5}} - \sqrt{1+\frac{2}{5}\sqrt{5}}\right)$$
$$- \frac{5}{4}\log 5 + \frac{3}{4}\sqrt{5}\,\log\frac{1}{2}(3+2\sqrt{5}) - 2\log 2;$$

$$\psi\left(\frac{7}{10}\right) = -\gamma + \frac{\pi}{2}\left(\sqrt{1+\frac{2}{5}\sqrt{5}} - 2\sqrt{1-\frac{2}{5}\sqrt{5}}\right)$$
$$- \frac{5}{4}\log 5 + \frac{3}{4}\sqrt{5}\,\log\frac{1}{2}(3+\sqrt{5}) - 2\log 2;$$

$$\psi\left(\frac{9}{10}\right) = -\gamma + \frac{\pi}{2}\left(\sqrt{1-\frac{2}{5}\sqrt{5}} + 2\sqrt{1+\frac{2}{5}\sqrt{5}}\right)$$
$$- \frac{5}{4}\log 5 - \frac{1}{4}\sqrt{5}\,\log\frac{1}{8}(147+65\sqrt{5}) - 2\log 2;$$

$$\psi\left(\frac{1}{12}\right) = -\gamma - \frac{\pi}{2}(2+\sqrt{3}) + \frac{\sqrt{3}}{2}\log(7-4\sqrt{3}) - \frac{3}{2}\log 3 - 3\log 2;$$

$$\psi\left(\frac{5}{12}\right) = -\gamma - \frac{\pi}{2}(2-\sqrt{3}) - \frac{\sqrt{3}}{2}\log(7-4\sqrt{3}) - \frac{3}{2}\log 3 - 3\log 2;$$

$$\psi\left(\frac{7}{12}\right) = -\gamma + \frac{\pi}{2}(2-\sqrt{3}) - \frac{\sqrt{3}}{2}\log(7-4\sqrt{3}) - \frac{3}{2}\log 3 - 3\log 2;$$

$$\psi\left(\frac{11}{12}\right) = -\gamma + \frac{\pi}{2}(2+\sqrt{3}) + \frac{\sqrt{3}}{2}\log(7-4\sqrt{3}) - \frac{3}{2}\log 3 - 3\log 2;$$

$$\psi\left(-\frac{1}{2}\right) = 2 - \gamma - 2\log 2;$$

$$\psi\left(-\frac{1}{3}\right) = 3 - \gamma + \frac{\sqrt{3}}{6}\pi - \frac{3}{2}\log 3;$$

$$\psi\left(-\frac{5}{6}\right) = \frac{6}{5} - \gamma - \frac{\sqrt{3}}{2}\pi - \frac{3}{2}\log 3 - 2\log 2.$$

The Polygamma functions

The *Polygamma functions* $\psi^{(n)}(z)$ $(n \in \mathbb{N})$ are defined by

(52) $$\psi^{(n)}(z) := \frac{d^{n+1}}{dz^{n+1}}\log\Gamma(z) = \frac{d^n}{dz^n}\psi(z) \quad (n \in \mathbb{N}_0;\ z \notin \mathbb{Z}_0^-).$$

In terms of the generalized (or Hurwitz) Zeta function $\zeta(s,a)$ (see Section 2.2), we can write

(53)
$$\psi^{(n)}(z) = (-1)^{n+1}\, n! \sum_{k=0}^{\infty} \frac{1}{(k+z)^{n+1}} = (-1)^{n+1}\, n!\, \zeta(n+1, z) \quad (n \in \mathbb{N};\ z \notin \mathbb{Z}_0^-),$$

which may be used to deduce the properties of $\psi^{(n)}(z)$ $(n \in \mathbb{N})$ from those of $\zeta(s,z)$ $(s = n+1;\ n \in \mathbb{N})$.

It is also easy to have the following expressions:

(54) $$\psi^{(n)}(z+m) - \psi^{(n)}(z) = (-1)^n\, n! \sum_{k=1}^{m} \frac{1}{(z+k-1)^{n+1}} \quad (m,\ n \in \mathbb{N}_0)$$

and

(55) $$\psi^{(n)}(z) - (-1)^n\, \psi^{(n)}(1-z) = -\pi \frac{d^n}{dz^n}\{\cot \pi z\} \quad (n \in \mathbb{N}_0),$$

it being understood (as usual) that $\psi^{(0)}(z) = \psi(z)$.

The asymptotic expansion for $\psi(z)$

From 1.1(54) we obtain the following asymptotic expansion for $\psi(z)$:

(56) $$\psi(z) = \log z - \frac{1}{2z} - \sum_{k=1}^{n} \frac{(-1)^{k-1}\, B_k}{2k\, z^{2k}} + O\left(z^{-2n-2}\right)$$
$$(|z| \to \infty;\ |\arg(z)| \leq \pi - \epsilon\ (0 < \epsilon < \pi);\ n \in \mathbb{N}_0).$$

Now we shall show [see Barnes (1899b)] that

(57) $$\sum_{k=1}^{n} \psi(k) = n\, \log n - n + \frac{1}{2} + O\left(n^{-1}\right) \quad (n \to \infty)$$

and

$$\sum_{k=1}^{n} \psi'(k) = \log n + 1 + \gamma + O\left(n^{-1}\right) \quad (n \to \infty), \tag{58}$$

where γ denotes the Euler-Mascheroni constant given in 1.1(3).

Indeed, setting $f(x) = 1/x$ and $f(x) = 1/x^2$ with $a = 1$ in the Euler-Maclaurin summation formula 1.3(68), we obtain

$$1 + \frac{1}{2} + \cdots + \frac{1}{n} = \gamma + \log n + \frac{1}{2\,n} - \frac{1}{8\,n^2} + \frac{15}{2\,n^4} - \cdots \tag{59}$$

and

$$1 + \frac{1}{2^2} + \cdots + \frac{1}{n^2} = \frac{\pi^2}{6} - \frac{1}{n} + \frac{1}{2\,n^2} - \frac{1}{6\,n^3} + \frac{1}{30\,n^5} - \cdots, \tag{60}$$

respectively. We can readily deduce from (60) that

$$\frac{1}{n^2} + \frac{1}{(n+1)^2} + \cdots = \frac{1}{n} + \frac{1}{2\,n^2} + O\left(n^{-3}\right) \quad (n \to \infty). \tag{61}$$

Since

$$\psi(m) = \frac{1}{1} + \frac{1}{2} + \cdots + \frac{1}{m-1} - \gamma \quad (m \in \mathbb{N} \setminus \{1\}), \tag{62}$$

we can also write (57) in the form:

$$\sum_{k=1}^{n} \psi(k) = -n\gamma + \frac{n-1}{1} + \frac{n-2}{2} + \cdots + \frac{1}{n-1}.$$

We thus obtain

$$\begin{aligned}\sum_{k=1}^{n} \psi(k) =& n\left(-\gamma + \frac{1}{1} + \frac{1}{2} + \cdots + \frac{1}{n-1}\right) - \frac{1}{1} - \frac{2}{2} - \cdots - \frac{n-1}{n-1} \\ =& n\left(-\gamma + \sum_{k=1}^{n-1} \frac{1}{k}\right) - n + 1,\end{aligned}$$

which, in view of (59), yields

$$\sum_{k=1}^{n} \psi(k) = n\left[-\gamma + \gamma + \log(n-1) + \frac{1}{2(n-1)} + O\left(n^{-2}\right)\right] - n + 1 \quad (n \to \infty),$$

which implies the result (57).

In order to prove (58), we consider (54) with $m = n = 1$:

$$\psi'(z+1) = \psi'(z) - \frac{1}{z^2}, \tag{63}$$

the repeated application of which gives

$$\psi'(m) = -\left[\frac{1}{1^2} + \frac{1}{2^2} + \cdots + \frac{1}{(m-1)^2}\right] + \psi'(1),$$

which may also follow directly from (54). Hence we have

$$\sum_{k=1}^{n} \psi'(k) = n\,\psi'(1) - \frac{n-1}{1^2} - \frac{n-2}{2^2} - \cdots - \frac{1}{(n-1)^2}. \tag{64}$$

Now apply the following special case of (53) when $n = z = 1$:

$$\psi'(1) = \zeta(2) = \frac{\pi^2}{6}, \tag{65}$$

in conjunction with (59), and (64) readily yields
(66)

$$\begin{aligned}\sum_{k=1}^{n} \psi'(k) =& n\,\frac{\pi^2}{6} - \left[\frac{1}{1^2} + \frac{1}{2^2} + \cdots + \frac{1}{(n-1)^2}\right] + \frac{1}{1^2} + \frac{2}{2^2} + \cdots + \frac{n-1}{(n-1)^2} \\ =& n\left[\frac{1}{n^2} + \frac{1}{(n+1)^2} + \cdots\right] + \gamma + \log(n-1) + O\left(n^{-1}\right) \quad (n \to \infty),\end{aligned}$$

which, by applying (61), produces the desired result (58).

1.3 The Multiple Gamma Functions

The double Gamma function Γ_2

The *double Gamma function* Γ_2 was defined and studied by Barnes (1899b, 1900, and 1901) and others in about 1900. Although this function did not appear in the tables of the most well-known special functions, yet it was cited in the exercises by Whittaker and Watson (1963, p. 264) and recorded also by Gradshteyn and Ryzhik (1980, p. 661, Entry 6.441(4); p. 937, Entry 8.333). Recently, this function was revived in the study of the determinants of the Laplacians on the n-dimensional unit sphere S^n [see Choi (1994a), Osgood *et al.* (1988), Quine and Choi (1996), Vardi (1988b), and Voros (1987)]. Shintani (1980) also used this function to prove the classical Kronecker limit formula. Its p-adic analytic extension appeared in a formula of Cassou-Noguès (1979) for the p-adic L-functions at the point 0. More recently, Choi *et al.* (1995, 1997, and 1999a) used this function to evaluate sums of series involving the Riemann Zeta function.

Matsumoto (1998), on the other hand, proved asymptotic expansions of the Barnes double Zeta function and the double Gamma function, and presented an application to the Hecke L-functions of real quadratic fields. Before their investigation by Barnes, these functions had been introduced in a different form by (for example) Hölder (1886), Alexeiewsky (1894), and Kinkelin (1860).

Here we summarize some properties of the double Gamma and related functions. We also introduce some mathematical constants associated with the double and triple Gamma functions.

Barnes (1899b) defined the double Gamma function $\Gamma_2 = 1/G$ satisfying each of the following properties:

(a) $G(z+1) = \Gamma(z)G(z)$ $(z \in \mathbb{C})$;

(b) $G(1) = 1$;

(c) Asymptotically,

(1)
$$\begin{aligned}\log G(z+n+2) =& \frac{n+1+z}{2}\log(2\pi) + \left[\frac{n^2}{2} + n + \frac{5}{12} + \frac{z^2}{2} + (n+1)z\right]\log n \\ & - \frac{3n^2}{4} - n - nz - \log A + \frac{1}{12} + O\left(n^{-1}\right) \quad (n \to \infty),\end{aligned}$$

where Γ is the familiar Gamma function introduced in Section 1.1 and A is called the Glaisher-Kinkelin constant defined by

(2)
$$\log A = \lim_{n\to\infty}\left\{\sum_{k=1}^{n} k\,\log\,k - \left(\frac{n^2}{2} + \frac{n}{2} + \frac{1}{12}\right)\log n + \frac{n^2}{4}\right\},$$

the numerical value of A being given by

$$A \cong 1.282427130\cdots.$$

From this definition, Barnes (1899b) deduced several explicit Weierstrass canonical product forms of the double Gamma function Γ_2, one of which is recalled here in the form:

(3)
$$\begin{aligned}\{\Gamma_2(z+1)\}^{-1} &= G(z+1) \\ &= (2\pi)^{\frac{1}{2}z}\exp\left(-\frac{1}{2}z - \frac{1}{2}(\gamma+1)z^2\right)\prod_{k=1}^{\infty}\left\{\left(1+\frac{z}{k}\right)^k\exp\left(-z+\frac{z^2}{2k}\right)\right\},\end{aligned}$$

where γ denotes the Euler-Mascheroni constant given by 1.1(3).

Barnes (1899b) also gave two more equivalent forms of the double Gamma function Γ_2:

(4)
$$\begin{aligned}&\{\Gamma_2(z+1)\}^{-1} = G(z+1) \\ &= (2\pi)^{\frac{1}{2}z}\exp\left(-\frac{1}{2}z(z+1) - \frac{1}{2}\gamma z^2\right)\prod_{k=1}^{\infty}\frac{\Gamma(k)}{\Gamma(z+k)}\exp\left[z\psi(k) + \frac{1}{2}z^2\,\psi'(k)\right];\end{aligned}$$

(5)
$$\begin{aligned}\{\Gamma_2(z+1)\}^{-1} = G(z+1) = (2\pi)^{\frac{1}{2}z}\exp\left[\left(\gamma - \frac{1}{2}\right)z - \left(\frac{\pi^2}{6} + 1 + \gamma\right)\frac{z^2}{2}\right]\Gamma(z+1) \\ \cdot\prod_{m=0}^{\infty}{\prod_{n=0}^{\infty}}'\left(1+\frac{z}{m+n}\right)\exp\left(-\frac{z}{m+n} + \frac{z^2}{2(m+n)^2}\right),\end{aligned}$$

where the prime denotes the exclusion of the case $n = m = 0$ and the Psi (or Digamma) function ψ is given by 1.2(1). Each form of these products is convergent for all finite values of $|z|$, by the Weierstrass factorization theorem [see Conway (1978, p. 170)].

The double Gamma function satisfies the following relations:

$$G(1)=1 \quad \text{and} \quad G(z+1)=\Gamma(z)G(z) \quad (z\in\mathbb{C}). \tag{6}$$

For sufficiently large real x and $a\in\mathbb{C}$, we have the Stirling formula for the G-function:

$$\begin{aligned}\log G(x+a+1)=&\frac{x+a}{2}\log(2\pi)-\log A+\frac{1}{12}-\frac{3x^2}{4}-ax\\&+\left(\frac{x^2}{2}-\frac{1}{12}+\frac{a^2}{2}+ax\right)\log x+O\left(x^{-1}\right) \quad (x\to\infty).\end{aligned} \tag{7}$$

The following special values of G [see Barnes (1899b)] may be recalled here:

$$G\left(\frac{1}{2}\right)=2^{\frac{1}{24}}\cdot\pi^{-\frac{1}{4}}\cdot e^{\frac{1}{8}}\cdot A^{-\frac{3}{2}}; \tag{8}$$

$$G(n+2)=1!\,2!\cdots n! \quad \text{and} \quad G(n+1)=\frac{(n!)^n}{1\cdot 2\cdot 3^2\cdot 4^3\cdots n^{n-1}} \quad (n\in\mathbb{N}). \tag{9}$$

We shall deduce only the expression (4) [see Barnes (1899b)]. Indeed, taking the logarithmic derivative on both sides of the fundamental functional relation in (6) with respect to z, we obtain

$$\frac{G'(z+1)}{G(z+1)}=\frac{\Gamma'(z)}{\Gamma(z)}+\frac{G'(z)}{G(z)},$$

whence

$$\frac{G'(z+n+2)}{G(z+n+2)}=\sum_{k=0}^{n}\frac{\Gamma'(z+k+1)}{\Gamma(z+k+1)}+\frac{G'(z+1)}{G(z+1)}.$$

For sufficiently small values of $|z|$, by Taylor's theorem, we have

$$\frac{\Gamma'(z+1+k)}{\Gamma(z+1+k)}=\frac{\Gamma'(1+k)}{\Gamma(1+k)}+z\,\frac{d}{dk}\frac{\Gamma'(1+k)}{\Gamma(1+k)}+\frac{z^2}{2!}\frac{d^2}{dk^2}\frac{\Gamma'(1+k)}{\Gamma(1+k)}+\cdots,$$

provided that the coefficients in the expansion are finite.

On the other hand, it follows from 1.1(2) or 1.2(3) that

$$-\frac{\Gamma'(1+z)}{\Gamma(1+z)}=\gamma+\sum_{m=1}^{\infty}\left(\frac{1}{z+m}-\frac{1}{m}\right),$$

which, upon differentiating r times with respect to z and setting $z=k$ in the resulting equation, yields

$$\frac{d^r}{dk^r}\frac{\Gamma'(1+k)}{\Gamma(1+k)}=(-1)^{r-1}r!\sum_{m=1}^{\infty}\frac{1}{(m+k)^{r+1}} \quad (r\in\mathbb{N}\setminus\{1\}),$$

so that

$$\frac{\Gamma'(z+1+k)}{\Gamma(z+1+k)} - \frac{\Gamma'(1+k)}{\Gamma(1+k)} - z\,\frac{d}{dk}\,\frac{\Gamma'(1+k)}{\Gamma(1+k)}$$
$$= \sum_{r=2}^{\infty}\left\{(-1)^{r-1}\sum_{m=1}^{\infty}\frac{1}{(m+k)^{r+1}}\,z^r\right\},$$

from which we find that

$$\sum_{k=0}^{n}\left\{\frac{\Gamma'(z+1+k)}{\Gamma(z+1+k)} - \frac{\Gamma'(1+k)}{\Gamma(1+k)} - z\,\frac{d}{dk}\,\frac{\Gamma'(1+k)}{\Gamma(1+k)}\right\}$$
$$= \sum_{r=2}^{\infty}(-1)^{r-1}\left\{\sum_{k=0}^{n}\sum_{m=1}^{\infty}\frac{1}{(m+k)^{r+1}}\,z^r\right\},$$

where we are not rearranging a double series, if n is not actually infinite.

Now it is not difficult to see from Eisenstein's theorem [see Forsyth (1900, p. 87)] that

$$\lim_{n\to\infty}\sum_{k=0}^{n}\sum_{m=1}^{\infty}\frac{1}{(m+k)^{r+1}}$$

is convergent when $r \geq 2$. And hence

$$\lim_{n\to\infty}\sum_{k=0}^{n}\left\{\frac{\Gamma'(z+1+k)}{\Gamma(z+1+k)} - \frac{\Gamma'(1+k)}{\Gamma(1+k)} - z\,\frac{d}{dk}\,\frac{\Gamma'(1+k)}{\Gamma(1+k)}\right\}$$

is finite when $|z| < 1$ $(z \in \mathbb{C})$.

We also have the following relation:

$$-\frac{G'(z+1)}{G(z+1)} = \sum_{k=0}^{n}\left\{\frac{\Gamma'(z+1+k)}{\Gamma(z+1+k)} - \frac{\Gamma'(1+k)}{\Gamma(1+k)} - z\,\frac{d}{dk}\,\frac{\Gamma'(1+k)}{\Gamma(1+k)}\right\}$$
$$- \frac{G'(z+n+2)}{G(z+n+2)} + \sum_{k=0}^{n}\left\{\frac{\Gamma'(1+k)}{\Gamma(1+k)} + z\,\frac{d}{dk}\,\frac{\Gamma'(1+k)}{\Gamma(1+k)}\right\}.$$

Therefore, for sufficiently small values of $|z|$, we may take

$$-\frac{G'(z+1)}{G(z+1)} = \alpha + 2\beta z$$
$$+\frac{d}{dz}\log\left[\prod_{k=0}^{\infty}\left\{\frac{\Gamma(z+1+k)}{\Gamma(1+k)}\exp\left(-z\,\frac{\Gamma'(1+k)}{\Gamma(1+k)} - \frac{z^2}{2}\,\frac{d}{dk}\,\frac{\Gamma'(1+k)}{\Gamma(1+k)}\right)\right\}\right], \tag{10}$$

provided that

$$-\lim_{n\to\infty}\frac{G'(z+n+2)}{G(z+n+2)} = -\alpha - 2\beta z + \lim_{n\to\infty}\sum_{k=0}^{n}\left\{\frac{\Gamma'(1+k)}{\Gamma(1+k)} + z\,\frac{d}{dk}\,\frac{\Gamma'(1+k)}{\Gamma(1+k)}\right\},$$

and provided also that the expression on the right-hand side tends to the same value as that obtainable from (1). Thus we have

$$\frac{G'(z+n+2)}{G(z+n+2)} = (n+1+z)\log n + \frac{1}{2}\log(2\pi) - n + O\left(n^{-1}\right) \quad (n\to\infty).$$

If, then, we can prove that, for suitable values of α and β,

$$\begin{aligned}&-\alpha-2\beta z+\sum_{k=0}^{n}\left(\frac{\Gamma'(1+k)}{\Gamma(1+k)}+z\,\frac{d}{dk}\,\frac{\Gamma'(1+k)}{\Gamma(1+k)}\right)\\ &\qquad=(n+1+z)\log n+\frac{1}{2}\log(2\pi)-n+O\left(n^{-1}\right)\quad(n\to\infty),\end{aligned}\tag{11}$$

we shall have shown that, for suitable values of α and β, (10) holds true from the fact [see Barnes (1899b, p. 269)] that $G'(z+1)/G(z+1)$ is the only solution of

$$f(z+1) = f(z) + \frac{\Gamma'(z)}{\Gamma(z)},$$

which has such an asymptotic expansion near at infinity.

Now the left-hand side of (11) may be written as

$$-\alpha - 2\beta z + \sum_{k=1}^{n+1}\left\{\psi(k) + z\,\psi'(k)\right\},$$

which, in view of 1.2(57) and 1.2(58), reduces at once to

$$(n+1)\log n - n + z\left(\log n + 1 + \gamma - 2\beta\right) + \frac{1}{2} - \alpha + O\left(n^{-1}\right) \quad (n\to\infty),$$

which, upon putting

$$\alpha = \frac{1}{2} - \frac{1}{2}\log(2\pi) \quad \text{and} \quad \beta = \frac{1+\gamma}{2},\tag{12}$$

becomes the right-hand side of (7).

If we now substitute the values of α and β from (12) into (10), we obtain the following expression:

$$\begin{aligned}-\frac{G'(z+1)}{G(z+1)} &= \frac{1}{2} - \frac{1}{2}\log(2\pi) + (1+\gamma)\,z\\ &+\frac{d}{dz}\log\left[\prod_{k=0}^{\infty}\left\{\frac{\Gamma(z+1+k)}{\Gamma(1+k)}\exp\left(-z\,\psi(1+k)-\frac{1}{2}z^2\,\psi'(1+k)\right)\right\}\right]\\ &\qquad(|z|<1;\ z\in\mathbb{C}),\end{aligned}$$

which, upon integrating with respect to z and obtaining the value of the constant of integration by making $z=0$, finally yields the desired expression (4) valid for

$|z| < 1$. In fact, (4) holds true for all finite values of $|z|$ by the principle of analytic continuation.

From (4) we may also evaluate a series involving the ψ-function:

$$\sum_{k=1}^{\infty} \left\{ \psi(k) + \frac{1}{2}\,\psi'(k) - \log\, k \right\} = 1 + \frac{\gamma}{2} - \frac{1}{2}\,\log\,(2\pi) \tag{13}$$

by taking logarithms on both sides of (4) and setting $z = 1$ in the resulting equation. In view of 1.2(53), the left-hand side of (13) can be written in its equivalent form:

$$\sum_{k=1}^{\infty} \left\{ \psi(k) + \frac{1}{2}\,\zeta(2,k) - \log\, k \right\} = 1 + \frac{\gamma}{2} - \frac{1}{2}\,\log\,(2\pi). \tag{14}$$

In view of their need in evaluating integrals involving the double Gamma function, it may be important to know other special values of the G–function. Here we give two special values of the G-function as an illustration. The following special value of the Γ-function is known [see Spiegel (1968, p.1)]:

$$\Gamma\left(\frac{1}{4}\right) \cong 3.62560\,99082\,21908\,\ldots. \tag{15}$$

The Catalan constant G is defined by

$$G := \frac{1}{2}\int_0^1 \mathbf{K}(k)dk = \sum_{m=0}^{\infty} \frac{(-1)^m}{(2m+1)^2} \cong 0.91596\,55941\,77219\,015\ldots, \tag{16}$$

where $\mathbf{K}$ is the complete elliptic integral of the first kind, given by [*cf.* Equation 1.4(33)]

$$\mathbf{K}(k) := \int_0^{\pi/2} \frac{dt}{\sqrt{1-k^2\sin^2 t}} \qquad (|k| < 1). \tag{17}$$

The following integral is known [see Gradshteyn and Ryzhik (1980, p. 526, Entry 4.224)]:

$$\int_0^{\pi/4} \log\sin t\, dt = -\frac{\pi}{4}\log 2 - \frac{1}{2}G. \tag{18}$$

Considering (6), (18), and (28), we obtain

$$G\left(\frac{3}{4}\right) = 2^{-\frac{1}{8}} \cdot \pi^{-\frac{1}{4}} \cdot e^{\frac{G}{2\pi}} \cdot \Gamma\left(\frac{1}{4}\right) G\left(\frac{1}{4}\right). \tag{19}$$

We also recall a duplication formula for the G-function [see Choi (1996b, p. 290)]:

$$G(z)\left\{G\left(z+\frac{1}{2}\right)\right\}^2 G(z+1) = e^{\frac{1}{4}} \cdot A^{-3} \cdot 2^{-2z^2+3z-\frac{11}{12}} \cdot \pi^{z-\frac{1}{2}} \cdot G(2z), \tag{20}$$

which is an obvious special case ($n = 2$) of the following multiplication formula given by Barnes (1899, p. 291):

$$\prod_{i=0}^{n-1}\prod_{j=0}^{n-1} G\left(z+\frac{i+j}{n}\right) = \mathcal{K}(n)\,(2\pi)^{\frac{1}{2}n(n-1)z}\, n^{nz-\frac{1}{2}n^2z^2}\, G(nz), \tag{21}$$

where, for convenience,

$$\mathcal{K}(n) := A^{1-n^2}\, e^{\frac{1}{12}(n^2-1)}\,(2\pi)^{-\frac{1}{2}(n-1)}\, n^{-\frac{5}{12}}.$$

Setting $z = \frac{1}{4}$ in (20), and using (6) and (8), we obtain

$$\Gamma\left(\frac{1}{4}\right)\left\{G\left(\frac{1}{4}\right)G\left(\frac{3}{4}\right)\right\}^2 = 2^{-\frac{1}{4}}\cdot\pi^{-\frac{1}{2}}\cdot e^{\frac{3}{8}}\cdot A^{-\frac{9}{2}}. \tag{22}$$

Combining (19) and (22), we obtain

$$G\left(\frac{1}{4}\right) = e^{\frac{3}{32}-\frac{G}{4\pi}}\cdot A^{-\frac{9}{8}}\left\{\Gamma\left(\frac{1}{4}\right)\right\}^{-\frac{3}{4}} \cong 0.293756\ldots \tag{23}$$

or, equivalently,

$$G\left(\frac{3}{4}\right) = 2^{-\frac{1}{8}}\cdot\pi^{-\frac{1}{4}}\cdot e^{\frac{3}{32}+\frac{G}{4\pi}}\cdot A^{-\frac{9}{8}}\left\{\Gamma\left(\frac{1}{4}\right)\right\}^{\frac{1}{4}} \cong 0.848718\ldots. \tag{24}$$

It follows from (23), (24), and (6) that

$$\frac{G\left(\frac{3}{4}\right)}{G\left(\frac{5}{4}\right)} = 2^{-\frac{1}{8}}\cdot\pi^{-\frac{1}{4}}\cdot e^{\frac{G}{2\pi}}. \tag{25}$$

Integral formulas involving the double Gamma function

We begin by recalling the following integral formula:

$$\int_0^z \pi t\cot\pi t\,dt = z\,\log(2\pi) + \log\frac{G(1-z)}{G(1+z)}, \tag{26}$$

which is due originally to Kinkelin (1860). Indeed, in view of (3), if we set

$$\Phi(z) := \frac{G(1+z)}{G(1-z)} = (2\pi)^z e^{-z}\prod_{k=1}^{\infty}\left\{\left(\frac{1+\frac{z}{k}}{1-\frac{z}{k}}\right)^k e^{-2z}\right\}$$

and differentiate logarithmically with respect to z, apply the known expansion [see Ahlfors (1979, p. 188)]:

$$\pi z\,\cot\pi z = 1 + 2\sum_{n=1}^{\infty}\frac{z^2}{z^2-n^2}\quad (z\notin\mathbb{Z}), \tag{27}$$

we shall readily obtain

$$\frac{d}{dz}\log\Phi(z)=\log(2\pi)-\pi z\cot\pi z\ \ (z\notin\mathbb{Z}),$$

which, upon integration, yields (26), since $\Phi(0)=1$.

Some simple consequences of Kinkelin's formula (26) are worthy of note here [see also Barnes (1899b, p. 279)]. First of all, by using integration by parts in (26), we have

$$\int_0^z \log\sin\pi t\,dt = z\log\left(\frac{\sin\pi z}{2\pi}\right)+\log\frac{G(1+z)}{G(1-z)}, \tag{28}$$

which, upon setting $t=\frac{1}{2}-u$ and replacing z by $\frac{1}{2}-z$, yields

(29)

$$\int_0^z \log\cos\pi t\,dt=\left(z-\frac{1}{2}\right)\log\left(\frac{\cos\pi z}{2\pi}\right)-\frac{1}{2}\log 2-\log\Gamma\left(\frac{1}{2}-z\right)+\log\frac{G\left(\frac{1}{2}+z\right)}{G\left(\frac{1}{2}-z\right)}.$$

Making use of (29), we obtain the following analogue of (26):

(30)

$$\int_0^z \pi t\tan\pi t\,dt=-\frac{1}{2}\log\left(\frac{\cos\pi z}{\pi}\right)-z\log(2\pi)-\log\Gamma\left(\frac{1}{2}-z\right)+\log\frac{G\left(\frac{1}{2}+z\right)}{G\left(\frac{1}{2}-z\right)},$$

which would follow also from (26) by setting $t=\frac{1}{2}-u$ and replacing z by $\frac{1}{2}-z$.

Combining (28) and (29), we readily have the integral formula:

$$\begin{aligned}\int_0^z \log\tan\pi t\,dt =&z\log\tan\pi z+\frac{1}{2}\log\frac{\cos\pi z}{\pi}+\log\Gamma\left(\frac{1}{2}-z\right)\\&+\log\frac{G(1+z)}{G(1-z)}-\log\frac{G\left(\frac{1}{2}+z\right)}{G\left(\frac{1}{2}-z\right)}.\end{aligned}\tag{31}$$

Similarly, by using various trigonometric identities, we can obtain the following integral formulas:

$$\begin{aligned}\int_0^z\left(\frac{\pi t}{\cos\pi t}\right)^2 dt =&\pi z^2\tan\pi z+\log\frac{\cos\pi z}{\pi}+2z\log(2\pi)\\&+2\log\Gamma\left(\frac{1}{2}-z\right)-2\log\frac{G\left(\frac{1}{2}+z\right)}{G\left(\frac{1}{2}-z\right)};\end{aligned}\tag{32}$$

$$\int_0^z\left(\frac{\pi t}{\sin\pi t}\right)^2 dt=-\pi z^2\cot\pi z+2z\log(2\pi)-2\log\frac{G(1-z)}{G(1+z)}. \tag{33}$$

$$\int_0^z \pi t\tan\pi t\,dt=\log\frac{G(1-z)}{G(1+z)}-\frac{1}{2}\log\frac{G(1-2z)}{G(1+2z)}, \tag{34}$$

which, in view of the known duplication formula (20) for the G-function, is the same as (31);

$$\text{(35)}\qquad \int_0^z \frac{\pi t}{\sin \pi t} dt = z\log(2\pi) + 4\log\frac{G\left(1-\frac{1}{2}z\right)}{G\left(1+\frac{1}{2}z\right)} - \log\frac{G(1-z)}{G(1+z)};$$

$$\text{(36)}\qquad \begin{aligned} \int_0^z (\pi t\,\tan \pi t)^2\,dt = &-\frac{\pi^2 z^3}{3} + \pi z^2 \tan \pi z + \log\frac{\cos \pi z}{\pi} \\ &+ 2z\log(2\pi) + 2\log\frac{G\left(\frac{3}{2}-z\right)}{G\left(\frac{1}{2}+z\right)}; \end{aligned}$$

$$\text{(37)}\qquad \int_0^z (\pi t\,\cot \pi t)^2\,dt = -\frac{\pi^2 z^3}{3} - \pi z^2 \cot \pi z + 2z\log(2\pi) + 2\log\frac{G(1-z)}{G(1+z)};$$

$$\text{(38)}\qquad \begin{aligned} \int_0^z \left(\frac{\pi t}{\sin \pi t}\right)^2 \cos \pi t\,dt &= 2\int_0^{z/2} \left(\frac{\pi t}{\sin \pi t}\right)^2 dt - 2\int_0^{z/2} \left(\frac{\pi t}{\cos \pi t}\right)^2 dt \\ &= -\frac{\pi z^2}{\sin \pi z} - 2\log\frac{\cos\frac{\pi z}{2}}{\pi} + 4\log\frac{G\left(1-\frac{z}{2}\right)}{G\left(1+\frac{z}{2}\right)} + 4\log\frac{G\left(\frac{1}{2}+\frac{z}{2}\right)}{G\left(\frac{3}{2}-\frac{z}{2}\right)}; \end{aligned}$$

$$\text{(39)}\qquad \int_0^z \frac{\pi t}{\sin \pi t\,\cos \pi t} dt = -\frac{1}{2}\log\frac{\cos \pi z}{\pi} + \log\frac{G(1-z)}{G(1+z)} + \log\frac{G\left(\frac{1}{2}+z\right)}{G\left(\frac{3}{2}-z\right)}.$$

Replacing t by at/π in (28), we obtain

$$\text{(40)}\qquad \int_0^z \log\sin at\,dt = z\log\frac{\sin az}{2\pi} + \frac{\pi}{a}\log\frac{G\left(1+\frac{a}{\pi}z\right)}{G\left(1-\frac{a}{\pi}z\right)},$$

which, in view of the trigonometric identity:

$$\cos bt - \cos at = 2\sin\left[\frac{1}{2}(a+b)t\right]\,\sin\left[\frac{1}{2}(a-b)t\right],$$

yields

$$\text{(41)}\qquad \begin{aligned} \int_0^z \log(\cos bt - \cos at)\,dt &= z\log\frac{\sin\left[\frac{1}{2}(a+b)z\right]\sin\left[\frac{1}{2}(a-b)z\right]}{2\pi^2} \\ &+ \frac{2\pi}{a+b}\log\frac{G\left(1+\frac{a+b}{2\pi}z\right)}{G\left(1-\frac{a+b}{2\pi}z\right)} + \frac{2\pi}{a-b}\log\frac{G\left(1+\frac{a-b}{2\pi}z\right)}{G\left(1-\frac{a-b}{2\pi}z\right)}. \end{aligned}$$

By differentiating Alexeiewsky's theorem [see Barnes (1899b, p. 281)]:

$$\text{(42)}\qquad \int_0^z \log\Gamma(t+1)\,dt = \frac{1}{2}[\log(2\pi)-1]z - \frac{z^2}{2} + z\log\Gamma(z+1) - \log G(z+1)$$

with respect to z, we obtain

$$\frac{d}{dz}\log G(z)=\frac{1}{2}[\log(2\pi)+1]-z+(z-1)\psi(z), \tag{43}$$

and so we have

$$\begin{aligned}&\frac{\partial}{\partial a}\log G(h(az))\\&=\left\{\frac{1}{2}[\log(2\pi)+1]-h(az)+[h(az)-1]\psi(h(az))\right\}\frac{\partial}{\partial a}h(az),\end{aligned} \tag{44}$$

where $h(z)$ is a function of z and a is a constant.

Differentiating each side of (41) with respect to a, and using (44) and 1.2(9), we obtain

$$\begin{aligned}\int_0^z \frac{t\sin at}{\cos bt-\cos at}dt&=\frac{2az}{a^2-b^2}\log(2\pi)\\&-\frac{2\pi}{(a+b)^2}\log\frac{G\left(1-\frac{a+b}{2\pi}z\right)}{G\left(1+\frac{a+b}{2\pi}z\right)}-\frac{2\pi}{(a-b)^2}\log\frac{G\left(1-\frac{a-b}{2\pi}z\right)}{G\left(1+\frac{a-b}{2\pi}z\right)}.\end{aligned} \tag{45}$$

Replacing t by $\frac{1}{\pi}\arctan at$ in (31), we obtain

$$\begin{aligned}\int_0^z \frac{\log t}{1+a^2t^2}dt&=\left(\frac{\log z}{a}\right)\arctan az+\frac{\pi}{2a}\log\left(\frac{\cos\arctan az}{\pi}\right)\\&+\frac{\pi}{a}\log\frac{G\left(1+\frac{1}{\pi}\arctan az\right)G\left(\frac{3}{2}-\frac{1}{\pi}\arctan az\right)}{G\left(1-\frac{1}{\pi}\arctan az\right)G\left(\frac{1}{2}+\frac{1}{\pi}\arctan az\right)}.\end{aligned} \tag{46}$$

Setting $a=1$ and $z=1$ in (46), and using 1.2(4), we obtain

$$\int_0^1 \frac{\log t}{1+t^2}dt=-G. \tag{47}$$

It is easy to verify that

$$\int_0^z \frac{a}{1+a^2+2a\cos t}dt=\frac{2a}{1-a^2}\arctan\left(\frac{1-a}{1+a}\tan\frac{1}{2}z\right)$$

and

$$\int_0^z \frac{\cos t}{1+a^2+2a\cos t}dt=\frac{z}{2a}-\frac{1+a^2}{a(1-a^2)}\arctan\left(\frac{1-a}{1+a}\tan\frac{1}{2}z\right).$$

Using these identities and applying the Leibniz rule [see De Lillo (1982, p. 665)] to the following integral and again integrating the resulting equation with respect to a from 1 to a, we obtain

$$\begin{aligned}\int_0^z \log\left(\frac{1+a^2+2a\cos t}{2+2\cos t}\right)dt&=z\log a-2\int_1^a \arctan\left(\frac{1-t}{1+t}\tan\frac{1}{2}z\right)\frac{dt}{t}\\&=(2z-\pi)\log a+2\int_1^a \arctan\left(\frac{t+\cos z}{\sin z}\right)\frac{dt}{t}.\end{aligned}$$

Equivalently, from (29) and the following integral:

$$\int_0^z \log(2+2\cos t)\,dt = (2z-2\pi)\log\left(\frac{\cos\left(\frac{1}{2}z\right)}{\pi}\right) + 4\pi\log\frac{G\left(\frac{1}{2}+\frac{z}{2\pi}\right)}{G\left(\frac{3}{2}-\frac{z}{2\pi}\right)},$$

we have

$$\begin{aligned}\int_0^z \log(1+a^2+2a\cos t)\,dt &= \pi\log a + (2z-2\pi)\log\left(\frac{a\cos\left(\frac{1}{2}z\right)}{\pi}\right)\\ &+ 4\pi\log\frac{G\left(\frac{1}{2}+\frac{z}{2\pi}\right)}{G\left(\frac{3}{2}-\frac{z}{2\pi}\right)} + 2\int_1^a \arctan\left(\frac{t+\cos z}{\sin z}\right)\frac{dt}{t}.\end{aligned}$$

Applying integration by parts to the last integral gives us the following equivalent form:

$$\begin{aligned}\int_0^z \log(1+a^2+2a\cos t)\,dt &= \pi\log a + (z-\pi)\log\left(\frac{a^2\cot\left(\frac{1}{2}z\right)}{2\pi^2}\right)\\ &+ 2\arctan\left(\frac{a+\cos z}{\sin z}\right)\log\left(\frac{a}{\sin z}\right) + 4\pi\log\frac{G\left(\frac{1}{2}+\frac{z}{2\pi}\right)}{G\left(\frac{3}{2}-\frac{z}{2\pi}\right)}\\ &- 2\int_{\cot\frac{1}{2}z}^{\frac{a+\cos z}{\sin z}} \frac{\log(t-\cot z)}{1+t^2}\,dt.\end{aligned} \tag{48}$$

Setting $z=\frac{1}{2}\pi$ in (48), and using 1.2(4), (46) and (47), we obtain

$$\begin{aligned}\int_0^{\pi/2} \log(1+a^2+2a\cos t)\,dt &= -\pi\log\left(\frac{\cos(\arctan a)}{\pi}\right)\\ &- 2\pi\log\frac{G\left(1+\frac{1}{\pi}\arctan a\right)G\left(\frac{3}{2}-\frac{1}{\pi}\arctan a\right)}{G\left(1-\frac{1}{\pi}\arctan a\right)G\left(\frac{1}{2}+\frac{1}{\pi}\arctan a\right)}.\end{aligned} \tag{49}$$

Integrating by parts and using (46), we have

$$\begin{aligned}\int_0^z \frac{\arctan at}{t}\,dt &= -\frac{\pi}{2}\log\left(\frac{\cos(\arctan az)}{\pi}\right)\\ &- \pi\log\frac{G\left(1+\frac{1}{\pi}\arctan az\right)G\left(\frac{3}{2}-\frac{1}{\pi}\arctan az\right)}{G\left(1-\frac{1}{\pi}\arctan az\right)G\left(\frac{1}{2}+\frac{1}{\pi}\arctan az\right)}.\end{aligned} \tag{50}$$

Replacing t by $\frac{1}{\pi}\arcsin at$ in (28), we get

$$\int_0^z \frac{\log t\,dt}{\sqrt{1-a^2t^2}} = \left(\frac{\arcsin az}{a}\right)\log\left(\frac{z}{2\pi}\right) + \frac{\pi}{a}\log\frac{G\left(1+\frac{1}{\pi}\arcsin az\right)}{G\left(1-\frac{1}{\pi}\arcsin az\right)}. \tag{51}$$

Using (51) and the following relation:

$$\int_0^z \frac{\arcsin at}{t}\,dt = (\arcsin az)\log z - a\int_0^z \frac{\log t\,dt}{\sqrt{1-a^2t^2}},$$

we obtain

$$\int_0^z \frac{\arcsin at}{t}\,dt = \log(2\pi)(\arcsin az) + \pi \log \frac{G\left(1 - \frac{1}{\pi}\arcsin az\right)}{G\left(1 + \frac{1}{\pi}\arcsin az\right)}. \tag{52}$$

Integrating by parts and using (46), we find that

$$\begin{aligned}
&\int_0^z \frac{(\arctan at)^2 dt}{t^2} \\
&= -\frac{(\arctan az)^2}{z} - \pi a \log\left(\frac{\cos(\arctan az)}{\pi}\right) - a\,\log(1 + a^2z^2)\,\arctan az \\
&+ 2\pi a \log \frac{G\left(1 - \frac{1}{\pi}\arctan az\right) G\left(\frac{1}{2} + \frac{1}{\pi}\arctan az\right)}{G\left(1 + \frac{1}{\pi}\arctan az\right) G\left(\frac{3}{2} - \frac{1}{\pi}\arctan az\right)} \\
&+ a^2 \int_0^z \frac{\log(1 + a^2t^2)}{1 + a^2t^2}\,dt.
\end{aligned} \tag{53}$$

Numerous special cases of the integral formulas (considered in this section) include the following results:

$$\int_0^{\pi/2} \left(\frac{x}{\sin x}\right)^2 \cos x\,dx = -\frac{\pi^2}{4} + 4G; \tag{54}$$

$$\int_0^{1/2} \log\Gamma(t+1)dt = -\frac{1}{2} - \frac{7}{24}\log 2 + \frac{1}{4}\log\pi + \frac{3}{2}\log A; \tag{55}$$

$$\int_0^{1/4} \log\Gamma(t+1)dt = -\frac{1}{4} - \frac{3}{8}\log 2 + \frac{1}{8}\log\pi + \frac{9}{8}\log A + \frac{G}{4\pi}; \tag{56}$$

$$\int_0^1 \frac{\arctan x}{x}\,dx = G; \tag{57}$$

$$\int_0^{1/\sqrt{2}} \frac{\arcsin x}{x}\,dx = \frac{\pi}{8}\log 2 + \frac{G}{2}; \tag{58}$$

$$\int_0^{\pi/4} x\cot x\,dx = \frac{\pi}{8}\log 2 + \frac{G}{2}; \tag{59}$$

$$\int_0^{\pi/4} \log\cos x\,dx = -\frac{\pi}{4}\log 2 + \frac{G}{2}; \tag{60}$$

$$\int_0^{\pi/4} x\tan x\,dx = -\frac{\pi}{8}\log 2 + \frac{G}{2}; \tag{61}$$

(62) $$\int_0^{\pi/4} \log\tan x\, dx = -G;$$

(63) $$\int_0^{\pi/4} \left(\frac{x}{\cos x}\right)^2 dx = \frac{\pi^2}{16} + \frac{\pi}{4}\log 2 - G;$$

(64) $$\int_0^{\pi/4} \left(\frac{x}{\sin x}\right)^2 dx = -\frac{\pi^2}{16} + \frac{\pi}{4}\log 2 + G;$$

(65) $$\int_0^{\pi/4} x^2\tan^2 x\, dx = -\frac{\pi^3}{192} + \frac{\pi^2}{16} + \frac{\pi}{4}\log 2 - G;$$

(66) $$\int_0^{\pi/4} x^2\cot^2 x\, dx = -\frac{\pi^3}{192} - \frac{\pi^2}{16} + \frac{\pi}{4}\log 2 + G;$$

(67) $$\int_0^{\pi/2} \frac{x}{\sin x} dx = 2G.$$

The evaluation of an integral involving $\log G(z)$

First we shall introduce two interesting mathematical constants, in addition to the Glaisher-Kinkelin constant A, by recalling the Euler-Maclaurin summation formula [*cf.* Hardy (1963, p. 318)]:

(68) $$\sum_{k=1}^{n} f(k) \sim C_0 + \int_a^n f(x)\, dx + \frac{1}{2} f(n) + \sum_{r=1}^{\infty} \frac{B_{2r}}{(2r)!} f^{(2r-1)}(n),$$

where C_0 is an arbitrary constant to be determined in each special case and

$$B_0 = 1,\ B_1 = -\frac{1}{2},\ B_2 = \frac{1}{6},\ B_4 = -\frac{1}{30},\ B_6 = \frac{1}{42},\ B_8 = -\frac{1}{30},\ B_{10} = \frac{5}{66}, \dots,$$

$$\text{and } B_{2n+1} = 0 \quad (n \in \mathbb{N})$$

are the Bernoulli numbers (see Section 1.6). For another useful summation formula, see Edwards (1954, p. 117).

Letting $f(x) = x^2\log x$ and $f(x) = x^3\log x$ in (68) with $a = 1$, we obtain

(69) $$\log B = \lim_{n\to\infty} \left[\sum_{k=1}^{n} k^2\log k - \left(\frac{n^3}{3} + \frac{n^2}{2} + \frac{n}{6}\right)\log n + \frac{n^3}{9} - \frac{n}{12}\right]$$

and

$$\log C = \lim_{n\to\infty}\left[\sum_{k=1}^{n} k^3 \log k - \left(\frac{n^4}{4}+\frac{n^3}{2}+\frac{n^2}{4}-\frac{1}{120}\right)\log n + \frac{n^4}{16}-\frac{n^2}{12}\right], \tag{70}$$

respectively; here B and C are constants whose approximate numerical values are given by

$$B \cong 1.03091\,675\ldots \tag{71}$$

and

$$C \cong 0.97955\,746\ldots. \tag{72}$$

The constants B and C were considered recently by Choi and Srivastava (1999a, p. 102; 2000a). See also Adamchik (1998, p. 199). Bendersky (1933) presented a set of constants including B and C.

It is known that [*cf.* Barnes (1899b, p. 288)]

$$\int_0^1 \log G(t+1)dt = \frac{1}{12}+\frac{1}{4}\log(2\pi) - 2\log A. \tag{73}$$

Setting $z = t$ in the logarithmic form of (3), and integrating both sides of the resulting equation from $t = 0$ to $t = \frac{1}{2}$, we obtain

$$\begin{aligned}\int_0^{\frac{1}{2}} \log G(t+1)dt &= \frac{1}{16}\log(2\pi) - \frac{\gamma}{48} - \frac{1}{12}\\ &+ \frac{1}{4}\sum_{k=1}^{\infty}\left[(4k^2+2k)\log(2k+1) - (4k^2+2k)\log(2k) - 2k - \frac{1}{2} + \frac{1}{12k}\right].\end{aligned} \tag{74}$$

Consider the sum S_n given by

$$\begin{aligned} S_n &:= \sum_{k=1}^{n}\left[(4k^2+2k)\log(2k+1) - (4k^2+2k)\log(2k)\right]\\ &= \sum_{k=1}^{n}\left[(2k+1)^2\log(2k+1) - (2k+1)\log(2k+1) - (2k)^2\log(2k) - 2k\log(2k)\right]\\ &= \left[\sum_{k=1}^{2n+1} k^2\log k - 2\sum_{k=1}^{n}(2k)^2\log(2k) - \sum_{k=1}^{2n+1} k\log k\right]\\ &= \left[\sum_{k=1}^{2n+1} k^2\log k - 8\sum_{k=1}^{n} k^2\log k - \sum_{k=1}^{2n+1} k\log k - 8\log 2\sum_{k=1}^{n} k^2\right],\end{aligned}$$

which, in view of (2) and (69), yields
(75)

$$\begin{aligned}S_n = &\left[\log B + \left\{\frac{(2n+1)^3}{3} + \frac{(2n+1)^2}{2} + \frac{(2n+1)}{6}\right\}\log(2n+1)\right.\\&- \frac{(2n+1)^3}{9} + \frac{(2n+1)}{12} - 8\log B - 8\left(\frac{n^3}{3} + \frac{n^2}{2} + \frac{n}{6}\right)\log n + \frac{8}{9}n^3 - \frac{8}{12}n\\&- \log A - \left\{\frac{(2n+1)^2}{2} + \frac{(2n+1)}{2} + \frac{1}{12}\right\}\log(2n+1)\\&\left.+\frac{(2n+1)^2}{4} - 8\log 2\frac{n(n+1)(2n+1)}{6}\right] + O\left(n^{-1}\right) \quad (n \to \infty)\\&= -\log A - 7\log B + \frac{2}{9} + \lim_{n\to\infty}\left[\left(\frac{8}{3}n^3 + 4n^2 + \frac{4}{3}n - \frac{1}{12}\right)\log(2n+1)\right.\\&- \left(\frac{8}{3}n^3 + 4n^2 + \frac{4}{3}n\right)\log n - \frac{1}{3}n^2 - \frac{1}{6}n\\&\left.- \left(\frac{8}{3}\log 2\right)n^3 - (4\log 2)n^2 - \left(\frac{4}{3}\log 2\right)n\right] + O\left(n^{-1}\right) \quad (n \to \infty).\end{aligned}$$

Now, substituting from (75) and recalling 1.1(3), and applying the following expansion:

$$\text{(76)} \qquad \log\left(1 + \frac{1}{2n}\right) = \frac{1}{2n} - \frac{1}{8n^2} + \frac{1}{24n^3} - \frac{d_n}{64n^4} \quad (0 < d_n < 1;\ n \in \mathbb{N}),$$

we evaluate the summation part of (74) as follows:

$$\begin{aligned}T_n := &-\log A - 7\log B + \frac{2}{9}\\&+\left[\left(\frac{8}{3}n^3 + 4n^2 + \frac{4}{3}n - \frac{1}{12}\right)\log(2n+1)\right.\\&- \left(\frac{8}{3}n^3 + 4n^2 + \frac{4}{3}n\right)\log n - \frac{1}{3}n^2 - \frac{1}{6}n\\&- \left(\frac{8}{3}\log 2\right)n^3 - (4\log 2)n^2 - \left(\frac{4}{3}\log 2\right)n\\&\left.-2\frac{n(n+1)}{2} - \frac{n}{2} + \frac{1}{12}(\gamma + \log n)\right] + O\left(n^{-1}\right) \quad (n \to \infty)\\&= \frac{\gamma}{12} - \log A - 7\log B - \frac{1}{12}\\&+\left[\left(\frac{8}{3}n^3 + 4n^2 + \frac{4}{3}n - \frac{1}{12}\right)\log 2\right.\\&+\left(\frac{8}{3}n^3 + 4n^2 + \frac{4}{3}n - \frac{1}{12}\right)\log\left(1 + \frac{1}{2n}\right)\\&\left.-\left(\frac{8}{3}\log 2\right)n^3 - \left(4\log 2 + \frac{4}{3}\right)n^2 - \left(\frac{4}{3}\log 2 + \frac{5}{3}\right)n\right] + O\left(n^{-1}\right) \quad (n \to \infty).\end{aligned}$$

Thus we have

$$\begin{aligned} T_n =& \frac{\gamma}{12} - \log A - 7\log B - \frac{1}{12} \\ &+ \left[\left(\frac{8}{3}n^3 + 4n^2 + \frac{4}{3}n - \frac{1}{12} \right) \log 2 + \frac{4}{3}n^2 + \frac{5}{3}n + \frac{5}{18} \right. \\ &\left. - \left(\frac{8}{3}\log 2 \right) n^3 - \left(4\log 2 + \frac{4}{3} \right) n^2 - \left(\frac{4}{3}\log 2 + \frac{5}{3} \right) n \right] + O\left(n^{-1}\right) \quad (n \to \infty), \end{aligned}$$

which, upon taking the limit as $n \to \infty$ and using (74), yields our desired result:

$$\int_0^{\frac{1}{2}} \log G(t+1)dt = \frac{1}{24}(\log 2 + 1) + \frac{1}{16}\log \pi - \frac{1}{4}\log A - \frac{7}{4}\log B \tag{77}$$

in terms of the mathematical constant B defined by (69).

Vignéras (1979) gave an integral representation for the double Gamma function Γ_2:

$$\begin{aligned} \log \Gamma_2(z+1) =& - \int_0^\infty \frac{e^{-t}}{t\left(1-e^{-t}\right)^2} \left(1 - zt - \frac{z^2t^2}{2} - e^{-zt} \right) dt \\ &+ (1+\gamma)\frac{z^2}{2} - \frac{3}{2}\log \pi \quad (\Re(z) > -1). \end{aligned} \tag{78}$$

The multiple Gamma functions

There are two known ways to define an n-ple Gamma functions Γ_n: Barnes (1904) [see also Vardi (1988b)] defined Γ_n by using the n-ple Hurwitz Zeta functions [see Choi and Quine (1992); also Seo *et al.* (1993)]; A recurrence formula of the Weierstrass canonical product forms of the n-ple Gamma functions Γ_n was given by Vignéras (1979) who used the theorem of Dufresnoy and Pisot (1963) which provides the existence, uniqueness, and expansion of the series of Weierstrass satisfying a certain functional equation.

By making use of the aforementioned Dufresnoy-Pisot theorem and starting with

$$f_1(x) = -\gamma x + \sum_{n=1}^{\infty} \left\{ \frac{x}{n} - \log\left(1 + \frac{x}{n}\right) \right\}, \tag{79}$$

Vignéras (1979) obtained a recurrence formula of Γ_n $(n \in \mathbb{N})$ which is given by

Theorem 1.3. *The n-ple Gamma functions Γ_n are defined by*

$$\Gamma_n(z) = \{G_n(z)\}^{(-1)^{n-1}} \quad (n \in \mathbb{N}), \tag{80}$$

where

$$G_n(z+1) = \exp\left(f_n(z)\right) \tag{81}$$

and the functions $f_n(z)$ are given by

$$f_n(z) = -x\,A_n(1) + \sum_{k=1}^{n-1} \frac{p_k(z)}{k!}\left[f_{n-1}^{(k)}(0) - A_n^{(k)}(1)\right] + A_n(z), \tag{82}$$

with

$$\begin{aligned} A_n(z) = \sum_{m\in\mathbb{N}_0{}^{n-1}\times\mathbb{N}} &\left[\frac{1}{n}\left(\frac{z}{L(m)}\right)^n - \frac{1}{n-1}\left(\frac{z}{L(m)}\right)^{n-1} + \cdots\right. \\ &\left. +(-1)^{n-1}\frac{z}{L(m)} + (-1)^n \log\left(1+\frac{z}{L(m)}\right)\right], \end{aligned} \tag{83}$$

where

$$L(m) = m_1 + m_2 + \cdots + m_n \quad if \quad m = (m_1, m_2, \ldots, m_n) \in \mathbb{N}_0{}^{n-1}\times\mathbb{N}$$

and the polynomials

$$p_n(z) = 1^n + 2^n + \cdots + (z-1)^n$$

satisfy the following relations:

$$p_n'(z) = B_n(z) \quad and \quad p_n(0) = 0, \tag{84}$$

$B_n(z)$ being the nth Bernoulli polynomials.

By analogy with the Bohr-Mollerup theorem (Theorem 1.1), which guarantees the uniqueness of the Gamma function Γ, one can give for the double Gamma function and (more generally) for the multiple Gamma functions of order n ($n \in \mathbb{N}$) a definition of Artin by means of the following theorem [see Vignéras (1979, p. 239)].

Theorem 1.4. *For all $n \in \mathbb{N}$, there exists a unique meromorphic function $G_n(z)$ satisfying each of the following properties:*
(a) $G_n(z+1) = G_{n-1}(z)\,G_n(z) \quad (z \in \mathbb{C})$;
(b) $G_n(1) = 1$;
(c) *For $x \geq 1$, $G_n(x)$ are infinitely differentiable and*

$$\frac{d^{n+1}}{dx^{n+1}}\{\log G_n(x)\} \geq 0;$$

(d) $G_0(x) = x$.

It may be remarked in passing that

$$G_1(z) = \Gamma_1(z) = \Gamma(z)$$

and the case $n = 2$ of (80) can readily be seen to produce the double Gamma function (5).

We observe that the function $\{\Gamma_n(z)\}^{-1}$ defined in (80) is an entire function of order n, all of whose zeros are the nonpositive integers $0, -1, -2, \ldots$, just as the Gamma function $\{\Gamma(z)\}^{-1}$, and the multiplicity of the zeros of $\{\Gamma_n(z)\}^{-1}$ at $z=-k$ $(k \in \mathbb{N}_0)$ is equal to the number of solutions of the equation:

$$L(m)=k+1 \quad \left(m \in {\mathbb{N}_0}^{n-1} \times \mathbb{N}\right).$$

The number of the solutions for $m \in {\mathbb{N}_0}^n$ is the coefficient of x^{k+1} in the Maclaurin series expansion of

$$(1-x)^{-n}, \qquad \text{that is,} \qquad \binom{n+k}{k+1}.$$

Therefore, the number of the solutions for $m \in {\mathbb{N}_0}^{n-1} \times \mathbb{N}$ is equal to

$$\binom{n+k}{k+1} - \binom{n+k-1}{k+1} = \binom{n+k-1}{n-1}.$$

We thus conclude that $\{\Gamma_n(z)\}^{-1}$ is an entire function with zeros at $z=-k$ $(k \in \mathbb{N}_0)$ whose multiplicities are

$$\binom{n+k-1}{n-1} \quad (n \in \mathbb{N};\ k \in \mathbb{N}_0). \tag{85}$$

The triple Gamma function Γ_3

When $n=3$ in (80), we readily obtain an explicit form of the *triple Gamma function* Γ_3:

$$\begin{aligned} \Gamma_3(1+z) &= G_3(1+z) \\ &= \exp\left[-\frac{1}{6}\left(\gamma+\frac{\pi^2}{6}+\frac{3}{2}\right)z^3+\frac{1}{4}\left(\gamma+\log(2\pi)+\frac{1}{2}\right)z^2+\Omega z\right] \\ &\cdot \prod_{m \in {\mathbb{N}_0}^2 \times \mathbb{N}} \left\{\left(1+\frac{z}{L(m)}\right)^{-1} \exp\left[\frac{z}{L(m)}-\frac{1}{2}\left(\frac{z}{L(m)}\right)^2+\frac{1}{3}\left(\frac{z}{L(m)}\right)^3\right]\right\}, \end{aligned} \tag{86}$$

where

$$\Omega = \frac{1}{12}\left(\frac{3}{2}-\gamma-3\log(2\pi)+\frac{\pi^2}{12}\right)+\frac{1}{2}\sum_{n=1}^{\infty}(-1)^n\frac{\zeta(n+2)}{(n+3)(n+4)}$$

and

$$L(m)=m_1+m_2+m_3 \quad \text{with} \quad m=(m_1,m_2,m_3) \in {\mathbb{N}_0}^2 \times \mathbb{N},$$

$\zeta(s)$ being the Riemann Zeta function (see Section 2.3). Now the infinite sum in Ω can be evaluated explicitly by using a known formula [see Choi and Srivastava (1997, p. 116, Eq. (2.63))]:

$$\sum_{k=3}^{\infty}(-1)^k\frac{\zeta(k)}{(k+1)(k+2)} = \frac{1}{2}+\frac{\gamma}{6}-\frac{\pi^2}{72}-2\log A. \tag{87}$$

We thus find that

$$\Omega = \frac{3}{8}-\frac{1}{4}\log(2\pi)-\log A \tag{88}$$

in terms of the Glaisher-Kinkelin constant A defined by (2).

If we set $n = 3$ in (85), we observe that $\{\Gamma_3(z)\}^{-1}$ is an entire function with zeros at $z = -k$ $(k \in \mathbb{N}_0)$ whose multiplicity is

$$\frac{1}{2}(k^2 + 3k + 2) \quad (k \in \mathbb{N}_0). \tag{89}$$

Furthermore, in view of (89), (86) can be written in the following equivalent form analogous to (3):

$$\begin{aligned}\Gamma_3(1+z) =& G_3(1+z) \\ =& e^{R(z)} \prod_{k=1}^{\infty} \left\{ \left(1+\frac{z}{k}\right)^{-\frac{1}{2}k(k+1)} \right. \\ &\left. \cdot \exp\left[\frac{1}{2}(k+1)z - \frac{1}{4}\left(1+\frac{1}{k}\right)z^2 + \frac{1}{6k}\left(1+\frac{1}{k}\right)z^3\right]\right\},\end{aligned} \tag{90}$$

where, for convenience,

$$R(z) := -\frac{1}{6}\left(\gamma + \frac{\pi^2}{6} + \frac{3}{2}\right)z^3 + \frac{1}{4}\left(\gamma + \log(2\pi) + \frac{1}{2}\right)z^2 + \left(\frac{3}{8} - \frac{1}{4}\log(2\pi) - \log A\right)z.$$

It follows that Γ_3 satisfies several basic properties and characteristics, which are summarized here in

Theorem 1.5. *The triple Gamma function* Γ_3 *is the unique meromorphic function satisfying each of the following properties*:

(a) $\Gamma_3(1) = 1$;

(b) $\Gamma_3(z+1) = G(z)\Gamma_3(z)$ $(z \in \mathbb{C})$;

(c) *For* $x \geq 1$, $\Gamma_3(x)$ *is infinitely differentiable and*

$$\frac{d^4}{dx^4}\{\log \Gamma_3(x)\} \geq 0.$$

Just as in (8), we now proceed to express the value of $\Gamma_3\left(\frac{1}{2}\right)$ in terms of the mathematical constants π, e, A, and B. We begin by recalling the following known results:

$$\frac{1}{2}\log(2\pi) = \lim_{n\to\infty}\left[\sum_{k=1}^{n}\log k - \left(n+\frac{1}{2}\right)\log n + n\right] \tag{91}$$

and

$$\log\left(1+\frac{1}{n}\right) = \frac{1}{n} - \frac{1}{2n^2} + \frac{1}{3n^3} + O\left(n^{-4}\right) \quad (n \to \infty). \tag{92}$$

By taking logarithms on both sides of (90), and setting $z = \frac{1}{2}$ in the resulting equation, if we make use of 1.1(3), we obtain

$$\begin{aligned}\log\Gamma_3\left(1+\frac{1}{2}\right) =& \frac{3}{16} - \frac{1}{16}\log(2\pi) - \frac{1}{2}\log A \\ &+ \lim_{n\to\infty}\left[-\sum_{k=1}^{n}\frac{k(k+1)}{2}\log\left(1+\frac{1}{2k}\right) + \frac{n^2}{8} + \frac{5}{16}n - \frac{1}{24}\log n\right].\end{aligned} \tag{93}$$

We first consider the following sum:

$$\begin{aligned}U_n :&= \sum_{k=1}^{n} \frac{k(k+1)}{2} \log\left(1+\frac{1}{2k}\right) \\ &= \frac{1}{2}\sum_{k=1}^{n} k(k+1)\log(2k+1) \\ &\quad - \frac{\log 2}{2}\sum_{k=1}^{n} k(k+1) - \frac{1}{2}\sum_{k=1}^{n} k^2\log k - \frac{1}{2}\sum_{k=1}^{n} k\log k \\ &= \frac{1}{8}\left(\sum_{k=1}^{n}(2k+1)^2\log(2k+1) - \sum_{k=1}^{n}\log(2k+1)\right) \\ &\quad - \frac{\log 2}{2}\sum_{k=1}^{n} k(k+1) - \frac{1}{2}\sum_{k=1}^{n} k^2\log k - \frac{1}{2}\sum_{k=1}^{n} k\log k\end{aligned}$$

and so

$$\begin{aligned}U_n &= \frac{1}{8}\left(\sum_{k=1}^{2n+1} k^2\log k - 4\sum_{k=1}^{n} k^2\log k - 4\log 2\sum_{k=1}^{n} k^2\right. \\ &\quad \left. - \sum_{k=1}^{2n+1}\log k + \sum_{k=1}^{n}\log k + n\log 2\right) \\ &\quad - \frac{\log 2}{2}\sum_{k=1}^{n} k(k+1) - \frac{1}{2}\sum_{k=1}^{n} k^2\log k - \frac{1}{2}\sum_{k=1}^{n} k\log k,\end{aligned}$$

which immediately leads us to

$$\begin{aligned}U_n =&\frac{1}{8}\sum_{k=1}^{2n+1} k^2\log k - \frac{1}{8}\sum_{k=1}^{2n+1}\log k - \sum_{k=1}^{n} k^2\log k - \frac{1}{2}\sum_{k=1}^{n} k\log k \\ &+ \frac{1}{8}\sum_{k=1}^{n}\log k - \left(\frac{n^3}{3}+\frac{3}{4}n^2+\frac{7}{24}n\right)\log 2.\end{aligned} \tag{94}$$

Upon substituting from (94) into (93), if we apply (2), (69), and (91), we obtain

$$\begin{aligned}\log\Gamma_3\left(1+\frac{1}{2}\right) &= \frac{3}{16} - \frac{1}{16}\log(2\pi) + \frac{7}{8}\log B \\ &\quad + \lim_{n\to\infty}\left[-\left(\frac{n^3}{3}+\frac{3}{4}n^2+\frac{7}{24}n-\frac{1}{16}\right)\log\left(1+\frac{1}{2n}\right)\right. \\ &\quad \left. +\frac{n^2}{6}+\frac{n}{3}-\frac{35}{288}+\frac{1}{16}\log 2\right] \\ &= \frac{3}{16} - \frac{1}{16}\log\pi + \frac{7}{8}\log B \\ &\quad + \lim_{n\to\infty}\left[-\frac{n^2}{6}-\frac{n}{3}-\frac{19}{288}+O\left(n^{-1}\right)+\frac{n^2}{6}+\frac{n}{3}-\frac{35}{288}\right] \\ &= -\frac{1}{16}\log\pi + \frac{7}{8}\log B,\end{aligned}$$

where we have also used (92) for the second equality. Thus we find that

$$\Gamma_3\left(1+\frac{1}{2}\right)=\pi^{-\frac{1}{16}}\cdot B^{\frac{7}{8}}, \tag{95}$$

which, in view of (8) and the assertion (b) of Theorem 1.5, yields

$$\Gamma_3\left(\frac{1}{2}\right)=2^{-\frac{1}{24}}\cdot\pi^{\frac{3}{16}}\cdot e^{-\frac{1}{8}}\cdot A^{\frac{3}{2}}\cdot B^{\frac{7}{8}}. \tag{96}$$

1.4 The Gaussian Hypergeometric Function and Its Generalization

The Gauss hypergeometric equation

The second-order linear ordinary differential equation:

$$z(1-z)\frac{d^2w}{dz^2}+[c-(a+b+1)\,z]\frac{dw}{dz}-abw=0 \tag{1}$$

or, equivalently,

$$\{\delta(\delta+c-1)-z(\delta+a)(\delta+b)\}w=0 \quad \left(\delta:=z\frac{d}{dz}\right), \tag{2}$$

in which a, b, and c are real or complex parameters, is called the *Gauss hypergeometric equation.* Its only singularities are at

$$z=0,\ 1,\ \infty;$$

each singularity is easily seen to be of *regular* kind.

The hypergeometric equation (1) or (2) is the most celebrated equation of the *Fuchsian class* which consists of differential equations whose only singularities (including the point at infinity) are regular singular points. Its importance stems, in part, from a well-known theorem that *every homogeneous linear differential equation of the second order, whose singularities (including the point at infinity) are regular and at most three in number, can be transformed into the hypergeometric equation.*

Power-series solutions of the hypergeometric equation (1) valid in the neighborhoods of the regular singular points $z=0$, 1, or ∞ can be developed by direct application of the classical method of Frobenius. Thus, if c is not an integer, the general solution of (1) valid in a neighborhood of the origin ($z=0$) is found to be

$$w=C_1\,{}_2F_1\,(a,\,b;\,c;\,z)+C_2\,z^{1-c}\,{}_2F_1\,(a-c+1,\,b-c+1;\,2-c;\,z)\quad (c\notin\mathbb{Z}), \tag{3}$$

where C_1 and C_2 are arbitrary constants, and (for convenience)

$$\begin{aligned}{}_2F_1\,(a,\,b;\,c;\,z):=&1+\frac{ab}{1\cdot c}z+\frac{a(a+1)b(b+1)}{1\cdot 2\cdot c(c+1)}z^2+\cdots\\ =&\sum_{n=0}^{\infty}\frac{(a)_n\,(b)_n}{(c)_n}\frac{z^n}{n!}\quad (c\notin\mathbb{Z}_0^-),\end{aligned} \tag{4}$$

in terms of the Pochhammer symbol $(\lambda)_n$ defined by 1.1(5).

Gauss's hypergeometric series

The infinite series in (4) obviously reduces to the elementary geometric series:

$$\sum_{n=0}^{\infty} z^n = 1 + z + z^2 + \cdots + z^n + \cdots \tag{5}$$

in its special cases when

$$\text{(i)} \quad a = c \quad \text{and} \quad b = 1 \quad \text{or} \quad \text{(ii)} \quad a = 1 \quad \text{and} \quad b = c. \tag{6}$$

Hence it is called the *hypergeometric series* or, more precisely, *Gauss's hypergeometric series* after the famous German mathematician, Carl Friedrich Gauss (1777–1855), who in the year 1812 introduced this series into analysis and gave the F–notation for it. (See also Equation 1.4(51) in which the general notation ${}_pF_q$ is introduced, p and q being any nonnegative integers.)

By d'Alembert's ratio test, it is easily seen that the hypergeometric series in (4) converges absolutely *within* the unit circle, that is, when $|z| < 1$, provided that the denominator parameter c is neither zero nor a negative integer. Notice, however, that if either or both of the numerator parameters a and b in (4) is zero or a negative integer, the hypergeometric series terminates, and the question of convergence does not enter into the discussion.

Further tests show that the hypergeometric series in (4), when $|z| = 1$ (that is, on the unit circle), is

(i) absolutely convergent if $\Re(c - a - b) > 0$;
(ii) conditionally convergent if $-1 < \Re(c - a - b) \leq 0 \quad \text{and} \quad z \neq 1$;
(iii) divergent if $\Re(c - a - b) \leq -1$.

As a matter of fact, in Case (i) we are led to the well-known *Gauss's summation theorem:*

$$\begin{gathered} {}_2F_1\,(a, b; c; 1) = \frac{\Gamma(c)\,\Gamma(c-a-b)}{\Gamma(c-a)\,\Gamma(c-b)} \\ (\Re(c-a-b) > 0;\ c \notin \mathbb{Z}_0^-). \end{gathered} \tag{7}$$

An obvious special case of (7) occurs when the numerator parameter a or b is a nonpositive integer $-n$, say. We thus have the *Chu-Vandermonde summation formula:*

$${}_2F_1\,(-n, b; c; 1) = \frac{(c-b)_n}{(c)_n} \quad (n \in \mathbb{N}_0;\ c \notin \mathbb{Z}_0^-), \tag{8}$$

which is, in fact, equivalent to *Vandermonde's convolution theorem:*

$$\begin{gathered} \sum_{k=0}^{n} \binom{\lambda}{k}\binom{\mu}{n-k} = \binom{\lambda+\mu}{n} = \sum_{k=0}^{n} \binom{\lambda}{n-k}\binom{\mu}{k} \\ (n \in \mathbb{N}_0;\ \lambda, \mu \in \mathbb{C}). \end{gathered} \tag{9}$$

For a number of summation theorems for the hypergeometric series (4) when z takes on other special values, see Bailey (1935, pp. 9–11), Erdélyi *et al.* (1953, pp. 104–105), Slater (1966, p. 243), and Luke (1975, pp. 271–273).

The hypergeometric series and its analytic continuation

We have seen that the hypergeometric series in (4) converges absolutely when $|z| < 1$ and thus defines a function:

$$ {}_2F_1\,(a,\, b;\, c;\, z) $$

which is analytic when $|z| < 1$, provided that c is neither zero nor a negative integer. This function is correspondingly called the *hypergeometric function* or Gauss's hypergeometric function. Indeed it is the only solution of the Fuchsian equation (1) that is analytic at the point $z = 0$ *and* assumes the value 1 at that point.

The hypergeometric function ${}_2F_1\,(a,\, b;\, c;\, z)$ can, in fact, be continued analytically outside the unit circle in a number of ways. If, for convenience, we use this same notation for the analytically continued function, one way is to employ *Euler's integral representation:*

$$ (10) \qquad \begin{gathered} {}_2F_1\,(a,\, b;\, c;\, z) = \frac{\Gamma(c)}{\Gamma(a)\,\Gamma(c-a)} \int_0^1 t^{a-1}\,(1-t)^{c-a-1}\,(1-zt)^{-b}\,dt \\ (\Re(c) > \Re(a) > 0;\ |\arg(1-z)| \leq \pi - \epsilon\ (0 < \epsilon < \pi)) \end{gathered} $$

or, equivalently,

$$ (11) \qquad \begin{gathered} {}_2F_1\,(a,\, b;\, c;\, z) = \frac{\Gamma(c)}{\Gamma(b)\,\Gamma(c-b)} \int_0^1 t^{b-1}\,(1-t)^{c-b-1}\,(1-zt)^{-a}\,dt \\ (\Re(c) > \Re(b) > 0;\ |\arg(1-z)| \leq \pi - \epsilon\ (0 < \epsilon < \pi)), \end{gathered} $$

since, by the series definition (4),

$$ (12) \qquad {}_2F_1\,(a,\, b;\, c;\, z) \equiv {}_2F_1\,(b,\, a;\, c;\, z). $$

An analytic continuation without such parametric constraints as in (10) and (11) can be provided by employing the following *Mellin-Barnes contour integral representation* of the hypergeometric function:

$$ (13) \qquad \begin{aligned} &\frac{\Gamma(a)\,\Gamma(b)}{\Gamma(c)}\,{}_2F_1\,(a,\, b;\, c;\, z) \\ &\quad = \frac{1}{2\pi i} \int_{-i\infty}^{i\infty} \frac{\Gamma(a+\zeta)\,\Gamma(b+\zeta)\,\Gamma(-\zeta)}{\Gamma(c+\zeta)}\,(-z)^{\zeta}\,d\zeta \\ &\quad (|\arg(-z)| \leq \pi - \epsilon\ (0 < \epsilon < \pi);\ a,\, b \notin \mathbb{Z}_0^-), \end{aligned} $$

where the path of integration is the imaginary axis (in the complex ζ-plane) which is indented, if necessary, to ensure that the poles of $\Gamma(a+\zeta)\,\Gamma(b+\zeta)$, *viz*

$$ (14) \qquad \zeta = -a - n \quad \text{and} \quad \zeta = -b - n \quad (n \in \mathbb{N}_0), $$

lie to the left of the path and the poles of $\Gamma(-\zeta)$, *viz*

$$\zeta = 0,\ 1,\ 2,\ \ldots\,, \tag{15}$$

lie to the right of the path.

The integrals in (10), (11), and (13) define analytic functions of z which are single-valued in the domain $|\arg(-z)| < \pi$, that is, in the whole z-plane with the exception of the points on the negative real axis. Since the Gaussian function ${}_2F_1\,(a,\ b;\ c;\ z)$ is defined by the hypergeometric series (4) everywhere within the unit circle $|z| = 1$ (including, for example, at points on the negative real axis from $z = 0$ to $z = 1 - \epsilon$, ϵ being an arbitrarily small positive number), we can use one or the other of these integrals to provide the analytic continuation of the hypergeometric function to the whole complex z-plane cut along the real axis from $z = 1$ to $z = \infty$. This analytic continuation of ${}_2F_1\,(a,\ b;\ c;\ z)$ is conveniently denoted by the same symbol ${}_2F_1\,(a,\ b;\ c;\ z)$ which will henceforth represent a branch (the *principal branch*) of the analytic function generated by the Gaussian hypergeometric series in (4).

By evaluating the contour integral in (13) using the sum of the residues of the integrand at the poles given by (14), it can be shown that

$$\begin{aligned}
&\frac{1}{\Gamma(c)}\,{}_2F_1\,(a,\ b;\ c;\ z)\\
&\quad= \frac{\Gamma(b-a)}{\Gamma(b)\,\Gamma(c-a)}\,(-z)^{-a}\,{}_2F_1\,\left(a,\ 1-c+a;\ 1-b+a;\ z^{-1}\right)\\
&\qquad+ \frac{\Gamma(a-b)}{\Gamma(a)\,\Gamma(c-b)}\,(-z)^{-b}\,{}_2F_1\,\left(b,\ 1-c+b;\ 1-a+b;\ z^{-1}\right)\\
&\qquad\qquad(|\arg(-z)| \leqq \pi - \epsilon\ (0 < \epsilon < \pi);\ a - b \notin \mathbb{Z}).
\end{aligned} \tag{16}$$

This is an important result for it may be used to provide the analytic continuation of ${}_2F_1\,(a,\ b;\ c;\ z)$ into the domain $|z| > 1$, that is, outside the circle of convergence of the hypergeometric series (4). It also shows that the function so defined possesses a branch point at infinity and that, asymptotically,

$$\begin{gathered}
{}_2F_1\,(a,\ b;\ c;\ z) \sim \mathcal{A}\,(-z)^{-a} + \mathcal{B}\,(-z)^{-b}\\
(|z| \to \infty;\ |\arg(-z)| \leqq \pi - \epsilon\ (0 < \epsilon < \pi)),
\end{gathered} \tag{17}$$

where, for convenience,

$$\mathcal{A} := \frac{\Gamma(c)\,\Gamma(b-a)}{\Gamma(b)\,\Gamma(c-a)} \quad\text{and}\quad \mathcal{B} := \frac{\Gamma(c)\,\Gamma(a-b)}{\Gamma(a)\,\Gamma(c-b)}. \tag{18}$$

Linear, quadratic, and cubic transformations

Pfaff-Kummer transformations:

$$\begin{gathered}
{}_2F_1\,(a,\ b;\ c;\ z) = (1-z)^{-a}\,{}_2F_1\,\left(a,\ c-b;\ c;\ \frac{z}{z-1}\right)\\
(c \notin \mathbb{Z}_0^-;\ |\arg(1-z)| \leqq \pi - \epsilon\ (0 < \epsilon < \pi));
\end{gathered} \tag{19}$$

(20)
$$ {}_2F_1(a,\, b;\, c;\, z) = (1-z)^{-b}\, {}_2F_1\left(b,\, c-a;\, c;\, \frac{z}{z-1}\right) $$
$$ (c \notin \mathbb{Z}_0^-;\ |\arg(1-z)| \leq \pi - \epsilon\ (0 < \epsilon < \pi)). $$

Euler's transformation:

(21)
$$ {}_2F_1(a,\, b;\, c;\, z) = (1-z)^{c-a-b}\, {}_2F_1(c-a,\, c-b;\, c;\, z) $$
$$ (c \notin \mathbb{Z}_0^-;\ |\arg(1-z)| \leq \pi - \epsilon\ (0 < \epsilon < \pi)). $$

The (Pfaff-Kummer) transformations (19) and (20) can be proven, for example, by setting $t = 1 - s$ in Euler's integral representations (10) and (11), respectively, and then appealing to the principle of analytic continuation. The Euler transformation (21) follows if we perform the transformations (19) and (20) successively.

For an extensive list of quadratic and cubic transformations of the Gaussian hypergeometric function, see Erdélyi *et al.* (1953, pp. 110–114) and Luke (1969, Vol. I, p. 92 *et seq.*).

Hypergeometric representations of elementary functions

(22)
$$ (1-z)^{-a} = \sum_{n=0}^{\infty} (a)_n \frac{z^n}{n!} = {}_2F_1(a,\, b;\, b;\, z); $$

(23)
$$ \log(1+z) = z\, {}_2F_1(1,\, 1;\, 2;\, -z); $$

(24)
$$ \log\left(\frac{1+z}{1-z}\right) = 2z\, {}_2F_1\left(\frac{1}{2},\, 1;\, \frac{3}{2};\, z^2\right); $$

(25)
$$ \sin^{-1} z = z\, {}_2F_1\left(\frac{1}{2},\, \frac{1}{2};\, \frac{3}{2};\, z^2\right); $$

(26)
$$ \tan^{-1} z = z\, {}_2F_1\left(\frac{1}{2},\, 1;\, \frac{3}{2};\, -z^2\right); $$

(27)
$$ \sinh^{-1} z = \log\left(z + \sqrt{1+z^2}\right) = z\, {}_2F_1\left(\frac{1}{2},\, \frac{1}{2};\, \frac{3}{2};\, -z^2\right); $$

(28)
$$ \left\{\frac{1}{2}\left(1+\sqrt{1-z}\right)\right\}^{1-2a} = {}_2F_1\left(a,\, a-\frac{1}{2};\, 2a;\, z\right) = \sqrt{1-z}\, {}_2F_1\left(a,\, a+\frac{1}{2};\, 2a;\, z\right); $$

(29) $$\left(1+\sqrt{z}\right)^{-2a}+\left(1-\sqrt{z}\right)^{-2a}=2\,{}_2F_1\left(a,\,a+\frac{1}{2};\,\frac{1}{2};\,z\right).$$

Hypergeometric representations of other functions

Legendre functions of the first and second kind:

(30) $$P_\nu^\mu(z)=\frac{1}{\Gamma(1-\mu)}\left(\frac{z+1}{z-1}\right)^{\frac{1}{2}\mu}\,{}_2F_1\left(-\nu,\,\nu+1;\,1-\mu;\,\frac{1-z}{2}\right);$$

(31) $$\begin{aligned}Q_\nu^\mu(z)=&\frac{\sqrt{\pi}\,e^{i\mu\pi}}{2^{\nu+1}}\,\frac{\Gamma(\mu+\nu+1)}{\Gamma\left(\nu+\frac{3}{2}\right)}\,\frac{\left(z^2-1\right)^{\frac{1}{2}\mu}}{z^{\mu+\nu+1}}\\&\cdot{}_2F_1\left[\frac{1}{2}(\mu+\nu+1),\,\frac{1}{2}(\mu+\nu)+1;\,\nu+\frac{3}{2};\,z^{-2}\right].\end{aligned}$$

The incomplete Beta function:

(32) $$B_z(\alpha,\beta)=\alpha^{-1}\,z^\alpha\,{}_2F_1\left(\alpha,\,1-\beta;\,\alpha+1;\,z\right).$$

Complete elliptic integrals of the first and second kind:

(33) $$\mathbf{K}(k)=\int_0^{\pi/2}\frac{d\theta}{\sqrt{1-k^2\sin^2\theta}}=\frac{1}{2}\pi\,{}_2F_1\left(\frac{1}{2},\,\frac{1}{2};\,1;\,k^2\right);$$

(34) $$\mathbf{E}(k)=\int_0^{\pi/2}\sqrt{1-k^2\sin^2\theta}\;d\theta=\frac{1}{2}\pi\,{}_2F_1\left(-\frac{1}{2},\,\frac{1}{2};\,1;\,k^2\right).$$

Jacobi polynomials:

(35) $$\begin{aligned}P_n^{(\alpha,\beta)}(z)=&(-1)^n\,P_n^{(\beta,\alpha)}(-z)\\=&\binom{\alpha+n}{n}\,{}_2F_1\left[\begin{matrix}-n,\,\alpha+\beta+n+1;\\ \alpha+1;\end{matrix}\,\frac{1-z}{2}\right].\end{aligned}$$

Gegenbauer (or ultraspherical) polynomials:

(36) $$\begin{aligned}C_n^\nu(z)=&\binom{n+\nu-\frac{1}{2}}{n}^{-1}\binom{n+2\nu-1}{n}\,P_n^{\left(\nu-\frac{1}{2},\nu-\frac{1}{2}\right)}(z)\\=&\binom{n+2\nu-1}{n}\,{}_2F_1\left[\begin{matrix}-n,\,2\nu+n;\\ \nu+\frac{1}{2};\end{matrix}\,\frac{1-z}{2}\right].\end{aligned}$$

Legendre (or spherical) polynomials:

(37) $$P_n(z)=P_n^{(0,0)}(z)=C_n^{\frac{1}{2}}(z)={}_2F_1\left(-n,\,n+1;\,1;\,\frac{1-z}{2}\right).$$

Tchebycheff polynomials of the first and second kind:

$$T_n(z) = \binom{n-\frac{1}{2}}{n}^{-1} P_n^{(-\frac{1}{2},-\frac{1}{2})}(z) = \frac{1}{2}\,n\,C_n^0(z) = {}_2F_1\left(-n,\, n;\, \frac{1}{2};\, \frac{1-z}{2}\right), \tag{38}$$

where

$$C_n^0(z) := \lim_{\nu\to 0}\left\{\nu^{-1}\, C_n^{\nu}(z)\right\}; \tag{39}$$

$$\begin{aligned} U_n(z) =& \frac{1}{2}\binom{n+\frac{1}{2}}{n+1}^{-1} P_n^{(\frac{1}{2},\frac{1}{2})}(z) = C_n^1(z) \\ =& (n+1)\,{}_2F_1\left(-n,\, n+2;\, \frac{3}{2};\, \frac{1-z}{2}\right). \end{aligned} \tag{40}$$

The confluent hypergeometric function

If, in Gauss's hypergeometric equation (1), we replace z by z/b, the resulting equation will have three regular singularities at

$$z = 0,\ b,\ \infty.$$

By letting $|b| \to \infty$, this transformed equation reduces to the form:

$$z\,\frac{d^2w}{dz^2} + (c-z)\,\frac{dw}{dz} - aw = 0 \tag{41}$$

or, equivalently,

$$\{\delta(\delta+c-1) - z(\delta+a)\}\,w = 0 \qquad \left(\delta := z\,\frac{d}{dz}\right). \tag{42}$$

Equation (41) or (42) has a regular singularity at $z = 0$, and an *irregular* singularity at $z = \infty$ which is formed by the *confluence* of two regular singularities at b and ∞ of Gauss's equation (1) with z replaced by z/b. Consequently, (41) or (42) is called the *confluent hypergeometric equation* or Kummer's differential equation after Ernst Eduard Kummer (1810–1893) who presented a detailed study of its solutions in 1836 [*cf.* Kummer (1836)].

The simplest solution of (1) or (2) is Kummer's function $F(a,\, c,\, z)$ which, in the notations of this section, is

$$\begin{aligned} {}_1F_1\,(a;\, c;\, z) :=& 1 + \frac{a}{1\cdot c}z + \frac{a(a+1)}{1\cdot 2\cdot c(c+1)}z^2 + \cdots \\ =& \sum_{n=0}^{\infty} \frac{(a)_n}{(c)_n}\,\frac{z^n}{n!} \qquad (c \notin \mathbb{Z}_0^-;\ |z| < \infty). \end{aligned} \tag{43}$$

Moreover, since

$$\lim_{|\lambda|\to\infty}\left\{(\lambda)_n\left(\frac{z}{\lambda}\right)^n\right\}=z^n=\lim_{|\mu|\to\infty}\left\{\frac{(\mu\, z)^n}{(\mu)_n}\right\}\tag{44}$$
$$(n\in\mathbb{N}_0;\ |z|<\infty),$$

we readily have

$$_1F_1\,(a;\ c;\ z)=\lim_{|b|\to\infty}\ _2F_1\left(a,\ b;\ c;\ \frac{z}{b}\right).\tag{45}$$

In view of the principle of confluence involved in (45), the solution (43) is also called the *confluent hypergeometric function.*

Other popular notations for the series solution (43) are (i) $M(a,\ c,\ z)$ used by Airey and Webb (1918) and (ii) $\Phi(a;\ c;\ z)$ or $\Phi(a,\ c,\ z)$ introduced by Humbert (1921a,b).

Important properties of Kummer's confluent hypergeometric function

The principle of confluence exhibited by (45) is useful in order to deduce properties of the confluent hypergeometric $_1F_1$ function from those of Gauss's hypergeometric $_2F_1$ function. Thus, from (10) and (20), we obtain

$$_1F_1\,(a;\ c;\ z)=\frac{\Gamma(c)}{\Gamma(a)\,\Gamma(c-a)}\int_0^1 t^{a-1}\,(1-t)^{c-a-1}\,e^{zt}\,dt\tag{46}$$
$$(\Re(c)>\Re(a)>0);$$

$$_1F_1\,(a;\ c;\ z)=e^z\ _1F_1\,(c-a;\ c;\ -z),\tag{47}$$

which is known as *Kummer's first transformation.* We also have *Kummer's second transformation*:

$$e^{-z}\ _1F_1\,(a;\ 2a;\ 2z)=\,_0F_1\left(-;\ a+\frac{1}{2};\ \frac{1}{4}z^2\right)\tag{48}$$
$$(2a\neq-1,\ -3,\ -5,\ \ldots),$$

where

$$\begin{aligned}_0F_1\,(-;\ c;\ z)&=\lim_{|a|\to\infty}\ _1F_1\,(a;\ c;\ z/a)\\&=\sum_{n=0}^{\infty}\frac{z^n}{n!\,(c)_n}\qquad(c\notin\mathbb{Z}_0^-;\ |z|<\infty).\end{aligned}\tag{49}$$

It follows immediately from the definition (43) that

$$_1F_1\,(a;\ a;\ z)=\,_0F_0\,(-;\ -;\ z)=e^z.\tag{50}$$

The generalized (Gauss and Kummer) hypergeometric function

A natural generalization of the hypergeometric functions ${}_2F_1$, ${}_1F_1$, *et cetera* (considered in this section) is accomplished by the introduction of an arbitrary number of numerator and denominator parameters. The resulting series:

$$(51)\qquad {}_pF_q\left[\begin{matrix}\alpha_1, \ldots, \alpha_p; \\ \beta_1, \ldots, \beta_q;\end{matrix}\; z\right] = \sum_{n=0}^{\infty} \frac{(\alpha_1)_n \cdots (\alpha_p)_n}{(\beta_1)_n \cdots (\beta_q)_n} \frac{z^n}{n!} = {}_pF_q(\alpha_1, \ldots, \alpha_p; \beta_1, \ldots, \beta_q; z),$$

where $(\lambda)_n$ is the Pochhammer symbol defined by 1.1(5), is known as the *generalized Gauss (and Kummer) series*, or simply, the *generalized hypergeometric series*. Here p and q are positive integers or zero (interpreting an empty product as 1), and we assume (for simplicity) that the variable z, the numerator parameters $\alpha_1, \ldots, \alpha_p$, and the denominator parameters $\beta_1, \ldots, \beta_q$ take on complex values, provided that no zeros appear in the denominator of (51), that is, that

$$(52)\qquad (\beta_j \notin \mathbb{Z}_0^-;\; j = 1, \ldots, q).$$

Thus, if a numerator parameter is a negative integer or zero, the ${}_pF_q$ series terminates in view of the identity 1.1(47), and we are led to a (generalized hypergeometric) polynomial of the type:

$$(53)\qquad \begin{aligned} {}_{p+1}F_q\left[\begin{matrix}-n, \alpha_1, \ldots, \alpha_p; \\ \beta_1, \ldots, \beta_q;\end{matrix}\; z\right] &= \sum_{k=0}^{n} \frac{(-n)_k(\alpha_1)_k \cdots (\alpha_p)_k}{(\beta_1)_k \cdots (\beta_q)_k} \frac{z^k}{k!} \\ &= \frac{(\alpha_1)_n \cdots (\alpha_p)_n}{(\beta_1)_n \cdots (\beta_q)_n}(-z)^n \\ &\cdot {}_{q+1}F_p\left[\begin{matrix}-n, 1-\beta_1-n, \ldots, 1-\beta_q-n; \\ 1-\alpha_1-n, \ldots, 1-\alpha_p-n;\end{matrix}\; \frac{(-1)^{p+q}}{z}\right] \qquad (n \in \mathbb{N}_0),\end{aligned}$$

where we have reversed the order of the terms of the polynomial by using 1.1(45) and 1.1(46).

Assuming that none of the numerator parameters is zero or a negative integer (otherwise the question of convergence will not arise), and with the usual restriction given by (52), the ${}_pF_q$ series in (51)
(i) converges for $|z| < \infty$ if $p \leq q$,
(ii) converges for $|z| < 1$ if $p = q + 1$, and
(iii) diverges for all z, $z \neq 0$, if $p > q + 1$.
Furthermore, if we set

$$(54)\qquad \omega := \sum_{j=1}^{q} \beta_j - \sum_{j=1}^{p} \alpha_j,$$

then it is known that the ${}_pF_q$ series in (51), with $p = q + 1$, is
I. absolutely convergent for $|z| = 1$ if $\Re(\omega) > 0$,
II. conditionally convergent for $|z| = 1$ $(z \neq 1)$ if $-1 < \Re(\omega) \leq 0$, and
III. divergent for $|z| = 1$ if $\Re(\omega) \leq -1$.

Analytic continuation of the generalized hypergeometric function

The generalized hypergeometric ${}_pF_q$ function is defined by the (absolutely) convergent series (51) whenever
(i) $p \leq q$ and $|z| < \infty$
or
(ii) $p = q + 1$ and $|z| < 1$.
As in the above-detailed case of the ${}_2F_1$ function, the generalized hypergeometric function ${}_pF_q$ (with $p = q + 1$) can be continued analytically to the domain $|\arg(1 - z)| < \pi$, that is, to the whole complex z-plane cut along the real axis from $z = 1$ to $z = \infty$, by using (51) in conjunction with the Mellin–Barnes contour integral representation:

(55)
$$\begin{aligned}&\frac{\Gamma(\alpha_1)\cdots\Gamma(\alpha_p)}{\Gamma(\beta_1)\cdots\Gamma(\beta_q)}\,{}_pF_q\left[\begin{matrix}\alpha_1, \dots, \alpha_p; \\ \beta_1, \dots, \beta_q;\end{matrix}\; z\right] \\ &= \frac{1}{2\pi i}\int_{-i\infty}^{i\infty} \frac{\Gamma(\alpha_1+\zeta)\cdots\Gamma(\alpha_p+\zeta)\,\Gamma(-\zeta)}{\Gamma(\beta_1+\zeta)\cdots\Gamma(\beta_q+\zeta)}(-z)^\zeta\, d\zeta \\ &(\alpha_j \notin \mathbb{Z}_0^-;\; j = 1, \dots, p;\; |\arg(-z)| < \pi),\end{aligned}$$

where the path of integration is the imaginary axis (in the complex ζ-plane) which is indented, if necessary, in oder to separate the poles of $\Gamma(\alpha_j + \zeta)$ $(j = 1, \dots, p)$ from those of $\Gamma(-\zeta)$.

Functions expressible in terms of the ${}_pF_q$ function

The following list is in addition to the important functions (considered above in this section) which can be expressed as a hypergeometric ${}_2F_1$ function.

Elementary functions:

(56)
$$e^z = {}_0F_0\,[-;-;\, z];$$

(57)
$$(1-z)^{-\alpha} = {}_1F_0\,[\alpha;-;\, z];$$

(58)
$$\cos z = {}_0F_1\left[-;\frac{1}{2};\, -\frac{1}{4}z^2\right];$$

(59)
$$\sin z = z\,{}_0F_1\left[-;\frac{3}{2};\, -\frac{1}{4}z^2\right].$$

Bessel functions:

(60)
$$\begin{aligned}J_\nu(z) &= \frac{\left(\frac{1}{2}z\right)^\nu}{\Gamma(\nu+1)}\,{}_0F_1\left[-;\nu+1;\, -\frac{1}{4}z^2\right] \\ &= \frac{\left(\frac{1}{2}z\right)^\nu}{\Gamma(\nu+1)}\exp(\pm iz)\,{}_1F_1\left[\nu+\frac{1}{2};\, 2\nu+1;\, \mp 2iz\right];\end{aligned}$$

$$I_\nu(z) = \frac{\left(\frac{1}{2}z\right)^\nu}{\Gamma(\nu+1)}\, {}_0F_1\left[-;\nu+1;\,\frac{1}{4}z^2\right] \tag{61}$$
$$= \frac{\left(\frac{1}{2}z\right)^\nu}{\Gamma(\nu+1)}\exp(\pm z)\, {}_1F_1\left[\nu+\frac{1}{2};\,2\nu+1;\,\mp 2z\right].$$

Lommel functions:

$$s_{\mu,\nu}(z) = \frac{z^{\mu+1}}{(\mu-\nu+1)(\mu+\nu+1)} \tag{62}$$
$$\cdot {}_1F_2\left[\begin{matrix} & 1; & \\ \frac{1}{2}(\mu-\nu+3),\ \frac{1}{2}(\mu+\nu+3); & & -\frac{1}{4}z^2\end{matrix}\right];$$

$$S_{\mu,\nu}(z) = s_{\mu,\nu}(z) + 2^{\mu-1}\,\Gamma\left(\frac{\mu-\nu+1}{2}\right)\Gamma\left(\frac{\mu+\nu+1}{2}\right) \tag{63}$$
$$\cdot\left\{\sin\left[\frac{1}{2}(\mu-\nu)\pi\right]J_\nu(z) - \cos\left[\frac{1}{2}(\mu-\nu)\pi\right]Y_\nu(z)\right\},$$

where $J_\nu(z)$ is given by (60), and

$$Y_\nu(z) = \csc(\nu\pi)\left[\cos(\nu\pi)J_\nu(z) - J_{-\nu}(z)\right]. \tag{64}$$

Struve functions:

$$\mathbf{H}_\nu(z) = \frac{\left(\frac{1}{2}z\right)^{\nu+1}}{\Gamma\left(\frac{3}{2}\right)\Gamma\left(\nu+\frac{3}{2}\right)}\, {}_1F_2\left[\begin{matrix} & 1; & \\ \frac{3}{2},\ \nu+\frac{3}{2}; & & -\frac{1}{4}z^2\end{matrix}\right] \tag{65}$$
$$= \frac{2^{1-\nu}}{\sqrt{\pi}\,\Gamma\left(\nu+\frac{1}{2}\right)}\, s_{\nu,\nu}(z)$$
$$= Y_\nu(z) + \frac{2^{1-\nu}}{\sqrt{\pi}\,\Gamma\left(\nu+\frac{1}{2}\right)}\, S_{\nu,\nu}(z);$$

$$\mathbf{L}_\nu(z) = \frac{\left(\frac{1}{2}z\right)^{\nu+1}}{\Gamma\left(\frac{3}{2}\right)\Gamma\left(\nu+\frac{3}{2}\right)}\, {}_1F_2\left[\begin{matrix} & 1; & \\ \frac{3}{2},\ \nu+\frac{3}{2}; & & \frac{1}{4}z^2\end{matrix}\right] \tag{66}$$
$$= \exp\left[-\frac{1}{2}(\nu+1)i\pi\right]\mathbf{H}_\nu(iz).$$

Whittaker function of the first kind:

$$M_{\kappa,\mu}(z) = z^{\mu+\frac{1}{2}}\exp\left(-\frac{1}{2}z\right)\, {}_1F_1\left(\mu-\kappa+\frac{1}{2};\,2\mu+1;\,z\right) \tag{67}$$
$$= z^{\mu+\frac{1}{2}}\exp\left(\frac{1}{2}z\right)\, {}_1F_1\left(\mu+\kappa+\frac{1}{2};\,2\mu+1;\,-z\right).$$

Incomplete Gamma function:

(68) $$\gamma(\alpha, z) = \alpha^{-1} z^{\alpha} {}_1F_1(\alpha; \alpha+1; -z).$$

Error function:

(69) $$\begin{aligned} \operatorname{Erf}(z) =& \frac{\sqrt{\pi}}{2} \operatorname{erf}(z) = \frac{1}{2}\gamma\left(\frac{1}{2}, z^2\right) \\ =& z\, {}_1F_1\left(\frac{1}{2}; \frac{3}{2}; -z^2\right). \end{aligned}$$

Hermite polynomials:

(70) $$\begin{aligned} H_n(z) =& \sum_{k=0}^{[n/2]} \frac{(-1)^k\, n!\, (2z)^{n-2k}}{k!\,(n-2k)!} \\ =& (2z)^n\, {}_2F_0\left(-\frac{1}{2}n, \frac{1}{2}-\frac{1}{2}n; -; -z^{-2}\right) \quad (n \in \mathbb{N}_0). \end{aligned}$$

Laguerre functions and polynomials:

(71) $$L_{\nu}^{(\alpha)}(z) = \frac{\Gamma(\nu+\alpha+1)}{\Gamma(\nu+1)\,\Gamma(\alpha+1)} {}_1F_1(-\nu; \alpha+1; z);$$

(72) $$\begin{aligned} L_n^{(\alpha)}(z) =& \sum_{k=0}^{n} \binom{n+\alpha}{n-k} \frac{(-z)^k}{k!} \\ =& \binom{n+\alpha}{n} {}_1F_1(-n; \alpha+1; z) \\ =& \frac{(-z)^n}{n!}\, {}_2F_0\left(-n, -\alpha-n; -; -\frac{1}{z}\right) \quad (n \in \mathbb{N}_0). \end{aligned}$$

Poisson–Charlier polynomials:

(73) $$\begin{aligned} c_n(x;\alpha) =& \sum_{k=0}^{n} (-1)^k \binom{n}{k}\binom{x}{k} k!\, \alpha^{-k} \quad (\alpha > 0;\, x \in \mathbb{N}_0) \\ =& n!\,(-\alpha)^{-n} L_n^{(x-n)}(\alpha) \\ =& (-\alpha)^{-n}\, n! \binom{x}{n} {}_1F_1(-n; x-n+1; \alpha). \end{aligned}$$

In an alternative notation, we have [see Szegö (1975, p. 35)]

(74) $$\begin{aligned} p_n(x) =& \alpha^{-\frac{1}{2}n} \frac{(x-n+1)_n}{\sqrt{n!}}\, {}_1F_1(-n; x-n+1; \alpha) \\ =& \alpha^{-\frac{1}{2}n} \sqrt{n!} \binom{x}{n} {}_1F_1(-n; x-n+1; \alpha). \end{aligned}$$

1.5 Stirling Numbers of the First and Second Kind

Stirling numbers of the first kind

The *Stirling numbers* $s(n,k)$ *of the first kind* are defined by the generating functions:

(1) $$z(z-1)\cdots(z-n+1)=\sum_{k=0}^{n} s(n,k)\,z^k$$

and

(2) $$\{\log(1+z)\}^k = k!\sum_{n=k}^{\infty} s(n,k)\,\frac{z^n}{n!} \quad (|z|<1).$$

We have the following recurrence relations satisfied by $s(n,k)$:

(3) $$s(n+1,k)=s(n,k-1)-n\,s(n,k) \quad (n\geq k\geq 1);$$

(4) $$\binom{k}{j}\,s(n,k)=\sum_{l=k-j}^{n-j}\binom{n}{l}\,s(n-l,j)\,s(l,k-j) \quad (n\geq k\geq j).$$

From the definition (1) of $s(n,k)$, the Pochhammer symbol in 1.1(5) can be written in the form:

(5) $$(z)_n = z(z+1)\cdots(z+n-1)=\sum_{k=0}^{n}(-1)^{n+k}\,s(n,k)\,z^k,$$

where $(-1)^{n+k}\,s(n,k)$ denotes the number of permutations of n symbols which has exactly k cycles.

It is not difficult to see also that

(6) $$s(n,0)=\begin{cases}1 & (n=0)\\ 0 & (n\in\mathbb{N}),\end{cases} \qquad s(n,n)=1,$$
$$s(n,1)=(-1)^{n+1}\,(n-1)!, \quad s(n,n-1)=-\frac{1}{2}n$$

and

(7) $$\sum_{k=1}^{n} s(n,k)=0 \quad (n\in\mathbb{N}\setminus\{1\}); \quad \sum_{k=0}^{n}(-1)^{n+k}\,s(n,k)=n!;$$
$$\sum_{j=k}^{n} s(n+1,j+1)\,n^{j-k}=s(n,k).$$

Yet another recursion formula for $s(n,k)$ is given by [see Shen (1995)]:

(8) $$(k-1)\,s(n,k)=-\sum_{m=1}^{k-1} s(n,k-m)\,C_m^n,$$

where

$$C_0^1 = 1,\ C_m^1 = 0\ (m \in \mathbb{N}),\quad \text{and}\quad C_m^n = \sum_{k=1}^{n-1} \frac{1}{k^m}\ (n \in \mathbb{N} \setminus \{1\}).$$

It readily follows from the recursion formula (8) that

$$(9)\quad \begin{aligned} s(n,2) &= (-1)^n (n-1)! \sum_{k=1}^{n-1} \frac{1}{k}; \\ s(n,3) &= (-1)^{n+1} \frac{(n-1)!}{2} \left\{ \left(\sum_{k=1}^{n-1} \frac{1}{k} \right)^2 - \sum_{k=1}^{n-1} \frac{1}{k^2} \right\}; \\ s(n,4) &= (-1)^n \frac{(n-1)!}{6} \left\{ \left(\sum_{k=1}^{n-1} \frac{1}{k} \right)^3 - 3 \left(\sum_{k=1}^{n-1} \frac{1}{k} \right) \left(\sum_{k=1}^{n-1} \frac{1}{k^2} \right) + 2 \sum_{k=1}^{n-1} \frac{1}{k^3} \right\}. \end{aligned}$$

In view of (5), by logarithmically differentiating 1.1(39) and then using 1.2(7), we obtain

$$(10)\quad \begin{aligned} \frac{d}{dz}(z)_n &= \sum_{k=1}^{n} (-1)^{n+k} k\, s(n,k)\, z^{k-1} \\ &= (z)_n \left(\sum_{k=1}^{n} \frac{1}{z+k-1} \right), \end{aligned}$$

which, upon employing Leibniz's rule for differentiation, yields a more general formula:

$$(11)\quad \begin{aligned} \frac{d^{j+1}}{dz^{j+1}}(z)_n &= \sum_{k=j+1}^{n} (-1)^{n+k+j+1} (-k)_{j+1}\, s(n,k)\, z^{k-j-1} \\ &= \sum_{l=0}^{j} (-1)^l \frac{j!}{(j-l)!} \left\{ \sum_{k=1}^{n} \frac{1}{(z+k-1)^{l+1}} \right\} \left\{ \frac{d^{j-l}}{dz^{j-l}}(z)_n \right\} \\ &\quad (j \in \mathbb{N}_0;\ n \in \mathbb{N}). \end{aligned}$$

For $j = 1$ and $j = 2$, (11) immediately yields

$$(12)\quad \begin{aligned} &\sum_{k=2}^{n} (-1)^{n+k} k(k-1)\, s(n,k)\, z^{k-2} \\ &\quad = (z)_n \left\{ \left(\sum_{k=1}^{n} \frac{1}{z+k-1} \right)^2 - \sum_{k=1}^{n} \frac{1}{(z+k-1)^2} \right\} \end{aligned}$$

and

$$(13)\quad \begin{aligned} &\sum_{k=3}^{n} (-1)^{n+k} k(k-1)(k-2)\, s(n,k)\, z^{k-3} \\ &= (z)_n \left[\left(\sum_{k=1}^{n} \frac{1}{z+k-1} \right)^3 - 3 \left(\sum_{k=1}^{n} \frac{1}{z+k-1} \right) \left\{ \sum_{k=1}^{n} \frac{1}{(z+k-1)^2} \right\} \right. \\ &\qquad \left. + \sum_{k=1}^{n} \frac{1}{(z+k-1)^3} \right]. \end{aligned}$$

Stirling numbers of the second kind

The *Stirling numbers* $S(n,k)$ *of the second kind* are defined by the generating functions:

$$z^n = \sum_{k=0}^{n} S(n,k)\, z(z-1)\cdots(z-k+1), \tag{14}$$

$$(e^z-1)^k = k! \sum_{n=k}^{\infty} S(n,k)\, \frac{z^n}{n!}, \tag{15}$$

and

$$(1-z)^{-1}(1-2z)^{-1}\cdots(1-kz)^{-1} = \sum_{n=k}^{\infty} S(n,k)\, z^{n-k} \quad (|z|<k^{-1}), \tag{16}$$

where $S(n,k)$ denotes the number of ways of partitioning a set of n elements into k non-empty subsets.

It is not difficult to see also that

$$\begin{gathered} S(n,0) = \begin{cases} 1 & (n=0) \\ 0 & (n \in \mathbb{N}), \end{cases} \\ S(n,1) = S(n,n) = 1, \quad \text{and} \quad S(n,n-1) = \binom{n}{2}. \end{gathered} \tag{17}$$

The recurrence relations for $S(n,k)$ are given by

$$S(n+1,k) = k\, S(n,k) + S(n,k-1) \quad (n \geq k \geq 1) \tag{18}$$

and

$$\binom{k}{j} S(n,k) = \sum_{i=k-j}^{n-j} \binom{n}{i} S(n-i,j)\, S(i,k-j) \quad (n \geq k \geq j). \tag{19}$$

The numbers $S(n,k)$ can be expressed in an explicit form:

$$S(n,k) = \frac{1}{k!} \sum_{j=0}^{k} (-1)^{k-j} \binom{k}{j} j^n. \tag{20}$$

Some additional properties of $S(n,k)$ are recalled here as follows:

$$\sum_{k=0}^{n} (-1)^{n-k}\, k!\, S(n,k) = 1; \tag{21}$$

$$\sum_{j=k}^{n} S(j-1,k-1)\, k^{n-j} = S(n,k); \tag{22}$$

(23) $$S(n,k)=\sum_{j=0}^{n-k}(-1)^j\binom{n-1+j}{n-k+j}\binom{2n-k}{n-k-j}S(n-k+j,j);$$

(24) $$\sum_{j=k}^{n} S(j,k)\,S(n,j) = \delta_{kn},$$

where δ_{mn} denotes the Kronecker delta defined by

(25) $$\delta_{mn}=\begin{cases}0 & (m\neq n),\\ 1 & (m=n).\end{cases}$$

1.6 Bernoulli and Euler Polynomials and Numbers

Bernoulli polynomials and numbers

The *Bernoulli polynomials* $B_n(x)$ are defined by the generating function:

(1) $$\frac{z\,e^{xz}}{e^z-1}=\sum_{n=0}^{\infty} B_n(x)\,\frac{z^n}{n!}\qquad (|z|<2\pi).$$

The numbers $B_n := B_n(0)$ are called the *Bernoulli numbers* generated by

(2) $$\frac{z}{e^z-1}=\sum_{n=0}^{\infty} B_n\,\frac{z^n}{n!}\qquad (|z|<2\pi).$$

It easily follows from (1) and (2) that

(3) $$B_n(x)=\sum_{k=0}^{n}\binom{n}{k}B_k\,x^{n-k}.$$

The Bernoulli polynomials $B_n(x)$ satisfy the difference equation:

(4) $$B_n(x+1)-B_n(x)=n\,x^{n-1}\qquad (n\in\mathbb{N}_0),$$

which yields

(5) $$B_n(0)=B_n(1)\qquad (n\in\mathbb{N}\setminus\{1\}).$$

Setting $x=1$ in (3), in view of (5), we have

(6) $$B_n=\sum_{k=0}^{n}\binom{n}{k}B_k,$$

which gives a recursion formula for computing Bernoulli numbers. The first few of the Bernoulli numbers are already listed with the Euler-Maclaurin summation formula 1.3(68), and (for the sake of completeness) we have the following list:

(7)
$$B_0=1,\ B_1=-\frac{1}{2},\ B_2=\frac{1}{6},\ B_4=-\frac{1}{30},\ B_6=\frac{1}{42},\ B_8=-\frac{1}{30},\ B_{10}=\frac{5}{66},$$
$$B_{12}=-\frac{691}{2730},\quad B_{14}=\frac{7}{6},\quad B_{16}=-\frac{3617}{510},\quad B_{18}=\frac{43867}{798},\quad B_{20}=-\frac{174611}{330},$$
$$B_{22}=\frac{854513}{138},\quad B_{24}=-\frac{236364091}{2730},\quad B_{26}=\frac{8553103}{6},\cdots,B_{2n+1}=0\quad (n\in\mathbb{N}).$$

The first few of the Bernoulli polynomials are given below:

(8)
$$B_0(x)=1,\quad B_1(x)=x-\frac{1}{2},\quad B_2(x)=x^2-x+\frac{1}{6},$$
$$B_3(x)=x^3-\frac{3}{2}x^2+\frac{1}{2}x,\quad B_4(x)=x^4-2x^3+x^2-\frac{1}{30},\cdots.$$

It is not difficult to derive the following identities for the Bernoulli polynomials:

(9)
$$B_n'(x)=n\,B_{n-1}(x)\quad (n\in\mathbb{N});$$

(10)
$$B_n(1-x)=(-1)^n\,B_n(x)\quad (n\in\mathbb{N}_0);$$

(11)
$$(-1)^n\,B_n(-x)=B_n(x)+n\,x^{n-1}\quad (n\in\mathbb{N}_0).$$

Multiplication formula:

(12)
$$B_n(mx)=m^{n-1}\sum_{k=0}^{m-1}B_n\left(x+\frac{k}{m}\right)\quad (n\in\mathbb{N}_0,\ m\in\mathbb{N}).$$

Addition formula:

(13)
$$B_n(x+y)=\sum_{k=0}^{n}\binom{n}{k}B_k(x)\,y^{n-k}\quad (n\in\mathbb{N}_0).$$

Integral formulas:

(14)
$$\int_x^y B_n(t)\,dt=\frac{B_{n+1}(y)-B_{n+1}(x)}{n+1};$$

(15)
$$\int_x^{x+1} B_n(t)\,dt=x^n;$$

(16)
$$\int_0^1 B_n(t)\,B_m(t)\,dt=(-1)^{n-1}\,\frac{m!\,n!}{(m+n)!}B_{m+n}\quad (m,\,n\in\mathbb{N}).$$

It follows from (14) and (15) that the finite sum of powers is expressed as Bernoulli polynomials and numbers:

(17)
$$\sum_{k=1}^{m}k^n=\frac{B_{n+1}(m+1)-B_{n+1}}{n+1}\quad (m,\,n\in\mathbb{N}).$$

By writing $2n$ for n in (3), we can deduce that

$$B_{2n}(x)+n\,x^{2n-1}=\sum_{k=0}^{n}\binom{2n}{2k}B_{2k}\,x^{2n-2k},$$

which, upon integrating from 0 to $\frac{1}{2}$, yields

$$\sum_{k=0}^{n} \frac{2^{2k} B_{2k}}{(2k)!(2n-2k+1)!} = \frac{1}{(2n)!} \quad (n \in \mathbb{N}_0), \tag{18}$$

where we have applied (14).

The generalized Bernoulli polynomials and numbers

The *generalized Bernoulli polynomials* $B_n^{(\alpha)}(x)$ of degree n in x are defined by the generating function:

$$\left(\frac{z}{e^z - 1}\right)^{\alpha} e^{xz} = \sum_{n=0}^{\infty} B_n^{(\alpha)}(x) \frac{z^n}{n!} \quad (|z| < 2\pi;\ 1^{\alpha} := 1) \tag{19}$$

for arbitrary (real or complex) parameter α. Clearly,

$$B_n^{(\alpha)}(x) = (-1)^n B_n^{(\alpha)}(\alpha - x), \tag{20}$$

so that

$$B_n^{(\alpha)}(\alpha) = (-1)^n B_n^{(\alpha)}(0) =: (-1)^n B_n^{(\alpha)} \tag{21}$$

in terms of the *generalized Bernoulli numbers* $B_n^{(\alpha)}$ defined by the generating function:

$$\left(\frac{z}{e^z - 1}\right)^{\alpha} = \sum_{n=0}^{\infty} B_n^{(\alpha)} \frac{z^n}{n!} \quad (|z| < 2\pi;\ 1^{\alpha} := 1). \tag{22}$$

It is easily observed that

$$B_n^{(1)}(x) = B_n(x) \quad \text{and} \quad B_n^{(1)} = B_n \quad (n \in \mathbb{N}_0). \tag{23}$$

From the generating function (19), it is fairly straightforward to deduce the addition theorem:

$$B_n^{(\alpha+\beta)}(x+y) = \sum_{k=0}^{n} \binom{n}{k} B_k^{(\alpha)}(x)\, B_{n-k}^{(\beta)}(y), \tag{24}$$

which, for $x = \beta = 0$, corresponds to the elegant representation:

$$B_n^{(\alpha)}(x) = \sum_{k=0}^{n} \binom{n}{k} B_k^{(\alpha)} x^{n-k} \tag{25}$$

for the generalized Bernoulli polynomials as a finite sum of the generalized Bernoulli numbers.

Srivastava *et al.* (1983, p. 442, Eqs. (4.4) and (4.5)) gave two new classes of addition theorems for the generalized Bernoulli polynomials:
(26)

$$B_n^{(\alpha+\lambda\gamma)}(x+\gamma y)=\sum_{k=0}^{n}\frac{\gamma+n}{\gamma+k}\binom{n}{k}B_k^{(\alpha-\lambda k)}(x-ky)\,B_{n-k}^{(\lambda k+\lambda\gamma)}(ky+\gamma y)\quad(\Re(\gamma)>0);$$

$$(27)\ B_n^{(\alpha+\beta+n+1)}(x+y+n)=\sum_{k=0}^{n}\binom{n}{k}B_k^{(\alpha+k+1)}(x+k)\,B_{n-k}^{(\beta+n-k+1)}(y+n-k).$$

Srivastava and Todorov (1988, p. 510, Eq. (3)) proved the following explicit formula for the generalized Bernoulli polynomials:

$$(28)\quad \begin{aligned}B_n^{(\alpha)}(x)=&\sum_{k=0}^{n}\binom{n}{k}\binom{\alpha+k-1}{k}\frac{k!}{(2k)!}\sum_{j=0}^{k}(-1)^j\binom{k}{j}j^{2k}\,(x+j)^{n-k}\\ &\cdot{}_2F_1[k-n,\,k-\alpha;\,2k+1;\,j/(x+j)],\end{aligned}$$

in terms of the Gaussian hypergeometric function (see Section 1.4). They also applied the representation (28) in order to derive certain interesting special cases considered earlier by Gould (1972) and Todorov (1985). Indeed, by the Chu-Vandermonde theorem 1.4(9), we have

$${}_2F_1(-N,\,b;\,c;\,1)=\binom{c-b+N-1}{N}\binom{c+N-1}{N}^{-1}\quad(N\in\mathbb{N}_0),$$

which, for $N=n-k$, $b=k-\alpha$, and $c=2k+1$, readily yields

$$(29)\qquad {}_2F_1(k-n,\,k-\alpha;\,2k+1;\,1)=\binom{\alpha+n}{n-k}\frac{(n-k)!\,(2k)!}{(n+k)!}\quad(0\leq k\leq n).$$

In view of (29), the special case of the Srivastava-Todorov formula (28) when $x=0$ gives the following representation for the generalized Bernoulli numbers:

$$(30)\qquad B_n^{(\alpha)}=\sum_{k=0}^{n}\binom{\alpha+n}{n-k}\binom{\alpha+k-1}{k}\frac{n!}{(n+k)!}\sum_{j=0}^{k}(-1)^j\binom{k}{j}j^{n+k}$$

or, equivalently,

$$(31)\qquad B_n^{(\alpha)}=\sum_{k=0}^{n}(-1)^k\binom{\alpha+n}{n-k}\binom{\alpha+k-1}{k}\frac{n!}{(n+k)!}\Delta^k\,0^{n+k},$$

where, for convenience,

$$(32)\qquad \Delta^k\,a^r=\Delta^k\,x^r\Big|_{x=a}=\sum_{j=0}^{k}(-1)^{k-j}\binom{k}{j}(a+j)^r,$$

Δ being the difference operator defined by [*cf.* Comtet (1974, p. 13 *et seq.*)]

(33) $$\Delta f(x) = f(x+1) - f(x),$$

so that, in general,

(34) $$\Delta^k f(x) = \sum_{j=0}^{k} (-1)^{k-j} \binom{k}{j} f(x+j).$$

Alternatively, since [Comtet (1974, p. 204, Theorem A); see also Eq. 1.5(20)]

(35) $$S(n,k) = \frac{1}{k!} \Delta^k 0^n,$$

where $S(n,k)$ denotes the Stirling number of the second kind defined by 1.5(14), that is, by

(36) $$z^n = \sum_{k=0}^{n} \binom{z}{k} k!\, S(n,k),$$

the representation (30) or (31) can be written also as [Todorov (1985, p. 665, Eq. (3))]

(37) $$B_n^{(\alpha)} = \sum_{k=0}^{n} (-1)^k \binom{\alpha+n}{n-k} \binom{\alpha+k-1}{k} \binom{n+k}{k}^{-1} S(n+k,k).$$

Formula (37) provides an interesting generalization of the following known result for the Bernoulli numbers B_n:

(38) $$B_n = \sum_{k=0}^{n} (-1)^k \binom{n+1}{k+1} \binom{n+k}{k}^{-1} \frac{\Delta^k 0^{n+k}}{k!},$$

which was considered, for example, by Gould (1972, p. 49, Eq. (17)).

Euler polynomials and numbers

The *Euler ploynomials* $E_n(x)$ and the *Euler numbers* E_n are defined by the following generating functions:

(39) $$\frac{2e^{xz}}{e^z+1} = \sum_{n=0}^{\infty} E_n(x) \frac{z^n}{n!} \quad (|z| < \pi);$$

(40) $$\frac{2e^z}{e^{2z}+1} = \operatorname{sech} z = \sum_{n=0}^{\infty} E_n \frac{z^n}{n!} \quad \left(|z| < \frac{\pi}{2}\right).$$

The following formulas are readily derivable from (39) and (40):

(41) $$E_n(x+1) + E_n(x) = 2\,x^n \quad (n \in \mathbb{N}_0);$$

(42) $$E_n'(x) = n\,E_{n-1}(x) \quad (n \in \mathbb{N});$$

(43) $$E_n(1-x) = (-1)^n\,E_n(x) \quad (n \in \mathbb{N}_0);$$

(44) $$(-1)^{n+1}\,E_n(-x) = E_n(x) - 2\,x^n \quad (n \in \mathbb{N}_0);$$

(45) $$E_n(x+y) = \sum_{k=0}^{n} \binom{n}{k} E_k(x)\,y^{n-k} \quad (n \in \mathbb{N}_0);$$

(46) $$E_n(x) = \sum_{k=0}^{n} \binom{n}{k} \frac{E_k}{2^k} \left(x - \frac{1}{2}\right)^{n-k} \quad (n \in \mathbb{N}_0),$$

which, upon taking $x = \frac{1}{2}$, yields

(47) $$E_n = 2^n\,E_n\left(\frac{1}{2}\right) \quad (n \in \mathbb{N}_0);$$

(48) $$\sum_{k=0}^{n} \binom{2n}{2k} E_{2k} = 0 \quad (n \in \mathbb{N}).$$

Multiplication formulas:

(49) $$E_n(mx) = m^n \sum_{k=0}^{m-1} (-1)^k\,E_n\left(x + \frac{k}{m}\right) \quad (n \in \mathbb{N}_0;\ m = 1,\ 3,\ 5,\ \ldots);$$

(50)
$$E_n(mx) = -\frac{2}{n+1}\,m^n \sum_{k=0}^{m-1} (-1)^k\,B_{n+1}\left(x + \frac{k}{m}\right) \quad (n \in \mathbb{N}_0;\ m = 2,\ 4,\ 6,\ \ldots).$$

Integral formulas:

(51) $$\int_x^y E_n(t)\,dt = \frac{E_{n+1}(y) - E_{n+1}(x)}{n+1} \quad (m,\ n \in \mathbb{N}_0);$$

(52)
$$\int_0^1 E_n(t)\,E_m(t)\,dt = (-1)^n\,4\,\left(2^{m+n+2} - 1\right) \frac{m!\,n!}{(m+n+2)!} B_{m+n+2} \quad (m,\ n \in \mathbb{N}_0).$$

An alternating finite sum of powers can be expressed as the Euler polynomials:

(53) $$\sum_{k=1}^{m} (-1)^{m-k}\,k^n = \frac{1}{2}\left\{E_n(m+1) + (-1)^m\,E_n(0)\right\} \quad (m,\ n \in \mathbb{N}).$$

Fourier series expansions of Bernoulli and Euler polynomials

By employing suitable contour integrations in complex function theory, we can obtain the following Fourier series expansions of Bernoulli and Euler polynomials:

$$B_{2n}(x) = \frac{(-1)^{n-1}\,2\,(2n)!}{(2\pi)^{2n}} \sum_{k=1}^{\infty} \frac{\cos 2k\pi x}{k^{2n}} \quad (n \in \mathbb{N};\ 0 \leq x \leq 1); \tag{54}$$

$$\begin{gathered} B_{2n-1}(x) = \frac{(-1)^{n}\,2\,(2n-1)!}{(2\pi)^{2n-1}} \sum_{k=1}^{\infty} \frac{\sin 2k\pi x}{k^{2n-1}} \\ (n = 1 \text{ and } 0 < x < 1;\ n \in \mathbb{N} \setminus \{1\} \text{ and } 0 \leq x \leq 1); \end{gathered} \tag{55}$$

$$\begin{gathered} E_{2n}(x) = \frac{(-1)^{n}\,4\,(2n)!}{\pi^{2n+1}} \sum_{k=0}^{\infty} \frac{\sin (2k+1)\pi x}{(2k+1)^{2n+1}} \\ (n = 0,\ 0 < x < 1;\ n \in \mathbb{N},\ 0 \leq x \leq 1); \end{gathered} \tag{56}$$

$$E_{2n-1}(x) = \frac{(-1)^{n}\,4\,(2n-1)!}{\pi^{2n}} \sum_{k=0}^{\infty} \frac{\cos (2k+1)\pi x}{(2k+1)^{2n}} \quad (n \in \mathbb{N};\ 0 \leq x \leq 1), \tag{57}$$

which, for $x = \frac{1}{2}$, yields

$$E_{2n+1} = 0 \quad (n \in \mathbb{N}_0), \tag{58}$$

where use is made of the relationship (47).

The first few of the Euler numbers E_n are given below:

$$E_0 = 1,\ E_2 = -1,\ E_4 = 5,\ E_6 = -61,\ E_8 = 1385,\ E_{10} = -50521,\ \ldots . \tag{59}$$

Relations between Bernoulli and Euler polynomials

The following relationships between the Bernoulli and Euler polynomials follow easily from the definitions (1) and (39):

$$\begin{aligned} E_n(x) &= \frac{2^{n+1}}{n+1} \left\{ B_{n+1}\left(\frac{x+1}{2}\right) - B_{n+1}\left(\frac{x}{2}\right) \right\} \\ &= \frac{2}{n+1} \left\{ B_{n+1}(x) - 2^{n+1} B_{n+1}\left(\frac{x}{2}\right) \right\} \quad (n \in \mathbb{N}_0), \end{aligned} \tag{60}$$

which, in view of (10), can also be written in the form:

$$E_n(1-x) = (-1)^n \frac{2}{n+1} \left[2^{n+1} B_{n+1}\left(\frac{x+1}{2}\right) - B_{n+1}(x) \right] \quad (n \in \mathbb{N}_0); \tag{61}$$

Two additional formulas involving these polynomials are given below:

(62) $$E_{n-2}(x) = 2\binom{n}{2}^{-1} \sum_{k=0}^{n-2} \binom{n}{k} \left(2^{n-k}-1\right) B_{n-k}\, B_k(x) \quad (n \in \mathbb{N}\setminus\{1\});$$

(63) $$B_n(x) = 2^{-n} \sum_{k=0}^{n} \binom{n}{k} B_{n-k}\, E_k(2x) \quad (n \in \mathbb{N}_0).$$

The generalized Euler polynomials and numbers

The *generalized Euler polynomials* $E_n^{(\alpha)}(x)$ and the *generalized Euler numbers* $E_n^{(\alpha)}$ are defined by the generating functions:

(64) $$\left(\frac{2}{e^z+1}\right)^{\alpha} \exp(xz) = \sum_{n=0}^{\infty} E_n^{(\alpha)}(x)\, \frac{z^n}{n!} \quad (|z| < \pi;\ 1^{\alpha} := 1)$$

and

(65) $$\left(\frac{2\, e^z}{e^{2z}+1}\right)^{\alpha} = \sum_{n=0}^{\infty} E_n^{(\alpha)}\, \frac{z^n}{n!} \quad \left(|z| < \frac{\pi}{2};\ 1^{\alpha} := 1\right)$$

for arbitrary (real or complex) parameter α. Clearly, we have

(66) $$E_n^{(1)}(x) = E_n(x) \quad \text{and} \quad E_n^{(1)} = E_n \quad (n \in \mathbb{N}_0).$$

It is easy to find from (64) and (65) that

(67) $$2^n\, E_n^{(\alpha)}\left(\frac{\alpha}{2}\right) = E_n^{(\alpha)} \quad (n \in \mathbb{N}_0),$$

which, for $\alpha = 1$, reduces to (47).

Srivastava *et al.* (1983, p. 443, Eq. (4.12)) proved the following interesting addition theorem for the generalized Euler polynomials:

(68)

$$E_n^{(\alpha+\lambda\gamma)}(x+\gamma y) = \sum_{k=0}^{n} \frac{\gamma+n}{\gamma+k} \binom{n}{k} E_k^{(\alpha-\lambda k)}(x-ky)\, E_{n-k}^{(\lambda k+\lambda\gamma)}(ky+\gamma y) \quad (\Re(\gamma) > 0).$$

Problems

1. Show that each form of the Gamma function defined by 1.1(1), 1.1(2), and 1.1(7) is analytic in its given domain.
[Whittaker and Watson (1963, Chapter 12); Rainville (1960, pp. 15–18)]

2. Prove the following series representations for the Euler-Mascheroni constant γ:

$$\gamma = 1 + \sum_{n=3}^{\infty} \frac{(-1)^n}{n} \left[\frac{\log(n-1)}{\log 2} \right]$$

and

$$\gamma = \sum_{n=1}^{\infty} \frac{(-1)^n}{n} \left[\frac{\log n}{\log 2} \right],$$

where the bracket denotes the greatest integer function.
[Gerst (1969); Sandham (1949)]

3. Deduce the explicit Weierstrass canonical product form 1.3(3) for the double Gamma function Γ_2 by using the three properties associated with Barnes's definition of Γ_2.
[Barnes (1899b, pp. 265–269); Whittaker and Watson (1963, p. 264)]

4. Prove the following generalization of Gauss's multiplication formula 1.1(51) for the Gamma function:

$$\prod_{\ell=0}^{m-1} \Gamma\left(kz + \frac{k\ell}{m}\right) = (2\pi)^{\frac{1}{2}(m-k)} \left(\frac{k}{m}\right)^{mkz + \frac{1}{2}(mk-m-k)} \cdot \prod_{n=0}^{k-1} \Gamma\left(mz + \frac{mn}{k}\right) \quad (k,\, m \in \mathbb{N}).$$

[Choi and Quine (1992, p. 131); Magnus *et al.* (1966, p. 3)]

5. For the Euler-Mascheroni constant γ, the sequence $\{s_n\}_{n=0}^{\infty}$ defined by

$$s_n := \left(1 + \frac{1}{2} + \cdots + \frac{1}{n-1}\right) - \log n \quad (n \in \mathbb{N} \setminus \{1\})$$

satisfies the asymptotic property:

$$s_n = \gamma + O\left(n^{-1}\right) \quad (n \to \infty).$$

Show that $\gamma - s_n$ $(n \in \mathbb{N} \setminus \{1\})$ can be expressed as follows as an infinite sum with rational terms:

$$\gamma - s_n = \frac{1}{n} \sum_{m=0}^{\infty} \frac{t_{m+2}}{\binom{m+n}{m}} \quad (n \in \mathbb{N} \setminus \{1\}),$$

where

$$t_{m+2} = -\frac{1}{(m+1)!} \int_0^1 (0-x)(1-x) \cdot \cdots \cdot (m-x)\, dx \quad (m \in \mathbb{N}_0).$$

[Elsner (1995, p. 1537); Jolley (1961, pp. 14–15)].

6. Prove the von Staudt-Clausen theorem (Theorem 6.1):

$$B_{2n} = I_n - \sum_{p-1|2n} \frac{1}{p} \quad (n \in \mathbb{N}),$$

where B_n is the Bernoulli number, I_n is an integer, and the sum is taken over all primes p such that $p-1$ divides $2n$.

[Carathéodory (1958, pp. 281–284); Apostol (1976, pp. 274–275)]

7. Prove the following rapidly converging series expansion for $\psi(z)$:

$$\psi(z) = \frac{1}{2}\log\left[(z+n)^2 + (z+n) + \frac{1}{3}\right] - \sum_{k=0}^{n} \frac{1}{z+k} - \frac{1}{9}\sum_{k=n+1}^{\infty} (k-n)A_k,$$

where

$$A_k := \int_0^1 (z+k+t)^{-2}\left[(z+k+t)^4 - \frac{1}{3}(z+k+t)^2 + \frac{1}{9}\right]^{-1} dt.$$

[Shafer and Lossers (1969)]

8. Let S_n denote the area of the sphere of radius 1 in $\mathbb{R}^n$. Prove that

$$S_n = \frac{\left[\int_{-\infty}^{\infty} e^{-t^2}\,dt\right]^n}{\int_0^{\infty} e^{-r^2} r^{n-1}\,dr} = \frac{2\left(\sqrt{\pi}\right)^n}{\Gamma\left(\frac{n}{2}\right)} \quad (n \in \mathbb{N}).$$

[Campbell (1966, pp. 126–128)]

9. Let

$$I_n := \int \cdots \int f\left(\sum_{k=1}^{n} t_k\right) t_1^{\alpha_1 - 1} \cdots t_n^{\alpha_n - 1}\,dt_1 \cdots dt_n$$

$$\left(\alpha_k > 0;\ t_k \geq 0;\ k = 1, \ldots, n;\ \sum_{k=1}^{n} t_k \leq 1\right),$$

where f is a continuous function. Prove that

$$I_n = \frac{\Gamma(\alpha_1)\cdots\Gamma(\alpha_n)}{\Gamma(\alpha_1 + \cdots + \alpha_n)} \int_0^1 f(\tau)\,\tau^{\Sigma_{k=1}^{n}\alpha_k - 1}\,d\tau.$$

[Whittaker and Watson (1963, p. 258); Campbell (1966, pp. 128–133)]

10. Prove that

$$\int_0^{\pi} (\sin\theta)^{\lambda-1} e^{-\mu\theta}\,d\theta = \frac{\pi\Gamma(\lambda)\,e^{-\frac{1}{2}\pi\mu}}{2^{\lambda-1}\,\Gamma\left(\frac{\lambda+i\mu+1}{2}\right)\Gamma\left(\frac{\lambda-i\mu+1}{2}\right)} \quad (\Re(\lambda) > 0;\ \mu \in \mathbb{C}).$$

[Nielsen (1965, p. 159)]

11. Show that

$$\int_{-\infty}^{\infty} \frac{e^{-2\pi izt}}{\cosh(\pi t)}dt = \frac{1}{\cosh(\pi z)} \qquad \left(-\frac{1}{2} < \Im(z) < \frac{1}{2}\right)$$

[*cf.* Equation 1.1(66)]

12. Making use of the Pochhammer symbol defined by 1.1(5), show that the Vandermonde convolution theorem 1.4(9) can be rewritten in the form:

$$\sum_{k=0}^{n} \frac{(\alpha)_k\,(\beta)_{n-k}}{k!\,(n-k)!} = \frac{(\alpha+\beta)_n}{n!} = \sum_{k=0}^{n} \frac{(\alpha)_{n-k}\,(\beta)_k}{(n-k)!\,k!}$$

$$(\alpha,\ \beta \in \mathbb{C};\ n \in \mathbb{N}_0)\,.$$

13. Show that Euler's transformation 1.4(21) is equivalent to the Pfaff-Saalschütz theorem:

$${}_3F_2\left[\begin{matrix} a,\ b,\ -n; \\ c,\ a+b-c-n+1; \end{matrix}\ 1\right] = \frac{(c-a)_n\,(c-b)_n}{(c)_n\,(c-a-b)_n}$$

$$\left(c \notin \mathbb{Z}_0^-\,;\ n \in \mathbb{N}_0\right).$$

14. Deduce Gauss's summation theorem 1.4(7) as a limit case of the Pfaff-Saalschütz theorem (see Problem 13 above).
[**Hint**: Let $n \to \infty$ and apply the asymptotic expansion 1.1(57)]

15. For the Bernoulli polynomials $B_n(x)$ generated by 1.6(1), show that

$$x^n = \sum_{k=0}^{n} \frac{1}{n+1}\binom{n+1}{k} B_k(x) \qquad (n \in \mathbb{N}_0)\,.$$

Hence (or otherwise) deduce the following representation for the Laguerre polynomials $L_n^{(\alpha)}(x)$ defined by 1.4(72):

$$L_n^{(\alpha)}(x) = \sum_{k=0}^{n} \frac{(-1)^k}{k!}\binom{n+\alpha}{n-k} B_k(x) \cdot {}_2F_2(-n+k,\ 1;\ \alpha+k+1,\ 2\,;\ 1).$$

[*Cf.* Magnus *et* al. (1966, p. 26); see also Popov (1998)]

16. Derive the following asymptotic expansion for the Psi (or Digamma) function:

$$\psi(z) = \log z - \frac{1}{2z} - \frac{1}{12z^2} + \frac{1}{120z^4} - \frac{1}{252z^6} + O\left(z^{-8}\right)$$

$$(|z| \to \infty;\ |\arg(z)| \le \pi - \epsilon\ (0 < \epsilon < \pi)).$$

17. For a, $b \in \mathbb{N}$, let (a, b) and $[a, b]$ denote the greatest common divisor and least common multiple, respectively, of a and b, and $\{x\} = x - [x]$ denotes the fractional part of x. Then show that, if $\Re(s) > \frac{1}{2}$,

$$\int_0^1 \zeta(1-s, \{ax\})\, \zeta(1-s, \{bx\})\, dx = \frac{2\, \{\Gamma(s)\}^2\, \zeta(2s)}{(2\pi)^{2s}} \left(\frac{(a,b)}{[a,b]}\right)^s,$$

where $\zeta(s, a)$ is the Hurwitz (or generalized) Zeta function defined by 2.2(1). Show also that

$$\int_0^1 B_n(\{ax\})\, B_n(\{bx\})\, dx = (-1)^{n-1} \frac{B_{2n}}{(2n)!} \left(\frac{(a,b)}{[a,b]}\right)^n \qquad (n \in \mathbb{N}).$$

[*Cf.* Mordell (1958b, p. 372); see also Mikolás (1957)]

18. Prove that

$$\int_0^{1/\sqrt{8}} \frac{t^2 \log t}{(t^2+1)^{\frac{1}{2}}}\, dt = \frac{1}{96} \left\{2\pi^2 + 30(\log 2)^2 - 39 \log 2 - 9\right\}.$$

[Knuth (1976, p. 138)]

19. Prove that

$$\int_0^\infty \frac{\cosh x \cdot \log x}{\cosh(2x) - \cos(2\pi a)}\, dx$$
$$= \frac{\pi}{2 \sin(\pi a)} \left\{\log \frac{\Gamma\left(\frac{1}{2}a + \frac{1}{2}\right)}{\Gamma\left(\frac{1}{2}a\right)} + \frac{1}{2} \log\left(2\pi \cot \frac{a\pi}{2}\right)\right\} \qquad (0 < a < 1).$$

[Williams and Zhang (1993, p. 44)]

20. Prove the following inequality:

$$\operatorname{cosec}^2 x - \frac{1}{2m+1} < \sum_{k=-m}^{m} \frac{1}{(x - k\pi)^2} < \operatorname{cosec}^2 x$$
$$\left(m \in \mathbb{N};\ 0 < |x| \le \frac{1}{2}\pi;\ x \in \mathbb{R}\right).$$

Also apply this inequality to show that

$$\int_0^\infty \left(\frac{\sin x}{x}\right)^2 dx = \frac{\pi}{2}.$$

[Neville (1951, pp. 629–630)]

21. Prove that

$$\int_0^\infty \frac{e^x \log x}{e^{2x} - 2\, e^x \cos(2\pi a) + 1}\, dx = \frac{\pi}{2} \frac{1}{\sin(2\pi a)} \log\left((2\pi)^{1-2a} \frac{\Gamma(1-a)}{\Gamma(a)}\right) \qquad (0 < a < 1).$$

[Zhang and Williams (1993c, p. 377)]

22. Prove that

$$\int_{\frac{\pi}{4}}^{\frac{\pi}{2}} \log\left(\log\left(\tan x\right)\right) dx = \frac{\pi}{2} \log\left(\frac{\Gamma\left(\frac{3}{4}\right)}{\Gamma\left(\frac{1}{4}\right)}\sqrt{2\pi}\right).$$

[*Cf.* Vardi (1988a, p. 308); see also Gradshteyn and Ryzhik (1980, p. 532)]

23. Prove the generalized Goldbach-Euler series:

$$\sum_{n=2}^{\infty}\sum_{k=0}^{\infty} \frac{1}{(pk+r)^n - 1} = \frac{1}{p}\left[\psi\left(\frac{r}{p}\right) - \psi\left(\frac{r-1}{p}\right)\right] + \frac{1}{r(r-1)}$$
$$(p \in \mathbb{N};\ r = p \text{ or } r = p+1),$$

where the case $r = 1 = p$ is excluded.

[Melzak (1973, p. 88)]

24. Let

$$P = \prod_{n=1}^{\infty} \frac{(n-a_1)(n-a_2)\cdots(n-a_k)}{(n-b_1)(n-b_2)\cdots(n-b_k)} \quad \left(\sum_{j=1}^{k} a_j = \sum_{j=1}^{k} b_j\right),$$

where no a_j or b_j is a positive integer. Show that

$$P = \prod_{j=1}^{k} \frac{\Gamma(1-b_j)}{\Gamma(1-a_j)}.$$

[*Cf.* Rainville (1960, pp. 13–15); see also Melzak (1973, p. 101)]

25. Let M_n and m_n denote the maximum and minimum of the Bernoulli polynomial $B_n(x)$ for $0 \leq x \leq 1$. Show that

$$\begin{aligned} M_{4n} &= B_{4n}\left(\tfrac{1}{2}\right) = \left(1 - 2^{1-4n}\right)|B_{4n}| \quad (n \in \mathbb{N}); \\ M_{4n+2} &= B_{4n+2}(0) = B_{4n+2} \quad (n \in \mathbb{N}_0); \\ m_{4n} &= B_{4n}(0) = -|B_{4n}| \quad (n \in \mathbb{N}); \\ m_{4n+2} &= B_{4n+2}\left(\tfrac{1}{2}\right) = -\left(1 - 2^{-1-4n}\right) B_{4n+2} \quad (n \in \mathbb{N}_0). \end{aligned}$$

[Lehmer (1940, pp. 534)]

26. Prove the following integrals:

$$\int_0^1 \frac{\log u}{u} \{\log(1+u)\}^2\, du = \mathcal{A} - \frac{\pi^4}{288}$$

and

$$\int_0^{\pi} u \left\{\log\left(2\cos\frac{u}{2}\right)\right\}^2 du = \mathcal{A} + \frac{31\,\pi^4}{480},$$

where

$$\mathcal{A}=\sum_{n=1}^{\infty}\frac{(-1)^{n-1}}{(n+1)^2}\left(\sum_{k=1}^{n}\frac{1}{k^2}\right).$$

[Rutledge and Douglass (1934, p. 30)]

27. Suppose that
(a) $f(z)$ is a meromorphic function of a complex variable z in $\mathbb{C}$;
(b) $f(z)$ is analytic and vanishes nowhere on $\mathbb{C}\setminus\mathbb{Z}_0^-$;
(c) $f(z)$ satisfies the functional equation:

$$n^{z-\frac{1}{2}}\prod_{k=0}^{n-1} f\left(\frac{z+k}{n}\right)=(2\pi)^{\frac{n-1}{2}}\,f(z)\quad\left(z\in\mathbb{C}\setminus\mathbb{Z}_0^-\right),$$

where n is an arbitrarily fixed positive integer greater than 1 and $n^{z-\frac{1}{2}}$ denotes the principal value $\exp\left[\left(z-\frac{1}{2}\right)\operatorname{Log} n\right]$. Then show that the function $f(z)$ must be of the form:

$$f(z)=\exp\left[a\left(z-\frac{1}{2}\right)+\frac{2m\pi i}{n-1}\right]\Gamma(z),$$

where a is an arbitrary complex constant and m is an arbitrary integer.

[*Cf.* Haruki (1986, p. 174); see also Theorem 1.1 (Bohr-Mollerup)]

28. Suppose that a function $f:\mathbb{R}^+\to\mathbb{R}$ satisfies the functional equation:

$$f(x+1)=x\,f(x)\quad\left(x\in\mathbb{R}^+\right)$$

and, moreover, that

$$\lim_{x\to\infty}\left(\frac{e}{x}\right)^x\sqrt{\frac{x}{2\pi}}\,f(x)=1.$$

Show that $f(x)=\Gamma(x)$ for all $x\in\mathbb{R}^+$.

[Kuczma (1968, p. 129)]

29. Suppose that a function $f:\mathbb{R}^+\to\mathbb{R}$ satisfies the following properties:
(a) $f(x+1)=x\,f(x)$;
(b) $\left(\frac{e}{x}\right)^x f(x)$ is decreasing for $x\in\mathbb{R}^+$;
(c) $f(1)=1$.
Show that $f(x)=\Gamma(x)$ for $x\in\mathbb{R}^+$.

[Anastassiadis (1957b, p. 117)]

30. Suppose that a function $f:\mathbb{R}^+\to\mathbb{R}^+$ satisfies the following properties:
(a) $f(x+1)=\frac{x}{x+y}\,f(x)\qquad(y>0)$;
(b) $f(x)$ is decreasing for $x\in\mathbb{R}^+$;
(c) $f(1)=\frac{1}{y}$.
Show that $f(x)=B(x,y)$ for $x\in\mathbb{R}^+$.

[Anastassiadis (1959, pp. 25-26)]

31. Prove that, for $p \geq 5$ and $p \in \mathbb{N}$,

$$B_p^{(p)} \equiv -\frac{1}{2} p^2 (p-1)! \pmod{p^5},$$

where $B_p^{(p)}$ denotes the generalized Bernoulli numbers defined by 1.6(22).

[Carlitz (1953, p. 112)]

32. Prove that, for $m \in \mathbb{N}$ and $i = \sqrt{-1}$,

$$\begin{aligned} &{}_2F_1(1,\, 1\,;\, 2m\,;\, i) \\ &\quad = (2m-1)\, i^m\, 2^{m-1} \left\{ \frac{\log 2}{2} - \frac{i\,\pi}{4} - \sum_{j=1}^{m-1} (-1)^j \frac{i^j}{2^j} \left(\frac{1}{2j} + \frac{1+i}{2j-1} \right) \right\} \end{aligned}$$

and

$$\begin{aligned} &{}_2F_1(1,\, 1\,;\, 2m+1\,;\, i) \\ &\quad = m(1+i)\, i^m\, 2^m \left\{ \frac{\log 2 - 1}{2} - i\frac{2-\pi}{4} - \sum_{j=1}^{m-1} (-1)^j \frac{i^j}{2^j} \left(\frac{1}{2j} + \frac{1-i}{4j+2} \right) \right\}. \end{aligned}$$

[Butzer and Hauss (1992b, p. 355)]

33. Prove that

$$\sum_{n=0}^{\infty} \frac{1}{(2n+1)^2 \binom{2n}{n}} = \frac{8}{3} G - \frac{\pi}{3} \log\left(2+\sqrt{3}\right).$$

[Borwein and Borwein (1987, p. 386)]

34. Prove that

$$\int_{-1}^{1} (2-2x)^\alpha (2+2x)^\beta \left(1-x^2\right)^{-\frac{1}{2}} dx = \frac{\Gamma\left(\alpha+\frac{1}{2}\right) \Gamma\left(\beta+\frac{1}{2}\right) \pi}{\Gamma(\alpha+\beta+1) \left[\Gamma\left(\frac{1}{2}\right)\right]^2}\, 2^{2\alpha+2\beta}$$

$$(\min\{\Re(\alpha), \Re(\beta)\} > -1).$$

[*Cf.* Askey (1980, p. 357)]

35. Prove that

$$\begin{aligned} \log \Gamma(z+1) = &\frac{1}{2}\left(z+\frac{1}{2}\right) \left\{ \log\left(z+\frac{1}{2}+\frac{1}{2\sqrt{3}}\right) + \log\left(z+\frac{1}{2}-\frac{1}{2\sqrt{3}}\right) \right\} \\ &- z - \frac{1}{2} + \frac{1}{2}\log(2\pi) + \int_0^\infty e^{-zt}\, \phi(t)\, dt, \end{aligned}$$

where, for convenience,

$$\begin{aligned} \phi(t) := &\frac{1}{t}\left(\frac{1}{2} - \frac{1}{t} + \frac{1}{e^t-1}\right) - \frac{1}{2t}\left\{1 - e^{-\frac{1}{2}t} \cosh\left(\frac{t}{2\sqrt{3}}\right)\right\} \\ &- \frac{d}{dt}\left\{ \frac{1 - e^{-\frac{1}{2}t} \cosh\left(\frac{t}{2\sqrt{3}}\right)}{t} \right\}; \end{aligned}$$

Moreover

$$\phi(t) = \frac{e^{-\frac{1}{2}t}}{t}\left\{\frac{1}{2}\operatorname{cosech}\left(\frac{1}{2}t\right) + \frac{1}{2\sqrt{3}}\sinh\left(\frac{t}{2\sqrt{3}}\right) - \frac{1}{t}\cosh\left(\frac{t}{2\sqrt{3}}\right)\right\}.$$

[*Cf.* Watson (1920, p. 5); see also Eq. 1.2 (29)]

36. Prove that, for fixed $p \in \mathbb{R}^+$,

$$\gamma = \lim_{n\to\infty} \frac{\sum_{k=0}^{\infty}\left(\frac{n^k}{k!}\right)^p (H_k - \log n)}{\sum_{k=0}^{\infty}\left(\frac{n^k}{k!}\right)^p},$$

where γ is the Euler-Mascheroni constant, and H_k denote the harmonic numbers defined by 3.2(36).

[Brent and McMillan (1980, pp. 310)]

37. Let $\mathbf{u}$ be the row vector $\{u_k = 1/k : k \in \mathbb{N}\}$ and let $\mathbf{M}$ be the matrix with entries $\{m_{ij} = 1/(i-j+1)$ if $j \leq i$, $m_{ij} = 0$ if $j > i : i, j \in \mathbb{N}\}$. Let $\mathbf{v}$ be the column vector $\{v_n = 1/(n+1) : n \in \mathbb{N}\}$. Show that the product $\mathbf{u}\left(\mathbf{M}^{-1}\mathbf{v}\right)$ exists (as a convergent series), and is equal to Euler-Mascheroni constant γ defined by 1.1(3).

[Kenter (1999, p. 452)]

38. Let $\Omega(z)$ be the function of the complex variable z defined by

$$\Omega(z) := \frac{\Gamma(z+a_1)\,\Gamma(z+a_2)}{\Gamma(z+b_1)\,\Gamma(z+b_2)}$$

in which a_1, a_2, b_1, b_2 are to be regarded as any constants (real or complex). Show that, for $|\arg z| < \pi$ and large $|z|$,

$$\Omega(z) \sim \left(\frac{1}{z}\right)^p \left\{1 + \frac{c_1}{z+1} + \frac{c_2}{(z+1)(z+2)} + \frac{c_3}{(z+1)(z+2)(z+3)} + \cdots\right\},$$

where the c_n are constants and $p = b_1 + b_2 - a_1 - a_2$.

[van Engen (1938, pp. 124–125)]

39. Prove the following Lipschitz summation formula:

$$\sum_{m=-\infty}^{\infty} (2m+\tau)^{-\lambda} = \frac{e^{-\pi i \lambda/2}\,\pi^{\lambda}}{\Gamma(\lambda)} \sum_{n=1}^{\infty} n^{\lambda-1}\, e^{\pi i \tau n} \quad (\Re(\tau) > 0;\ \lambda > 1).$$

[Knopp (1970, p. 65)]

40. Prove the following series-integral representation of the Catalan constant G:

$$G = \frac{1}{2}\sum_{n=0}^{\infty}(-1)^n \left(\int_0^{\frac{\pi}{2}} \cos^n t\, dt\right)^2.$$

[Catalan (1883, p. 51)]

41. Prove that

$$\frac{1}{2}\psi'(x)\,\psi'''(x) < (\psi''(x))^2 \quad (x > 0),$$

where ψ denotes the Psi (or Digamma) function.

[*Cf.* English and Rousseau (1997, p. 432); see also Alzer and Wells (1998, p. 1459)]

42. A function f is said to be completely monotonic on an interval I, if $f \in C^\infty(I)$ and

$$(-1)^k f^{(k)}(x) \geq 0 \qquad (*)$$

for all $x \in I$ and for all $k \in \mathbb{N}_0$. If the inequality $(*)$ is strict for all $x \in I$ and for all $k \in \mathbb{N}_0$, then f is said to be strictly completely monotonic on I. Consider the function

$$F_n(x;c) := \left(\psi^{(n)}(x)\right)^2 - c\,\psi^{(n-1)}(x)\,\psi^{(n+1)}(x)$$
$$\left(x \in \mathbb{R}^+;\ c \in \mathbb{R};\ n \in \mathbb{N} \setminus \{1\}\right).$$

Let $n \in \mathbb{N} \setminus \{1\}$, and let α, $\beta \in \mathbb{R}$. Show that the function:

$$x \mapsto F_n(x;\alpha)$$

is strictly completely monotonic on $(0,\infty)$ if and only if

$$\alpha \leq \frac{n-1}{n},$$

and that the function:

$$x \mapsto -F_n(x;\beta)$$

is strictly completely monotonic on $(0,\infty)$ if and only if

$$\beta \geq \frac{n}{n+1}.$$

[Alzer and Wells (1998, p. 1460)]

43. Let the function $g(x)$ be defined by

$$g(x) := x^2\,\psi'(1+x) - x\,\psi(1+x) + \log\Gamma(x+1) \quad (x > -1).$$

Show that $g(x)$ strictly decreases from ∞ to 0 on $(-1,0]$ and strictly increases from 0 to ∞ on $[0,\infty)$.

[Elbert and Laforgia (2000, Theorem 2)]

44. Prove that

$$\frac{x^{x-1}\,e^y}{y^{y-1}\,e^x} \leq \frac{\Gamma(x)}{\Gamma(y)} \leq \frac{x^{x-\frac{1}{2}}\,e^y}{y^{y-\frac{1}{2}}\,e^x} \quad (x \geq y > 1).$$

[Kečkić and Vasić (1971, p. 107)]

45. Prove that

(a) For n an odd positive integer,

$$G = -T\left(\frac{1}{4}\right) = \frac{2n+1}{n+1}\sum_{j=1}^{n}(-1)^j\, T\left(\frac{2j-1}{8n+4}\right);$$

(b) For n an even positive integer,

$$G = -T\left(\frac{1}{4}\right) = \frac{2n+1}{n}\sum_{j=1}^{n}(-1)^{j+1}\, T\left(\frac{2j-1}{8n+4}\right),$$

where, for convenience, T is defined by

$$T(r) := \int_0^{r\pi} \log(\tan\theta)\, d\theta \qquad \left(0 \leq r \leq \frac{1}{2}\right).$$

[Bradley (1999, pp. 164–165)]

46. For the Stirling numbers $s(n,k)$ defined by 1.5(1), show that

(a) $$\sum_{n=0}^{\infty} \frac{(-1)^n}{(n+k)(a)_n} s(n+k,k) = -\frac{(1-a)_k}{k!}\psi^{(k)}(a-k);$$

(b) $$\sum_{n=0}^{\infty} \frac{(-1)^n}{(n+k-1)(a)_n} s(n+k,k) = \frac{(-a)_k}{a(k-1)!}\psi^{(k-1)}(a-k).$$

Also deduce the special cases of *each* of these summation formulas when

$$a = k+1 \quad \text{and} \quad a = k+2.$$

[*Cf.* Jordan (1965, p. 343); see also Hansen (1975, p. 348)]

47. Verify that the special case of Problem 46(a) above when $a = k+1$ is *precisely* the relationship 3.5(16).

[*Cf.* Problem 5 (Chapter 3)]

CHAPTER 2

THE ZETA AND RELATED FUNCTIONS

This chapter aims at providing a self-contained theory of the Zeta and related functions, which will be required in each of the next chapters. We first introduce (and investigate the various properties and relationships satisfied by) the multiple Hurwitz Zeta function $\zeta_n(s,a)$ $(n \in \mathbb{N})$ and consider its relatively more familiar special case when $n = 1$, that is, the Hurwitz (or generalized) Zeta function. We then deal rather systematically with the Riemann Zeta function which (for the *main* purpose of this book) happens to be the most important member of the significantly large family of Zeta functions considered in this chapter. Other functions (introduced in this chapter) include the Polylogarithm functions, Legendre's Chi function, Clausen's integral (or function), the Hurwitz-Lerch Zeta function, and so on.

2.1 Multiple Hurwitz Zeta Functions

Barnes (1904) introduced and studied the generalized multiple Hurwitz Zeta function $\zeta_n\left(s, a \,|\, w_1, \ldots, w_n\right)$ defined, for $\Re(s) > n$, by the n-ple series

$$(1) \qquad \zeta_n\left(s, a \,|\, w_1, \ldots, w_n\right) := \sum_{m_1, \ldots, m_n = 0}^{\infty} \frac{1}{(a+\Omega)^s} \qquad (\Re(s) > n;\; n \in \mathbb{N}),$$

where $\Omega = m_1 w_1 + \cdots + m_n w_n$ and the general conditions for a and the parameters $w_1, \ldots, w_n$ are given in Barnes (1904) who used it in the study of the multiple Gamma functions (see Section 1.3). In this section we consider only the simple case of (1) when

$$w_j = 1 \, (j = 1, \ldots, n;\; j, n \in \mathbb{N})$$

and

$$(2) \qquad \zeta_n\left(s, a\right) := \sum_{k_1, \ldots, k_n = 0}^{\infty} \left(a + k_1 + \cdots + k_n\right)^{-s} \qquad (\Re(s) > n;\; a \notin \mathbb{Z}_0^-),$$

which is also referred to as *n-ple* (or, simply, *multiple*) *Hurwitz Zeta function.* We shall give a rather detailed investigation of the properties and characteristics of the function $\zeta_n\left(s, a\right)$ in (2), including its analytic continuation.

The analytic continuation of $\zeta_n\left(s, a\right)$

In order to give the analytic continuation of $\zeta_n\left(s, a\right)$, we first recall a criterion for the convergence of an n-ple series attributed to Eisenstein in the work of Forsyth (1900) who gave the proof for the double series. Here we shall give a proof of the general case:

Theorem 2.1. *The n-ple series:*

$$\sum_{m_1,\ldots,m_n=-\infty}^{\infty}{}' \left(m_1^2+\cdots+m_n^2\right)^{-\mu} \tag{3}$$

converges if $\mu > \frac{1}{2}n$, *where the prime denotes the exclusion of the case when* $m_j = 0$ $(j = 1, \ldots, n)$.

Proof. Let the series be arranged in the partial series: For this purpose, choose integers k_j such that

$$2^{k_j} \leq m_j < 2^{k_j+1} \quad (1 \leq j \leq n_j;\ k_j \in \mathbb{N}_0).$$

Let

$$D := \{(m_1,\cdots,m_n) \,|\, 2^{k_j} \leq m_j < 2^{k_j+1};\ 1 \leq j \leq n\}$$

and let $I_j := [2^{k_j}, 2^{k_j+1})$ $(1 \leq j \leq n)$ be half-open intervals. Since $2^{k+1} - 2^k = 2^k$, the number of integers m_j in I_j is 2^{k_j} $(1 \leq j \leq n)$. So the number of n-tuples in D is

$$2^{k_1}\cdots 2^{k_n} = 2^{k_1+\cdots+k_n} = 2^{nx},$$

where

$$x := \frac{k_1+\cdots+k_n}{n}.$$

Note that, for $(m_1,\cdots,m_n) \in D$,

$$\sum_{j=1}^{n} 2^{2k_j} \leq m_1^2+\cdots+m_n^2 < \sum_{j=1}^{n} 2^{2k_j+2}$$

and, by comparing the arithmetic and geometric means of 2^{2k_j} $(1 \leq j \leq n)$, we also have

$$2^{2x} \leq \frac{1}{n}\sum_{j=1}^{n} 2^{2k_j} \leq m_1^2+\cdots+m_n^2.$$

Now, for any $(m_1,\cdots,m_n)$ in D and any positive real μ, we have

$$\frac{1}{(m_1^2+\cdots+m_n^2)^{\mu}} \leq \frac{1}{2^{2x\mu}}.$$

Then the sum of all terms as $1/(m_1^2+\cdots+m_n^2)^{\mu}$ in D is less than or equal to

$$\frac{2^{nx}}{2^{2x\mu}} = \frac{1}{2^{x(2\mu-n)}} = \frac{1}{2^{\frac{2\mu-n}{n}k_1}} \cdots \frac{1}{2^{\frac{2\mu-n}{n}k_n}}.$$

Let $k = \max\{k_j \,|\, 1 \leq j \leq n\}$. We find that the sum of all the partial series is less than or equal to

$$\begin{aligned}\sum_{k_1=0}^{k}\cdots\sum_{k_n=0}^{k} \frac{1}{2^{\frac{2\mu-n}{n}k_1}}\cdots\frac{1}{2^{\frac{2\mu-n}{n}k_n}} &= \left(\sum_{k_1=0}^{k}\frac{1}{2^{\frac{2\mu-n}{n}k_1}}\right)^n \\ &= \left[\frac{1-2^{-(k+1)\frac{2\mu-n}{n}}}{1-\frac{1}{2^{\frac{2\mu-n}{n}}}}\right]^n.\end{aligned} \tag{4}$$

Taking the limit in (4) as $k \to \infty$, we observe that the n-ple series in (4) converges if $2\mu - n > 0$, that is, if $\mu > \frac{1}{2}n$, and the sum is

$$\left(\frac{1}{1-\frac{1}{2^{\frac{2\mu-n}{n}}}}\right)^n = \frac{2^{2\mu-n}}{\left(2^{\frac{2\mu-n}{n}}-1\right)^n}.$$

This completes the proof of Theorem 2.1.

Let $s = \sigma + it$ ($\sigma,\ t \in \mathbb{R}$). First, for convergence, we consider $\zeta_n(s,a)$ in (2) for the case when $a > 0$:

$$\zeta_n(s,a) = \sum_{k_1,\cdots,k_n=0}^{\infty} (a+k_1+\cdots+k_n)^{-s} \quad (\Re(s) = \sigma > n;\ a > 0). \tag{5}$$

Theorem 2.2. *The series for $\zeta_n(s,a)$ in (5) converges absolutely for $\sigma > n$. The convergence is uniform in every half-plane $\sigma \geq n+\delta$ $(\delta > 0)$, so $\zeta_n(s,a)$ is an analytic function of s in the half-plane $\sigma > n$.*

Proof. Observe that, for $\sigma > 0$,

$$\begin{aligned}\sum_{k_1,\cdots,k_n=0}^{\infty}{}' (k_1+\cdots+k_n)^{-\sigma} &= \sum_{k_1,\cdots,k_n=0}^{\infty}{}' [(k_1+\cdots+k_n)^2]^{-\frac{1}{2}\sigma} \\ &\leq \sum_{k_1,\cdots,k_n=0}^{\infty}{}' (k_1^2+\cdots+k_n^2)^{-\frac{1}{2}\sigma},\end{aligned}$$

in which the prime denotes the exclusion of the case when $k_j = 0$ $(1 \leq j \leq n)$ and that the last series converges for $\sigma > n$ by Eisenstein's theorem (Theorem 2.1). Thus all statements in Theorem 2.2 follow from the inequality:

$$\begin{aligned}\sum_{k_1,\cdots,k_n=0}^{\infty} |(a+k_1+\cdots+k_n)^{-s}| &= \sum_{k_1,\cdots,k_n=0}^{\infty} (a+k_1+\cdots+k_n)^{-\sigma} \\ &\leq \sum_{k_1,\cdots,k_n=0}^{\infty} (a+k_1+\cdots+k_n)^{-n-\delta}.\end{aligned}$$

We now choose to recall a convergence theorem concerning term-by-term integration of monotonic sequences of functions, which is due to Lévi [see Apostol (1974, p. 268, Theorem 10.25)].

Theorem 2.3. *Let $L(I)$ denote the set of all Lebesgue-integrable functions on an interval I. Let $\{g_n\}$ be a sequence of functions in $L(I)$ such that*

(a) *each g_n is nonnegative almost everywhere on I*

and

(b) *the series $\sum_{n=1}^{\infty} \int_I g_n$ converges.*

Then the series $\sum_{n=1}^{\infty} g_n$ converges almost everywhere on I to a sum function g in $L(I)$. Moreover,

$$\int_I g = \int_I \sum_{n=1}^{\infty} g_n = \sum_{n=1}^{\infty} \int_I g_n.$$

Next we present an integral representation of $\zeta_n(s,a)$, which is given by

Theorem 2.4. *If* $\Re(s)=\sigma>n$, *then*

$$\Gamma(s)\,\zeta_n(s,a)=\int_0^\infty \frac{x^{s-1}e^{-ax}}{(1-e^{-x})^n}dx \quad (\Re(s)>n;\ n\in\mathbb{N}). \tag{6}$$

Proof. It follows from 1.1(1) that, for $\Re(s)=\sigma>0$,

$$\Gamma(s)=\int_0^\infty x^{s-1}e^{-x}dx.$$

First we keep s real, $s>n$ $(n\in\mathbb{N})$, and then extend the result to complex s by analytic continuation. In this Eulerian integral for $\Gamma(s)$ we set

$$x=(a+k_1+\cdots+k_n)t \quad (k_j\in\mathbb{N}_0;\ j=1,\ldots,n),$$

and we find that

$$\Gamma(s)=(a+k_1+\cdots+k_n)^s\int_0^\infty e^{-(a+k_1+\cdots+k_n)t}\ t^{s-1}dt$$

so that

$$(a+k_1+\cdots+k_n)^{-s}\,\Gamma(s)=\int_0^\infty e^{-(k_1+\cdots+k_n)t}e^{-at}t^{s-1}dt.$$

Summing over all $k_j\in\mathbb{N}_0$ $(1\leq j\leq n)$, we obtain

$$\zeta_n(s,a)\,\Gamma(s)=\sum_{k_1,\cdots,k_n=0}^{\infty}\int_0^\infty e^{-(k_1+\cdots+k_n)t}e^{-at}t^{s-1}\,dt,$$

the series on the right being convergent if $s>n$.

Now we wish to interchange the order of summation and integration. The simplest way to justify this process is to regard the integral as a Lebesgue integral. Since the integrand is nonnegative, Lévi's convergence theorem (Theorem 2.3) implies that the series:

$$\sum_{k_1,\cdots,k_n=0}^{\infty}\int_0^\infty e^{-(a+k_1+\cdots+k_n)t}e^{-at}t^{s-1}dt$$

converges almost everywhere to a sum function which itself is Lebesgue-integrable on $[0,\infty)$ and that

$$\zeta_n(s,a)\,\Gamma(s)=\int_0^\infty \sum_{k_1,\cdots,k_n=0}^{\infty} e^{-(k_1+\cdots+k_n)t}e^{-at}t^{s-1}dt.$$

But, if $t>0$, we have $0<e^{-t}<1$, and hence

$$\sum_{k=0}^{\infty}e^{-kt}=\frac{1}{1-e^{-t}},$$

the series being a geometric series. We, therefore, have

$$\sum_{k_1,\cdots,k_n=0}^{\infty} e^{-(k_1+\cdots+k_n)t} e^{-at} t^{s-1} = \frac{e^{-at} t^{s-1}}{(1-e^{-t})^n}$$

almost everywhere on $[0,\infty)$; in fact, everywhere except at 0, so we get

$$\zeta_n(s,a)\,\Gamma(s) = \int_0^\infty \frac{e^{-at} t^{s-1}}{(1-e^{-t})^n} dt.$$

This proves (6) for real s and $s > n$ $(n \in \mathbb{N})$.

To extend this integral representation to all complex $s = \sigma + it$ with $\sigma > n$, we note that both functions on the left-hand side of (6) are analytic for $\sigma > n$. To show that the right member is also analytic, we assume that $n + \delta \leq \sigma \leq c$, where $c > n$ and $\delta > 0$, and write

$$\begin{aligned}\int_0^\infty \left| \frac{e^{-at} t^{s-1}}{(1-e^{-t})^n} \right| dt &= \int_0^\infty \frac{e^{-at} t^{\sigma-1}}{(1-e^{-t})^n} dt \\ &= \left(\int_0^1 + \int_1^\infty \right) \frac{e^{-at} t^{\sigma-1}}{(1-e^{-t})^n} dt.\end{aligned}$$

If $0 \leq t \leq 1$, we have $t^{\sigma-n} \leq t^\delta$, and if $t \geq 1$, we have $t^{\sigma-n} \leq t^{c-n}$. Also, since $e^t - 1 \geq t$ for $t \geq 0$, we have

$$\begin{aligned}\int_0^1 \frac{e^{-at} t^{\sigma-1}}{(1-e^{-t})^n} dt &\leq \int_0^1 \frac{e^{(n-a)t} t^{\delta+n-1}}{(e^t-1)^n} dt \\ &\leq \begin{cases} e^{n-a} \displaystyle\int_0^1 t^{\delta-1}\, dt = \frac{e^{n-a}}{\delta} & \text{if} \quad 0 < a \leq n, \\ \displaystyle\int_0^1 t^{\delta-1} dt = \frac{1}{\delta} & \text{if} \quad a > n \end{cases}\end{aligned}$$

and

$$\int_1^\infty \frac{e^{-at} t^{\sigma-1}}{(1-e^{-t})^n} dt \leq \int_0^\infty \frac{e^{-at} t^{c-1}}{(1-e^{-t})^n} dt = \Gamma(c)\,\zeta_n(c,a).$$

This shows that the integral in (6) converges uniformly in every strip $n+\delta \leq \sigma \leq c$, where $\delta > 0$, and therefore represents an analytic function in every such strip; hence also in the half-plane $\sigma > n$. Therefore, by the principle of analytic continuation, (6) holds true for all s with $\Re(s) = \sigma > n$ $(n \in \mathbb{N})$.

In order to extend $\zeta_n(s,a)$ to the half-plane on the left of the line $\sigma = n$, we derive another representation in terms of a contour integral. The contour C is essentially a Hankel's loop [*cf.*, *e.g.*, Whittaker and Watson (1963, p. 245)], which starts from ∞ along the upper side of the positive real axis, encircles the origin once in the positive (counter-clockwise) direction, and then returns to ∞ along the lower side of the positive real axis.

The loop C is composed of three parts C_1, C_2, and C_3, where C_2 is a positively-oriented circle of radius $c < 2\pi$ about the origin, and C_1 and C_3 are the upper and lower edges of a cut in the complex z-plane along the positive real axis, traversed as described above. Thus we can use the parametrizations $-z = re^{-\pi i}$ on C_1 and $-z = re^{\pi i}$ on C_3, where r varies from c to ∞.

Theorem 2.5. *If $a > 0$, then the function defined by the contour integral:*

$$I_n(s,a) = -\frac{1}{2\pi i}\int_{\infty}^{(0+)} \frac{(-z)^{s-1}e^{-az}}{(1-e^{-z})^n}\,dz. \tag{7}$$

is an entire function of s. Moreover,

$$\zeta_n(s,a) = \Gamma(1-s)\,I_n(s,a) \quad (\Re(s) = \sigma > n). \tag{8}$$

Proof. Here $(-z)^s$ means $r^s e^{-\pi is}$ on C_1 and $r^s e^{\pi is}$ on C_3. We consider an arbitrary compact disk $|s| \le M$ and prove that the integrals along C_1 and C_3 converge uniformly on every such disk. Since the integrand in (7) is an entire function of s, this will prove that $I_n(s,a)$ is also an entire function of s for $a > 0$. Along C_1 we have, for $r \ge 1$,

$$|(-z)^{s-1}| = r^{\sigma-1}\left|e^{-\pi i(\sigma-1+it)}\right| = r^{\sigma-1}\,e^{\pi t} \le r^{M-1}e^{\pi M},$$

since $|s| \le M$. Similarly, along C_3 we have, for $r \ge 1$,

$$\left|(-z)^{s-1}\right| = r^{\sigma-1}|e^{\pi i\,(\sigma-1+it)}| = r^{\sigma-1}e^{-\pi t} \le r^{M-1}e^{\pi M}.$$

Hence, on either C_1 or C_3, we have, for $r \ge 1$,

$$\left|\frac{(-z)^{s-1}e^{-az}}{(1-e^{-z})^n}\right| \le \frac{r^{M-1}e^{\pi\,M}e^{-ar}}{(1-e^{-r})^n} \le \frac{e^{\pi\,M}}{(1-e^{-1})^n}\cdot r^{M-1}\cdot e^{-ar}.$$

But $\int_c^\infty r^{M-1}\,e^{-ar}\,dr$ converges if $c > 0$; this shows that the integrals along C_1 and C_3 converge uniformly on every compact disk $|s| \le M$, and hence $I_n(s,a)$ is an entire function of s. To prove (8), we write

$$-2\pi i\,I_n(s,a) = \left(\int_{C_1} + \int_{C_2} + \int_{C_3}\right)(-z)^{s-1}g(-z)\,dz,$$

where

$$g(-z) = \frac{e^{-az}}{(1-e^{-z})^n}.$$

On C_1 and C_3 we have $g(-z) = g(-r)$, and on C_2 we write $-z = ce^{i\theta}$, where θ varies from 2π to 0. This gives us

$$\begin{aligned}
-2\pi i\,I_n(s,a) &= \int_\infty^c r^{s-1}\,e^{-\pi i\,(s-1)}\,g(-r)\,dr \\
&\quad - i\int_{2\pi}^0 c^{s-1}e^{(s-1)i\theta}ce^{i\theta}g(ce^{i\theta})\,d\theta \\
&\quad + \int_c^\infty r^{s-1}e^{\pi i\,(s-1)}g(-r)\,dr \\
&= -\,2i\sin(\pi s)\int_c^\infty r^{s-1}g(-r)\,dr - ic^s\int_{2\pi}^0 e^{is\theta}\,g(ce^{i\theta})\,d\theta.
\end{aligned} \tag{9}$$

Dividing both sides of (9) by $-2i$, we obtain

$$\pi I_n(s,a) = \sin(\pi s) I_1(s,c) + I_2(s,c),$$

where

$$I_1(s,c) = \int_c^\infty r^{s-1} g(-r)\,dr \quad \text{and} \quad I_2(s,c) = \frac{c^s}{2}\int_{2\pi}^0 e^{is\theta} g(ce^{i\theta})\,d\theta.$$

Now let $c \to 0$. We find that, in view of (6),

$$\lim_{c\to 0} I_1(s,c) = \int_0^\infty \frac{r^{s-1}e^{-ar}}{(1-e^{-r})^n}\,dr = \Gamma(s)\,\zeta_n(s,a) \quad (\sigma = \Re(s) > n).$$

We next show that

$$\lim_{c\to 0} I_2(s,c) = 0.$$

To do this, we note that $g(-z)$ is analytic in $|z| < 2\pi$, except for a pole of order n at $z = 0$. Therefore, $z^n g(-z)$ is analytic everywhere inside $|z| < 2\pi$, and hence is bounded there, that is, $|g(-z)| \leq A/|z|^n$, where $|z| = c < 2\pi$ and A is a positive constant. We thus have

$$|I_2(s,c)| \leq \frac{c^\sigma}{2}\int_0^{2\pi} e^{-t\theta}\cdot\frac{A}{c^n}\,d\theta \leq \pi A\, e^{2\pi|t|}\, c^{\sigma-n}.$$

If $\sigma > n$ and $c \to 0$, we find that $I_2(s,c) \to 0$; hence

$$\pi I_n(s,a) = \sin(\pi s)\,\Gamma(s)\,\zeta_n(s,a), \tag{10}$$

which is, in terms of 1.1(12), seen to be equivalent to (8).

In the equation (8), valid for $\sigma > n$, the function $I_n(s,a)$ is an entire function of s, and $\Gamma(1-s)$ is analytic for every complex s for $s \in \mathbb{C}\setminus\mathbb{N}$. We, therefore, can use this equation to define $\zeta_n(s,a)$ for $\sigma \leq n$, that is, outside $\sigma > n$ as desired.

Definition 2.1. If $\Re(s) = \sigma \leq n$, we define $\zeta_n(s,a)$ by

$$\zeta_n(s,a) := \Gamma(1-s)\, I_n(s,a), \tag{11}$$

where $I_n(s,a)$ is given in (7).

This equation (11) provides the analytic continuation of $\zeta_n(s,a)$ to the whole complex s-plane.

Theorem 2.6. *The function $\zeta_n(s,a)$ defined by* (11) *is analytic for all s except for simple poles at $s = k$ $(1 \leq k \leq n)$, with their respective residues given by*

$$\operatorname*{Res}_{s=k} \zeta_n(s,a) = \frac{1}{(n-k)!(k-1)!}\lim_{z\to 0}\frac{d^{n-k}}{dz^{n-k}}\frac{z^n e^{-az}}{(1-e^{-z})^n} \quad (k = 1, \ldots, n;\ n \in \mathbb{N}). \tag{12}$$

In particular, when $s = n$, its residue is $1/(n-1)!$.

Proof. Since $I_n(s,a)$ is an entire function of s for $a > 0$, the only possible singularities of $\zeta_n(s,a)$ are the poles of $\Gamma(1-s)$. Since $1/\Gamma(1-s)$ has simple zeros at $s \in \mathbb{N}$, $\Gamma(1-s)$ has simple poles at $s \in \mathbb{N}$. But Theorem 2.2 shows that $\zeta_n(s,a)$ is analytic in $\Re(s) = \sigma > n$, and so $s = 1, \cdots, n$ are the only poles of $\zeta_n(s,a)$.

We next prove the assertion (12). If s is any integer, say $s = k$, the integrand in the contour integral for $I_n(s,a)$ in (7) takes the same values on both C_1 and C_3, and hence the integrals along C_1 and C_3 cancel, giving us

$$\begin{aligned} I_n(k,a) &= -\frac{1}{2\pi i}\int_{C_2} \frac{(-z)^{k-1}e^{-az}}{(1-e^{-z})^n}\,dz \\ &= -\operatorname*{Res}_{z=0} f(z), \end{aligned}$$

where, for convenience,

$$f(z) := \frac{(-z)^{k-1}e^{-az}}{(1-e^{-z})^n}.$$

We observe that, for $1 \leq k \leq n$, the function $f(z)$ has a pole of order $n+1-k$ at $z = 0$. We, therefore, have

$$I_n(k,a) = \frac{(-1)^k}{(n-k)!}\lim_{z\to 0}\frac{d^{n-k}}{dz^{n-k}}\frac{z^n e^{-az}}{(1-e^{-z})^n}. \tag{13}$$

To find the residue of $\zeta_n(s,a)$ at $s = k$ $(1 \leq k \leq n)$, by using 1.1(12), we compute the limit:

$$\begin{aligned} \lim_{s\to k}(s-k)\zeta_n(s,a) &= \lim_{s\to k}(s-k)\Gamma(1-s)I_n(s,a) \\ &= \frac{\pi\, I_n(k,a)}{\Gamma(k)}\cdot\lim_{s\to k}\frac{s-k}{\sin(\pi s)} \\ &= \frac{I_n(k,a)}{(-1)^k(k-1)!}, \end{aligned}$$

which, by virtue of (13), immediately yields (12).

Relationship between $\zeta_n(s,x)$ and $B_n^{(\alpha)}(x)$

By using the multiple Hurwitz Zeta function, Choi (1996a) derived the following explicit formula for $B_n^{(\alpha)}(x)$ different from 1.6(28):

$$B_{n+k}^{(n)}(x) = n\binom{n+k}{n}\sum_{j=0}^{n-1}(-1)^j\frac{B_{k+j+1}(x)}{k+j+1}\sum_{\ell=j}^{n-1}\binom{\ell}{j}s(n,\ell+1)x^{\ell-j} \tag{14}$$

in terms of the Stirling numbers $s(n,k)$ of the first kind (see Section 1.5).

The value of $\zeta_n(-\ell,x)$ can be calculated explicitly for $\ell \in \mathbb{N}_0$. Taking $s = -\ell$ in the relation (8) with a replaced by x, we find that

$$\zeta_n(s,-\ell) = \Gamma(1+\ell)I_n(-\ell,x) = \ell!\, I_n(-\ell,x). \tag{15}$$

Now, from the definition (19) of the generalized Bernoulli polynomials $B_n^{(\alpha)}(x)$, we have

$$\begin{aligned}
I_n(-\ell,x) &= -\frac{1}{2\pi i}\int_C \frac{(-z)^{-\ell-1}e^{-xz}}{(1-e^{-z})^n}\,dz \\
&= -\operatorname*{Res}_{z=0}\frac{(-z)^{-\ell-1}e^{-xz}}{(1-e^{-z})^n} \\
&= (-1)^\ell \operatorname*{Res}_{z=0} z^{-n-\ell-1}\frac{z^n e^{(n-x)z}}{(e^z-1)^n} \\
&= (-1)^\ell \operatorname*{Res}_{z=0} z^{-n-\ell-1}\sum_{k=0}^{\infty} B_k^{(n)}(n-x)\frac{z^k}{k!} \\
&= (-1)^\ell \frac{B_{n+\ell}^{(n)}(n-x)}{(n+\ell)!},
\end{aligned}$$

which, in view of 1.6(20) and (15), yields the desired relationship:

(16) $$\zeta_n(-\ell,x) = (-1)^n \frac{\ell!}{(n+\ell)!}B_{n+\ell}^{(n)}(x) \quad (\ell \in \mathbb{N}_0).$$

Setting $n=1$ in (16), we have the well-known result:

(17) $$\zeta(-\ell,x) = -\frac{B_{\ell+1}(x)}{\ell+1} \quad (\ell \in \mathbb{N}_0),$$

where $\zeta(s,x) := \zeta_1(s,x)$ is the Hurwitz (or generalized) Zeta function (see Section 2.2).

The number of solutions of

$$k_1+\cdots+k_n = k \quad (k\in\mathbb{N}_0;\ (k_1,\ldots,k_n)\in {\mathbb{N}_0}^n)$$

is equal to the coefficient of x^k in the Maclaurin series expansion of $(1-x)^{-n}$:

(18) $$(1-x)^{-n} = \sum_{k=0}^{\infty}\binom{-n}{k}(-x)^k = \sum_{k=0}^{\infty}\binom{k+n-1}{n-1}x^k.$$

The multiple Hurwitz Zeta function in (5) can thus be expressed as a simple series:

(19) $$\zeta_n(s,x) = \sum_{k=0}^{\infty}\binom{k+n-1}{n-1}(x+k)^{-s}.$$

From 1.5(1) or 1.5(5), we find that

(20) $$\binom{k+n-1}{n-1} = \frac{1}{(n-1)!}\sum_{j=0}^{n-1}|s|(n,j+1)k^j,$$

where $|s|(n,k) := (-1)^{n+k}\, s(n,k)$ are often called the *unsigned* or *absolute* Stirling numbers of the first kind. Combining (19) and (20), we have

$$\zeta_n(s,x) = \frac{1}{(n-1)!}\sum_{k=0}^{\infty}\left(\sum_{j=0}^{n-1}|s|(n,j+1)k^j\right)(x+k)^{-s},$$

which, by virtue of the identity:

$$k^\ell = \{(-x)+(x+k)\}^\ell = \sum_{j=0}^{\ell}\binom{\ell}{j}(-x)^{\ell-j}(x+k)^j,$$

yields

$$\begin{aligned}\zeta_n(s,x) &= \frac{1}{(n-1)!}\left[\sum_{\ell=0}^{n-1}|s|(n,\ell+1)\left\{\sum_{j=0}^{\ell}\binom{\ell}{j}(-x)^{\ell-j}\right\}\right]\left(\sum_{k=0}^{\infty}\frac{1}{(x+k)^{s-j}}\right)\\ &= \frac{1}{(n-1)!}\sum_{\ell=0}^{n-1}|s|(n,\ell+1)\sum_{j=0}^{\ell}\binom{\ell}{j}(-x)^{\ell-j}\zeta(s-j,x).\end{aligned}$$

Next it is easy to show that $\zeta_n(s,x)$ is expressible as a finite combination of the generalized Zeta function $\zeta(s,x)$ with polynomial coefficients in x:

$$\zeta_n(s,x) = \sum_{j=0}^{n-1} p_{n,j}(x)\zeta(s-j,x), \tag{21}$$

where

$$p_{n,j}(x) = \frac{1}{(n-1)!}\sum_{\ell=j}^{n-1}(-1)^{n+1-j}\binom{\ell}{j}s(n,\ell+1)x^{\ell-j}. \tag{22}$$

We shall now find $p_{n,j}(x)$ in (22) as a polynomial in x of degree $n-1-j$ with rational coefficients.

Since $\zeta(s,x)$ can be continued analytically to a meromorphic function having a simple pole at $s=1$ with its residue 1, the representation (21) shows that $\zeta_n(s,x)$ is analytic for all s except for simple poles only at $s=k$ $(k=1,\ldots,n;\ n\in\mathbb{N})$ with their respective residues given by

$$\operatorname*{Res}_{s=k}\zeta_n(s,x) = p_{n,k-1}(x) \quad (k=1,\ldots,n;\ n\in\mathbb{N}). \tag{23}$$

In view of 1.5(3) and 1.5(6), $\zeta_n(s,x)$ can be expressed explicitly for the first few values of n:

$$\begin{aligned}\zeta_2(s,x) =& (1-x)\zeta(s,x)+\zeta(s-1,x),\\ \zeta_3(s,x) =& \frac{1}{2}\left(x^2-3x+2\right)\zeta(s,x)+\left(\frac{3}{2}-x\right)\zeta(s-1,x)+\frac{1}{2}\zeta(s-2,x),\\ \zeta_4(s,x) =& \frac{1}{6}\left\{\left(-x^3+6x^2-11x+6\right)\zeta(s,x)+\left(3x^2-12x+11\right)\zeta(s-1,x)\right.\\ &\left.-(3x-6)\zeta(s-2,x)+\zeta(s-3,x)\right\}.\end{aligned} \tag{24}$$

Letting $s = -\ell$ in (21) and applying (16), we obtain the desired formula (14). Also, upon setting $x = n$ $(n \in \mathbb{N})$ in (14) and making use of 1.6(21), we find that

$$B_{n+k}^{(n)} = (-1)^{n+k} n \binom{n+k}{n} \sum_{j=0}^{n-1} (-1)^j \frac{B_{k+j+1}(n)}{k+j+1} \sum_{\ell=j}^{n-1} \binom{\ell}{j} s(n,\ell+1) n^{\ell-j}. \tag{25}$$

Now we express $\operatorname{Res}_{s=k} \zeta_n(s,x)$ in (12) in a more recognizable form:

$$\operatorname*{Res}_{s=k} \zeta_n(s,x) = \frac{B_{n-k}^{(n)}(n-x)}{(n-k)!(k-1)!}. \tag{26}$$

Indeed, it follows from 1.6(19) and (12) that

$$\begin{aligned}\lim_{s\to k} (s-k)\zeta_n(s,x) &= \frac{1}{(n-k)!(k-1)!} \lim_{z\to 0} \frac{d^{n-k}}{dz^{n-k}} \sum_{j=0}^{\infty} B_j^{(n)}(n-x) \frac{z^j}{j!} \\ &= \frac{B_{n-k}^{(n)}(n-x)}{(n-k)!(k-1)!}.\end{aligned}$$

The Vardi-Barnes multiple Gamma functions

Vardi (1988, p. 498) gave another expression for the multiple Gamma functions $\Gamma_n(a)$ (see Section 1.3) whose general form was also studied by Barnes (1904):

$$\Gamma_n(a) = \left[\prod_{m=1}^{n} R_{n-m+1}^{(-1)^m \binom{a}{m-1}} \right] G_n(a) \quad (n \in \mathbb{N}), \tag{27}$$

where

$$G_n(a) := \exp\left[\zeta_n'(0,a)\right] \quad \text{with} \quad \zeta_n'(s,a) = \frac{\partial}{\partial s} \zeta_n(s,a)$$

and

$$R_m := \exp\left(\sum_{k=1}^{m} \zeta_k'(0,1) \right) \quad \text{with } R_0 = 1. \tag{28}$$

In particular, the special cases of (27) when $n = 1$ and $n = 2$ give other forms of the simple and double Gamma functions $\Gamma_1 = \Gamma$ and Γ_2:

$$\begin{aligned}\Gamma(a) &= \exp\left[-\zeta'(0) + \zeta'(0,a)\right] \\ &= \sqrt{2\pi} \exp\left[\zeta'(0,a)\right],\end{aligned} \tag{29}$$

where $\zeta(s) := \zeta(s,1)$ is the Riemann Zeta function (see Section 2.3);

$$\Gamma_2(a) = A\,(2\pi)^{\frac{1}{2}-\frac{1}{2}a} \exp\left(-\frac{1}{12} + \zeta_2'(0,a) \right), \tag{30}$$

where we have used (24) and the known identity [see Voros (1987, p. 462, Eq. (A.11))]:

$$\log A = \frac{1}{12} - \zeta'(-1). \tag{31}$$

Here we can give another proof of the multiplication formula for Γ_2 (see 1.3(21)) different from that of Barnes (1899b) by using (30) [see Choi and Quine (1992)]. We consider

$$\sum_{\ell=0}^{n-1}\sum_{j=0}^{n-1}\zeta_2\left(s,a+\frac{\ell+j}{n}\right)=\sum_{\ell=0}^{n-1}\sum_{j=0}^{n-1}\sum_{k_1,\,k_2=0}^{\infty}\left(a+\frac{\ell+j}{n}+k_1+k_2\right)^{-s}$$

$$=n^s\sum_{k_1,\,k_2=0}^{\infty}\sum_{\ell=0}^{n-1}\sum_{j=0}^{n-1}(na+\ell+j+nk_1+nk_2)^{-s}$$

$$=n^s\sum_{k_1,\,k_2=0}^{\infty}(na+k_1+k_2)^{-s}=n^s\,\zeta_2(s,na),$$

which, upon differentiating with respect to s, yields

$$\sum_{\ell=0}^{n-1}\sum_{j=0}^{n-1}\zeta_2'\left(s,a+\frac{\ell+j}{n}\right)=(\log n)\,n^s\,\zeta_2(s,na)+n^s\,\zeta_2'(s,na).$$

By virtue of (30), we readily obtain the following multiplication formula for Γ_2:

$$\prod_{\ell=0}^{n-1}\prod_{j=0}^{n-1}\Gamma_2\left(a+\frac{\ell+j}{n}\right)=C(n)\,(2\pi)^{-\frac{1}{2}n(n-1)a}\,n^{-\frac{n^2a^2}{2}+na}\,\Gamma_2(na), \tag{32}$$

where

$$C(n):=A^{n^2-1}\cdot e^{\frac{1}{12}(1-n^2)}\cdot(2\pi)^{\frac{1}{2}(n-1)}\cdot n^{\frac{5}{12}}.$$

An interesting identity is also obtained from (32):

$$\prod_{\ell=0}^{n-1}\prod_{j=0}^{n-1}{}'\,\Gamma_2\left(\frac{\ell+j}{n}\right)=\frac{C(n)}{n}, \tag{33}$$

where the prime denotes the exclusion of the case when $\ell=0=j$.

2.2 The Hurwitz (or Generalized) Zeta Function

The *Hurwitz* (or *generalized*) *Zeta function* $\zeta(s,a)$ is defined by

$$\zeta(s,a):=\sum_{k=0}^{\infty}(k+a)^{-s}\qquad(\Re(s)>1;\ a\notin\mathbb{Z}_0^-). \tag{1}$$

It is easy to see that $\zeta(s,a)=\zeta_1(s,a)$ for the case when $n=1$ in 2.1(2). Thus we can deduce many properties of $\zeta(s,a)$ from those of $\zeta_n(s,a)$ in Section 2.1. Indeed, the series for $\zeta(s,a)$ in (1) converges absolutely for $\Re(s)=\sigma>1$. The convergence is uniform in every half-plane $\sigma\geq 1+\delta$ $(\delta>0)$, so $\zeta(s,a)$ is an analytic function of s in the half-plane $\Re(s) = \sigma > 1$. Setting $n = 1$ in 2.1(6), we have the integral representation:

(2)
$$\Gamma(s)\,\zeta(s,a) = \int_0^\infty \frac{x^{s-1}\,e^{-ax}}{1-e^{-x}}\,dx = \int_0^\infty \frac{x^{s-1}\,e^{-(a-1)x}}{e^x - 1}\,dx$$
$$= \int_0^1 \frac{x^{a-1}}{1-x}\left(\log\frac{1}{x}\right)^{s-1} dx \quad (\Re(s) > 1;\ \Re(a) > 0).$$

Moreover, $\zeta(s,a)$ can be continued meromorphically to the whole complex s-plane (except for a simple pole at $s = 1$ with its residue 1) by means of the contour integral representation (see Theorem 2.5):

(3)
$$\zeta(s,a) = -\frac{\Gamma(1-s)}{2\pi i}\int_C \frac{(-z)^{s-1}\,e^{-az}}{1-e^{-z}}\,dz,$$

where the contour C is the Hankel loop of Theorem 2.5. The connection between $\zeta(s,a)$ and the Bernoulli polynomials $B_n(x)$ is also given in 2.1(17).

From the definition (1) of $\zeta(s,a)$, it easily follows that

(4)
$$\zeta(s,a) = \zeta(s, n+a) + \sum_{k=0}^{n-1} (k+a)^{-s} \quad (n \in \mathbb{N});$$

(5)
$$\zeta\left(s, \frac{1}{2}a\right) - \zeta\left(s, \frac{1}{2}a + \frac{1}{2}\right) = 2^s \sum_{n=0}^{\infty} (-1)^n\,(a+n)^{-s}.$$

Hurwitz's formula for $\zeta(s,a)$

The series expression $\zeta(s,a)$ was originally meaningful for $\sigma > 1$ $(s = \sigma + it)$. Hurwitz obtained another series representation for $\zeta(s,a)$ valid in the half-plane $\sigma < 0$:

(6)
$$\zeta(1-s,a) = \frac{\Gamma(s)}{(2\pi)^s}\left\{e^{-\frac{1}{2}\pi i s}\,L(a,s) + e^{\frac{1}{2}\pi i s}\,L(-a,s)\right\}$$
$$(0 < a \leq 1,\ \sigma = \Re(s) > 1;\ 0 < a < 1,\ \sigma > 0),$$

where the function $L(x,s)$ is defined by

(7)
$$L(x,s) := \sum_{n=1}^{\infty} \frac{e^{2\pi i n x}}{n^s} \quad (x \in \mathbb{R};\ \sigma = \Re(s) > 1),$$

which is often referred to as the *periodic* (or *Lerch*) *Zeta function.*

We note that the Dirichlet series in (7) is a periodic function of x with period 1 and that $L(1,s) = \zeta(s)$, the Riemann Zeta function (see Section 2.3). The series in (7) converges absolutely for $\sigma > 1$. Yet, if $x \notin \mathbb{Z}$, the series can also be seen to converge conditionally for $\sigma > 0$. So the formula (6) is also valid for $\sigma > 0$ if $a \neq 1$.

We observe that the function $L(x, s)$ in (7) is a linear combination of the Hurwitz Zeta functions when x is a rational number. Indeed, setting $x = p/q$ $(1 \leq p \leq q;\ p,\ q \in \mathbb{N})$ in (7), the terms in (7) can be rearranged according to the residue classes mod q by letting

$$n = kq + r \quad (1 \leq r \leq q;\ k \in \mathbb{N}_0),$$

which gives us, for $\sigma > 1$,

$$\begin{aligned} L\left(\frac{p}{q}, s\right) &= \sum_{n=1}^{\infty} \frac{1}{n^s} \exp\left(\frac{2\pi i n p}{q}\right) = \sum_{r=1}^{q} \sum_{k=0}^{\infty} \frac{\exp\left(\frac{2\pi i r p}{q}\right)}{(kq+r)^s} \\ &= \frac{1}{q^s} \sum_{r=1}^{q} \exp\left(\frac{2\pi i r p}{q}\right) \sum_{k=0}^{\infty} \frac{1}{\left(k + \frac{r}{q}\right)^s} \\ &= \frac{1}{q^s} \sum_{r=1}^{q} \exp\left(\frac{2\pi i r p}{q}\right) \zeta\left(s, \frac{r}{q}\right). \end{aligned}$$

Therefore, if we take $a = p/q$ in the Hurwitz formula (6), we obtain:

$$\zeta\left(1 - s, \frac{p}{q}\right) = \frac{2\,\Gamma(s)}{(2\pi q)^s} \sum_{r=1}^{q} \cos\left(\frac{\pi s}{2} - \frac{2\pi r p}{q}\right) \zeta\left(s, \frac{r}{q}\right) \tag{8}$$

$$(1 \leq p \leq q;\ p,\ q \in \mathbb{N}),$$

which holds true, by the principle of analytic continuation, for all admissible values of $s \in \mathbb{C}$.

Hermite's formula for $\zeta(s, a)$

We first recall *Plana's summation formula:*

$$\sum_{k=m}^{n} f(k) = \frac{1}{2}[f(m) + f(n)] + \int_m^n f(\tau)\, d\tau - 2 \int_0^{\infty} \frac{q(m,t) - q(n,t)}{e^{2\pi t} - 1}\, dt, \tag{9}$$

where $f(z)$ is a bounded analytic function in $m \leq \Re(z) \leq n$ and

$$q(\lambda, t) = \frac{1}{2i}[f(\lambda + it) - f(\lambda - it)] \quad (i = \sqrt{-1}).$$

Choose the function $f(z)$ in (9) as follows:

$$f(z) = (a + z)^{-s} \quad (s = \sigma + it;\ |\arg(a + z)| < \pi). \tag{10}$$

Then it is easy to see that

$$q(x, y) = -\left[(a + x)^2 + y^2\right]^{-\frac{1}{2}s} \sin\left[s \arctan\left(\frac{y}{x + a}\right)\right].$$

In view of the elementary inequality:

$$\left|\arctan\left(\frac{y}{x+a}\right)\right| \leq \min\left\{\frac{\pi}{2}, \frac{|y|}{x+a}\right\},$$

we obtain

$$(11)\qquad \begin{aligned} |q(x,y)| &\leq \left[(a+x)^2+y^2\right]^{\frac{1}{2}-\frac{1}{2}\sigma} \left|y^{-1}\right| \sinh\left(\frac{1}{2}\pi|s|\right) && (|y|>a); \\ |q(x,y)| &\leq \left[(a+x)^2+y^2\right]^{-\frac{1}{2}\sigma} \left|\sinh\left(\frac{y|s|}{x+a}\right)\right| && (|y|<a). \end{aligned}$$

Making use of (11), it is easily seen that the integral:

$$\int_0^\infty q(x,y)\left(e^{2\pi y}-1\right)^{-1} dy \quad (\sigma>0)$$

converges when $x \geq 0$ and tends to 0 as $x \to \infty$. Also the improper integral:

$$\int_0^\infty (a+x)^{-s} dx \quad (\sigma = \Re(s) > 1)$$

converges. Therefore, if $\sigma > 1$, it is valid to make $n \to \infty$ $(m = 0)$ in (9) with the function $f(z)$ in (10). Thus we readily obtain *Hermite's formula* for $\zeta(s,a)$:

$$(12)\quad \zeta(s,a) = \frac{1}{2}a^{-s} + \frac{a^{1-s}}{s-1} + 2\int_0^\infty \left(a^2+y^2\right)^{-\frac{1}{2}s}\left\{\sin\left(s\arctan\frac{y}{a}\right)\right\}\frac{dy}{e^{2\pi y}-1}.$$

Note that the integral involved in (12) converges for all admissible values of $s \in \mathbb{C}$. Moreover, the integral is an entire function of s. A special case of the formula (12) when $a = 1$ is attributed to Jensen.

Setting $s = 0$ in (12), we have

$$(13)\qquad \zeta(0,a) = \frac{1}{2} - a,$$

which is also obtained from 2.1(17) in view of 1.6(8). If we set $z = s$ in 1.2(30) and differentiate the resulting equation with respect to s, we find that

$$(14)\qquad \psi(s) = \log s - \frac{1}{2s} - 2\int_0^\infty \frac{t\,dt}{(t^2+s^2)\left(e^{2\pi t}-1\right)} \quad (\Re(s)>0).$$

Taking the limit in (12) as $s \to 1$, by virtue of the uniform convergence of the integral in (12), we get

$$\lim_{s\to 1}\left\{\zeta(s,a) - \frac{1}{s-1}\right\} = \lim_{s\to 1}\frac{a^{1-s}-1}{s-1} + \frac{1}{2a} + 2\int_0^\infty \frac{y\,dy}{(a^2+y^2)\left(e^{2\pi y}-1\right)},$$

which, in view of (14), yields

$$(15)\qquad \lim_{s\to 1}\left\{\zeta(s,a) - \frac{1}{s-1}\right\} = -\frac{\Gamma'(a)}{\Gamma(a)} = -\psi(a).$$

Differentiating (12) with respect to s and setting $s = 0$ in the resulting equation, we have

$$\left\{\frac{d}{ds}\zeta(s,a)\right\}_{s=0} = \left(a - \frac{1}{2}\right)\log a - a + 2\int_0^\infty \frac{\arctan\left(\frac{y}{a}\right)}{e^{2\pi y}-1}\,dy, \tag{16}$$

which, by virtue of 1.2(30), yields

$$\left.\frac{d}{ds}\zeta(s,a)\right|_{s=0} = \log\Gamma(a) - \frac{1}{2}\log(2\pi), \tag{17}$$

which is equivalent to the identity 2.1(29). In addition to (17), it is easy to find from the definition (1) of $\zeta(s,a)$ that

$$\frac{\partial}{\partial a}\zeta(s,a) = -s\,\zeta(s+1,a). \tag{18}$$

The respective special cases of (15) and (17) when $a = 1$, by means of 1.2(4) and 1.1(13), become

$$\lim_{s\to 1}\left\{\zeta(s,a) - \frac{1}{s-1}\right\} = \lim_{\epsilon\to 0}\left\{\zeta(1+\epsilon,a) - \frac{1}{\epsilon}\right\} = \gamma \tag{19}$$

and

$$\zeta'(0) = -\frac{1}{2}\log(2\pi), \tag{20}$$

where $\zeta(s)$ is the Riemann Zeta function (see Section 2.3).

Further integral representations for $\zeta(s,a)$

In addition to (12), some known integral representations of $\zeta(s,a)$ are recalled here:

$$\zeta(s,a) = \frac{1}{\Gamma(s)}\int_0^\infty \frac{t^{s-1}\exp[-(a-1)t]}{e^t-1}\,dt \qquad (\Re(s) > 1;\ \Re(a) > 0), \tag{21}$$

which is the same as (2);

$$\zeta(s,a) = \frac{1}{2}a^{-s} - \frac{a^{1-s}}{1-s} + \frac{1}{\Gamma(s)}\int_0^\infty \left(\frac{1}{e^t-1} - \frac{1}{t} + \frac{1}{2}\right)e^{-at}\,t^{s-1}\,dt \qquad (\Re(s) > -1;\ \Re(a) > 0); \tag{22}$$

$$\zeta(s,a) = \frac{\pi\,2^{s-2}}{s-1}\int_0^\infty \left[t^2 + (2a-1)^2\right]^{\frac{1}{2}(1-s)} \frac{\cos\left[(s-1)\arctan\left(\frac{t}{2a-1}\right)\right]}{\cosh^2\left(\frac{1}{2}\pi t\right)}\,dt \qquad \left(\Re(a) > \frac{1}{2}\right); \tag{23}$$

$$
\begin{aligned}
\zeta(s,a) =& \cos\left(\frac{1}{2}\pi s\right)\sin(2\pi a)\int_0^\infty \frac{t^{-s}}{\cosh(2\pi t)-\cosh(2\pi a)}\,dt \\
&+ \sin\left(\frac{1}{2}\pi s\right)\int_0^\infty \frac{t^{-s}\left[\cosh(2\pi a)-e^{-2\pi t}\right]}{\cosh(2\pi t)-\cosh(2\pi a)}\,dt \\
&(\Re(s)<1 \text{ when } 0<\Re(a)<1;\ \Re(s)<0 \text{ when } a=1);
\end{aligned}
\tag{24}
$$

$$
\begin{aligned}
\zeta(s,a) =& a^{-s} + \sum_{k=0}^{n} \frac{\Gamma(k+s-1)}{\Gamma(s)}\frac{B_k}{k!}\,a^{-k-s+1} \\
&+ \frac{1}{\Gamma(s)}\int_0^\infty \left(\frac{1}{e^t-1}-\sum_{k=0}^{n}\frac{B_k}{k!}\,t^{k-1}\right)e^{-at}\,t^{s-1}\,dt \\
&(\Re(s)>-(2n-1);\ \Re(a)>0;\ n\in\mathbb{N}_0).
\end{aligned}
\tag{25}
$$

Some applications of the derivative formula (17)

We begin by giving another proof of the derivative formula (17). Indeed, by the analytic continuation of $\zeta(s,a)$ and the special case of (4) when $n=1$, we observe that

$$
\zeta(s,a+1)=\zeta(s,a)-a^{-s}, \tag{26}
$$

which, upon differentiating with respect to s and setting $s=0$ in the resulting equation, yields

$$
\zeta'(0,a+1)=\zeta'(0,a)+\log a.
$$

Let $f(a):=\exp[\zeta'(0,a)]$. We then have $f(a+1)=a\,f(a)$ $(a>0)$ and

$$
\frac{d^2}{da^2}\log f(a)=\frac{d^2}{da^2}\frac{d}{ds}\zeta(s,a)\bigg|_{s=0}=\sum_{k=0}^{\infty}\frac{1}{(k+a)^2}>0 \quad (a>0),
$$

which implies that $f(a)$ is logarithmically convex on $(0,\infty)$. Thus, by appealing to the Bohr-Mollerup theorem (Thorem 1.1), we obtain, for some constant C,

$$
f(a)=C\,\Gamma(a),
$$

which, for $a=1$, yields

$$
C=\exp[\zeta'(0)]=(2\pi)^{-\frac{1}{2}}.
$$

This completes our second proof of the derivative formula (17).

Many authors gave seemingly different proofs of Stirling's formula 1.1(52) [see (for example) Blyth and Pathak (1986), Choi (1994b), Diaconis and Freeman (1986), and Patin (1989)]. Here, by taking the limit in (16) as $a\to\infty$, we have

$$
\lim_{a\to\infty}\left\{\zeta'(0,a)+a+\frac{1}{2}\log a-a\,\log a\right\}=0, \tag{27}
$$

which, upon taking the exponential and using (17), immediately yields Stirling's formula 1.1(53).

Combining formulas (17) and 1.1(62), we obtain a formula for the Beta function $B(\alpha,\beta)$:

$$B(\alpha,\beta) = (2\pi)^{\frac{1}{2}} \exp\left[\zeta'(0,\alpha) + \zeta'(0,\beta) - \zeta'(0,\alpha+\beta)\right], \tag{28}$$

where $\zeta'(s,a) = \frac{\partial}{\partial s}\zeta(s,a)$. Applying the formula (16) and the following trigonometric identities:

$$\begin{aligned} \arctan a + \arctan b &= \arctan \frac{a+b}{1-ab} \quad (ab < 1), \\ \arctan a - \arctan b &= \arctan \frac{a-b}{1+ab} \quad (ab > -1) \end{aligned} \tag{29}$$

to (28), we can readily deduce an integral representation of $B(\alpha,\beta)$ [*cf.* Choi and Nam (1995)]:

$$B(\alpha,\beta) = \frac{\alpha^{\alpha-\frac{1}{2}}\,\beta^{\beta-\frac{1}{2}}}{(\alpha+\beta)^{\alpha+\beta-\frac{1}{2}}}(2\pi)^{\frac{1}{2}} e^{I(\alpha,\beta)} \quad (\alpha > 0,\ \beta > 0), \tag{29}$$

where, for convenience,

$$I(\alpha,\beta) := 2\rho \int_0^\infty \arctan\left\{\frac{(t^3+t)\rho^3}{\alpha\beta(\alpha+\beta)}\right\} \frac{dt}{e^{2\pi t\rho}-1} \quad \left(\rho^2 := \alpha^2+\alpha\beta+\beta^2\right).$$

Differentiating both sides of (22) with respect to s and letting $s = 0$ in the resulting equation, we obtain

$$\zeta'(0,a) = \left(a-\frac{1}{2}\right)\log a - a + \int_0^\infty \left(\frac{1}{e^t-1} - \frac{1}{t} + \frac{1}{2}\right)\frac{e^{-at}}{t}\,dt \qquad (\Re(a) > 0). \tag{31}$$

In a similar manner, (31) leads us to another integral representation of $B(\alpha,\beta)$ [*cf.* Choi and Nam (1995)]:

$$B(\alpha,\beta) = \frac{\alpha^{\alpha-\frac{1}{2}}\beta^{\beta-\frac{1}{2}}}{(\alpha+\beta)^{\alpha+\beta-\frac{1}{2}}}(2\pi)^{\frac{1}{2}} e^{J(\alpha,\beta)} \quad (\alpha > 0;\ \beta > 0), \tag{32}$$

where, for convenience,

$$J(\alpha,\beta) := \int_0^\infty \left(\frac{1}{e^t-1} - \frac{1}{t} + \frac{1}{2}\right)\left(e^{-\alpha t} + e^{-\beta t} - e^{-(\alpha+\beta)t}\right)\frac{dt}{t}.$$

Another form for $\Gamma_2(a)$

From 2.1(24) and 2.1(30), by virtue of (17), we obtain another form for the double Gamma function $\Gamma_2(a)$:

$$\Gamma_2(a) = A\left\{\Gamma(a)\right\}^{1-a} \exp\left[-\frac{1}{12} + \zeta'(-1,a)\right] \quad (a > 0), \tag{33}$$

where $\zeta'(s,a) = \frac{\partial}{\partial s}\zeta(s,a)$.

In addition to the integral representation 1.3(78), we can express $\log\Gamma_2(a)$ as improper integrals in many ways. For example, we give two integral representations for $\log\Gamma_2(a)$:

$$\begin{aligned}\log\Gamma_2(a) = & -\frac{1}{12} + \log A - \frac{a^2}{4} + \left(\frac{1}{2}a^2 - \frac{1}{2}a\right)\log a + (1-a)\log\Gamma(a) \\ & + 2\int_0^\infty \left\{\frac{1}{2}(a^2+t^2)^{\frac{1}{2}} \sin\left(\arctan\frac{t}{a}\right)\log(a^2+t^2)\right. \\ & \left. + (a^2+t^2)^{\frac{1}{2}}\cos\left(\arctan\frac{t}{a}\right)\arctan\frac{t}{a}\right\}\frac{dt}{e^{2\pi t}-1} \quad (\Re(a)>0);\end{aligned} \tag{34}$$

$$\begin{aligned}\log\Gamma_2(a) = & \log A - \frac{a^2}{4} + \left(\frac{a^2}{2} - \frac{a}{2} + \frac{1}{12}\right)\log a + (1-a)\log\Gamma(a) \\ & - \int_0^\infty \left(\frac{1}{e^t-1} - \frac{1}{t} + \frac{1}{2} - \frac{t}{12}\right)\frac{e^{-at}}{t^2}\,dt \quad (\Re(a)>0).\end{aligned} \tag{35}$$

Indeed, by differentiating Hermite's formula (12) for $\zeta(s,a)$ with respect to s, and letting $s \to -1$ in the resulting equation, and applying (17) and the identity for $\zeta'(-1,a)$, we readily obtain (34). On the other hand, setting $n = 2$ in (25), we have

$$\begin{aligned}\zeta(s,a) = & \frac{sa^{-s-1}}{12} + \frac{a^{-s}}{2} + \frac{a^{1-s}}{s-1} \\ & + \frac{1}{\Gamma(s)}\int_0^\infty \left(\frac{1}{e^t-1} - \frac{1}{t} + \frac{1}{2} - \frac{t}{12}\right)t^{s-1}e^{-at}dt \\ & (\Re(s) > -3;\ \Re(a) > 0).\end{aligned} \tag{36}$$

Employing the same technique as in getting (34) by making use of (36) and considering the following identities:

$$\left.\frac{d}{ds}\frac{1}{\Gamma(s)}\right|_{s=-1} = -1 \qquad \text{and} \qquad \left.\frac{1}{\Gamma(s)}\right|_{s=-1} = 0,$$

we obtain (35).

Glaisher (1877, p. 47) expressed the Glaisher-Kinkelin constant A given in 1.3(2) as an integral:

$$A = 2^{\frac{7}{36}}\pi^{-\frac{1}{6}}\exp\left(\frac{1}{3} + \frac{2}{3}\int_0^{\frac{1}{2}} \log\Gamma(t+1)\,dt\right). \tag{37}$$

By setting $a = 1$ in (34) and (35), we can also obtain integral representations of $\log A$:

$$\begin{aligned}\log A = & \frac{1}{3} - 2\int_0^\infty \left\{\frac{1}{2}(1+t^2)^{\frac{1}{2}}\sin(\arctan t)\log(1+t^2)\right. \\ & \left. + (1+t^2)^{\frac{1}{2}}\cos(\arctan t)\arctan t\right\}\frac{dt}{e^{2\pi t}-1}\end{aligned} \tag{38}$$

and

$$\log A = \frac{1}{4} + \int_0^\infty \left(\frac{1}{e^t - 1} - \frac{1}{t} + \frac{1}{2} - \frac{t}{12} \right) \frac{e^{-t}}{t^2}\, dt. \tag{39}$$

The formula (35) can be used to obtain an asymptotic formula for $\log \Gamma_2(a)$ by first observing that, for some $M > 0$ and for all $a > 0$,

$$\left| \int_0^\infty \left(\frac{1}{e^t - 1} - \frac{1}{t} + \frac{1}{2} - \frac{t}{12} \right) \frac{e^{-at}}{t^2}\, dt \right| < \frac{M}{a}.$$

Thus, by employing 1.1(54), we have

$$\begin{aligned} \log \Gamma_2(a) = {}& \log A + \frac{3a^2}{4} - a - \frac{1}{12} - \left(\frac{a^2}{2} - a + \frac{5}{12} \right) \log a \\ & + \frac{1}{2}(1-a)\log(2\pi) + O\left(a^{-1}\right) \quad (a \to \infty;\ a > 0), \end{aligned} \tag{40}$$

which may be compared with 1.3(7).

2.3 The Riemann Zeta Function

The *Riemann Zeta function* $\zeta(s)$ is defined by

$$\zeta(s) := \begin{cases} \displaystyle\sum_{n=1}^\infty \frac{1}{n^s} = \frac{1}{1-2^{-s}} \sum_{n=1}^\infty \frac{1}{(2n-1)^s} & (\Re(s) > 1) \\ \displaystyle\left(1 - 2^{1-s}\right)^{-1} \sum_{n=1}^\infty \frac{(-1)^{n-1}}{n^s} & (\Re(s) > 0;\ s \neq 1). \end{cases} \tag{1}$$

It is easy to see from the definitions (1) and 2.2(1) that

$$\zeta(s) = \zeta(s,1) = (2^s - 1)^{-1}\, \zeta\left(s, \frac{1}{2}\right) = 1 + \zeta(s,2) \tag{2}$$

and

$$\zeta(s) = \frac{1}{m^s - 1} \sum_{j=1}^{m-1} \zeta\left(s, \frac{j}{m}\right) \quad (m \in \mathbb{N} \setminus \{1\}). \tag{3}$$

In view of (2), we can deduce many properties of $\zeta(s)$ from those of $\zeta(s,a)$ given in Section 2.2. In fact, the series $\zeta(s) = \sum_{n=1}^\infty n^{-s}$ in (1) represents an analytic function of s in the half-plane $\Re(s) = \sigma > 1$. Setting $a = 1$ in 2.2(2), we have an integral representation of $\zeta(s)$ in the form:

$$\begin{aligned} \Gamma(s)\,\zeta(s) &= \int_0^\infty \frac{x^{s-1} e^{-x}}{1 - e^{-x}}\, dx = \int_0^\infty \frac{x^{s-1}}{e^x - 1}\, dx \\ &= \int_0^1 \frac{1}{1-x} \left(\log \frac{1}{x} \right)^{s-1} dx \quad (\Re(s) > 1). \end{aligned} \tag{4}$$

Furthermore, just as $\zeta(s,a)$, $\zeta(s)$ can be continued meromorphically to the whole complex s-plane (except for a simple pole at $s=1$ with its residue 1) by means of the contour integral representation:

$$\zeta(s) = -\frac{\Gamma(1-s)}{2\pi i} \int_C \frac{(-z)^{s-1}\, e^{-z}}{1-e^{-z}}\, dz, \tag{5}$$

where the contour C is the Hankel loop of Theorem 2.5.

Now 2.2(19) and (5), together, imply that the Laurent series of $\zeta(s)$ in a neighborhood of its pole $s=1$ has the form:

$$\zeta(s) = \frac{1}{s-1} + \gamma + \sum_{n=1}^{\infty} a_n\,(s-1)^n, \tag{6}$$

where γ is the Euler-Mascheroni constant given in 1.1(3) and a_n is also expressed as [see Ivić (1985, pp. 4–6)]:

$$a_n = \lim_{m\to\infty} \left[\sum_{k=1}^{m} \frac{(\log k)^n}{k} - \frac{(\log m)^{n+1}}{n+1}\right] \qquad (n \in \mathbb{N}). \tag{7}$$

The Riemann Zeta function $\zeta(s)$ in (1) plays a central rôle in the applications of complex analysis to number theory. The number-theoretic properties of $\zeta(s)$ are exhibited by the following result known as *Euler's formula*, which gives a relationship between the set of primes and the set of positive integers:

$$\zeta(s) = \prod_p \left(1-p^{-s}\right)^{-1} \qquad (\Re(s) > 1), \tag{8}$$

where the product is taken over all primes.

From 2.2(4), we have [*cf.* Equation (2) for the *special* cases when $n=0$ and $n=1$]

$$\zeta(s) = \zeta(s, n+1) + \sum_{k=1}^{n} k^{-s} \qquad (n \in \mathbb{N}_0). \tag{9}$$

The connection between $\zeta(s)$ and the Bernoulli numbers is given as follows:

$$\zeta(-n) = \begin{cases} -\dfrac{1}{2} & (n=0) \\[2mm] -\dfrac{B_{n+1}}{n+1} & (n \in \mathbb{N}), \end{cases} \tag{10}$$

which is deduced by setting $x=1$ in 2.1(17) and using 1.6(5).

Riemann's functional equation for $\zeta(s)$

The special case of 2.2(8) when $p=1=q$ yields *Riemann's functional equation* for $\zeta(s)$:

$$\zeta(1-s) = 2(2\pi)^{-s}\,\Gamma(s)\,\cos\left(\frac{1}{2}\pi s\right)\zeta(s) \tag{11}$$

or, equivalently,

$$\zeta(s) = 2(2\pi)^{s-1}\,\Gamma(1-s)\,\sin\left(\frac{1}{2}\pi s\right)\,\zeta(1-s). \tag{12}$$

Taking $s = 2n+1$ ($n \in \mathbb{N}$) in (11), the factor $\cos\left(\frac{1}{2}\pi s\right)$ vanishes and we find that

$$\zeta(-2n) = 0 \quad (n \in \mathbb{N}), \tag{13}$$

which are often referred to as the *trivial zeros* of $\zeta(s)$. The equation (13) can also be proved by combining (10) and 1.6(7).

By using the Legendre duplication formula 1.1(49) and 1.1(12), it is not difficult to see that the functional equation (11) or (12) can be written in a simpler form:

$$\Phi(s) = \Phi(1-s), \tag{14}$$

where the function $\Phi(s)$ is defined by

$$\Phi(s) := \pi^{-\frac{1}{2}s}\,\Gamma\left(\frac{1}{2}s\right)\,\zeta(s). \tag{15}$$

The function $\Phi(s)$ has simple poles at $s = 0$ and $s = 1$. According to Riemann, in order to remove these poles, we multiply $\Phi(s)$ by $\frac{1}{2}s(1-s)$ and define

$$\xi(s) := \frac{1}{2}\,s(1-s)\,\Phi(s), \tag{16}$$

which is an entire function of s and satisfies the functional equation:

$$\xi(s) = \xi(1-s). \tag{17}$$

Setting $s = 2n$ ($n \in \mathbb{N}$) in (11) and applying (10), we have the well-known identity:

$$\zeta(2n) = (-1)^{n+1}\,\frac{(2\pi)^{2n}}{2\,(2n)!}B_{2n} \quad (n \in \mathbb{N}_0), \tag{18}$$

which, in view of 1.6(7), enables us to list the following special values:

$$\begin{aligned} &\zeta(2) = \frac{\pi^2}{6}, \quad \zeta(4) = \frac{\pi^4}{90}, \quad \zeta(6) = \frac{\pi^6}{945}, \\ &\zeta(8) = \frac{\pi^8}{9450}, \quad \zeta(10) = \frac{\pi^{10}}{93555}, \ldots . \end{aligned} \tag{19}$$

We may recall here a known recursion formula for $\zeta(2n)$ [see also Section 4.1]:

$$\zeta(2n) = \frac{2}{2n+1}\sum_{k=1}^{n-1}\zeta(2k)\zeta(2n-2k) \quad (n \in \mathbb{N}\setminus\{1\}), \tag{20}$$

which can also be used to evaluate $\zeta(2n)$ $(n \in \mathbb{N}\setminus\{1\})$.

We get no information about $\zeta(2n+1)$ $(n \in \mathbb{N})$ from Riemann's functional equation, since both members of (11) vanish upon setting $s = 2n+1$ $(n \in \mathbb{N})$. In fact, until now no simple formula analogous to (18) is known for $\zeta(2n+1)$ or even for any special case such as $\zeta(3)$. It is not even known whether $\zeta(2n+1)$ is rational or irrational, except that the irrationality of $\zeta(3)$ was proved recently by Apéry (1979). Instead, a known integral formula for $\zeta(2n+1)$ is recalled here:

$$\zeta(2n+1) = \frac{(-1)^{n+1}\,(2\pi)^{2n+1}}{2(2n+1)!} \int_0^1 B_{2n+1}(t)\,\cot(\pi t)\,dt \quad (n \in \mathbb{N}). \tag{21}$$

It is readily seen that $\zeta(s) \neq 0$ $(\Re(s) = \sigma \geq 1)$, and (12) shows that $\zeta(s) \neq 0$ $(\sigma \leq 0)$ except for the trivial zeros in (13). Furthermore, in view of the second series definition of $\zeta(s)$ in (1), we find that $\zeta(s) < 0$ $(s \in \mathbb{R};\ 0 < s < 1)$. The assertion that all the *non-trivial* zeros of $\zeta(s)$ have real part $\frac{1}{2}$ is popularly known as the *Riemann hypothesis* which was conjectured (but not proven) in the memoir of Riemann (1859). This hypothesis is still one of the most challenging mathematical problems today [see Edwards (1974)].

It is easy to derive from (12) and (13) that [*cf.*, *e.g.*, Srivastava (1999, p. 387, Eq. (1.15))]

$$\zeta'(-2n) = \lim_{\epsilon \to 0} \frac{\zeta(-2n+\epsilon)}{\epsilon} = (-1)^n \frac{(2n)!}{2(2\pi)^{2n}} \zeta(2n+1) \quad (n \in \mathbb{N}). \tag{22}$$

Relationship between $\zeta(s)$ and the mathematical constants B and C

Just as $\log A$ in 2.1(31), we can also express $\log B$ and $\log C$ introduced in Section 1.3 as special cases of $\zeta'(s)$. Indeed, using the Euler-Maclaurin summation formula 1.3(68), we can obtain a number of analytical representations of $\zeta(s)$, such as [*cf.* Hardy (1949, p. 333)]

(23)
$$\zeta(s) = \lim_{n \to \infty} \left\{ \sum_{k=1}^{n} k^{-s} - \frac{n^{1-s}}{1-s} - \frac{1}{2} n^{-s} \right\} \quad (\Re(s) > -1),$$

(24)
$$\zeta(s) = \lim_{n \to \infty} \left\{ \sum_{k=1}^{n} k^{-s} - \frac{n^{1-s}}{1-s} - \frac{1}{2} n^{-s} + \frac{1}{12} s n^{-s-1} \right\} \quad (\Re(s) > -3),$$

and

$$\begin{aligned}\zeta(s) = \lim_{n \to \infty} \Bigg\{ &\sum_{k=1}^{n} k^{-s} - \frac{n^{1-s}}{1-s} - \frac{1}{2} n^{-s} + \frac{1}{12} s n^{-s-1} \\ &- \frac{1}{720} s(s+1)(s+2) n^{-s-3} \Bigg\} \quad (\Re(s) > -5).\end{aligned} \tag{25}$$

Now it is not difficult to express the mathematical constants B and C as follows:

$$\log B = -\zeta'(-2) \tag{26}$$

and

$$\log C = -\zeta'(-3) - \frac{11}{720}, \tag{27}$$

respectively.

It is clear from 1.3(70) that $\log C$ must be the finite part of the divergent sum $\sum k^3 \log k$ according to some regularization; hence $\log C$ must be related to $\zeta'(s)$ for some special value of s. By differentiating both sides of (25) with respect to s and letting $s = -3$, we obtain

(28)

$$-\zeta'(-3) = \lim_{n\to\infty} \left\{ \sum_{k=1}^{n} k^3 \log k - \left(\frac{n^4}{4} + \frac{n^3}{2} + \frac{n^2}{4} - \frac{1}{120} \right) \log n + \frac{n^4}{16} - \frac{n^2}{12} \right\} + \frac{11}{720},$$

which, when compared with 1.3(70), yields the desired expression (27). The special case of (22) when $n = 1$ also shows that, in view of (26),

$$\log B = \frac{\zeta(3)}{4\pi^2}. \tag{29}$$

Integral representations for $\zeta(s)$

In addition to (4) and (5), in terms of (2), we find from Section 2.2 that

$$\zeta(s) = \frac{1}{\Gamma(s)} \int_0^\infty \frac{t^{s-1}}{e^t - 1} dt \quad (\Re(s) > 1); \tag{30}$$

$$\zeta(s) = \frac{1}{2} + \frac{1}{s-1} + \frac{1}{\Gamma(s)} \int_0^\infty \left(\frac{1}{e^t - 1} - \frac{1}{t} + \frac{1}{2} \right) e^{-t}\, t^{s-1}\, dt \quad (\Re(s) > -1); \tag{31}$$

$$\zeta(s) = \sin\left(\frac{\pi}{2}s\right) \int_0^\infty \frac{\cosh(2\pi) - e^{-2\pi t}}{\cosh(2\pi t) - \cosh(2\pi)} \frac{dt}{t^s} \quad (\Re(s) < 0); \tag{32}$$

$$\begin{aligned}\zeta(s) ={}& 1 + \frac{1}{s-1} + \sum_{k=1}^{n} \frac{\Gamma(k+s-1)}{\Gamma(s)} \frac{B_k}{k!} \\ &+ \frac{1}{\Gamma(s)} \int_0^\infty \left(\frac{1}{e^t - 1} - \sum_{k=0}^{n} \frac{B_k}{k!} t^{k-1} \right) e^{-t}\, t^{s-1}\, dt \\ &(\Re(s) > -(2n-1);\ n \in \mathbb{N}_0);\end{aligned} \tag{33}$$

$$\zeta(s) = \frac{1}{2} + \frac{1}{s-1} + 2\int_0^\infty \frac{\sin(s \arctan t)}{(1+t^2)^{\frac{1}{2}s}} \frac{dt}{e^{2\pi t} - 1}. \tag{34}$$

Now we recall some summation formulas analogous to the infinite series version of Plana's summation formula 2.2(9). To do this, let $z = \tau + it$,

$$\frac{1}{e^{-2\pi iz} - 1} = X(\tau, t) + iY(\tau, t)$$

and

$$\frac{1}{e^{-\pi iz} - e^{\pi iz}} = X_1(\tau, t) + iY_1(\tau, t),$$

whence

$$X(\tau, t) = \frac{\cos(2\pi\tau) - e^{-2\pi t}}{e^{2\pi t} - 2\cos(2\pi\tau) + e^{-2\pi t}},$$

$$Y(\tau, t) = \frac{\sin(2\pi\tau)}{e^{2\pi t} - 2\cos(2\pi\tau) + e^{-2\pi t}},$$

$$X_1(\tau, t) = \frac{\cos(\pi\tau)\ (e^{\pi t} - e^{-\pi t})}{e^{2\pi t} - 2\cos(2\pi\tau) + e^{-2\pi t}},$$

and

$$Y_1(\tau, t) = \frac{\sin(\pi\tau)\,(e^{\pi t} + e^{-\pi t})}{e^{2\pi t} - 2\cos(2\pi\tau) + e^{-2\pi t}}.$$

On the other hand, we let

$$p(\tau, t) := \frac{1}{2}[f(\tau + it) + f(\tau - it)]$$

and

$$q(\tau, t) := \frac{1}{2i}[f(\tau + it) - f(\tau - it)],$$

so that, by appealing to the Reflection Principle, $p(\tau, t)$ and $q(\tau, t)$ represent, respectively, the real and imaginary parts of the analytic function $f(\tau + it)$ whose domain is symmetric with respect to the τ-axis, in case the function $f(\tau + i0)$ is real. Then we have the summation formulas [see Lindelöf (1947, Chapter III)]:

$$\sum_{n=m}^{\infty} f(n) = \int_{\alpha}^{\infty} f(\tau)\,d\tau - 2\int_0^{\infty} Q(\alpha, t)\,dt, \tag{35}$$

where, for convenience,

$$Q(\tau, t) := p(\tau, t)\,Y(\tau, t) + q(\tau, t)\,X(\tau, t); \tag{36}$$

$$\sum_{n=m}^{\infty} f(n) = -\frac{1}{2\pi i}\int_{\alpha - i\infty}^{\alpha + i\infty} \left(\frac{\pi}{\sin(\pi z)}\right)^2 F(z)\,dz, \tag{37}$$

F being a primitive function of f;

$$\sum_{n=m}^{\infty} (-1)^n f(n) = -\int_{\alpha - i\infty}^{\alpha + i\infty} \frac{f(z)\,dz}{e^{\pi iz} - e^{-\pi iz}} = -2\int_0^{\infty} Q_1(\alpha, t)\,dt, \tag{38}$$

where, for convenience,

$$Q_1(\tau,t) := p(\tau,t)\,Y_1(\tau,t) + q(\tau,t)\,X_1(\tau,t), \tag{39}$$

α being any number between $m-1$ and m in both (37) and (39).

Moreover, the formulas (35), (37), and (38) are valid if the function f satisfies each of the following conditions:

(a) $f(z)$ is analytic for $\Re(z) > 0$ $(z = \tau + it)$;

(b) $\lim_{t\to\infty} \exp(-2\pi|t|)\, f(\tau+it) = 0$ uniformly for $0 \le \tau < \infty$;

(c)

$$\lim_{\tau\to\infty} \int_{-\infty}^{\infty} \exp(-2\pi|t|)\, |f(\tau+it)|\, dt = 0.$$

Now we set $f(z) = z^{-s}$, with $m = 1$ and $\alpha = \frac{1}{2}$, in the three summation formulas just introduced. We thus obtain

$$\begin{aligned} p(\tau,t) &= (\tau^2+t^2)^{-\frac{1}{2}s} \cos\left(s \arctan\frac{t}{\tau}\right), \\ q(\tau,t) &= -\,(\tau^2+t^2)^{-\frac{1}{2}s} \sin\left(s \arctan\frac{t}{\tau}\right). \end{aligned} \tag{40}$$

We also find from (35) that

$$\zeta(s) = \frac{2^{s-1}}{s-1} - 2^s \int_0^\infty \left(1+t^2\right)^{-\frac{1}{2}s} \sin(s \arctan t) \frac{dt}{e^{\pi t}+1}. \tag{41}$$

Setting $z = \frac{1}{2} + it$ in (37), we have

$$\zeta(s) = \frac{\pi\, 2^{s-2}}{s-1} \int_0^\infty \left(1+t^2\right)^{\frac{1-s}{2}} \frac{\cos[(s-1)\arctan t]}{\left[\cosh\left(\frac{1}{2}\pi t\right)\right]^2}\, dt. \tag{42}$$

Starting with the second part of the definition (1) of $\zeta(s)$ and applying (38), we also get

$$\zeta(s) = \frac{2^{s-1}}{1-2^{1-s}} \int_0^\infty \left(1+t^2\right)^{-\frac{1}{2}s} \frac{\cos(s \arctan t)}{\cosh\left(\frac{1}{2}\pi t\right)}\, dt. \tag{43}$$

Formulas (34), (41), (42), and (43) are due to Jensen. It is observed that the definite integrals appearing in these four expressions represent analytic functions for all bounded values of s. In fact, anyone of these expressions gives the analytic continuation of $\zeta(s)$ in the whole complex s-plane except for a simple pole at $s = 1$ with its residue 1, just as it is with the contour integral representation (5).

Some other integral representations of $\zeta(s)$ are recalled here:

$$\zeta(s) = \frac{1}{s-1} + \frac{\sin(\pi s)}{\pi} \int_0^\infty \left[\log(1+t) - \psi(1+t)\right] \frac{dt}{t^s}, \tag{44}$$

(45)
$$\begin{aligned}\zeta(s+1) =&\frac{\sin(\pi s)}{\pi s} \int_0^\infty \psi'(1+t)]\,\frac{dt}{t^s}\\ =&\frac{\sin(\pi s)}{\pi s} \int_0^\infty [\psi(1+t)+\gamma]\,\frac{dt}{t^{1+s}},\end{aligned}$$

and

(46)
$$\zeta(s+m) = (-1)^{m-1}\,\frac{\Gamma(s)\,\sin(\pi s)}{\pi\,\Gamma(s+m)} \int_0^\infty \psi^{(m)}(1+t)\,\frac{dt}{t^s} \quad (m \in \mathbb{N}),$$

all three of which are due to de Bruijn (1937) and are valid for $0 < \Re(s) < 1$. Moreover,

(47)
$$\zeta(s) =\frac{1}{s-1} + \frac{\sin(\pi s)}{\pi(s-1)} \int_0^\infty \left(\psi'(1+t) - \frac{1}{1+t}\right) \frac{dt}{t^{1-s}}$$
$$(0 < \Re(s) < 2;\ s \neq 1);$$

(48)
$$\zeta(s) = \frac{1}{\Gamma(s+1)} \int_0^\infty \frac{e^t\,t^s}{(e^t-1)^2}\,dt \quad (\Re(s) > 1);$$

(49)
$$\zeta(s) = \frac{\left(1-2^{1-s}\right)^{-1}}{\Gamma(s)} \int_0^\infty \frac{t^{s-1}}{e^t+1}\,dt \quad (\Re(s) > 0);$$

(50)
$$\zeta(s) = \frac{\left(1-2^{-s}\right)^{-1}}{2\,\Gamma(s)} \int_0^\infty \frac{t^{s-1}}{\sinh t}\,dt \quad (\Re(s) > 1);$$

(51)
$$\zeta(s) = \frac{\left(1-2^{1-s}\right)^{-1}}{2\,\Gamma(s+1)} \int_0^\infty \frac{t^s\,e^t}{(e^t+1)^2}\,dt \quad (\Re(s) > 0);$$

(52)
$$\zeta(s) = \frac{\pi^{\frac{1}{2}s}}{s(s-1)\,\Gamma\left(\frac{1}{2}s\right)} + \frac{\pi^{\frac{1}{2}s}}{\Gamma\left(\frac{1}{2}s\right)} \int_1^\infty \left(t^{\frac{1}{2}(1-s)} + t^{\frac{1}{2}s}\right) \frac{\omega(t)}{t}\,dt,$$

which is due to Riemann and where

(53)
$$\omega(t) := \sum_{n=1}^\infty \exp\left(-n^2\,\pi\,t\right).$$

A summation identity for $\zeta(n)$

By using a known transformation formula and some reciprocity relations, Sitaramachandrarao and Sivaramasarma (1979) proved the following well-known summation identity for $\zeta(n)$, which was first given by Euler in 1775 (*cf.*, *e.g.*, Berndt (1985, p. 252, Eq. (9.5)):

(54)
$$2\sum_{q=1}^\infty \frac{1}{q^n} \left(\sum_{k=1}^{q-1} \frac{1}{k}\right) = n\,\zeta(n+1) - \sum_{k=1}^{n-2} \zeta(n-k)\,\zeta(k+1) \quad (n \in \mathbb{N}\setminus\{1\}),$$

where an empty sum is understood (as usual) to be nil. The formula (54) contains many (known or new) special cases. Here we give an elementary proof of (54) [*cf.* Williams (1953)]. Let us first consider the sum:

$$\begin{aligned}\sum_{k=1}^{n-2} \zeta(n-k)\zeta(k+1) &= \zeta(2)\zeta(n-1) + \zeta(3)\zeta(n-2) + \cdots + \zeta(n-1)\zeta(2) \\ &= \sum_{p=1}^{\infty}\sum_{q=1}^{\infty}\left\{\frac{1}{p^2}\cdot\frac{1}{q^{n-1}} + \frac{1}{p^3}\cdot\frac{1}{q^{n-2}} + \cdots + \frac{1}{p^{n-1}}\cdot\frac{1}{q^2}\right\} \\ &= \lim_{N\to\infty} S_N,\end{aligned}$$

where, for convenience,

$$S_N := \sum_{p=1}^{N}\sum_{q=1}^{N}\left\{\frac{1}{p^2}\cdot\frac{1}{q^{n-1}} + \frac{1}{p^3}\cdot\frac{1}{q^{n-2}} + \cdots + \frac{1}{p^{n-1}}\cdot\frac{1}{q^2}\right\} \quad (N \in \mathbb{N}). \tag{55}$$

Observing that the sum within braces in (55) is a geometric series, and taking care of the exceptional case when $p = q$, the series (55) becomes

$$\sum_{q=1}^{N}\left\{\sum_{\substack{p=1\\p\neq q}}^{N}\frac{1}{p-q}\left(\frac{q^{1-n}}{p} - \frac{p^{1-n}}{q}\right) + (n-2)\frac{1}{q^{n+1}}\right\}. \tag{56}$$

Now consider the first double sum in (56):

$$\begin{aligned}&\sum_{q=1}^{N}\sum_{\substack{p=1\\p\neq q}}^{N}\frac{1}{p-q}\left(\frac{q^{1-n}}{p} - \frac{p^{1-n}}{q}\right) \\ &= \sum_{q=1}^{N}\sum_{\substack{p=1\\p\neq q}}^{N}\frac{1}{p-q}\cdot\frac{q^{1-n}}{p} + \sum_{p=1}^{N}\sum_{\substack{q=1\\q\neq p}}^{N}\frac{1}{q-p}\cdot\frac{p^{1-n}}{q}.\end{aligned} \tag{57}$$

Inverting the order of summation in the second term on the right-hand side of (57), it is fairly straightforward to write (57) in the form:

$$\sum_{q=1}^{N}\sum_{\substack{p=1\\p\neq q}}^{N}\frac{q^{1-n}}{p(p-q)} + \sum_{p=1}^{N}\sum_{\substack{q=1\\q\neq p}}^{N}\frac{p^{1-n}}{q(q-p)} = 2\sum_{q=1}^{N}\frac{1}{q^{n-1}}\sum_{\substack{p=1\\p\neq q}}^{N}\frac{1}{p(p-q)}. \tag{58}$$

Combining (55) through (58), we find that

$$\begin{aligned}&\sum_{q=1}^{N}\sum_{p=1}^{N}\left\{\frac{1}{p^2}\cdot\frac{1}{q^{n-1}} + \frac{1}{p^3}\cdot\frac{1}{q^{n-2}} + \cdots + \frac{1}{p^{n-1}}\cdot\frac{1}{q^2}\right\} \\ &= (n-2)\sum_{q=1}^{N}\frac{1}{q^{n+1}} + 2\sum_{q=1}^{N}\frac{1}{q^{n-1}}\sum_{\substack{p=1\\p\neq q}}^{N}\frac{1}{p(p-q)}.\end{aligned} \tag{59}$$

Now consider

$$
(60) \quad \begin{aligned}
-q \sum_{\substack{p=1 \\ p \neq q}}^{N} \frac{1}{p(p-q)} &= \sum_{\substack{p=1 \\ p \neq q}}^{N} \left(\frac{1}{p} - \frac{1}{p-q} \right) \\
&= \sum_{p=1}^{q-1} \frac{1}{q-p} - \sum_{p=q+1}^{N} \frac{1}{p-q} + \sum_{p=1}^{N} \frac{1}{p} - \frac{1}{q} \\
&= \sum_{p=1}^{q-1} \frac{1}{p} - \sum_{p=1}^{N-q} \frac{1}{p} + \sum_{p=1}^{N} \frac{1}{p} - \frac{1}{q} \\
&= -\frac{1}{q} + \sum_{p=1}^{q-1} \frac{1}{p} + \sum_{p=N-q+1}^{N} \frac{1}{p},
\end{aligned}
$$

which, when substituted into (59), yields

$$
(61) \quad \begin{aligned}
&\sum_{q=1}^{N} \sum_{p=1}^{N} \left\{ \frac{1}{p^2} \cdot \frac{1}{q^{n-1}} + \frac{1}{p^3} \cdot \frac{1}{q^{n-2}} + \cdots + \frac{1}{p^{n-1}} \cdot \frac{1}{q^2} \right\} \\
&= n \sum_{q=1}^{N} \frac{1}{q^{n+1}} - 2 \sum_{q=1}^{N} \frac{1}{q^n} \left(\sum_{k=1}^{q-1} \frac{1}{k} \right) - 2 \sum_{q=1}^{N} \frac{1}{q^n} \left(\sum_{k=N-q+1}^{N} \frac{1}{k} \right).
\end{aligned}
$$

Finally, we consider the inequality:

$$
0 \leq \sum_{k=N-q+1}^{N} \frac{1}{k} = \frac{1}{N-q+1} + \frac{1}{N-q+2} + \cdots + \frac{1}{N} \leq \frac{q}{N-q+1},
$$

so that

$$
(62) \quad \begin{aligned}
0 &< \sum_{q=1}^{N} \frac{1}{q^n} \left(\sum_{k=N-q+1}^{N} \frac{1}{k} \right) < \sum_{q=1}^{N} \frac{1}{q^{n-1}} \cdot \frac{1}{N-q+1} \\
&\leq \sum_{q=1}^{N} \frac{1}{q(N-q+1)} \qquad (n \in \mathbb{N} \setminus \{1\}) \\
&= \frac{1}{N+1} \sum_{q=1}^{N} \left(\frac{1}{q} + \frac{1}{N-q+1} \right) = \frac{2}{N+1} \sum_{q=1}^{N} \frac{1}{q} \\
&< \frac{2}{N+1} (1 + \log N) \to 0 \quad \text{as } N \to \infty.
\end{aligned}
$$

By taking the limit in (61) as $N \to \infty$, and applying (62), we complete the proof of the summation identity (54).

2.4 Polylogarithm Functions

The so-called *Polylogarithm functions* have been studied rather extensively in the works of (for example) Lewin (1958, 1981), who was exceedingly fascinated by the following integral formula even during his school days:

$$\int_0^{\frac{1}{2}(\sqrt{5}-1)} \frac{\log(1-t)}{t}\, dt = \left\{\log\left(\frac{\sqrt{5}-1}{2}\right)\right\}^2 - \frac{\pi^2}{10}. \tag{1}$$

Many of the earlier authors who studied these functions include Euler, Abel, Legendre, Kummer, Spence, and so on. These functions have been found to be useful in physics, electronics, quantum eletrodynamics, etc., and are closely related to many other mathematical functions such as the Clausen integral (or function) $\mathrm{Cl}_2(x)$ given by

$$\mathrm{Cl}_2(x) = x\log\pi - x\log\sin\left(\frac{x}{2}\right) + 2\pi\log\frac{G\left(1-\frac{x}{2\pi}\right)}{G\left(1+\frac{x}{2\pi}\right)}, \tag{2}$$

where G is the Barnes G-function studied in Section 1.3.

Throughout this work, we choose to follow the notations and conventions used by Lewin (1958, 1981).

The Dilogarithm function

The *Dilogarithm function* $\mathrm{Li}_2(z)$ is defined by

$$\begin{aligned} \mathrm{Li}_2(z) &:= \sum_{n=1}^{\infty} \frac{z^n}{n^2} \quad (|z| \leq 1) \\ &= -\int_0^z \frac{\log(1-t)}{t}\, dt. \end{aligned} \tag{3}$$

It follows from (3) that

$$\frac{d}{dx}\mathrm{Li}_2\left(-\frac{1}{x}\right) = \frac{\log(1+x)-\log x}{x},$$

the integration of which yields

$$\mathrm{Li}_2\left(-\frac{1}{x}\right) + \mathrm{Li}_2(-x) = 2\,\mathrm{Li}_2(-1) - \frac{1}{2}(\log x)^2, \tag{4}$$

where the constant of integration is determined by taking $x = 1$.

The special case of (4) when $x = -1 = e^{i\pi}$ gives

$$2\,\mathrm{Li}_2(1) = 2\,\mathrm{Li}_2(-1) + \frac{\pi^2}{2}. \tag{5}$$

It is easy to see also that

$$\mathrm{Li}_2(-1) = -\frac{1}{2}\mathrm{Li}_2(1), \tag{6}$$

which, in view of (5), yields

$$\mathrm{Li}_2(1) = \frac{1}{1^2} + \frac{1}{2^2} + \frac{1}{3^2} + \cdots = \frac{\pi^2}{6} = \zeta(2), \tag{7}$$

where $\zeta(s)$ is the Riemann Zeta function (see Section 2.3). It follows from (6) and (7) that

$$\mathrm{Li}_2(-1) = -\frac{\pi^2}{12}. \tag{8}$$

By taking $x = y\, e^{i\pi}$ in (4), we obtain

$$\mathrm{Li}_2\,(y) + \mathrm{Li}_2\left(\frac{1}{y}\right) = \frac{\pi^2}{3} - \frac{1}{2}\,(\log\, y)^2 - i\pi\, \log y \quad (y > 1). \tag{9}$$

From (3), we also find that

$$\begin{aligned}\mathrm{Li}_2(z) &= -\,\log z\, \log(1-z) - \int_0^z \frac{\log t}{1-t}\, dt \\ &= -\,\log z\, \log(1-z) - \mathrm{Li}_2(1-z) + \mathrm{Li}_2(1),\end{aligned}$$

whence

$$\mathrm{Li}_2(z) + \mathrm{Li}_2(1-z) = \frac{\pi^2}{6} - \log z\, \log(1-z). \tag{10}$$

By choosing a suitable argument, and then employing differentiation and integration, we have Landen's formulas [see Lewin (1958, p. 5)]:

$$\mathrm{Li}_2\,(x) + \mathrm{Li}_2\left(\frac{x}{x-1}\right) = -\frac{1}{2}\,\{\log\,(1-x)\}^2 \quad (x < 1) \tag{11}$$

and

$$\begin{aligned}\mathrm{Li}_2\,(x) + \mathrm{Li}_2\left(\frac{x}{x-1}\right) =& \frac{\pi^2}{2} - 2i\pi\, \log x + i\pi\, \log(x-1) \\ & - \frac{1}{2}\,\{\log\,(x-1)\}^2 \quad (x > 1).\end{aligned} \tag{12}$$

A factorization formula for $\mathrm{Li}_2\,(x)$ is given by

$$\frac{1}{n}\,\mathrm{Li}_2\,(x^n) = \sum_{k=0}^{n-1} \mathrm{Li}_2\,(\omega^k\, x) \quad \left(\omega := \exp\left(\frac{2\pi i}{n}\right)\right), \tag{13}$$

which follows from the elementary identity:

$$1 - x^n = \prod_{k=0}^{n-1} (1 - \omega^k\, x) \quad \left(\omega := \exp\left(\frac{2\pi i}{n}\right)\right).$$

The special case of (13) when $n = 2$ gives us

$$\frac{1}{2}\operatorname{Li}_2\left(x^2\right) = \operatorname{Li}_2(x) + \operatorname{Li}_2(-x). \tag{14}$$

By eliminating $\operatorname{Li}_2(x)$ between (11) and (14), we get

$$\operatorname{Li}_2\left(\frac{x}{x-1}\right) + \frac{1}{2}\operatorname{Li}_2\left(x^2\right) - \operatorname{Li}_2(-x) = -\frac{1}{2}\left\{\log(1-x)\right\}^2. \tag{15}$$

Setting $x = \frac{3-\sqrt{5}}{2}$ and $x = \frac{\sqrt{5}-1}{2}$ in (11) yields the following results:

$$\operatorname{Li}_2\left(\frac{1-\sqrt{5}}{2}\right) = -\frac{\pi^2}{15} + \frac{1}{2}\left\{\log\left(\frac{\sqrt{5}-1}{2}\right)\right\}^2 \tag{16}$$

and

$$\operatorname{Li}_2\left(-\frac{1+\sqrt{5}}{2}\right) = -\frac{\pi^2}{10} + \frac{1}{2}\left\{\log\left(\frac{\sqrt{5}+1}{2}\right)\right\}^2, \tag{17}$$

respectively.

The *Legendre's Chi function* $\chi_2(x)$ is defined by

$$\begin{aligned} \chi_2(x) :=& \sum_{n=1}^{\infty} \frac{x^{2n-1}}{(2n-1)^2} \qquad (-1 \leq x \leq 1) \\ =& \frac{1}{2}\left\{\operatorname{Li}_2(x) - \operatorname{Li}_2(-x)\right\}, \end{aligned} \tag{18}$$

which, in view of (3), immediately becomes

$$\chi_2(x) = \frac{1}{2}\int_0^x \log\left(\frac{1+t}{1-t}\right)\frac{dt}{t}. \tag{19}$$

From (19), one readily finds the *Euler-Legendre-Landen* identity:

$$\chi_2\left(\frac{1-x}{1+x}\right) + \chi_2(x) = \frac{\pi^2}{8} + \frac{1}{2}\log x \,\log\left(\frac{1+x}{1-x}\right). \tag{20}$$

Putting $(1-x)/(1+x) = x$ gives $x^2 + 2x - 1 = 0$ or $x = \sqrt{2} - 1 = \tan\left(\frac{1}{8}\pi\right)$. We thus find from (20) that

$$\chi_2\left(\tan\frac{\pi}{8}\right) = \frac{\pi^2}{16} - \frac{1}{4}\left[\log\left(\tan\frac{\pi}{8}\right)\right]^2. \tag{21}$$

Upon integrating by parts twice, it is easily observed that

(22)

$$\int_0^x t^{\alpha}\operatorname{Li}_2(t)\,dt = \frac{x^{\alpha+1}}{\alpha+1}\operatorname{Li}_2(x) + \frac{x^{\alpha+1}-1}{(\alpha+1)^2}\log(1-x) - \frac{1}{(\alpha+1)^2}\int_0^x \frac{1-t^{\alpha+1}}{1-t}\,dt.$$

In view of 1.2(3), the integral formula (22) with $x = 1$ assumes the form:

$$\int_0^1 t^{\alpha-1} \operatorname{Li}_2(t)\, dt = \frac{\pi^2}{6\alpha} - \frac{\psi(\alpha+1)+\gamma}{\alpha^2}, \tag{23}$$

which, for $\alpha = n \in \mathbb{N}_0$, yields

$$\int_0^1 t^n \operatorname{Li}_2(t)\, dt = \frac{\pi^2}{6(n+1)} - \frac{1}{(n+1)^2} \sum_{k=1}^{n+1} \frac{1}{k}. \tag{24}$$

Nielsen [see Lewin (1958, pp. 20–21)] applied (24) to show that

$$\int_0^1 f(tx) \operatorname{Li}_2(t)\, dt = \frac{\pi^2}{6x} \int_0^x f(t)\, dt - \frac{1}{x} \sum_{n=0}^{\infty} \frac{a_n x^{n+1}}{(n+1)^2} \left(\sum_{k=1}^{n+1} \frac{1}{k} \right), \tag{25}$$

where $f(x) := \sum_{n=0}^{\infty} a_n x^n$. Take $a_n = n+1$, so that $f(t) = (1-t)^{-2}$. By making use of the well-known expansion:

$$\frac{1}{2} \{\log(1-x)\}^2 = \sum_{n=1}^{\infty} \frac{x^{n+1}}{n+1} \left(\sum_{k=1}^{n} \frac{1}{k} \right),$$

we find the following integral equation involving the Dilogarithm function:

$$\int_0^1 \frac{\operatorname{Li}_2(t)}{(1-xt)^2}\, dt = \frac{\pi^2}{6(1-x)} - \frac{\operatorname{Li}_2(x)}{x} - \frac{\{\log(1-x)\}^2}{2x} \qquad (-1 \leq x \leq 1), \tag{26}$$

whose special cases when $x = -1$ and $x = \frac{1}{2}$ become

$$\int_0^1 \frac{\operatorname{Li}_2(t)}{(1+t)^2}\, dt = \frac{1}{2} (\log 2)^2 \tag{27}$$

and

$$\int_0^1 \frac{\operatorname{Li}_2(t)}{(2-t)^2}\, dt = \frac{\pi^2}{24}, \tag{28}$$

respectively.

If z is a pure imaginary number in (3), say $z = iy$ $(y \in \mathbb{R})$, then

$$\operatorname{Li}_2(iy) = \frac{1}{4} \operatorname{Li}_2(-y^2) + i \operatorname{Ti}_2(y) \qquad (y \in \mathbb{R}), \tag{29}$$

where the new function $\operatorname{Ti}_2(x)$ is defined by

$$\operatorname{Ti}_2(x) := \sum_{n=1}^{\infty} \frac{(-1)^{n+1} x^{2n-1}}{(2n-1)^2} \qquad (x \leq 1). \tag{30}$$

From (29), we also see that

$$\mathrm{Ti}_2(y) = \frac{1}{2i}\left[\mathrm{Li}_2(iy) - \mathrm{Li}_2(-iy)\right]. \tag{31}$$

It follows from (18) that

$$\mathrm{Ti}_2(y) = -i\,\chi_2(iy). \tag{32}$$

In view of the following series expansion of $\tan^{-1} x$:

$$\tan^{-1} x = \sum_{n=1}^{\infty} (-1)^{n+1} \frac{x^{2n-1}}{2n-1} \qquad (x \leq 1), \tag{33}$$

the definition (30) can be written in the form:

$$\mathrm{Ti}_2(x) = \int_0^x \frac{\tan^{-1} t}{t}\,dt \quad \text{or} \quad \mathrm{Ti}_2(x) = \int_0^x \left(\tan^{-1} t\right) \frac{d(\log t)}{dt}\,dt. \tag{34}$$

If one takes (34) as a definition of $\mathrm{Ti}_2(x)$, the domain of $\mathrm{Ti}_2(x)$ may be extended to ∞.

Some known formulas and special values for $\mathrm{Ti}_2(x)$ are recalled here:

$$\mathrm{Ti}_2(x) - \mathrm{Ti}_2\left(\frac{1}{x}\right) = \mathrm{sgn}(x)\,\frac{\pi}{2}\,\log|x| \qquad (x \in \mathbb{R}), \tag{35}$$

where $\mathrm{sgn}(x)$ is defined by

$$\mathrm{sgn}(x) := \begin{cases} 1 & (x > 0), \\ -1 & (x < 0); \end{cases} \tag{36}$$

$$\mathrm{Ti}_2(1) = G, \tag{37}$$

where G is the Catalan constant given in 1.3(16);

$$\frac{1}{2}\,\mathrm{Ti}_2\left(\frac{2x}{1-x^2}\right) = \mathrm{Ti}_2(x) + \mathrm{Ti}_2(-x,1) - \mathrm{Ti}_2(x,1), \tag{38}$$

where $\mathrm{Ti}_2(x,a)$ is defined by

$$\mathrm{Ti}_2(x,a) := \int_0^x \frac{\tan^{-1} t}{a+t}\,dt; \tag{39}$$

$$\begin{aligned} \frac{1}{3}\,\mathrm{Ti}_2\left(\frac{3x-x^3}{1-3x^2}\right) =& \mathrm{Ti}_2(x) + \mathrm{Ti}_2\left(\frac{1-\sqrt{3}\,x}{\sqrt{3}+x}\right) - \mathrm{Ti}_2\left(\frac{1+\sqrt{3}\,x}{\sqrt{3}-x}\right) \\ &+ \frac{\pi}{6}\log\left(\frac{\sqrt{3}+x}{1-\sqrt{3}x}\cdot\frac{1+\sqrt{3}x}{\sqrt{3}-x}\right) \quad \left(-\frac{1}{\sqrt{3}} < x < \frac{1}{\sqrt{3}}\right), \end{aligned} \tag{40}$$

which, upon setting $x = \tan\theta$, yields a trigonometric form:

$$\begin{aligned}\frac{1}{3}\,\mathrm{Ti}_2(\tan 3\theta) =& \mathrm{Ti}_2(\tan\theta) + \mathrm{Ti}_2\left[\tan\left(\frac{\pi}{6}-\theta\right)\right] - \mathrm{Ti}_2\left[\tan\left(\frac{\pi}{6}+\theta\right)\right] \\ &+ \frac{\pi}{6}\log\left[\frac{\tan\left(\frac{\pi}{6}+\theta\right)}{\tan\left(\frac{\pi}{6}-\theta\right)}\right].\end{aligned} \tag{41}$$

Clausen's integral (or function)

From the series definition in (3) of the Dilogarithm function $\mathrm{Li}_2(z)$, we have

$$\mathrm{Li}_2\left(e^{i\theta}\right) = \sum_{n=1}^{\infty}\frac{\cos n\theta}{n^2} + i\sum_{n=1}^{\infty}\frac{\sin n\theta}{n^2}. \tag{42}$$

From the integral definition in (3) of $\mathrm{Li}_2(z)$, we also have

$$\mathrm{Li}_2\left(e^{i\theta}\right) - \mathrm{Li}_2(1) = -\int_1^{\exp(i\theta)}\frac{\log(1-t)}{t}\,dt,$$

which, upon putting $t = e^{i\eta}$, yields

$$\begin{aligned}\mathrm{Li}_2\left(e^{i\theta}\right) - \mathrm{Li}_2(1) &= -i\int_0^{\theta}\log\left(1-e^{i\eta}\right)d\eta \\ &= -i\int_0^{\theta}\log\left[2\sin\left(\frac{1}{2}\eta\right)\exp\left(-\frac{1}{2}(\pi-\eta)\right)i\right]d\eta \\ &= -i\int_0^{\theta}\log\left[2\sin\left(\frac{1}{2}\eta\right)\right]d\eta + \frac{1}{4}(\pi-\theta)^2 - \frac{1}{4}\pi^2 \quad (0 \leqq \theta \leqq 2\pi).\end{aligned} \tag{43}$$

Comparing the real parts in (42) and (43) gives us a well-known Fourier series:

$$\sum_{n=1}^{\infty}\frac{\cos n\theta}{n^2} = \frac{\pi^2}{6} - \frac{\theta(2\pi-\theta)}{4} \quad (0 \leqq \theta \leqq 2\pi). \tag{44}$$

On the other hand, if we compare the imaginary parts in (42) and (43), we obtain *Clausen's integral* (or *function*) $\mathrm{Cl}_2(\theta)$ defined by [*cf.* Equation (2) above]

$$\mathrm{Cl}_2(\theta) := \sum_{n=1}^{\infty}\frac{\sin n\theta}{n^2} = -\int_0^{\theta}\log\left[2\sin\left(\frac{1}{2}\eta\right)\right]d\eta. \tag{45}$$

Equation (43) can, therefore, be written in the form:

$$\mathrm{Li}_2\left(e^{i\theta}\right) = \frac{\pi^2}{6} - \frac{\theta(2\pi-\theta)}{4} + i\,\mathrm{Cl}_2(\theta) \quad (0 \leqq \theta \leqq 2\pi). \tag{46}$$

Some known properties and special values of the Clausen integral (or function) include the periodic properties:

$$\mathrm{Cl}_2(2n\pi \pm \theta) = \mathrm{Cl}_2(\pm\theta) = \pm\mathrm{Cl}_2(\theta), \tag{47}$$

which, for $n = 1$ and with θ replaced by $\pi + \theta$, yields

$$\mathrm{Cl}_2(\pi + \theta) = -\mathrm{Cl}_2(\pi - \theta). \tag{48}$$

From the series definition (45), it is obvious that

$$\mathrm{Cl}_2(n\pi) = 0 \quad (n \in \mathbb{Z}), \tag{49}$$

which, for $n = 1$, gives

$$\int_0^{\pi} \log\left(2 \sin \frac{1}{2}\theta\right) d\theta = 0 \quad \text{and} \quad \int_0^{\pi/2} \log(\sin\theta)\, d\theta = -\frac{1}{2}\pi \log 2. \tag{50}$$

Setting $\theta = \frac{1}{2}\pi$ in the series definition (45), and using (48), we find that

$$\mathrm{Cl}_2\left(\frac{1}{2}\pi\right) = G = -\mathrm{Cl}_2\left(\frac{3}{2}\pi\right), \tag{51}$$

where G is the Catalan constant given in 1.3(16);

$$\frac{1}{2}\,\mathrm{Cl}_2(2\theta) = \mathrm{Cl}_2(\theta) - \mathrm{Cl}_2(\pi - \theta); \tag{52}$$

From (45), by employing integration by parts, we have

$$\mathrm{Cl}_2(\theta) = -\theta \log\left(\sin \frac{1}{2}\theta\right) + \int_0^{\theta} \frac{1}{2}\eta \cot\left(\frac{1}{2}\eta\right) d\eta. \tag{53}$$

If we substitute the known formula:

$$\cot\left(\frac{1}{2}\theta\right) = \frac{1}{\frac{1}{2}\theta} - \sum_{k=1}^{\infty} \frac{\theta}{k^2\pi^2 - \frac{1}{4}\theta^2} \tag{54}$$

into (53) and evaluate the resulting integral, we obtain

$$\mathrm{Cl}_2(\theta) = -\theta \log\left(\sin \frac{1}{2}\theta\right) + \theta + \sum_{n=1}^{\infty} \left[2\theta - 2n\pi \log\left(\frac{2n\pi + \theta}{2n\pi - \theta}\right)\right] \quad (0 \leq \theta < 2\pi); \tag{55}$$

$$\mathrm{Ti}_2(\tan\theta) = \theta \log(\tan\theta) + \frac{1}{2}\mathrm{Cl}_2(2\theta) + \frac{1}{2}\mathrm{Cl}_2(\pi - 2\theta), \tag{56}$$

whose interesting special case when $\theta = \frac{1}{12}\pi$ becomes

$$\mathrm{Cl}_2\left(\frac{\pi}{6}\right) + \mathrm{Cl}_2\left(\frac{5\pi}{6}\right) = \frac{4}{3}G. \tag{57}$$

The Trilogarithm function

The *Trilogarithm function* $\mathrm{Li}_3(z)$ is defined by

$$\begin{aligned}\mathrm{Li}_3(z) &:= \sum_{n=1}^{\infty} \frac{z^n}{n^3} \quad (|z| \leq 1) \\ &= \int_0^z \frac{\mathrm{Li}_2(t)}{t}\, dt.\end{aligned} \tag{58}$$

An obvious special case of (58) is

$$\mathrm{Li}_3(1) = \zeta(3), \tag{59}$$

where $\zeta(s)$ is the Riemann Zeta function (see Section 2.3).

The following simple functional relationship for $\mathrm{Li}_3(x)$ would follow if we divide both sides of (14) by x and then integrate each term:

$$\frac{1}{4}\,\mathrm{Li}_3\left(x^2\right) = \mathrm{Li}_3\left(x\right) + \mathrm{Li}_3\left(-x\right), \tag{60}$$

which, for $x = 1$, yields

$$\mathrm{Li}_3(-1) = -\frac{3}{4}\,\mathrm{Li}_3(1) \tag{61}$$

Similarly, we find from (4) and (9) that

$$\mathrm{Li}_3\left(-x\right) - \mathrm{Li}_3\left(-\frac{1}{x}\right) = -\frac{\pi^2}{6}\log x - \frac{1}{6}\left(\log x\right)^3 \tag{62}$$

and

$$\mathrm{Li}_3\left(y\right) - \mathrm{Li}_3\left(\frac{1}{y}\right) = \frac{\pi^2}{3}\log y - \frac{1}{6}\left(\log y\right)^3 - \frac{1}{2}i\pi(\log y)^2 \quad (y > 1), \tag{63}$$

respectively.

Landen's formula for $\mathrm{Li}_3\left(x\right)$ [see Lewin (1958, p. 138)] is recalled here as follows:

$$\begin{aligned}\mathrm{Li}_3\left(\frac{x}{x-1}\right) &+ \mathrm{Li}_3\left(1-x\right) + \mathrm{Li}_3\left(x\right) = \mathrm{Li}_3\left(1\right) + \frac{\pi^2}{6}\log(1-x) \\ &- \frac{1}{2}\log x\left[\log\left(1-x\right)\right]^2 + \frac{1}{6}\left[\log\left(1-x\right)\right]^3 \quad (0 < x < 1),\end{aligned} \tag{64}$$

whose domain can be extended by using (63) with $y = 1 - x$:

$$\begin{aligned}\mathrm{Li}_3\left(\frac{x}{x-1}\right) &+ \mathrm{Li}_3\left(\frac{1}{1-x}\right) + \mathrm{Li}_3\left(x\right) = \mathrm{Li}_3\left(1\right) - \frac{\pi^2}{6}\log(1-x) \\ &- \frac{1}{2}\log(-x)\left[\log\left(1-x\right)\right]^2 + \frac{1}{3}\left[\log\left(1-x\right)\right]^3 \quad (x < 0).\end{aligned} \tag{65}$$

Taking $x = \frac{1}{2}$ in (64) or $x = -1$ in (65), we obtain

$$2\,\mathrm{Li}_3\left(\frac{1}{2}\right) + \mathrm{Li}_3(-1) = \mathrm{Li}_3(1) - \frac{\pi^2}{6}\log 2 + \frac{1}{3}(\log 2)^3,$$

which, upon eliminating $\mathrm{Li}_3(-1)$ with the aid of (61), yields

(66) $$\mathrm{Li}_3\left(\frac{1}{2}\right) = \frac{7}{8}\mathrm{Li}_3(1) - \frac{\pi^2}{12}\log 2 + \frac{1}{6}(\log 2)^3.$$

Setting $1 - x = x/(1-x)$, we have

$$(1-x)^2 = x \quad \text{or} \quad x^2 - 3x + 1 = 0,$$

which gives the value:

(67) $$x = \frac{3-\sqrt{5}}{2} = 4\sin^2\left(\frac{\pi}{10}\right) \quad (x < 1).$$

Now, writing $1-x$ for x in (60), and evaluating (64) with x given by (67), we obtain

(68) $$\mathrm{Li}_3\left(\frac{3-\sqrt{5}}{2}\right) = \frac{4}{5}\mathrm{Li}_3(1) + \frac{\pi^2}{15}\log\left(\frac{3-\sqrt{5}}{2}\right) - \frac{1}{12}\left[\log\left(\frac{3-\sqrt{5}}{2}\right)\right]^3.$$

An interesting definite integral of $\mathrm{Li}_3(1)$ was given by Hjortnaes (1954):

(69) $$\mathrm{Li}_3(1) = 10\int_0^{\log\left(\frac{1+\sqrt{5}}{2}\right)} t^2 \coth t\,dt,$$

which, upon writing $\coth t$ in exponential form and putting $y = 1 - e^{-2t}$, yields

(70) $$\mathrm{Li}_3(1) = \frac{5}{4}\int_0^{\frac{\sqrt{5}-1}{2}} [\log(1-y)]^2\left(\frac{2}{y} + \frac{1}{1-y}\right)dy.$$

Integrating (70) by parts and expressing it in terms of trilogarithms, we arrive once again at (68).

The Polylogarithm functions

The *Polylogarithm function* $\mathrm{Li}_n(z)$ is defined by

(71) $$\begin{aligned}\mathrm{Li}_n(z) &:= \sum_{k=1}^{\infty}\frac{z^k}{k^n} \quad (|z| \le 1;\ n \in \mathbb{N}\setminus\{1\}) \\ &= \int_0^z \frac{\mathrm{Li}_{n-1}(t)}{t}\,dt \quad (n \in \mathbb{N}\setminus\{1,\,2\}).\end{aligned}$$

Clearly, we have

(72) $$\mathrm{Li}_n(1) = \zeta(n) \quad (n \in \mathbb{N}\setminus\{1\}),$$

in terms of the Riemann Zeta function $\zeta(s)$ (see Section 2.3). Setting $z = iy$ $(y \in \mathbb{R})$ in (71), we have

$$\mathrm{Li}_n(iy) = \frac{1}{2^n}\mathrm{Li}_n\left(-y^2\right) + i\,\mathrm{Ti}_n(y) \quad (y \in \mathbb{R}), \tag{73}$$

where $\mathrm{Ti}_n(x)$ is the *inverse tangent integral of order* n defined by

$$\mathrm{Ti}_n(x) := \sum_{k=1}^{\infty} (-1)^{k+1} \frac{x^{2k-1}}{(2k-1)^n} \quad (|x| \leq 1). \tag{74}$$

It is obvious that

$$\mathrm{Ti}_n(x) = \int_0^x \frac{\mathrm{Ti}_{n-1}(t)}{t}\, dt, \tag{75}$$

which, in conjunction with (34) for $\mathrm{Ti}_2(x)$, extends the domain of $\mathrm{Ti}_n(x)$.

If $z = re^{i\theta}$, one defines

$$\mathrm{Li}_n\left(re^{i\theta}\right) := \mathrm{Li}_n(r, \theta) + i\left[\mathrm{Ti}_n(\varrho) - \mathrm{Ti}_n(\varrho, \tan\theta)\right], \tag{76}$$

where $\varrho = r\,\sin\theta/(1 - r\,\cos\theta)$. Then it is readily seen that

$$\mathrm{Li}_n(r, \theta) = \int_0^r \frac{\mathrm{Li}_{n-1}(t, \theta)}{t}\, dt \tag{77}$$

and

$$\mathrm{Ti}_n(\varrho, \tan\theta) = \int_0^\varrho \left(\frac{\mathrm{Ti}_{n-1}(t)}{t + \tan\theta} + \frac{\mathrm{Ti}_{n-1}(t, \tan\theta)}{t} - \frac{\mathrm{Ti}_{n-1}(t, \tan\theta)}{t + \tan\theta}\right) dt. \tag{78}$$

The *generalized Clausen function* $\mathrm{Cl}_n(\theta)$ is defined by

$$\mathrm{Cl}_n(\theta) := \begin{cases} \displaystyle\sum_{k=1}^{\infty} \frac{\sin k\theta}{k^n} & (n \text{ is even}), \\ \displaystyle\sum_{k=1}^{\infty} \frac{\cos k\theta}{k^n} & (n \text{ is odd}). \end{cases} \tag{79}$$

From these definitions, we have

$$\mathrm{Cl}_{2n+1}(\theta) = \mathrm{Li}_{2n+1}(1) - \int_0^\theta \mathrm{Cl}_{2n}(t)\, dt \tag{80}$$

and

$$\mathrm{Cl}_{2n}(\theta) = \int_0^\theta \mathrm{Cl}_{2n-1}(t)\, dt. \tag{81}$$

The *Log-Sine integral* $\mathrm{Ls}_n(\theta)$ *of order* n is defined by

$$\mathrm{Ls}_n(\theta) := -\int_0^\theta \left(\log\left|2\sin\frac{1}{2}t\right|\right)^{n-1} dt \quad (n \in \mathbb{N}\setminus\{1\}), \tag{82}$$

whose special case *only* when $n = 2$ satisfies the relationship [*cf.* Equation (45)]:

$$\mathrm{Ls}_2(\theta) = \mathrm{Cl}_2(\theta). \tag{83}$$

The *generalized Log-Sine integral of order* n *and index* m is defined by

$$\mathrm{Ls}_n^{(m)}(\theta) := -\int_0^\theta t^m \left(\log\left|2\sin\frac{1}{2}t\right|\right)^{n-m-1} dt. \tag{84}$$

The *associated Clausen function* $\mathrm{Gl}_n(\theta)$ *of order* n is defined by

$$\mathrm{Gl}_n(\theta) := \begin{cases} \displaystyle\sum_{k=1}^\infty \frac{\cos k\theta}{k^n} & (n \text{ is even}), \\ \displaystyle\sum_{k=1}^\infty \frac{\sin k\theta}{k^n} & (n \text{ is odd}), \end{cases} \tag{85}$$

which also satisfies the following obvious relationships analogous to (80) and (81):

$$\mathrm{Gl}_{2n}(\theta) = \mathrm{Li}_{2n}(1) - \int_0^\theta \mathrm{Gl}_{2n-1}(t)\, dt \tag{86}$$

and

$$\mathrm{Gl}_{2n+1}(\theta) = \int_0^\theta \mathrm{Gl}_{2n}(t)\, dt, \tag{87}$$

each of which was studied extensively by Glaisher (1898).

Dividing both sides of (4) by x and integrating from 1 to x, we obtain [*cf.* Equation (62)]

$$\mathrm{Li}_3(-x) - \mathrm{Li}_3\left(-\frac{1}{x}\right) = -\frac{1}{3!}(\log x)^3 + 2\,\mathrm{Li}_2(-1)\log x. \tag{88}$$

By repeating this process $n-2$ times, we finally arrive at

$$\begin{aligned} \mathrm{Li}_n(-x) + (-1)^n \mathrm{Li}_n\left(-\frac{1}{x}\right) &= -\frac{1}{n!}(\log x)^n \\ &+ 2\sum_{k=1}^{\left[\frac{1}{2}n\right]} \frac{(\log x)^{n-2k}}{(n-2k)!}\mathrm{Li}_{2k}(-1) \quad (n \in \mathbb{N}\setminus\{1\}), \end{aligned} \tag{89}$$

which was attributed to Jonquière by Nielsen (1923), although equivalent formulas were given much earlier by Spence and Kummer [see Lewin (1958, p. 173)].

Taking $x = e^{i\pi}$ in (89), we find that

$$[1 + (-1)^n] \operatorname{Li}_n(1) = -\frac{i^n \pi^n}{n!} + 2 \sum_{k=1}^{[\frac{1}{2}n]} \frac{i^{n-2k} \pi^{n-2k}}{(n-2k)!} \operatorname{Li}_{2k}(-1). \tag{90}$$

Since the left-hand side of (90) vanishes when n is odd, the evaluation of $\operatorname{Li}_{2n+1}(1)$ is not possible from (90). Yet, using the following known relationship between $\operatorname{Li}_n(1)$ and $\operatorname{Li}_n(-1)$:

$$\operatorname{Li}_n(-1) = -\left(1 - 2^{1-n}\right) \operatorname{Li}_n(1), \tag{91}$$

and writing $2n$ for n in (90), we readily get

$$\frac{2(-1)^{n-1}\left(2^{2n}-1\right)}{(2n)!} \mathrm{L}_{2n} = \frac{1}{(2n)!} - 2 \sum_{k=1}^{n-1} \frac{(-1)^{k-1}\left(2^{2k-1}-1\right)}{(2k)!\,(2n-2k)!} \mathrm{L}_{2k}, \tag{92}$$

where, for the sake of brevity,

$$\mathrm{L}_{2k} := (2k)! \pi^{-2k} 2^{1-2k} \operatorname{Li}_{2k}(1). \tag{93}$$

By using the generating function 1.6(2) of the Bernoulli numbers, one can also obtain

$$\operatorname{Li}_{2n}(-1) = (-1)^n \left(2^{2n-1} - 1\right) B_{2n} \frac{\pi^{2n}}{(2n)!} \quad (n \in \mathbb{N}) \tag{94}$$

and

$$\operatorname{Li}_{2n}(1) = (-1)^{n+1} \frac{(2\pi)^{2n}}{2(2n)!} B_{2n} \quad (n \in \mathbb{N}), \tag{95}$$

which is precisely the same as 2.3(18).

Although there is no result like (94) when the order of the Polylogarithm function is odd, yet there is an analogous relation involving the Euler numbers E_n given in Section 1.6. Dividing (35) by x and integrating from 1 to x would yield

$$\operatorname{Ti}_3(x) + \operatorname{Ti}_3\left(\frac{1}{x}\right) = 2\operatorname{Ti}_3(1) + \frac{1}{4}\pi(\log x)^2 \quad (x > 0). \tag{96}$$

Repeated applications of this process lead us to *Spence's formula:*

$$\begin{aligned} \operatorname{Ti}_n(x) + (-1)^{n-1} \operatorname{Ti}_n\left(\frac{1}{x}\right) &= \frac{1}{2}\pi \frac{(\log x)^{n-1}}{(n-1)!} \\ &+ 2 \sum_{k=1}^{[\frac{1}{2}(n-1)]} \frac{(\log x)^{n-2k-1}}{(n-2k-1)!} \operatorname{Ti}_{2k+1}(1) \quad (x > 0). \end{aligned} \tag{97}$$

Next we find from (73) that

$$\operatorname{Li}_{2n+1}(i) - \operatorname{Li}_{2n+1}(-i) = 2i \operatorname{Ti}_{2n+1}(1). \tag{98}$$

Setting $x = \exp\left(\frac{1}{2}\pi i\right)$ and writing $2n+1$ for n in (89), we obtain

(99)

$$\mathrm{Li}_{2n+1}(-i) - \mathrm{Li}_{2n+1}(i) = (-1)^n i \left[-\frac{\pi^{2n+1}}{(2n+1)!\,2^{2n+1}} + 2\pi^{2n+1} \sum_{k=1}^{n} \frac{B_{2k}\left(2^{2k-1}-1\right)}{(2k)!\,(2n-2k+1)!\,2^{2n-2k+1}} \right].$$

Consider

$$\frac{2e^{\frac{1}{2}z}}{e^z+1} = 2\frac{e^{\frac{1}{2}z}-e^{-\frac{1}{2}z}}{2}\left(\frac{z}{e^z-1} - \frac{1}{2}\frac{2z}{e^{2z}-1}\right), \tag{100}$$

which can be expanded by 1.6(2) and 1.6(40) in powers of z. Upon equating the coefficients of z^{2n}, we find that

$$\frac{E_{2n}}{(2n)!\,2^{2n+1}} = \left[\frac{1}{(2n+1)!\,2^{2n+1}} - 2\sum_{k=1}^{n} \frac{B_{2k}\left(2^{2k-1}-1\right)}{(2k)!\,(2n-2k+1)!\,2^{2n-2k+1}} \right]. \tag{101}$$

Now a combination of (98), (99), and (101) leads us to the well-known Euler series:

$$\mathrm{Ti}_{2n+1}(1) = \sum_{k=1}^{\infty} (-1)^{k+1} \frac{1}{(2k-1)^{2n+1}} = \frac{(-1)^n \pi^{2n+1}}{(2n)!\,2^{2n+2}} E_{2n}, \tag{102}$$

which, in view of 1.6(59), gives

$$\mathrm{Ti}_3(1) = \frac{\pi^3}{32}, \quad \mathrm{Ti}_5(1) = \frac{5\pi^5}{1536}, \quad \mathrm{Ti}_7(1) = \frac{61\,\pi^7}{184320}, \cdots. \tag{103}$$

Starting with the known formula for $\mathrm{Li}_2(x,\theta)$:

$$\mathrm{Li}_2(r,\theta) + \mathrm{Li}_2\left(\frac{1}{r},\theta\right) = 2\,\mathrm{Gl}_2(\theta) - \frac{1}{2}(\log r)^2, \tag{104}$$

and following a procedure analogous to that in proving (89), one obtains

$$\begin{aligned} \mathrm{Li}_n(r,\theta) + (-1)^n \mathrm{Li}_n\left(\frac{1}{r},\theta\right) &= -\frac{1}{n!}(\log r)^n \\ &+ 2\sum_{k=1}^{\left[\frac{1}{2}n\right]} \frac{(\log r)^{n-2k}}{(n-2k)!} \mathrm{Gl}_{2k}(\theta) \quad (n \in \mathbb{N}\setminus\{1\}), \end{aligned} \tag{105}$$

which, upon setting $r = e^{i\pi}$, noting that $\mathrm{Li}_n(-r,\theta) = \mathrm{Li}_n(r,\pi-\theta)$, and writing $2n$ for n, gives

$$\mathrm{Gl}_{2n}(\pi-\theta) = \frac{(-1)^{n-1}\pi^{2n}}{2(2n)!} + \sum_{k=1}^{n} \frac{(-1)^{n-k}\pi^{2n-2k}}{(2n-2k)!} \mathrm{Gl}_{2k}(\theta). \tag{106}$$

The following relationship between $\mathrm{Gl}_{2n}(\theta)$ and $\mathrm{Gl}_{2n}(\pi-\theta)$ can readily be obtained from the series definition (85):

$$\mathrm{Gl}_{2n}(\pi-\theta) + \mathrm{Gl}_{2n}(\theta) = \frac{1}{2^{2n-1}} \mathrm{Gl}_{2n}(2\theta). \tag{107}$$

The generating function 1.6(1) for the Bernoulli polynomials can be rewritten in the form:

$$\frac{te^{xt}}{e^t - 1} = 1 + t \sum_{n=1}^{\infty} \frac{2t\cos(2n\pi x) - 4n\pi\sin(2n\pi x)}{t^2 + 4n^2\pi^2}. \tag{108}$$

Making use of the expansion:

$$\frac{1}{t^2 + 4n^2\pi^2} = \frac{1}{4n^2\pi^2} \sum_{k=0}^{\infty} (-1)^k \left(\frac{t}{2n\pi}\right)^{2k} \qquad (|t| < 2n\pi;\ n \in \mathbb{N})$$

and rearranging the double series in (108) in powers of t, the coefficients are easily expressible in terms of $\mathrm{Gl}_n(2\pi x)$ defined by (85). Thus, employing 1.6(1) on the left-hand side of (108) and equating the coefficients of the same powers on both sides, we find that

$$\mathrm{Gl}_n(2\pi x) = (-1)^{1+\left[\frac{1}{2}n\right]} 2^{n-1} \pi^n \frac{B_n(x)}{n!} \qquad (0 \leq x \leq 1;\ n \in \mathbb{N} \setminus \{1\}), \tag{109}$$

which gives us the following special cases:

$$\begin{aligned}
\mathrm{Gl}_2(\theta) &= \frac{1}{4}(\pi - \theta)^2 - \frac{\pi^2}{12}, \\
\mathrm{Gl}_3(\theta) &= \frac{1}{12}\theta(\pi - \theta)(2\pi - \theta), \\
\mathrm{Gl}_4(\theta) &= \frac{1}{90}\pi^4 - \frac{1}{12}\theta^2\left(\pi^2 - \pi\theta + \frac{1}{4}\theta^2\right), \\
\mathrm{Gl}_5(\theta) &= \frac{1}{720}\theta(\pi - \theta)(2\pi - \theta)\left(4\pi^2 + 6\pi\theta - 3\theta^2\right).
\end{aligned} \tag{110}$$

The Log-Sine integrals

A recurrence relationship for $\mathrm{Ls}_n(\pi)$ defined by (82) is given by

$$\begin{aligned}
&\frac{(-1)^m}{m!}\mathrm{Ls}_{m+2}(\pi) = \pi\left(1 - 2^{-m}\right)\zeta(m+1) \\
&\quad + \sum_{k=2}^{m-1} \frac{(-1)^{k+1}}{k!}\left(1 - 2^{k-m}\right)\zeta(m-k+1)\,\mathrm{Ls}_{k+1}(\pi) \qquad (m \in \mathbb{N}).
\end{aligned} \tag{111}$$

In order to derive the recurrence relation (111), we let

$$\begin{aligned}
I &:= \int_0^{\pi} \exp\left[x \log\left(2\sin\frac{1}{2}\theta\right)\right] d\theta \\
&= \sum_{n=0}^{\infty} \int_0^{\pi} \frac{x^n}{n!}\left[\log\left(2\sin\frac{1}{2}\theta\right)\right]^n d\theta \\
&= -\sum_{n=0}^{\infty} \frac{x^n}{n!}\mathrm{Ls}_{n+1}(\pi).
\end{aligned} \tag{112}$$

Alternatively, in view of 1.1(64), we have

$$\begin{aligned} I &= \int_0^\pi \left[\exp\left\{\log\left(2\sin\frac{1}{2}\theta\right)\right\}\right]^x d\theta \\ &= \int_0^\pi 2^x \sin^x\left(\frac{1}{2}\theta\right) d\theta \\ &= 2^x\sqrt{\pi}\,\frac{\Gamma\left(\frac{1}{2}+\frac{1}{2}x\right)}{\Gamma\left(1+\frac{1}{2}x\right)}. \end{aligned}$$

By the Legendre duplication formula 1.1(49) for Γ, we thus obtain

(113) $$I = \pi\, y,$$

where, for convenience,

$$y := \frac{\Gamma(1+x)}{\left[\Gamma\left(1+\frac{1}{2}x\right)\right]^2}.$$

Making use of the following notations:

$$D^n f(x) = \frac{d^n}{dx^n} f(x) = f^{(n)}(x) \quad \text{and} \quad D_0^n f(x) = \frac{d^n}{dx^n} f(x)\bigg|_{x=0} = f^{(n)}(0),$$

differentiating (112) n times, and setting $x = 0$, we find that

(114) $$\mathrm{Ls}_{n+1}(\pi) = -D_0^n I = -\pi\, D_0^n\, y,$$

which, by virtue of 1.2(53), yields

(115) $$D_0^n \log y = (-1)^n (n-1)!\,\left(1-2^{1-n}\right)\zeta(n) \quad (n \in \mathbb{N}\setminus\{1\}),$$

so that we have

(116) $$\log y =: f(x) = \sum_{n=2}^\infty \frac{x^n}{n}(-1)^n\left(1-2^{1-n}\right)\zeta(n),$$

since

$$\log y|_{x=0} = 0 = D_0^1 \log y.$$

From (114), we get $\mathrm{Ls}_{n+1}(\pi) = -\pi\, D_0^n \exp(f(x))$. Let $y_m := D^m e^f$. Then

$$y_{m+1} = D^{m+1} e^f = D^m\left(De^f\right) = D^m\left(f'\,e^f\right).$$

Hence, using Leibniz's rule for differentiation, we have

(117) $$y_{m+1} = \sum_{k=0}^m \binom{m}{k} f^{(m-k+1)}\, y_k,$$

which, upon considering

$$y_m|_{x=0} = -\frac{1}{\pi}\,\mathrm{Ls}_{m+1}(\pi)$$

in (114), yields

(118) $$\mathrm{Ls}_{m+2}(\pi) = -\pi\, f_0^{(m+1)} + \sum_{k=1}^m \binom{m}{k} f_0^{(m-k+1)}\,\mathrm{Ls}_{k+1}(\pi).$$

Now, setting

$$\mathrm{Ls}_2(\pi) = 0, \quad f_0' = 0, \quad \text{and} \quad f_0^{(m)} = (-1)^m (m-1)!\,\left(1-2^{1-m}\right)\zeta(m) \quad (m \in \mathbb{N}\setminus\{1\})$$

in (118), we immediately arrive at the desired recurrence relation (111).

Some simple consequences of (111) are presented below:

$$\mathrm{Ls}_2(\pi) = -\int_0^{\pi} \log\left(2\sin\frac{1}{2}\theta\right) d\theta = 0,$$

$$\mathrm{Ls}_3(\pi) = -\int_0^{\pi} \left[\log\left(2\sin\frac{1}{2}\theta\right)\right]^2 d\theta = -\frac{1}{12}\pi^3,$$

$$\mathrm{Ls}_4(\pi) = -\int_0^{\pi} \left[\log\left(2\sin\frac{1}{2}\theta\right)\right]^3 d\theta = \frac{3}{2}\pi\,\zeta(3),$$

$$\mathrm{Ls}_5(\pi) = -\int_0^{\pi} \left[\log\left(2\sin\frac{1}{2}\theta\right)\right]^4 d\theta = -\frac{19}{240}\pi^5,$$

$$\mathrm{Ls}_6(\pi) = -\int_0^{\pi} \left[\log\left(2\sin\frac{1}{2}\theta\right)\right]^5 d\theta = \frac{45}{2}\pi\,\zeta(5) + \frac{5}{4}\pi^3\,\zeta(3),$$

$$\mathrm{Ls}_7(\pi) = -\int_0^{\pi} \left[\log\left(2\sin\frac{1}{2}\theta\right)\right]^6 d\theta = -\frac{45}{2}\pi\,\{\zeta(3)\}^2 - \frac{275}{1344}\pi^7,$$

$$\mathrm{Ls}_8(\pi) = -\int_0^{\pi} \left[\log\left(2\sin\frac{1}{2}\theta\right)\right]^7 d\theta = \frac{2835}{4}\pi\,\zeta(7) + \frac{315}{8}\pi^3\,\zeta(5) + \frac{133}{20}\pi^5\,\zeta(3),$$

$$\mathrm{Ls}_9(\pi) = -\int_0^{\pi} \left[\log\left(2\sin\frac{1}{2}\theta\right)\right]^8 d\theta = -\frac{24177}{26880}\pi^9 - 1890\,\pi\,\zeta(3)\,\zeta(5) - \frac{105}{2}\pi^3\,\{\zeta(3)\}^2.$$

2.5 Hurwitz–Lerch Zeta Functions

The *Hurwitz-Lerch Zeta function* $\Phi(z,s,a)$ is defined by

$$\Phi(z,s,a) := \sum_{n=0}^{\infty} \frac{z^n}{(n+a)^s} \tag{1}$$

$$\left(a \in \mathbb{C}\setminus\mathbb{Z}_0^-;\ s \in \mathbb{C} \text{ when } |z|<1;\ \Re(s)>1 \text{ when } |z|=1\right),$$

which satisfies the obvious functional relation:

$$\Phi(z,s,a) = z^n\,\Phi(z,s,n+a) + \sum_{k=0}^{n-1} \frac{z^k}{(k+a)^s} \qquad \left(n\in\mathbb{N};\ a\in\mathbb{C}\setminus\mathbb{Z}_0^-\right). \tag{2}$$

By writing the Eulerian integral 1.1(1) in the form:

$$\Gamma(z) = s^z \int_0^{\infty} e^{-st}\,t^{z-1}\,dt \quad (\Re(z)>0;\ \Re(s)>0), \tag{3}$$

we can deduce the following integral representation from (1):

$$\Phi(z,s,a) = \frac{1}{\Gamma(s)}\int_0^{\infty} \frac{t^{s-1}\,e^{-at}}{1-ze^{-t}}\,dt = \frac{1}{\Gamma(s)}\int_0^{\infty} \frac{t^{s-1}\,e^{-(a-1)t}}{e^t - z}\,dt \tag{4}$$

$$(\Re(a)>0;\ |z|\le 1,\ z\ne 1,\ \Re(s)>0;\ z=1,\ \Re(s)>1)$$

by noting that

$$\frac{z^n}{(n+a)^s} = \frac{1}{\Gamma(s)} \int_0^\infty e^{-at} t^{s-1} \left(ze^{-t}\right)^n dt \quad (\Re(a) > 0;\ \Re(s) > 0). \tag{5}$$

If use is made of the infinite-series version of Plana's summation formula 2.2(9) [*cf.* Lindelöf (1905, p. 61); see also Erdélyi *et al.* (1953, p. 22)] and the definition (1), another definite integral representation of $\Phi(z,s,a)$ is obtained in the form:

$$\begin{aligned}\Phi(z,s,a) &= \frac{1}{2a^s} + \int_0^\infty \frac{z^t}{(t+a)^s}\,dt \\ &- 2\int_0^\infty \sin\left[t\log z - s\tan^{-1}\left(\frac{t}{a}\right)\right]\left(t^2+a^2\right)^{-\frac{1}{2}s}\frac{dt}{e^{2\pi t}-1} \quad (\Re(a) > 0),\end{aligned} \tag{6}$$

which, for $z=1$, immediately reduces to Hermite's formula 2.2(12) for $\zeta(s,a)$.

By setting $z = e^{i\theta}$ in (4) and using (3), we get *Lipschitz's formula:*

$$2\,\Gamma(s)\sum_{n=1}^\infty \frac{e^{in\theta}}{(n+a)^s} = \int_0^\infty e^{-at}\,t^{s-1}\frac{e^{i\theta}-e^{-t}}{\cosh t - \cos\theta}\,dt \tag{7}$$
$$(0<\theta<2\pi;\ \Re(a) > -1;\ \Re(s) > 0).$$

Several contour integral and series representations of $\Phi(z,s,a)$ include [see Erdélyi *et al.* (1953, pp. 28-31)]:

$$\Phi(z,s,a) = -\frac{\Gamma(1-s)}{2\pi i} + \int_C \frac{(-t)^{s-1}\,e^{-at}}{1-z\,e^{-t}}\,dt \tag{8}$$
$$(\Re(a) > 0;\ |\arg(-t)| \leq \pi),$$

where the contour C is the Hankel loop of Theorem 2.5, which obviously does not enclose any pole of the integrand, that is, $t = \log z \pm 2n\pi i \ \ (n \in \mathbb{N}_0)$;

$$\Phi(z,s,a) = \frac{\Gamma(1-s)}{z^a}\sum_{n=-\infty}^\infty (-\log z + 2n\pi i)^{s-1}\,e^{2n\pi i a} \tag{9}$$
$$(0 < a \leq 1;\ \Re(s) < 0;\ |\arg(-\log z + 2n\pi i)| \leq \pi);$$

Lerch's transformation formula:

$$\begin{aligned}\Phi(z,s,a) =& \frac{i(2\pi)^{s-1}\,\Gamma(1-s)}{z^a}\left\{e^{-\frac{1}{2}i\pi s}\,L\left(-a,1-s,\frac{\log z}{2\pi i}\right)\right. \\ & \left. - e^{i\pi\left(\frac{1}{2}s+2a\right)}\,L\left(a,1-s,1-\frac{\log z}{2\pi i}\right)\right\},\end{aligned} \tag{10}$$

where the so-called Lipschitz-Lerch transcendent $L(x,s,a)$ is defined by

$$L(x,s,a) := \sum_{n=0}^\infty \frac{\exp(2\pi i n x)}{(n+a)^s} = \Phi\left(e^{2\pi i x}, s, a\right) \tag{11}$$
$$(a \in \mathbb{C}\setminus\mathbb{Z}_0^-;\ x \in \mathbb{R}\setminus\mathbb{Z} \text{ and } \Re(s) > 0;\ x \in \mathbb{Z} \text{ and } \Re(s) > 1);$$

Replacing s by $1-s$ in 2.2(6), we obtain another equivalent form of Hurwitz's formula 2.2(6):

$$\zeta(s,a) = 2(2\pi)^{s-1}\,\Gamma(1-s)\sum_{n=1}^{\infty} n^{s-1}\sin\left(2n\pi a+\frac{\pi s}{2}\right) \tag{12}$$
$$(\Re(s)<0;\ 0<a\leq 1);$$

By applying the binomial theorem to $(-\log z \pm 2n\pi i)^{s-1}$ in (9), and making use of (12), 1.1(12), and 1.1(41), we readily obtain the following formulas [*cf.* Erdélyi *et al.* (1953, p. 29, Eq. (8))]:

$$\Phi(z,s,a) = \frac{\Gamma(1-s)}{z^a}\left(\log\frac{1}{z}\right)^{s-1} + \frac{1}{z^a}\sum_{k=0}^{\infty}\zeta(s-k,a)\,\frac{(\log z)^k}{k!} \tag{13}$$
$$(|\log z|<2\pi;\ s\notin\mathbb{N};\ a\notin\mathbb{Z}_0^-);$$

(14)

$$\Phi(z,m,a) = \frac{1}{z^a}\left\{\sum_{\substack{k=0\\k\neq m-1}}^{\infty}\zeta(m-k,a)\,\frac{(\log z)^k}{k!} + \frac{(\log z)^{m-1}}{(m-1)!}\left[\psi(m)-\psi(a)-\log\left(\log\frac{1}{z}\right)\right]\right\}$$
$$(m\in\mathbb{N}\setminus\{1\};\ |\log z|<2\pi;\ a\notin\mathbb{Z}_0^-);$$

By setting $s=-m$ $(m\in\mathbb{N})$ in (13) and using 2.1(17), Φ can also be expressed in terms of the Bernoulli polynomials as follows:

$$\Phi(z,-m,a) = \frac{m!}{z^a}\left(\log\frac{1}{z}\right)^{-m-1} - \frac{1}{z^a}\sum_{k=0}^{\infty}\frac{B_{m+k+1}(a)\,(\log z)^k}{k!(m+k+1)} \tag{15}$$
$$(|\log z|<2\pi).$$

The special case of (1) when $s=1$ yields

$$\Phi(z,1,a) = \sum_{n=0}^{\infty}\frac{z^n}{n+a} = a^{-1}\,{}_2F_1(1,\,a;\,1+a;\,z) \quad (|z|<1), \tag{16}$$

and, more generally, we have

$$\Phi(z,k,a) = a^{-k}\,{}_{k+1}F_k(1,\,a,\,\ldots,\,a;\,a+1,\,\ldots,a+1;\,z) \quad (k\in\mathbb{N}_0), \tag{17}$$

where ${}_pF_q$ denotes the generalized hypergeometric function (see Section 1.4).

It follows from (13) and (16) that

$$\lim_{z\to 1}\left\{(1-z)^{1-s}\,\Phi(z,s,a)\right\} = \Gamma(1-s) \quad (\Re(s)<1) \tag{18}$$

and

$$\lim_{z\to 1}\left\{\frac{\Phi(z,1,1)}{\log(1-z)}\right\} = -1. \tag{19}$$

The *Jonquière function* $F(z,s)$ is defined by

$$F(z,s) := \sum_{k=1}^{\infty} \frac{z^k}{k^s} = z\,\Phi(z,s,1), \tag{20}$$

whose many properties can be deduced from those of the Φ-function. For example, Lerch's transformation formula (10) readily yields *Jonquière's formula*:

$$F(z,s) + e^{is\pi} F\left(\frac{1}{z}, s\right) = \frac{(2\pi)^s}{\Gamma(s)} e^{\frac{1}{2}i\pi s} \zeta\left(1-s, \frac{\log z}{2\pi i}\right). \tag{21}$$

Moreover, we obtain

$$F(z,-m) = (-1)^{m+1} F\left(\frac{1}{z}, -m\right) \qquad (m \in \mathbb{N}) \tag{22}$$

and

$$F(z,m) + (-1)^m F\left(\frac{1}{z}, m\right) = -\frac{2\pi i}{m!} B_m\left(\frac{\log z}{2\pi i}\right) \qquad (m \in \mathbb{N}\setminus\{1\}), \tag{23}$$

both of which provide the analytic continuation of $F(z,s)$ in (20) outside its circle of convergence $|z| = 1$.

By taking $a = 1$ in (15) and using 1.6(5), we get

$$\begin{gathered} F(z,-m) = m!\left(\log\frac{1}{z}\right)^{-m-1} - \sum_{k=0}^{\infty} \frac{B_{m+k+1}}{k!(m+k+1)} (\log z)^k \\ (m \in \mathbb{N};\ |\log z| < 2\pi). \end{gathered} \tag{24}$$

In addition to the Jonquière function $F(z,s)$ in (20), various special cases of $\Phi(z,s,a)$ are listed below:

$$\zeta(s) = \Phi(1,s,1); \tag{25}$$

$$\zeta(s,a) = \Phi(1,s,a); \tag{26}$$

$$\mathrm{Li}_2(z) = z\Phi(z,2,1); \tag{27}$$

$$\begin{gathered} \mathrm{Li}_s(z) := \sum_{k=1}^{\infty} \frac{z^k}{k^s} = z\Phi(z,s,1) = F(z,s) \\ (s \in \mathbb{C} \text{ and } |z| < 1;\ \Re(s) > 1 \text{ and } |z| = 1). \end{gathered} \tag{28}$$

The Taylor series expansion of the Lipschitz-Lerch transcendent $L(x,s,a)$

The function $L(x,s,a)$ defined by (11) was first studied by Lipschitz (1857, 1889) and Lerch (1887) in connection with Dirichlet's famous theorem on primes in arithmetic progressions. For $x \in \mathbb{Z}$, (11) reduces immediately to the Hurwitz Zeta function $\zeta(s,a)$ (see Section 2.2). Since $L(x,s,a)$ is a special case of $\Phi(z,s,a)$, many properties of this function can be obtained from those of $\Phi(z,s,a)$.

The Lerch functional equation for $L(x,s,a)$ can be deduced from (10):

(29)

$$L(x,1-s,a) = \frac{\Gamma(s)}{(2\pi)^s}\left\{e^{\pi i\left(\frac{1}{2}s-2ax\right)}\,L(-a,s,x) + e^{\pi i\left(-\frac{1}{2}s+2a(1-x)\right)}\,L(a,s,1-x)\right\}$$

$$(0<x<1;\ 0<a<1).$$

Several interesting proofs of (29) were given by Apostol (1951) and Berndt (1972a).

The function $L(x,s,a)$ satisfies the differential-difference equations:

(30) $$\frac{\partial L(x,s,a)}{\partial a} = -s\,L(x,s+1,a)$$

and

(31) $$\frac{\partial L(x,s,a)}{\partial x} + 2\pi i a\,L(x,s,a) = 2\pi i\,L(x,s-1,a).$$

Klusch (1992) gave a Taylor series expansion of the function $L(x,s,a+\xi)$ in the neighborhood of $\xi=0$ in the form: For fixed $a \in \mathbb{R}^+$,

(32) $$\sum_{k=0}^{\infty}(-1)^k\frac{(s)_k}{k!}\,L(x,s+k,a)\,\xi^k = L(x,s,a+\xi) \quad (|\xi|<a),$$

which provides the analytic continuation of $L(x,s,a)$ to the whole complex s-plane. Klusch (1992) also applied (32) to derive several summation formulas which will be investigated systematically in Chapter 3.

By setting $\xi=-t$ in (32), we get

(33) $$\sum_{k=0}^{\infty}\frac{(s)_k}{k!}\,L(x,s+k,a)\,t^k = L(x,s,a-t) \quad (|t|<a),$$

which, upon replacing k by $k+2$ and s by $s-1$, dividing the resulting equation by t^2, and differentiating both sides with respect to t, yields the following *corrected* version of one of Klusch's results [Klusch (1992, p. 517, Eq. (3.3))]:

(34)

$$\sum_{k=1}^{\infty}\frac{k}{(k+2)!}\,\frac{\Gamma(s+k+1)}{\Gamma(s)}\,L(x,s+k+1,a)\,t^{k-1}$$

$$= t^{-2}\left\{L(x,s,a-t)+L(x,s,a)\right\} - \frac{2t^{-3}}{s-1}\left\{L(x,s-1,a-t)-L(x,s-1,a)\right\}$$

$$(|t|<a;\ t\neq 0).$$

Evaluation of $L(x,-n,a)$

Since $L(x,s,a)=\zeta(s,a)$ $(x\in\mathbb{Z})$ whose properties are studied in Section 2.2, in what follows we assume that x is not an integer. By making use of the classical method presented in Section 2.1, $L(x,s,a)$ can be extended to the whole complex s-plane by means of the contour integral [see Apostol (1951)]:

$$I(x,s,a)=\frac{1}{2\pi i}\int_{C_1}\frac{z^{s-1}\,e^{az}}{1-e^{z+2\pi ix}}\,dz, \tag{35}$$

where the contour C_1 is a loop which begins at $-\infty$, encircles the origin once in the positive (counter-clockwise) direction, and returns to $-\infty$. Since $I(x,s,a)$ is an entire function of s, and since

$$L(x,s,a)=\Gamma(1-s)\,I(x,s,a), \tag{36}$$

we can use (36) to continue $L(x,s,a)$ analytically to the whole complex s-plane as an entire function of s $(x\notin\mathbb{Z})$.

Apostol (1951) computed recursively the values of $L(x,-n,a)$ $(n\in\mathbb{N}_0;\ x\notin\mathbb{Z})$ by applying (31) and (35). Thus he obtained [*cf.* Apostol (1951)]

$$L(x,0,a)=\frac{1}{1-\exp(2\pi ix)}=\frac{i}{2}\cot(\pi x)+\frac{1}{2}, \tag{37}$$

$$L(x,-1,a)=\frac{a}{2}[i\cot(\pi x)+1]-\frac{1}{4}\csc^2(\pi x), \tag{38}$$

and

$$L(x,-2,a)=\frac{a^2}{2}\left[i\cot(\pi x)+\frac{1}{4}\right]-\frac{a}{2}\csc^2(\pi x)-\frac{i}{4}\cot(\pi x)\csc^2(\pi x), \tag{39}$$

and

$$L(x,-n,a)=-\frac{\mathcal{B}_{n+1}\left(a,e^{2\pi ix}\right)}{n+1}\qquad(n\in\mathbb{N}_0), \tag{40}$$

where the $\mathcal{B}_n(a,\alpha)$ are defined by means of the generating function:

$$\frac{z\,e^{az}}{\alpha\,e^z-1}:=\sum_{n=0}^{\infty}\mathcal{B}_n\left(a,\alpha\right)\frac{z^n}{n!}\qquad(|z+\log\alpha|<2\pi). \tag{41}$$

The special case of $\mathcal{B}_n\left(a,\alpha\right)$ when $\alpha=1$ becomes the familiar Bernoulli polynomial $B_n(a)$ (see Section 1.6). Under the assumption $\alpha\neq1$, Apostol (1951) gave the main properties of $\mathcal{B}_n\left(a,\alpha\right)$ including, for example, the summation formula:

$$\mathcal{B}_n\left(a,\alpha\right)=\sum_{k=0}^{n}\binom{n}{k}\mathcal{B}_k\left(\alpha\right)a^{n-k}\qquad(n\in\mathbb{N}_0;\ \mathcal{B}_k\left(\alpha\right):=\mathcal{B}_k\left(0,\alpha\right)), \tag{42}$$

from which it is seen that $\mathcal{B}_n(a,\alpha)$ are polynomials in a, satisfying the difference equation:

$$\alpha\,\mathcal{B}_n(a+1,\alpha) - \mathcal{B}_n(a,\alpha) = n\,a^{n-1} \quad (n \in \mathbb{N}), \tag{43}$$

which yields the following special cases:

$$\alpha\,\mathcal{B}_1(1,\alpha) = 1 + \mathcal{B}_1(\alpha) \tag{44}$$

and

$$\alpha\,\mathcal{B}_n(1,\alpha) = \mathcal{B}_n(\alpha) \quad (n \in \mathbb{N}\setminus\{1\}). \tag{45}$$

Equations (44), (45), and (42) with $a = 1$, together, aid us in computing $\mathcal{B}_n(\alpha)$ recursively. Thus we have

$$\begin{aligned} &\mathcal{B}_0(\alpha) = 0, \quad \mathcal{B}_1(\alpha) = \frac{1}{\alpha-1}, \quad \mathcal{B}_2(\alpha) = -\frac{2\alpha}{(\alpha-1)^2}, \quad \mathcal{B}_3(\alpha) = \frac{3\alpha(\alpha+1)}{(\alpha-1)^3}, \\ &\mathcal{B}_4(\alpha) = -\frac{4\alpha\left(\alpha^2+4\alpha+1\right)}{(\alpha-1)^4}, \quad \mathcal{B}_5(\alpha) = \frac{5\alpha\left(\alpha^3+11\alpha^2+11\alpha+1\right)}{(\alpha-1)^5}, \\ &\mathcal{B}_6(\alpha) = -\frac{6\alpha\left(\alpha^4+26\alpha^3+66\alpha^2+26\alpha+1\right)}{(\alpha-1)^6}, \end{aligned} \tag{46}$$

and (in general)

$$\mathcal{B}_n(\alpha) = \frac{n\alpha}{(\alpha-1)^n}\sum_{k=1}^{n-1}(-1)^k\,k!\,\alpha^{k-1}\,(\alpha-1)^{n-1-k}\,S(n-1,k), \tag{48}$$

in terms of the Stirling numbers of the second kind (see Section 1.5).

We recall here several further properties of the functions $\mathcal{B}_n(a,\alpha)$:

$$\frac{\partial^p}{\partial a^p}\mathcal{B}_n(a,\alpha) = \frac{n!}{(n-p)!}\mathcal{B}_{n-p}(a,\alpha) \quad (p = 0,\,1,\,\ldots,\,n); \tag{49}$$

$$\mathcal{B}_n(a+b,\alpha) = \sum_{k=0}^{n}\binom{n}{k}\mathcal{B}_k(a,\alpha)\,b^{n-k}; \tag{50}$$

$$\int_a^b \mathcal{B}_n(t,\alpha)\,dt = \frac{\mathcal{B}_{n+1}(b,\alpha) - \mathcal{B}_{n+1}(a,\alpha)}{n+1}; \tag{51}$$

$$\sum_{k=0}^{m-1} k^n = \frac{\alpha-1}{n+1}\sum_{k=1}^{m}\mathcal{B}_{n+1}(k,\alpha) + \frac{\mathcal{B}_{n+1}(m,\alpha) - \mathcal{B}_{n+1}(\alpha)}{n+1}, \tag{52}$$

which obviously generalizes the familiar result 1.6(17).

Problems

1. Show that

$$\begin{aligned}\zeta'\left(-2k+1,\frac{p}{q}\right)=&\frac{1}{2k}\,[\psi(2k)-\log(2\pi q)]\;B_{2k}\left(\frac{p}{q}\right)-\frac{q^{-2k}}{2k}\,[\psi(2k)-\log(2\pi)]\;B_{2k}\\&+\frac{(-1)^{k+1}\,\pi}{(2\pi q)^{2k}}\sum_{n=1}^{q-1}\sin\left(\frac{2\pi pn}{q}\right)\,\psi^{(2k-1)}\left(\frac{n}{q}\right)\\&+\frac{(-1)^{k+1}\,2\cdot(2k-1)!}{(2\pi q)^{2k}}\sum_{n=1}^{q-1}\cos\left(\frac{2\pi pn}{q}\right)\,\zeta'\left(2k,\frac{n}{q}\right)\\&+\frac{\zeta'(-2k+1)}{q^{2k}}\qquad(p<q;\;p,\,q,\,k\in\mathbb{N}).\end{aligned}$$

[Miller and Adamchik (1998, p. 203, Proposition 1)]

2. Prove that there exist constants D_k defined by

$$\log D_k:=\lim_{n\to\infty}\left(\sum_{m=1}^{n}m^k\,\log m-p(n,k)\right)\qquad(k\in\mathbb{N}_0),$$

where the definition of $p(n,k)$ in Adamchik (1998, p. 198, Eq. (20)) is *corrected* here as follows:

$$\begin{aligned}p(n,k):=&\frac{n^k}{2}\log n+\frac{n^{k+1}}{k+1}\left(\log n-\frac{1}{k+1}\right)\\&+k!\sum_{j=1}^{k}\frac{n^{k-j}\,B_{j+1}}{(j+1)!\,(k-j)!}\left[\log n+(1-\delta_{kj})\sum_{\ell=1}^{j}\frac{1}{k-\ell+1}\right]\end{aligned}$$

and δ_{kj} is the Kronecker symbol defined by 1.5(25).

[Bendersky (1933, pp. 273–275); Adamchik (1998, p. 198)]

3. For the constants D_k $(k\in\mathbb{N}_0)$ defined in Problem 2, show that

$$D_0=(2\pi)^{\frac{1}{2}},\quad D_1=A,\quad D_2=B,\quad\text{and}\quad D_3=C,$$

and

$$\log\,D_k=\frac{B_{k+1}\,H_k}{k+1}-\zeta'(-k)\qquad(k\in\mathbb{N}_0),$$

where B_n are the Bernoulli numbers and H_n are the harmonic numbers, and the mathematical constants A, B, and C are given as in Section 1.3.

[Adamchik (1998, pp. 198-199)]

4. Prove that

$$\int_0^{\pi}\left(\log\left[\left(2\sin\tfrac{1}{2}t\right)\left(2\cos\tfrac{1}{2}t\right)^p\right]\right)^3\,dt=-2\pi\zeta(3)\left[\tfrac{3}{4}\left(1+p^3\right)-\tfrac{3}{8}\left(p+p^2\right)\right].$$

[Lewin (1981, p. 220)]

5. The Barnes double Zeta function $\zeta_2(v;\alpha,w)$ is the meromorphic continuation of the series defined by

$$\zeta_2(v;\alpha,w) := \sum_{m=0}^{\infty}\sum_{n=0}^{\infty}(\alpha+m+nw)^{-v} \quad (\alpha>0;\ w>0;\ v\in\mathbb{C}).$$

The functions $\rho_2(1,w)$ and $\Gamma_2(\alpha,(1,w))$ are defined by

$$\log\rho_2(1,w) := -\lim_{\alpha\to 0}\left[\zeta_2'(0;\alpha,w)+\log\alpha\right]$$

and

$$\Gamma_2(\alpha,(1,w)) := \rho_2(1,w)\exp\left[\zeta_2'(0;\alpha,w)\right],$$

respectively.

(a) Prove that, for any $N\in\mathbb{N}$ and $\Re(v)>-N+1$ $(v\in\mathbb{C}\setminus[\mathbb{Z}_0^-\cup\{1,\ 2\}])$,

$$\begin{aligned}\zeta_2(v;\alpha,w) =& \zeta(v,\alpha)+\frac{\zeta(v-1)}{v-1}\,w^{1-v}\\ &+\sum_{n=0}^{N-1}\binom{-v}{n}\zeta(-n,\alpha)\zeta(v+n)w^{-v-n}+R_N(v;\alpha,w)\end{aligned}$$

with the estimate:

$$R_N(v;\alpha,w)=O\left(w^{-\Re(v)-N}\right),$$

where the O-constant depends on v, N, and α.

(b) Show also that, for $N\in\mathbb{N}\setminus\{1,\ 2\}$,

$$\begin{aligned}\zeta_2'(0;\alpha,w) =& -\frac{1}{12}w\log w+\left[\frac{1}{12}-\zeta'(-1)\right]w\\ &+\frac{1}{2}\left(\frac{1}{2}-\alpha\right)\log w+\log\Gamma(\alpha)+\left(\frac{1}{2}\alpha-\frac{3}{4}\right)\log(2\pi)\\ &+\zeta(-1,\alpha)\,w^{-1}(\log w-\gamma)\\ &+\sum_{n=2}^{N-1}\frac{(-1)^n}{n}\zeta(-n,\alpha)\,\zeta(n)\,w^{-n}+R_N'(0;\alpha,w)\end{aligned}$$

and

$$\begin{aligned}\log\Gamma_2(\alpha,(1,w)) =& \frac{1}{2}\alpha\log\left(2\pi w^{-1}\right)+\log\Gamma(\alpha)\\ &+\left[\zeta(-1,\alpha)-\zeta(-1)\right]w^{-1}(\log w-\gamma)\\ &+\sum_{n=2}^{N-1}\frac{(-1)^n}{n}\left[\zeta(-n,\alpha)-\zeta(n)\right]\zeta(n)\,w^{-n}+S_N(\alpha,(1,w))\end{aligned}$$

with the estimates:

$$R'_N(0;\alpha,w) = O\left[w^{-N}(|\log w|+1)\right]$$

and

$$S_N(\alpha,(1,w)) = O\left(w^{-N}(|\log w|+1)\right],$$

where the O-constants depend on N and α and the prime in $\zeta_2'(v;\alpha,w)$ denotes differentiation with respect to v.

[Matsumoto (1998 and 2001)]

6. For the Bernoulli numbers and polynomials, show that

$$\text{(a)} \qquad B_{2n} = 4n\,(-1)^{n+1}\int_0^\infty \frac{x^{2n-1}}{e^{2\pi x}-1}dx \quad (n\in\mathbb{N});$$

$$\text{(b)} \qquad |B_{2n}(x)| \leq |B_{2n}| \quad (n\in\mathbb{N};\ 0\leq x\leq 1).$$

7. Show that

$$\zeta(s) = \frac{1}{(1-2^{1-s})\,\Gamma(s)}\int_0^\infty \frac{x^{s-1}}{e^x+1}\,dx \quad (\Re(s)>0;\ s\neq 1).$$

[See, *e.g.*, Chen (1993, p. 263)]

8. Prove that

$$\int_m^{m+n+\frac{1}{2}} \log\Gamma(t)\,dt = -\sum_{k=1}^{m-1} k\,\log k + \frac{1}{2}\sum_{k=1}^{m+n-1}(2k+1)\log(2k+1)$$
$$-\frac{1}{2}(m+n)^2(1+\log 2) + \frac{m(m-1)}{2} - \frac{3}{2}\zeta'(-1)$$
$$+\left(-\frac{3}{2}+\frac{1}{2}\log 2\right)\zeta(-1) + \left(n+\frac{1}{2}\right)\log\sqrt{2\pi} \quad (m\in\mathbb{N};\ n\in\mathbb{N}_0).$$

[Elizalde *et al.* (1994, p. 19)]

9. Prove that

$$\int_0^z t^n\,\psi(t)\,dt = (-1)^{n-1}\zeta'(-n) + \frac{(-1)^n}{n+1}B_{n+1}H_n$$
$$-\sum_{k=0}^{n}(-1)^k\binom{n}{k}\frac{z^{n-k}}{k+1}B_{k+1}(z)\,H_k + \sum_{k=0}^{n}(-1)^k\binom{n}{k}z^{n-k}\zeta'(-k,z)$$
$$(n\in\mathbb{N}_0;\ \Re(z)>0).$$

[Adamchik (1998, p. 197)]

10. Prove that

$$\frac{z^a}{\Gamma(1-s)}\Phi(z,s,a) = \left(\log\frac{1}{z}\right)^{s-1} + \sum_{k=0}^{\infty}\binom{s-1}{2k}(\log z)^{2k}\frac{\zeta(s-2k,a)}{\Gamma(2k+1-s)}$$
$$-\sum_{k=0}^{\infty}\binom{s-1}{2k+1}(\log z)^{2k+1}\frac{\zeta(s-2k-1,a)}{\Gamma(2k+2-s)} \quad (|\log z|<2\pi;\ s\notin\mathbb{N};\ a\notin\mathbb{Z}_0^-).$$

Also deduce the relatively more familiar result 2.5(13).

11. Prove that

$$E_{2n-1}(0) = \frac{4(-1)^n}{(2\pi)^{2n}}(2n-1)!\,\left(2^{2n}-1\right)\zeta(2n) \quad (n \in \mathbb{N}),$$

where $E_n(x)$ denotes the Euler polynomials (see Section 1.6).

[*Cf.* Srivastava (1999a, p. 390)]

12. Let $f : \mathbb{N} \times \mathbb{N} \to \mathbb{R}$ be a function. For a fixed $n \in \mathbb{N}$, let

$$T_n := \{(k,j) \in \mathbb{N} \times \mathbb{N} \mid 1 \leq k,\ j \leq n;\ k+j \geq n+1\}$$

and

$$S_n := \sum_{(x,y)\in T_n} f(x,y),$$

the sum being taken over all pairs $(x,y) \in T_n$. Show that

$$S_n = \sum_{k=1}^{n} f(k,k) + \sum_{k=2}^{n}\sum_{j=1}^{k-1}\{f(j,k)+f(k,j)-f(j,k-j)\}.$$

[Sitaramachandrarao and Sivaramasarma (1979, p. 603)]

13. Prove that

$$2\sum_{q=1}^{\infty}\frac{1}{q^n}\left(\sum_{\substack{k=1\\(k,q)=1}}^{q-1}\frac{1}{k}\right) = n+2-\frac{1}{\zeta(n+1)}\sum_{j=1}^{n-2}\zeta(n-j)\,\zeta(j+1) \quad (n \in \mathbb{N}\setminus\{1\}),$$

where an empty sum is understood (as usual) to be nil, and $(k,q) = 1$ denotes that k and q are relatively prime.

[*Cf.* Equation 2.3(54); see also Sitaramachandrarao and Sivaramasarma (1979, p. 602)]

14. Prove that

$$\sum_{k=1}^{\infty}\frac{H_k}{k^{2n+1}} = \frac{1}{2}\sum_{j=2}^{2n}(-1)^j\,\zeta(j)\,\zeta(2n-j+2) \quad (n \in \mathbb{N}),$$

and

$$\sum_{k=1}^{\infty}\frac{H_k}{k^n} = \left(1+\frac{n}{2}\right)\zeta(n+1) - \frac{1}{2}\sum_{j=2}^{n-1}\zeta(j)\,\zeta(n-j+1) \quad (n \in \mathbb{N}\setminus\{1\}),$$

where H_k are the harmonic numbers defined by 3.2(36).

[Georghiou and Philippou (1983, p. 29); see also Equation 2.3(54)]

15. For bounded maps $f,\, g : \mathbb{N} \to \mathbb{C}$, let

$$S_{f,g}(u,v;w) = \sum_{r,k=1}^{\infty} \frac{f(r+k)\,g(k)}{r^u\,k^u\,(r+k)^w} \qquad (u,\, v \in \mathbb{N};\; w \in \mathbb{R}^+)$$

and

$$C_{f,g}(x,y) = \sum_{r=2}^{\infty} \frac{f(r)}{r^x} \sum_{k=1}^{r-1} \frac{g(k)+g(r-k)}{k^y} \qquad (x>1;\; y \in \mathbb{N}).$$

Show that

$$\begin{aligned} S_{f,g}(u,v;w) &+ S_{f,g}(v,u;w) \\ &= \sum_{j=0}^{u-1} \binom{v+j-1}{v-1} C_{f,g}(v+w+j,u-j) \\ &\quad + \sum_{j=0}^{v-1} \binom{u+j-1}{u-1} C_{f,g}(u+w+j,v-j). \end{aligned}$$

[Subbarao and Sitaramachandrarao (1985, p. 246)]

16. Prove that

$$\sum_{r,k=1}^{\infty} \frac{(-1)^{r+k}}{rk\,(r+k)} = \frac{1}{4}\,\zeta(3)$$

and

$$\sum_{r,k=1}^{\infty} \frac{(-1)^{k-1}}{rk\,(r+k)} = \frac{5}{8}\,\zeta(3).$$

[Subbarao and Sitaramachandrarao (1985, p. 247)]

17. Prove that

$$\begin{aligned} \sum_{k=1}^{\infty} \frac{H_k^{(2)}}{k^{2n+1}} &= \zeta(2)\,\zeta(2n+1) - \frac{(n+2)(2n+1)}{2}\zeta(2n+3) \\ &\quad + 2\sum_{j=2}^{n+1} (j-1)\zeta(2j-1)\,\zeta(2n+4-2j) \quad (n \in \mathbb{N}), \end{aligned}$$

where $H_n^{(m)}$ denotes the generalized harmonic number defined by

$$H_0^{(m)} = 0 \quad \text{and} \quad H_n^{(m)} = \sum_{\ell=1}^{n} \ell^{-m} \qquad (m,\, n \in \mathbb{N}).$$

[Georghiou and Philippou (1983, p. 35)]

18. Prove that

$$\sum_{m,\,n=0}^{\infty}{}' \frac{1}{m^6\,(m^2+n^2)} = \frac{13\,\pi^8}{28350} \quad \text{and} \quad \sum_{m,\,n=0}^{\infty}{}' \frac{1}{m^2\,(m^2-mn+n^2)} = \frac{\sqrt{3}\,\pi^4}{30},$$

where the prime denotes that the summations are taken over all pairs of integers except $(0,0)$.

[Smart (1973, p. 10)]

19. Prove that

$$\sum_{k=1}^{\infty} \frac{1}{k} \left(H_{2k-1} - H_k - \log 2\right) = (\log 2)^2 - \frac{\pi^2}{6}.$$

[Knuth (1976, p. 138)]

20. Let a, b, $c \in \mathbb{R}$ with $a > 0$ and $d = b^2 - 4ac < 0$. The Epstein Zeta function is given by

$$Z(s) := \frac{1}{2} \sum{}' \left(a m^2 + b mn + c n^2\right)^{-s} \quad (\Re(s) > 1),$$

where the prime denotes that the summation is to be taken over all pairs (m, n) of integers other than the pair $(0,0)$. Let $k \in \mathbb{R}^+$ be a number satisfying

$$k^2 = \frac{c}{a} - \left(\frac{b}{2a}\right)^2 \quad (a,\ b,\ c \in \mathbb{R};\ a > 0)$$

and suppose that the Bessel function $K_\nu(z)$ is defined by

$$\begin{aligned} K_\nu(z) &= \frac{1}{2} \int_0^\infty e^{-\frac{z}{2}\left(u+\frac{1}{u}\right)}\, u^{\nu-1}\, du = \frac{1}{2} \int_{-\infty}^\infty e^{-z\cosh t}\, e^{\nu t}\, dt \\ &= \int_0^\infty e^{-z\cosh t} \cosh(\nu t)\, dt \quad \left(\nu \in \mathbb{C};\ |\arg z| < \frac{\pi}{2}\right). \end{aligned}$$

Also let

$$\sigma_\nu(n) := \sum_{d|n} d^\nu = \sum_{d|n} \left(\frac{n}{d}\right)^\nu \quad (\nu \in \mathbb{C};\ n \in \mathbb{N}).$$

Show that

$$\begin{aligned} a^s\, Z(s) =& \zeta(2s) + k^{1-2s}\, \zeta(2s-1) \frac{\Gamma\left(s - \frac{1}{2}\right) \Gamma\left(\frac{1}{2}\right)}{\Gamma(s)} \\ &+ \frac{\pi^s}{\Gamma(s)} k^{\frac{1}{2}-s}\, H(s) \quad (\Re(s) > 1), \end{aligned}$$

where

$$H(s) = 4 \sum_{n=1}^{\infty} n^{s-\frac{1}{2}}\, \sigma_{1-2s}(n) \cos\left(\frac{n\pi b}{a}\right) K_{s-\frac{1}{2}}(2\pi k n).$$

Show also that $H(s)$ is an entire function of s such that

$$H(s) = H(1-s).$$

[Bateman and Grosswald (1964, p. 366)]

21. Show that

$$(cz+d)^{2n}\,A\left(\frac{az+b}{cz+d},-2n\right)=A(z,-2n)+\frac{1}{2}\left\{1-(cz+d)^{2n}\right\}\zeta(2n+1)$$
$$+\frac{(-1)^n\,(2\pi)^{2n+1}\,i}{2\,(2n+2)!}\sum_{j=1}^{c}\sum_{k=0}^{2n+2}\binom{2n+2}{k}B_k\left(\frac{j}{c}\right)\bar{B}_{2n-k+2}\left(\frac{jd}{c}\right)\{-(cz+d)\}^{k-1}$$
$$(c,\,n\in\mathbb{N}),$$

where $A(z,-2n)$ denotes a Lambert series in the variable $e^{2\pi iz}$ defined by

$$A(z,-2n):=\sum_{k=1}^{\infty}k^{-2n-1}\frac{e^{2\pi ikz}}{1-e^{2\pi ikz}}$$

and $\bar{B}_n(x):=B_n\,(x-[x])$ ($[x]$ being, as usual, the greatest integer in x).

[*Cf.* Berndt (1973, p. 505); see also Apostol (1950, p. 153)]

22. Show that

$$2^{k-s}\left\{\zeta\left(s,\frac{a}{2}\right)-\zeta\left(s,\frac{a+1}{2}\right)\right\}=\sum_{\ell=1}^{k}(-1)^{\ell-1}\sum_{j=0}^{\ell-1}(-1)^j\binom{k}{j}(\ell-1-j+a)^{-s}$$
$$+\sum_{n=0}^{\infty}(-1)^{n+k}\sum_{j=0}^{k}(-1)^j\binom{k}{j}(n+k-j+a)^{-s}\quad(k\in\mathbb{N};\ \Re(s)>-k),$$

where $\zeta(s,a)$ $(a>0)$ is the Hurwitz (or generalized) Zeta function defined by 2.2(1).

[Balakrishnan (1988, p. 205)]

23. Show that

$$\int_0^{\frac{\pi}{3}}\sum_{k=0}^{m-2}\frac{(-1)^k}{2k+1}\binom{2m-2}{2k}\left(\frac{\theta}{2}\right)^{2k+1}\left\{\log\left(2\,\sin\frac{\theta}{2}\right)\right\}^{2m-2k-2}d\theta$$
$$=\frac{(-1)^m\,\pi^{2m}}{8m(2m-1)}\left[\left(\frac{1}{6}\right)^{2m-1}-2\left(1-\frac{1}{2^{2m-1}}\right)B_{2m}\right]\quad(m\in\mathbb{N}\setminus\{1\})$$

and

$$\int_0^{\frac{\pi}{3}}\sum_{k=0}^{m-1}(-1)^k\binom{2m}{2k}\left(\frac{\theta}{2}\right)^{2k}\left\{\log\left(2\,\sin\frac{\theta}{2}\right)\right\}^{2m-2k}d\theta$$
$$=\frac{(-1)^m\,\pi^{2m+1}}{2^{2m+2}}\left(E_{2m}-\frac{1}{(2m+1)\,3^{2m}}\right)\quad(m\in\mathbb{N}),$$

where B_n and E_n are the Bernoulli and Euler numbers, respectively, given in Section 1.6.

[Zhang and Williams (1995, p. 282)]

24. Show that

$$\sum_{n=1}^{\infty} \frac{1}{n^{m+2}\binom{2n}{n}} = \frac{(-1)^m\, 2^m}{m!} \int_0^{\frac{\pi}{3}} \theta \left\{\log\left(2 \sin\frac{\theta}{2}\right)\right\}^m d\theta \quad (m \in \mathbb{N}_0)$$

and

$$\sum_{n=1}^{\infty} \frac{(-1)^{n-1}}{n^{m+2}\binom{2n}{n}} = \frac{(-1)^m\, 2^m}{m!} \int_0^{2\log\tau} \theta \left\{\log\left(2 \sinh\frac{\theta}{2}\right)\right\}^m d\theta \quad (m \in \mathbb{N}_0).$$

[Zhang and Williams (1995, pp. 286 and 288)]

25. Show that

$$\int_0^{\frac{\pi}{3}} \left[\left\{\log\left(2 \sin\frac{\theta}{2}\right)\right\}^4 - \frac{3\,\theta^2}{2}\left\{\log\left(2 \sin\frac{\theta}{2}\right)\right\}^2\right] d\theta = \frac{253\,\pi^5}{2^3\cdot 3^4\cdot 5}$$

and

$$\int_0^{\frac{\pi}{3}} \left[\theta\left\{\log\left(2 \sin\frac{\theta}{2}\right)\right\}^4 - \frac{\theta^3}{2}\log\left(2 \sin\frac{\theta}{2}\right)\right] d\theta = \frac{313\,\pi^6}{2^4\cdot 3^6\cdot 5\cdot 7}.$$

[Zucker (1985, pp. 98 and 99)]

26. Prove each of the following claims:

(a) For all $a_1, a_2, \ldots$

$$\sum_{k=1}^{\infty} \frac{a_1\, a_2\, \cdots\, a_{k-1}}{(x+a_1)\cdots(x+a_k)} = \frac{1}{x}.$$

(b)

$$\zeta(3) =: \sum_{n=1}^{\infty} \frac{1}{n^3} = \frac{5}{2}\sum_{n=1}^{\infty} \frac{(-1)^{n-1}}{n^3\binom{2n}{n}}.$$

(c) Consider the recursion:

$$n^3\, u_n + (n-1)^3\, u_{n-2} = \left(34\,n^3 - 51\,n^2 + 27\,n - 5\right)\, u_{n-1} \quad (n \in \mathbb{N}\setminus\{1\}).$$

Let $\{b_n\}$ be the sequence defined by $b_0 = 1$, $b_1 = 5$, and $b_n = u_n$ for all $n \in \mathbb{N}\setminus\{1\}$; then the b_n all are integers. Let $\{a_n\}$ be the sequence defined by

$a_0 = 0$, $a_1 = 6$, and $a_n = u_n$ for all $n \in \mathbb{N}\setminus\{1\}$; then the a_n are rational numbers with denominator dividing $2[1, 2, \ldots, n]^3$ (here $[1, 2, \ldots, n]$ is the lowest common multiple) of $1, \ldots, n$).

(d) $a_n/b_n \to \zeta(3)$ as $n \to \infty$; indeed the convergence is so fast as to prove that $\zeta(3)$ cannot be rational. To be precise, for all integers p, q with q sufficiently large relative to $\epsilon > 0$,

$$\left|\zeta(3) - \frac{p}{q}\right| > \frac{1}{q^{\theta+\epsilon}} \quad \text{and} \quad \theta = 13.41782\ldots,$$

which proves the irrationality of $\zeta(3)$.

[*Cf.* van der Poorten (1979, p. 195); see also Equation 4.2(2) and Apéry (1979)]

27. Prove the following inequalities for $n,\ p \in \mathbb{N}$:

$$0 \leq \left(p+\frac{1}{2}\right)\sum_{k=1}^{n}\frac{1}{k^{2p}} - \sum_{j=1}^{p-1}\left(\sum_{k=1}^{n}\frac{1}{k^{2p-2j}}\right)\left(\sum_{k=1}^{n}\frac{1}{k^{2j}}\right) < \frac{8}{n}$$

and

$$\left|\sum_{k=1}^{n}\frac{1}{(2k-1)^2} - 2\left(\sum_{k=1}^{n}\frac{(-1)^{k+1}}{2k-1}\right)^2\right| < \frac{6}{n}.$$

[Hovstad (1972, p. 93)]

28. Show that the first inequality of Problem 27 implies that

$$\sum_{k=1}^{\infty}\frac{1}{k^{2p}} = \mathcal{R}_p\left(\sum_{k=1}^{\infty}\frac{1}{k^2}\right)^p \qquad (p \in \mathbb{N}),$$

where $\mathcal{R}_p$ is a rational number satisfying the recurrence formula:

$$\mathcal{R}_1 = 1 \quad \text{and} \quad \left(p+\frac{1}{2}\right)\mathcal{R}_p = \sum_{j=1}^{p-1}\mathcal{R}_j\,\mathcal{R}_{p-j} \quad (p \in \mathbb{N}\setminus\{1\}).$$

[Hovstad (1972, p. 93)]

29. Prove that $\zeta(s)$ has the factorization:

$$\zeta(s) = \frac{e^{bs}}{2\,(s-1)\,\Gamma\left(1+\frac{1}{2}s\right)}\prod_{\rho}\left[\left(1-\frac{s}{\rho}\right)e^{\frac{s}{\rho}}\right],$$

where

$$b = -1 - \frac{\gamma}{2} + \log(2\pi);$$

γ is the Euler-Mascheroni constant, and the product is taken over all the so-called non-trivial zeros of $\zeta(s)$.

[*Cf.* Titchmarsh (1951, pp. 30–31); see also Melzak (1973, p. 111)]

30. Prove that

$$\zeta\left(\frac{1}{2}, a+1\right) = \sum_{n=0}^{\infty}\frac{(2n)!}{2^{2n}\,(n!)^2}\,\zeta\left(n+\frac{1}{2}\right)(-a)^n \quad (|a|<1)$$

and

$$\zeta\left(\frac{1}{2}, na\right) = n^{-\frac{1}{2}}\sum_{k=0}^{n-1}\zeta\left(\frac{1}{2}, a+\frac{k}{n}\right) \quad (n \in \mathbb{N}).$$

[Powell (1952, p. 117)]

31. Let

$$(s-1)\,\zeta(s,a) = 1 + \sum_{n=0}^{\infty} \gamma_n(a)\,(s-1)^{n+1} \quad (0 < a \leq 1;\ n \in \mathbb{N}_0)\,.$$

Show that

$$\gamma_n(a) = \frac{(-1)^n}{n!} \lim_{m \to \infty} \left(\sum_{k=0}^{m} \frac{\{\log(k+a)\}^n}{k+a} - \frac{\{\log(m+a)\}^{n+1}}{n+1} \right),$$

which, for $a = 1$, gives the coefficients in the Laurent expansion of $\zeta(s)$ about $s = 1$ [*cf.* Equation 2.3(7)].

[Berndt (1972b, p. 152)]

32. Prove that

$$\limsup_{t \to \infty} \frac{|\zeta(1+it)|}{\log(\log t)} \geq e^{\gamma},$$

where γ is the Euler-Mascheroni constant defined by 1.1(3).

[Titchmarsh (1927, p. 79)]

33. Prove that

$$\mathrm{Li}_m(e^z) = \sum_{\substack{n=0 \\ n \neq m-1}}^{\infty} \zeta(m-n)\,\frac{z^n}{n!}$$

$$+ \left(1 + \frac{1}{2} + \cdots + \frac{1}{m-1} - \log(-z)\right) \frac{z^{m-1}}{(m-1)!} \quad (m \in \mathbb{N};\ |z| < 2\pi).$$

[Cohen *et al.* (1992, p. 26)]

34. Prove that

$$\zeta(5) = \frac{5}{2} \sum_{n=1}^{\infty} \left(\sum_{k=1}^{n-1} \frac{1}{k^2} - \frac{4}{5}\,\frac{1}{n^2} \right) \frac{(-1)^n}{n^3 \binom{2n}{n}}.$$

[van der Poorten (1979, p. 274)]

35. Let K be a normal extension field of degree n over the rational number field $\mathbb{Q}$. Denote by O_K the integer ring of K. Let $I(K)$ be the set of all non-zero ideals of O_K, $N_{\mathfrak{a}}$ the absolute norm of an ideal $\mathfrak{a} \in I(K)$, and $\mathrm{Tr}\,\alpha$ the trace of $\alpha \in K$ over $\mathbb{Q}$. We assume that $[K : \mathbb{Q}] = n > 1$, let

$$O_{K,0} = \{\alpha \in O_K \mid \mathrm{Tr}\,\alpha = 0\};$$
$$TK = \min\{\mathrm{Tr}\,\alpha \mid \alpha \in O_K,\ \mathrm{Tr}\,\alpha > 0\}.$$

For $\mathfrak{a} \in I(K)$ and $\mathfrak{a}_0 = O_{K,0} \cap \mathfrak{a}$, define

$$T(\mathfrak{a}) = \min\left\{ \frac{\mathrm{Tr}\alpha}{T(K)} \mid \alpha \in \mathfrak{a},\ \mathrm{Tr}\alpha > 0 \right\};$$
$$N_0(\mathfrak{a}) = \#\{O_{K,0}/\mathfrak{a}_0\},$$

where $O_{K,0}/\mathfrak{a}_0$ is the quotient of $\mathbb{Z}$-module $O_{K,0}$ by submodule $\mathfrak{a}_0$. For any $x \in \mathbb{R}^+$, let

$$\begin{aligned} j_K(x) &= \#\{\mathfrak{a} \in I(K) \mid N\mathfrak{a} \le x\}; \\ j_{K,0}(x) &= \#\{\mathfrak{a} \in I(K) \mid N_0(\mathfrak{a}) \le x\} \\ &= \#\left\{\mathfrak{a} \in I(K) \mid \frac{N\mathfrak{a}}{T(\mathfrak{a})} \le x\right\}. \end{aligned}$$

Assume that k is a positive integer and $n = 2k+1$. Let q be a prime number and $q \equiv 1 (\mathrm{mod}\ n)$. Suppose that K is the subfield of the qth cyclotomic field C_q such that $K/\mathbb{Q}$ is a tamely ramified cyclic extension of degree n. Show that

$$\zeta(2k+1) = \frac{2\,q^n - q^{n-1} - q}{q^n - 1}\,\zeta(2k) \lim_{x \to +\infty} \frac{j_k(x)}{j_{K,0}(x)}.$$

[Lan (1999, p. 273)]

36. Prove that

$$\sum_{n=1}^{\infty} \frac{(H_n)^3}{n^5} - \frac{11}{4} \sum_{n=1}^{\infty} \frac{H_n^{(2)}}{n^6} = \frac{469}{32}\zeta(8) - 16\,\zeta(3)\,\zeta(5) + \frac{3}{2}\,\zeta(2)\,\{\zeta(3)\}^2$$

and

$$\begin{aligned} \sum_{n=1}^{\infty} \frac{(H_n)^3}{n^7} - \frac{13}{4} \sum_{n=1}^{\infty} \frac{H_n^{(2)}}{n^8} &= \frac{561}{20}\zeta(10) - \frac{47}{4}\,\{\zeta(5)\}^2 \\ &\quad - \frac{49}{2}\,\zeta(7)\,\zeta(3) + 3\,\zeta(2)\,\zeta(3)\,\zeta(5) + \frac{15}{4}\,\{\zeta(3)\}^2\,\zeta(4). \end{aligned}$$

[Flajolet and Salvy (1998, p. 27)]

37. Prove that, for an odd weight $m = p + q$,

$$\begin{aligned} \sum_{n=1}^{\infty} \frac{H^{(p)}(n)}{n^q} =& \zeta(m)\left(\frac{1}{2} - \frac{(-1)^p}{2}\binom{m-1}{p} - \frac{(-1)^p}{2}\binom{m-1}{q}\right) \\ &+ \frac{1-(-1)^p}{2}\,\zeta(p)\,\zeta(q) + (-1)^p \sum_{k=1}^{[p/2]} \binom{m-2k-1}{q-1}\,\zeta(2k)\,\zeta(m-2k) \\ &+ (-1)^p \sum_{k=1}^{[q/2]} \binom{m-2k-1}{p-1}\,\zeta(2k)\,\zeta(m-2k), \end{aligned}$$

where $\zeta(1)$ should be interpreted as 0 wherever it occurs.

[*Cf.* Borwein *et al.* (1995, p. 278); see also Flajolet and Salvy (1998, p. 22)]

38. Prove that

$$\sum_{n=1}^{\infty} \sum_{k=1}^{2n-1} \frac{(-1)^{n+k}}{n^2\,k} = \pi\,G - \frac{27}{16}\,\zeta(3)$$

and

$$\sum_{n=1}^{\infty} \sum_{k=1}^{2n-1} \frac{(-1)^{n-1}}{n^2\, k} = \pi\, G - \frac{29}{16}\, \zeta(3),$$

where G denotes the Catalan constant defined by 1.3(16).

[Sitaramachandrarao (1987, p. 13)]

39. Prove that, for $m \in \mathbb{N}$ and $\Re(a) > 0$,

$$\begin{aligned}\zeta'(-m,a) &= \frac{\partial}{\partial z}\, \zeta(z,a)\bigg|_{z=-m} \\ &= \frac{1}{m+1}\, a^{m+1} \log a - \frac{1}{(m+1)^2}\, a^{m+1} - \frac{1}{2}\, a^m \log a \\ &\quad + \frac{m}{12}\, a^{m-1} \log a + \frac{1}{12}\, a^{m-1} + \sum_{k=1}^{\infty} \alpha_{2k}\, a^{-(2k-m+1)},\end{aligned}$$

where

$$\alpha_{2k} := \begin{cases} \dfrac{B_{2k+2}}{2k+2} \left[\dbinom{m}{2k+1} \log a + \displaystyle\sum_{j=0}^{2k} \dbinom{m}{j} \frac{(-1)^j}{2k-j+1} \right] & (2k \leq m-1), \\ \dfrac{B_{2k+2}}{2k+2} \displaystyle\sum_{j-0}^{m} \dbinom{m}{j} \frac{(-1)^j}{2k-j+1} & (2k \geq m). \end{cases}$$

[Elizalde (1986, p. 349)]

40. Define constants Φ_k, for $k > 1$ an odd integer, by

$$\Phi_k = -\frac{2}{\pi} \sum_{\substack{d=1 \\ d\,\text{odd}}}^{k-2} (-1)^{\frac{d-1}{2}}\, \frac{\pi^d}{d!}\, \zeta(k-d+1).$$

Show that, for real $r \geq 1$ and $s > 1$,

$$\begin{aligned}\zeta(r,s) &= -\frac{1}{2}\, \zeta(r+s) \\ &\quad + \sum_{\substack{k=3 \\ k\,\text{odd}}}^{\infty} \Phi_k \sum_{\substack{j=0 \\ j\,\text{even}}}^{k-1} \binom{k}{j}\, \eta(r-j)\, \eta(s-k+j),\end{aligned}$$

where the Eta function is defined by

$$\eta(s) := \left(1 - 2^{1-s}\right)\, \zeta(s).$$

[Crandall and Buhler (1994, p. 279)]

41. Let $f(s)$ be a function defined by the Dirichlet series as follows:

$$f(s) := \sum_{n=2}^{\infty} n^{-s} \sum_{k<n} k^{-1} \quad (\Re(s) = \sigma > 1).$$

Show that $f(s)$ is analytic in the whole complex s-plane except at simple poles $s=0$ and $s=1-2a$ $(a \in \mathbb{N})$ with residues given by

$$\operatorname*{Res}_{s=0} f(s) = -\frac{1}{2}$$

and

$$\operatorname*{Res}_{s=1-2a} f(s) = -\frac{B_{2a}}{2a} \quad (a \in \mathbb{N}),$$

and a pole of order 2 at $s=1$ with residue given by

$$\operatorname*{Res}_{s=1} f(s) = \gamma,$$

where B_{2a} are the Bernoulli numbers defined by 1.6(2) and γ is the Euler-Mascheroni constant defined by 1.1(3).

[Matsuoka (1982, p. 399)]

42. By using the definitions and restrictions as in Problem 20 of Chapter 2 with an *additional* condition $c > 0$, show that

$$Z(s) = \zeta(2s)\, a^{-s} + \frac{2^{2s-1}\, a^{s-1}\, \sqrt{\pi}}{\Gamma(s)\, (-d)^{s-\frac{1}{2}}} \zeta(2s-1)\, \Gamma\left(s - \frac{1}{2}\right) + Q(s),$$

where

$$Q(s) = \frac{2\pi^s \cdot 2^{s-\frac{1}{2}}}{\sqrt{a}\,\Gamma(s) \cdot (-d)^{\frac{s}{2}-\frac{1}{4}}} \sum_{n=1}^{\infty} n^{s-\frac{1}{2}}\, \sigma_{1-2s}(n) \cos\left(\frac{n\pi b}{a}\right) \cdot \int_0^{\infty} u^{s-\frac{3}{2}} \exp\left(-\frac{\pi n\sqrt{-d}}{2a}\left(u + u^{-1}\right)\right) du.$$

[Selberg and Chowla (1967, p. 87)]

43. Let $r,\ s \in \mathbb{N}_0$ with $r > s$. Show that

$$\int_0^1 \int_0^1 \frac{x^r\, y^s}{1 - xy}\, dx dy$$

is a rational number whose denominator is a divisor of d_r^2, where d_r denotes the lowest common multiple of $1,\ 2,\ \ldots,\ r$.

[Beukers (1979, p. 268)]

44. Prove that

$$\zeta(3) = \frac{1}{4} \sum_{n=1}^{\infty} (-1)^{n-1} \frac{56\, n^2 - 32\, n + 5}{(2n-1)^2} \frac{1}{\binom{3n}{n}\binom{2n}{n} n^3}$$

and

$$\zeta(3) = \sum_{n=0}^{\infty} \frac{(-1)^n}{72\binom{4n}{n}\binom{3n}{n}} \frac{5265\, n^4 + 13878\, n^3 + 13761\, n^2 + 6120\, n + +1040}{(4n+1)(4n+3)(n+1)(3n+1)^2\,(3n+2)^2}.$$

[Amdeberhan (1996, p. 2)]

45. Prove that

$$\sum_{n=1}^{\infty}\frac{(-1)^n}{n^2}\sum_{k=1}^{n}\frac{1}{k^2}=-\frac{1}{6}\left[24\,\mathrm{Li}_4\left(\frac{1}{2}\right)+21\,\zeta(3)\,\log 2+(\log 2)^4\right]$$
$$+\frac{\pi^2}{6}(\log 2)^2+\frac{17}{480}\pi^4,$$

where $\mathrm{Li}_4(z)$ denotes the Tetralogarithm (see Section 2.4).

[Daudé *et al.* (1997, p. 421)]

46. Prove that

$$\sum_{n=1}^{\infty}(-1)^n\frac{(\log(2n+1))^2}{2n+1}=\frac{\pi}{4}\left\{2\left[\zeta''\left(0,\tfrac{1}{4}\right)-\zeta''\left(0,\tfrac{3}{4}\right)\right]\right.$$
$$\left.-4[\gamma+\log(2\pi)]\log\left(\frac{\Gamma\left(\frac{1}{4}\right)}{\Gamma\left(\frac{3}{4}\right)}\right)+[\gamma+\log(2\pi)]^2+\frac{\pi^2}{12}\right\}$$

and

$$\int_0^{\infty}\frac{(\log t)^2}{\cosh t}dt=\frac{\pi}{2}\left\{2\left[\zeta''\left(0,\tfrac{1}{4}\right)-\zeta''\left(0,\tfrac{3}{4}\right)\right]\right.$$
$$\left.-4\,\log(2\pi)\,\log\left(\frac{\Gamma\left(\frac{1}{4}\right)}{\Gamma\left(\frac{3}{4}\right)}\right)+[\log(2\pi)]^2+\frac{1}{4}\pi^2\right\}.$$

[Shail (2000)]

47. For an integer $a\geq -1$, prove the following asymptotic expansion:

$$G_a(x):=\sum_{n\geq a+1}^{\infty}\zeta(n-a)\frac{(-x)^n}{n!}$$
$$=-\frac{(-x)^{a+1}}{(a+1)!}\{\log x-\psi(a+2)+\gamma\}$$
$$+\sum_{n=0}^{a}(-x)^n\left(\sum_{k=0}^{n}\frac{(a-k)!}{(a+1)!(n-k)!}-\frac{1}{n!(a+1-n)}\right)+O\left(\frac{1}{x}\right)$$
$$(x\to\infty).$$

[Buschman and Srivastava (1993, p. 296)]

CHAPTER 3

SERIES INVOLVING ZETA FUNCTIONS

The main purpose of this chapter is to present a rather extensive collection of closed-form sums of series involving the Zeta functions. Many of these summation formulas will find their applications in Chapter 5 (especially Section 5.3) in the evaluations of the determinants of the Laplacians for the n-dimensional sphere $\mathbf{S}^n$ with the standard metric. We begin this chapter by presenting an interesting historical introduction to the remarkably widely investigated subject of closed-form summation of series involving the Zeta functions. Among the various methods and techniques used in the vast literature on the subject, we give here reasonably detailed accounts of those using the binomial theorem, generating functions, multiple Gamma functions, and hypergeometric identities. The last section (Section 3.6) deals with other methods based (for example) on the Weierstrass canonical product form for the Gamma function and higher-order derivatives of the Gamma function.

3.1 Historical Introduction

A rather classical (over two centuries old) theorem of Christian Goldbach (1690–1764), which was stated in a letter dated 1729 from Goldbach to Daniel Bernoulli (1700–1782), was revived in 1986 by Shallit and Zikan (1986) as the following problem:

$$\sum_{\omega \in \mathcal{S}} (\omega - 1)^{-1} = 1, \tag{1}$$

where $\mathcal{S}$ denotes the set of all nontrivial integer kth powers, that is,

$$\mathcal{S} := \left\{ n^k \mid n, k \in \mathbb{N} \setminus \{1\} \right\}. \tag{2}$$

In terms of the Riemann Zeta function $\zeta(s)$ defined by 2.3(1), Goldbach's theorem (1) assumes the elegant form [*cf.* Shallit and Zikan (1986, p. 403)]:

$$\sum_{k=2}^{\infty} \{\zeta(k) - 1\} = 1 \tag{3}$$

or, equivalently,

$$\sum_{k=2}^{\infty} \mathcal{F}(\zeta(k)) = 1, \tag{4}$$

where $\mathcal{F}(x) := x - [x]$ denotes the *fractional* part of $x \in \mathbb{R}$. As a matter of fact, it is fairly straightforward to observe also that

$$\sum_{k=2}^{\infty} (-1)^k \mathcal{F}(\zeta(k)) = \frac{1}{2}, \tag{5}$$

(6) $$\sum_{k=1}^{\infty} \mathcal{F}(\zeta(2k)) = \frac{3}{4}, \quad \text{and} \quad \sum_{k=1}^{\infty} \mathcal{F}(\zeta(2k+1)) = \frac{1}{4}.$$

Another remarkable result involving the Riemann's ζ-function is the following series representation for $\zeta(3)$:

(7) $$\zeta(3) = \frac{\pi^2}{7}\left\{1 - 4\sum_{k=1}^{\infty} \frac{\zeta(2k)}{(2k+1)(2k+2)\,2^{2k}}\right\}$$

or, equivalently,

(8) $$\zeta(3) = -\frac{4\pi^2}{7}\sum_{k=0}^{\infty} \frac{\zeta(2k)}{(2k+1)(2k+2)\,2^{2k}},$$

since $\zeta(0) = -\frac{1}{2}$ [see Equation 2.3(10)]. The series representation (7) is contained in a 1772 paper by Leonhard Euler (1707–1783) [see, *e.g.*, Ayoub (1974, pp. 1084–1085)]. It was rediscovered by Ramaswami (1934) and (more recently) by Ewell (1990). [See also Srivastava (1988a, p. 7, Equation (2.23)) where Euler's result (7) was reproduced *actually* from the work of Ramaswami (1934)]. Numerous further series representations for $\zeta(3)$, which are analogous to (7) or (8), can be found in the works of Wilton (1922–1923), Zhang and Williams (1993a and 1993b), Cvijović and Klinowski (1997), Chen and Srivastava (1998), Srivastava (1997, 1998, 1999a, and 2000a), Srivastava and Tsumura (2000a,b), and others [*cf.*, *e.g.*, Tsumura (1994, p. 384) and Ewell (1995, p. 1004); see also Berndt (1989)].

A considerably large variety of methods were used in the aforementioned works, and also in the works of (among others) Jensen (1887), Dinghas (1974), Srivastava (1986, 1987, 1988a, and 1988b), Klusch (1992), and Choi *et al.* (1995), Choi and Srivastava (1997, 1999a, 2000a, 2000b, and 2000d), Choi and Seo (1999), and Kanemitsu *et al.* (2000), dealing with summation of series involving the Riemann ζ-function and its such extensions as the (Hurwitz's) generalized Zeta function $\zeta(s,a)$ defined by 2.2(1).

3.2 Use of the Binomial Theorem

Landau's formula [*cf.* Landau (1953, p. 274, Eq.(3)); see also Titchmarsh (1986, p. 33, Eq. (2.14.1))]:

(1) $$\zeta(s) = 1 + \frac{1}{s-1} - \sum_{k=1}^{\infty} \frac{(s)_k}{(k+1)!}\{\zeta(s+k) - 1\}$$

is capable of providing the analytic continuation of $\zeta(s)$ over the whole complex s-plane; here $(s)_k$ denotes the Pochhammer symbol defined by 1.1(5).

Another formula, which can also be used in a similar way, is attributed to Ramaswami (1934, p. 166) [see also Titchmarsh (1986, p. 33, Eq. (2.14.2))]:

(2) $$(1 - 2^{1-s})\zeta(s) = \sum_{k=1}^{\infty} \frac{(s)_k}{k!}\frac{\zeta(s+k)}{2^{s+k}}.$$

Motivated by these well-known results (1) and (2) in the theory of the Riemann Zeta function $\zeta(s)$, Singh and Verma (1983) derived the following infinite series involving $\zeta(s)$:

$$(3)\qquad \zeta(s)=\frac{1}{2}+\frac{1}{s-1}+\frac{1}{2}\sum_{k=1}^{\infty}(-1)^{k-1}\frac{k\cdot(s)_{k+1}}{(k+2)!}\zeta(s+k+1)\quad(\Re(s)<1),$$

and

$$(4)\qquad \zeta(s)=1+\frac{1}{2^{s+1}}\frac{s+3}{s-1}+\frac{1}{2}\sum_{k=1}^{\infty}(-1)^{k-1}\frac{k\cdot(s)_{k+1}}{(k+2)!}\{\zeta(s+k+1)-1\}.$$

The proofs of (3) and (4) by Singh and Verma (1983, Sections 2 and 3) depend rather heavily upon the integral representation [see Titchmarsh (1986, p. 14, Eq. (2.1.4)]:

$$(5)\qquad \zeta(s)=\frac{1}{2}+\frac{1}{s-1}-s\int_{1}^{\infty}\left(x-[x]-\frac{1}{2}\right)\frac{dx}{x^{s+1}}\quad(\Re(s)>-1).$$

Srivastava (1987) gave relatively simple proofs of (3) and (4) *without* using the integral representation (5). He was thus led naturally to an interesting unification (and generalization) of (3) and (4) involving the Hurwitz (or generalized) Zeta function $\zeta(s,a)$ (see Section 2.2). The elementary techniques employed in Srivastava (1987) are shown to apply also to the derivation of numerous other results for $\zeta(s,a)$ including, for example, some useful analogues of (1) and (2).

The derivation of Srivastava's unification (and generalization) of (3) and (4) is based simply upon the familiar binomial expansion [see Equation 1.4(22)]:

$$(6)\qquad \sum_{k=0}^{\infty}\frac{(\lambda)_k}{k!}z^k=(1-z)^{-\lambda}\quad(|z|<1).$$

Indeed it follows readily from (6) and the definition 2.2(1) of the generalized Zeta function $\zeta(s,a)$ that [*cf.*, *e.g.*, Wilton (1922–1923, p. 90, Eq. (1))]

$$(7)\qquad \sum_{k=0}^{\infty}\frac{(\lambda)_k}{k!}\zeta(\lambda+k,a)\,t^k=\zeta(\lambda,a-t)\quad(|t|<|a|),$$

which holds true, by the principle of analytic continuation of $\zeta(s,a)$, for all values of $\lambda\neq 1$. Observe that (7) is the special case of 2.5(33) when $x\in\mathbb{Z}$.

Now replace the summation index in (7) by $k+2$, set $\lambda=s-1$, and divide both sides of the resulting equation by t^2. We thus observe from (7) that

$$(8)\qquad \sum_{k=0}^{\infty}\frac{(s-1)_{k+2}}{(k+2)!}\zeta(s+k+1,a)t^k=\{\zeta(s-1,a-t)-\zeta(s-1,a)\}t^{-2}-(s-1)\zeta(s,a)t^{-1}\quad(0<|t|<|a|).$$

Differentiating both sides of (8) with respect to t, and noticing from the definition 1.1(5) and 2.2(18) that

(9) $(s-1)_{k+2} = (s-1)(s)_{k+1}$ and $\dfrac{\partial}{\partial t}\{\zeta(s-1, a-t)\} = (s-1)\zeta(s, a-t),$

we have

$$\begin{aligned}\sum_{k=1}^{\infty} \frac{k \cdot (s)_{k+1}}{(k+2)!}\zeta(s+k+1, a)t^{k-1} &= \{\zeta(s, a-t) + \zeta(s, a)\}t^{-2} \\ &- \frac{2}{s-1}\{\zeta(s-1, a-t) - \zeta(s-1, a)\}t^{-3} \quad (0 < |t| < |a|),\end{aligned} \tag{10}$$

which can also be obtained as the special case of 2.5(34) when $x \in \mathbb{Z}$.

For $t = -1$, (10) readily yields the desired unification (and generalization) of (3) and (4) in the form [see Srivastava (1987, p. 49, Eq. (2.6))]:

$$\zeta(s, a) = a^{-s}\left(\frac{1}{2} + \frac{a}{s-1}\right) + \frac{1}{2}\sum_{k=1}^{\infty}(-1)^{k-1}\frac{k \cdot (s)_{k+1}}{(k+2)!}\zeta(s+k+1, a), \tag{11}$$

provided that the series converges.

In the special case of (11) when $a = 1$, the series converges if $\Re(s) < 1$, and we immediately obtain (3). Furthermore, in view of 2.3(2) and the special case of 2.3(9) when $n - 1$, the formula (11) with $a = 2$ is precisely the same as the known result (4).

It may be of interest to remark here that alternative proofs of the well-known results (1) and (2), based upon the integral representation 2.3(30), were given by Menon (1961). As a matter of fact, both (1) and (2) can also be deduced fairly easily from (7).

Replacing the summation index n in (7) by $n+1$, and setting $\lambda = s-1$, we have

$$\begin{aligned}&\sum_{k=0}^{\infty} \frac{(s)_k}{(k+1)!}\zeta(s+k, a)t^{k+1} \\ &\quad = \frac{1}{s-1}\{\zeta(s-1, a-t) - \zeta(s-1, a)\} \quad (|t| < |a|),\end{aligned} \tag{12}$$

where we have made use of the first identity in (9).

By virtue of the definition 2.2(1), (12) with $t = 1$ assumes the form:

$$\zeta(s, a) = \frac{(a-1)^{1-s}}{s-1} - \sum_{k=1}^{\infty} \frac{(s)_k}{(k+1)!}\zeta(s+k, a), \tag{13}$$

which, in view of 2.3(9) when $n = 1$, yields Landau's formula (1) for $a = 2$. On the other hand, (12) with $t = \frac{1}{2}$ (and s replaced by $s+1$) or (7) with $t = \frac{1}{2}$ (and $\lambda = s$) similarly yields

$$\zeta(s, 2a-1) - 2^{1-s}\zeta(s, a) = \sum_{k=1}^{\infty} \frac{(s)_k}{k!}\frac{\zeta(s+k, a)}{2^{s+k}}, \tag{14}$$

which leads us immediately to Ramaswami's formula (2) upon setting $a = 1$.

For $t = -1$, (12) yields

$$\zeta(s,a) = \frac{a^{1-s}}{s-1} + \sum_{k=1}^{\infty}(-1)^{k-1}\frac{(s)_k}{(k+1)!}\zeta(s+k,a), \tag{15}$$

and (12) with $t = -\frac{1}{2}$ (and s replaced by $s+1$) or (7) with $t = -\frac{1}{2}$ (and $\lambda = s$) gives

$$\zeta(s,2a) - 2^{1-s}\zeta(s,a) = -\sum_{k=1}^{\infty}(-1)^{k-1}\frac{(s)_k}{k!}\frac{\zeta(s+k,a)}{2^{s+k}}. \tag{16}$$

Formulas (15) with $a = 2$ and (16) with $a = 1$ provide interesting analogues of (1) and (2), respectively; in fact, this indicated analogue of (2) [that is, (16) with $a = 1$] was also given by Ramaswami (1934, p. 166). It is not difficult to deduce (11) as a natural consequence of (15).

Numerous other consequences of the general results (7), (11), and (12) can be deduced by suitably specializing the parameter t in a manner detailed above.

In terms of the binomial coefficient 1.1(40), (6) and (7) with $a = 2$ are written in the following equivalent forms:

$$\sum_{k=0}^{\infty}\binom{\lambda+k-1}{k}t^k = (1-t)^{-\lambda} \quad (|t| < 1) \tag{17}$$

and

$$\sum_{k=0}^{\infty}\binom{\lambda+k-1}{k}\{\zeta(\lambda+k)-1\}t^k = \zeta(\lambda,2-t) \quad (|t| < 2), \tag{18}$$

respectively. This last identity (18) is, in fact, equivalent to [*cf.* Ramanujan (1916–1917, p. 78, Eq. (15)); Apostol (1954, p. 240, Eq. (7))]

$$\sum_{k=0}^{\infty}\binom{\lambda+k-1}{k}\zeta(\lambda+k)t^k = \zeta(\lambda,1-t) \quad (|t| < 1). \tag{19}$$

For fixed $\lambda \neq 1$, the series in (18) and (19) converge absolutely for $|t| < 2$ and $|t| < 1$, respectively. Thus, by the principle of analytic continuation, the formulas (18) and (19) are valid for all values of $\lambda \neq 1$.

Formula (18) provides a unification (and generalization) of 3.1(4) and 3.1(5), and indeed also of a fairly large number of other summation formulas scattered in the literature. For example, in view of the relationships 2.3(1) and 2.2(4), (18) with $t = 1$ gives us [*cf.* Hansen (1975, p. 356, Eq. (54.4.1))]

$$\sum_{k=1}^{\infty}\binom{\lambda+k-1}{k}\{\zeta(\lambda+k)-1\} = 1 \tag{20}$$

which generalizes 3.1(4), and a special case of (18) when $t=-1$ yields [*cf.* Hansen (1975, p. 356, Eq. (54.4.2))]

$$\sum_{k=1}^{\infty}(-1)^{k-1}\binom{\lambda+k-1}{k}\{\zeta(\lambda+k)-1\}=2^{-\lambda} \tag{21}$$

which generalizes 3.1(5).

Several additional consequences of the general summation formulas (18) and (19) are worthy of note. First of all, replace the summation index k in (18) by $k+1$, and set $\lambda=s-1$, so that

$$\sum_{k=0}^{\infty}\binom{s+k-1}{k+1}\{\zeta(s+k)-1\}t^{k+1}=\zeta(s-1,2-t)-\zeta(s-1)+1 \quad (|t|<2) \tag{22}$$

which, for $t=1$, reduces immediately to the following alternative form of (20)

$$\sum_{k=0}^{\infty}\binom{s+k-1}{k+1}\{\zeta(s+k)-1\}=1. \tag{23}$$

Now it follows from the definition 1.1(40) of the binomial coefficient that

$$\binom{s+k-1}{k+1}=\frac{(s-1)(s)_k}{(k+1)!} \quad (k\in\mathbb{N}_0). \tag{24}$$

Thus the formula (23) can easily be rewritten as Landau's formula (1).

For $t=-1$, (22) readily yields [see Srivastava (1988b, p. 131, Eq. (2.4))]:

$$\zeta(s)=1+\frac{1}{2^{s-1}}\frac{1}{s-1}+\sum_{k=1}^{\infty}(-1)^{k-1}\frac{(s)_k}{(k+1)!}\{\zeta(s+k)-1\}, \tag{25}$$

which provides an interesting companion of Landau's formula (1).

Setting $t=\frac{1}{2}$ in (22), and making use of 2.2(4) with $a=\frac{1}{2}$ and $n=1$, we obtain another series representation for $\zeta(s)$ [see Srivastava (1988a, p. 5, Eq. (2.13))]:

$$\zeta(s)=\frac{2^s-1}{2^s-2}+\frac{1}{2^s-2}\sum_{k=1}^{\infty}\frac{(s)_k}{k!2^k}\{\zeta(s+k)-1\}. \tag{26}$$

In their special cases when $s=2$, (1) and (25) reduce simply to the summation formulas (4) and (5), respectively, while (26) similarly yields the elegant sum:

$$\sum_{k=2}^{\infty}\frac{k-1}{2^k}\{\zeta(k)-1\}=\frac{\pi^2}{8}-1. \tag{27}$$

Next we turn to the summation formula (19) which (for $\lambda=s-1$ and with k replaced by $k+1$) assumes the form:

$$\sum_{k=0}^{\infty}\binom{s+k-1}{k+1}\zeta(s+k)t^{k+1}=\zeta(s-1,1-t)-\zeta(s-1) \quad (|t|<1). \tag{28}$$

In view of the identities in 2.3(1) and (24), a special case of (28) when $t = \frac{1}{2}$ readily yields Ramaswami's result (2).

In case we add (19) to itself (with t replaced by $-t$), we obtain the summation formula [*cf.* Hansen (1975, p. 357, Eq. (54.6.3))]:

$$(29)\quad \sum_{k=0}^{\infty} \binom{\lambda+2k-1}{2k} \zeta(\lambda+2k)t^{2k} = \frac{1}{2}\{\zeta(\lambda-1, 1-t) + \zeta(\lambda, 1+t)\} \quad (|t| < 1)$$

while a similar subtraction yields

$$(30)\quad \sum_{k=0}^{\infty} \binom{\lambda+2k}{2k+1} \zeta(\lambda+2k+1)t^{2k+1} = \frac{1}{2}\{\zeta(\lambda, 1-t) - \zeta(\lambda, 1+t)\} \quad (|t| < 1).$$

Various interesting special cases of (29) and (30) are given in the literature. In particular, the special cases of (29) when $t = \frac{1}{2}$, $t = \frac{1}{3}$, and $t = \frac{1}{6}$ were considered by Ramaswami (1934, p. 167, Eqs. (1), (3), and (4)) who also gave a special case of (30) when $t = \frac{1}{2}$ [*cf.* Ramaswami (1934, p. 167, Eq. (2))], and by Apostol (1954) who proved various generalizations of Ramaswami's results.

By assigning suitable numerical values to the variable s in some of the aforementioned special cases of (29) and (30), Ramaswami (1934) also evaluated a number of special sums including various formulas in Section 3.4 and 3.1(7). Wilton (1922–1923, p. 92) showed that

$$(31)\quad \sum_{k=1}^{\infty} \frac{(2k-1)!}{(2k+3)!} \frac{\zeta(2k)}{2^{2k}} = \frac{\zeta(3)}{2\pi^2} + \frac{1}{12}\log\pi - \frac{11}{72}$$

and

$$(32)\quad \sum_{k=1}^{\infty} \frac{(2k)!}{(2k+3)!} \frac{\zeta(2k+1)}{2^{2k+1}} = \frac{3}{8} - \frac{1}{6}\log\pi - \frac{1}{4}\gamma + \frac{1}{12}\log 2 + \frac{\zeta'(2)}{\pi^2}.$$

Many closed-form evaluations of series involving the Zeta function were investigated systematically by Srivastava (1988a) who gave a rather detailed discussion about their derivations and relationships (if any) among them. As a matter of fact, most of the series identities considered by Srivastava (1988a) will be included in our extensive list given in Section 3.4.

We record here the following identity involving the Bernoulli polynomials:

$$(33)\quad \frac{B_{n+1}(a+t)}{n+1} = \sum_{k=0}^{n} \binom{n}{k} \frac{B_{k+1}(a)}{k+1} t^{n-k} + \frac{t^{n+1}}{n+1} \quad (n \in \mathbb{N}_0),$$

which is, in fact, an immediate consequence of the known formula 1.6(13).

It follows from 2.2(15) that the Laurent series expansion of $\zeta(s, a)$ at $s = 1$ is of the form

$$(34)\quad \zeta(s, a) = \frac{1}{s-1} - \psi(a) + \sum_{n=1}^{\infty} c_n (s-1)^n,$$

where the coefficients c_n are constants to be determined and the Psi (or Digamma) function $\psi(z)$ is given in Section 1.2.

Making use of (33), (34), and several results presented in previous sections, we shall now prove a general series identity which is contained in [see Choi and Srivastava (2000b)]

Theorem 3.1. *For every nonnegative integer* n,

$$\sum_{k=2}^{\infty} \frac{\zeta(k,\, a)}{(k)_{n+1}} t^{n+k} = \frac{(-1)^n}{n!} \left[\zeta'(-n,\, a-t) - \zeta'(-n,\, a)\right]$$

$$+ \sum_{k=1}^{n} \frac{(-1)^{n+k}}{n!} \binom{n}{k} \left[(H_n - H_{n-k})\, \zeta(k-n,\, a) - \zeta'(k-n,\, a)\right] t^k \tag{35}$$

$$+ \left[H_n + \psi(a)\right] \frac{t^{n+1}}{(n+1)!} \qquad (|t| < |a|;\ n \in \mathbb{N}_0),$$

where H_n *denotes the harmonic numbers defined by*

$$H_n := \sum_{j=1}^{n} \frac{1}{j} \tag{36}$$

and it is understood (as usual) that an empty sum is nil.

Proof. Upon transposing the first $n+2$ terms from $k=0$ to $k=n+1$ in (7) $(\lambda = s)$ to the right-hand side, if we divide both sides of the resulting equation by $s+n$, we get

$$\sum_{k=n+2}^{\infty} (s)_n\, (s+n+1)_{k-n-1}\, \zeta(s+k,\, a)\, \frac{t^k}{k!} = \frac{g_n(s,t,a)}{s+n} \qquad (|t| < |a|;\ n \in \mathbb{N}_0), \tag{37}$$

where, for convenience,

$$g_n(s,t,a) := \zeta(s,\, a-t) - \sum_{k=0}^{n+1} \frac{(s)_k}{k!}\, \zeta(s+k,\, a)\, t^k. \tag{38}$$

Now we shall show that

$$\lim_{s \to -n} g_n(s,t,a) = 0. \tag{39}$$

Since $\zeta(s+n+1,\, a)$ has a simple pole at $s=-n$ with its residue 1, we find from (34) that

$$\lim_{s \to -n} (s+n)\, \zeta(s+n+1,\, a) = 1. \tag{40}$$

By rewriting (38) in the form:

$$g_n(s,t,a) = \zeta(s,\, a-t) - \sum_{k=0}^{n} \frac{(s)_k}{k!}\, \zeta(s+k,\, a)\, t^k - (s)_n\, (s+n)\, \zeta(s+n+1,\, a)\, \frac{t^{n+1}}{(n+1)!},$$

if we take the limit as $s \to -n$ with the aid of (40) and make use of the elementary identity 1.1(40), we obtain

$$\lim_{s\to -n} g_n(s,t,a) = \zeta(-n,\, a-t) - \sum_{k=0}^{n} (-1)^{n-k} \binom{n}{k} \zeta(-k,\, a)\, t^{n-k} + \frac{(-1)^{n+1}}{n+1}\, t^{n+1},$$

which, in view of the relationship 2.1(17), can be put in its equivalent form:

$$\lim_{s\to -n} g_n(s,t,a) = -\frac{B_{n+1}(a-t)}{n+1} + \sum_{k=0}^{n} (-1)^{n-k} \binom{n}{k} \frac{B_{k+1}(a)}{k+1}\, t^{n-k} + \frac{(-1)^{n+1}}{n+1}\, t^{n+1},$$

from which our assertion (39) follows easily by applying (33). Thus, by l'Hôpital's rule, we have

$$\lim_{s\to -n} \frac{g_n(s,t,a)}{s+n} = \lim_{s\to -n} \frac{\partial}{\partial s}\{g_n(s,t,a)\}. \tag{41}$$

Next, by appealing to (38) and 1.5(10), we observe that

$$\begin{aligned} \frac{\partial}{\partial s}\{g_n(s,t,a)\} &= \zeta'(s,\, a-t) - \zeta'(s,\, a) \\ &\quad - \sum_{k=1}^{n} h(s,a,k)\frac{t^k}{k!} - h(s,a,n+1)\frac{t^{n+1}}{(n+1)!}, \end{aligned} \tag{42}$$

where, for convenience,

$$h(s,a,k) := (s)_k \left[\left(\sum_{l=1}^{k} \frac{1}{s+l-1} \right) \zeta(s+k,\, a) + \zeta'(s+k,\, a) \right]. \tag{43}$$

It follows from (43) that

$$\begin{aligned} \lim_{s\to -n} h(s,a,k) &= -\,(-n)_k \left[(H_n - H_{n-k})\, \zeta(k-n,\, a) - \zeta'(k-n,\, a) \right] \\ &\qquad (k = 1, \ldots, n) \end{aligned} \tag{44}$$

and

$$\begin{aligned} &\lim_{s\to -n} h(s,a,n+1) \\ &\quad = \lim_{s\to -n} \left[\left(\sum_{j=0}^{n} \frac{(s)_n}{s+j} \right) (s+n)\, \zeta(s+n+1,\, a) + (s)_{n+1}\, \zeta'(s+n+1,\, a) \right], \end{aligned}$$

which, upon writing

$$\sum_{j=0}^{n} \frac{(s)_n}{s+j} = \sum_{j=0}^{n-1} \frac{(s)_n}{s+j} + \frac{(s)_n}{s+n},$$

reduces to the form:

$$\lim_{s\to -n} h(s,a,n+1) = -\sum_{j=0}^{n-1} \frac{(-n)_n}{n-j} + \lim_{s\to -n} \phi(s,a,n), \tag{45}$$

where, for convenience,

$$\phi(s,a,n) := (s)_n\,\zeta(s+n+1,\,a) + (s)_{n+1}\,\zeta'(s+n+1,\,a). \tag{46}$$

Now, by virtue of (34), we readily have

$$\begin{aligned}\lim_{s\to -n}\phi(s,a,n) &= \lim_{s\to -n}(s)_n\left[-\psi(a) + \sum_{j=1}^{\infty} jc_j\,(s+n)^j\right]\\ &= -(-n)_n\,\psi(a)\\ &= (-1)^{n+1}\,n!\,\psi(a).\end{aligned} \tag{47}$$

It follows from (45) and (47) that

$$\lim_{s\to -n} h(s,a,n+1) = (-1)^{n+1}\,n!\,\left[H_n + \psi(a)\right], \tag{48}$$

where H_n denotes the harmonic numbers defined by (36).

Making use of (44) and (48), we find from (42) that

$$\begin{aligned}&\lim_{s\to -n}\frac{\partial}{\partial s}\{g_n(s,t,a)\} = \zeta'(-n,\,a-t) - \zeta'(-n,\,a)]\\ &+\sum_{k=1}^{n}(-1)^k\binom{n}{k}\left[(H_n - H_{n-k})\,\zeta(k-n,\,a) - \zeta'(k-n,\,a)\right]t^k\\ &+\frac{(-1)^n}{n+1}\left[H_n + \psi(a)\right]t^{n+1}.\end{aligned} \tag{49}$$

Finally, since

$$\lim_{s\to -n}\sum_{k=n+2}^{\infty}(s)_n\,(s+n-1)_{k-n-1}\,\zeta(s+k,a)\frac{t^k}{k!} = (-1)^n\,n!\sum_{k=2}^{\infty}\frac{\zeta(k,\,a)}{(k)_{n+1}}t^{k+n}, \tag{50}$$

by equating the second members of (49) and (50), we are led immediately to our desired series identity (35). This evidently completes the proof of Theorem 3.1.

We deduce another interesting identity by suitably combining the special cases of (35) when $a=1$ and $a=2$. By applying 2.2(4), 1.2(4), and 1.2(7), we thus find that

$$\begin{aligned}\sum_{k=1}^{\infty}\frac{t^{n+k}}{(k)_{n+1}} &= \frac{(-1)^{n+1}}{n!}\,(1-t)^n\,\log(1-t)\\ &+\sum_{k=1}^{n}\frac{(-1)^{n+k}}{n!}\binom{n}{k}(H_n - H_{n-k})\,t^k \quad (|t|<1;\ n\in\mathbb{N}_0),\end{aligned} \tag{51}$$

which, in the special case when $n=2$, is a known result recorded (for example) by Hansen (1975, p. 37, Entry (5.7.40); p. 74, Entry (5.16.26)).

We shall deal with other interesting applications of Theorem 3.1 in Section 3.4.

3.3 Use of Generating Functions

Adamchik and Srivastava (1998) proposed and developed a novel method of evaluation of the sums of series involving Zeta functions by using generating functions. The key ingredients in their approach happen to include the familiar integral representations 2.2(21) and 2.3(30). This method has already been implemented in *Mathematica* (Version 3.0).

Consider the sum:

$$\Omega(a) = \sum_{k=1}^{\infty} f(k)\,\zeta(k+1,a) \qquad (\Re(a) > 0), \tag{1}$$

where the sequence $\{f(n)\}_{n=1}^{\infty}$ is assumed to possess a generating function:

$$F(t) = \sum_{k=1}^{\infty} f(k)\,\frac{t^k}{k!} \tag{2}$$

and

$$f(n) = O\left(\frac{1}{n}\right) \qquad (n \to \infty). \tag{3}$$

Upon replacing the Hurwitz (or generalized) Zeta function in (1) by its integral representation given by 2.2(21) with $s = k+1$, if we invert the order of summation and integration, we obtain

$$\Omega(a) = \int_0^{\infty} F(t)\,\frac{e^{-(a-1)t}}{e^t - 1}\,dt, \tag{4}$$

where we have also made use of the generating function (2).

Thus the problem of summation of series of the type (1) has been reduced formally to that of integration in (4). Although the integral in (4) appears to be fairly involved for symbolic integration, yet it may be possible to reduce it to 2.2(21) or 2.3(30) (or another known integral), especially when $F(t)$ is a power, exponential, trigonometric or hyperbolic function.

The first example (Proposition 1 below) would illustrate how this technique actually works [see Adamchik and Srivastava (1998, p. 135, Proposition 1)]

Proposition 3.2. *Let n be a positive integer. Then*

$$\sum_{k=1}^{\infty} \frac{(-1)^k}{k}\,\{\zeta(nk) - 1\} = \log\left(\prod_{j=0}^{n-1} \Gamma\left(2 - (-1)^{(2j+1)/n}\right)\right). \tag{5}$$

Remark 1. In Proposition 1, as also in Equations (7), (8), and (10) below, it is tacitly assumed that

$$-1 = e^{\log(-1)},$$

where $\log z$ denotes the principal branch of the logarithmic function in the complex z-plane for which

$$-\pi < \arg(z) \leq \pi \qquad (z \neq 0).$$

Proof. Denote, for convenience, the left-hand side of the summation formula (5) by $\Theta(n)$. Then, in view of the special case of 2.3(9) when $n = 1$, we can apply the integral representation 2.2(21) with $s = nk$ and $a = 2$. Upon inverting the order of summation and integration, which can be justified by the absolute convergence of the series and the integral involved, we thus find that

$$\Theta(n) = n \int_0^\infty \frac{dt}{te^t(e^t - 1)} \sum_{k=1}^\infty \frac{(-t^n)^k}{\Gamma(nk+1)}. \tag{6}$$

By recognizing the series in (6) as a trigonometric function of order n [see, for details, Erdélyi *et al.* (1955, Section 18.2)] or, alternatively, by using an easily derivable special case of a known result in Hansen (1975, p. 207, Entry (10.49.1)), we have

$$\sum_{k=1}^\infty \frac{(-t^n)^k}{\Gamma(nk+1)} = -1 + \frac{1}{n} \sum_{j=1}^n \exp\left(t(-1)^{(2j+1)/n}\right). \tag{7}$$

Substituting from (7) into (6), and inverting the order of summation and integration once again, we obtain

$$\begin{aligned}\Theta(n) = \lim_{r \to 1-} \Bigg\{ -n \int_0^\infty \frac{dt}{t^r\, e^t(e^t - 1)} \\ + \sum_{j=1}^n \int_0^\infty \frac{\exp\left(-\left[1 - (-1)^{(2j+1)/n}\right] t\right)}{t^r(e^t - 1)}\, dt \Bigg\},\end{aligned} \tag{8}$$

where the parameter $r < 1$ is inserted with a view to providing convergence of the integrals at their lower terminal $t = 0$.

Finally, we evaluate each integral in (8) by means of 2.2(21) and proceed to the limit as $r \to 1-$. Indeed, by making use of the known behavior of $\zeta(1 - r, a)$ near $r = 1$ for fixed a [*cf.*, *e.g.*, Erdélyi *et al.* (1953, p. 26)]:

$$\zeta(1 - r, a) \sim \frac{1}{2} - a + (1 - r) \log\left(\frac{\Gamma(a)}{\sqrt{2\pi}}\right) \qquad (r \to 1-;\ a \text{ fixed}), \tag{9}$$

we arrive at the right-hand side of the assertion (5).

In precisely the same manner, we can prove a *mild* generalization of Proposition 3.2, which we state here as [see Adamchik and Srivastava (1998, p. 136, Proposition 2)]

Proposition 3.3. *Let n be a positive integer. Then*

$$\begin{aligned}\sum_{k=1}^\infty \frac{(-1)^k}{k} \zeta(nk, a) = -n \log \Gamma(a) \\ + \log\left(\prod_{j=0}^{n-1} \Gamma\left(a - (-1)^{(2j+1)/n}\right)\right) \qquad (\Re(a) \geq 1).\end{aligned} \tag{10}$$

Next we prove [see Adamchik and Srivastava (1998, p. 136, Proposition 3)]

Proposition 3.4. *Let n be a positive integer. Then, in terms of the Bernoulli numbers B_n defined by* 1.6(2) *and the Stirling numbers $S(n,k)$ defined by* 1.5(14),

$$\begin{aligned}\sum_{k=2}^{\infty}(-1)^k\,\{\zeta(k)-1\}\,k^n &= -1+\frac{1-2^{n+1}}{n+1}\,B_{n+1}\\ &\quad -\sum_{k=1}^{n}(-1)^k\,k!\,\zeta(k+1)S(n+1,k+1).\end{aligned}\tag{11}$$

Proof. Making use of the integral representation 2.2(21) with $a=2$, if we invert the order of summation and integration, and then evaluate the resulting integral and sum, we find that

$$\begin{aligned}\Lambda(n) &:= \sum_{k=2}^{\infty}(-1)^k\,\{\zeta(k)-1\}\,k^n\\ &= \lim_{r\to 1}\left(r\,\frac{d}{dr}\right)^n \{r\,[\psi(r+2)-\psi(2)]\}.\end{aligned}\tag{12}$$

Since

$$\left(r\,\frac{d}{dr}\right)^r \{r\,f(r)\} = \sum_{k=0}^{n} S(n+1,k+1)\,f^{(k)}(r)\,r^{k+1}\tag{13}$$

and [*cf.* Equation 1.2(53)]

$$\psi^{(k)}(3) = (-1)^k\,k!\,\left\{1+2^{-k-1}-\zeta(k+1)\right\}\qquad (k\in\mathbb{N}),\tag{14}$$

we find from (12) that

$$\begin{aligned}\Lambda(n) &= -1-\sum_{k=1}^{n}(-1)^k\,k!\,\zeta(k+1)\,S(n+1,k+1)\\ &\quad +\sum_{k=0}^{n}(-1)^k\,k!\,2^{-k-1}\,S(n+1,k+1).\end{aligned}\tag{15}$$

In order to complete the proof of Proposition 3.4, we thus need to show that

$$\sum_{k=0}^{n}(-1)^k\,k!\,2^{-k-1}\,S(n+1,k+1) = \frac{1-2^{n+1}}{n+1}\,B_{n+1}\tag{16}$$

in terms of the Bernoulli numbers B_n defined by 1.6(2). In fact, it is known that [Hansen (1975, p. 351, Entry (52.2.36))]

$$\sum_{k=1}^{n}(-1)^k\,k!\,2^{-k}\,S(n,k) = \frac{2}{n+1}\,(1-2^{n+1})\,B_{n+1},\tag{17}$$

which, in view of the recurrence relation 1.5(18) with k replaced by $k+1$, yields the desired identity (16).

Similarly, we can prove [see Adamchik and Srivastava (1998, p. 137, Proposition 4)]

Proposition 3.5. *Let n be a positive integer. Then*

$$\sum_{k=2}^{\infty} \{\zeta(k)-1\}\, k^n = 1 + \sum_{k=1}^{n} k!\, \zeta(k+1)\, S(n+1,k+1). \tag{18}$$

The following list provides further summation formulas involving series of Zeta functions, which can be derived by applying the foregoing technique [see Adamchik and Srivastava (1998, pp. 138–139)]:

$$\sum_{k=2}^{\infty} \frac{k^2}{k+1} \{\zeta(k)-1\} = \frac{3}{2} - \frac{\gamma}{2} + \frac{\pi^2}{6} - \frac{1}{2}\log(2\pi); \tag{19}$$

$$\sum_{k=2}^{\infty} \frac{k^2}{k+1} \{\zeta(2k+1)-1\} = \frac{9}{16} - \gamma + \log 2 - \frac{1}{2}\zeta(3); \tag{20}$$

$$\sum_{k=1}^{\infty} \{\zeta(4k)-1\} = \frac{7}{8} - \frac{\pi}{4}\coth\pi; \tag{21}$$

$$\sum_{k=1}^{\infty} (-1)^k \{\zeta(4k)-1\} = 1 + \frac{\pi}{2\sqrt{2}} \frac{\sin(\pi\sqrt{2}) + \sinh(\pi\sqrt{2})}{\cos(\pi\sqrt{2}) - \cosh(\pi\sqrt{2})}; \tag{22}$$

$$\sum_{k=1}^{\infty} \{\zeta(4k)-1\}\, z^{4k} = \frac{3z^4-1}{2(z^4-1)} - \frac{\pi z}{4}\{\cot(\pi z) + \coth(\pi z)\} \qquad (|z|<2), \tag{23}$$

which contains both (21) and (22) as limiting cases;

$$\begin{aligned}\sum_{k=1}^{\infty} \{\zeta(2k)-1\}\, \sin k &= -\frac{1}{2}\cot\left(\frac{1}{2}\right) \\ &+ \frac{\pi}{2}\, \frac{\sin\left(\frac{1}{2}\right)\sin\left(2\pi\cos\left(\frac{1}{2}\right)\right) - \cos\left(\frac{1}{2}\right)\sinh\left(2\pi\sin\left(\frac{1}{2}\right)\right)}{\cos\left(2\pi\cos\left(\frac{1}{2}\right)\right) - \cosh\left(2\pi\sin\left(\frac{1}{2}\right)\right)};\end{aligned} \tag{24}$$

$$\sum_{k=1}^{\infty} \binom{p+k}{k} \zeta(p+k+1,a)\, z^k = \frac{(-1)^p}{p!} \left\{\psi^{(p)}(a) - \psi^{(p)}(a-z)\right\} \tag{25}$$

$$(p \in \mathbb{N};\ \Re(a) > 0;\ |z| < |a|);$$

$$\sum_{k=2}^{\infty} \frac{k}{2^k} \zeta\left(k+1, \frac{3}{4}\right) = 8G \tag{26}$$

and

$$\sum_{k=2}^{\infty} \frac{k}{2^k} \zeta\left(k+1, \frac{5}{4}\right) = 8(1-G), \tag{27}$$

where G denotes Catalan's constant defined by 1.3(16).

Remark 2. Since the right-hand side of the relationship 1.2(53) [with $n!$ replaced by $\Gamma(n+1)$] is well-defined for $n \in \mathbb{C} \setminus \{-1\}$, the summation formula (25) may be put in a slightly more general form [*cf.* Wilton (1922–1923); see also Srivastava (1988a, p. 137, Equation (6.6))]:

$$\sum_{k=0}^{\infty} \binom{s+k-1}{k} \zeta(s+k, a)\, z^k = \zeta(s, a-z) \tag{28}$$
$$\left(s \in \mathbb{C} \setminus \{1\};\ a \neq \mathbb{Z}_0^-;\ |z| < |a|\right),$$

which, in view of 1.1(40), is equivalent to 3.2(7) for $\lambda = s$ and $t = z$.

The foregoing method has been implemented in *Mathematica* (Version 3.0), as we remarked at the beginning of this section.

Series involving Polygamma functions

The generalized harmonic numbers $H_n^{(s)}$ defined by [*cf.* Graham *et al.* (1989)]

$$H_n^{(s)} := \sum_{k=1}^{n} \frac{1}{k^s} \qquad (n \in \mathbb{N},\ s \in \mathbb{C}), \tag{29}$$

which, by virtue of 2.3(9), can be written in the form:

$$H_n^{(s)} = \zeta(s) - \zeta(s, n+1) \qquad (\Re(s) > 1;\ n \in \mathbb{N}). \tag{30}$$

In view of the relationships 1.2(53) and (30), the foregoing techniques can be applied also to series involving Polygamma functions and generalized harmonic numbers. We first state [see Adamchik and Srivastava (1998, p. 139, Proposition 5)]

Proposition 3.6. *Let p be a positive integer. Then, in terms of the function $\Phi(z, s, a)$ defined by* 2.5(1),

$$\sum_{k=1}^{\infty} \psi^{(p)}(a+k)\, z^k = \frac{(-1)^{p+1}\, p!}{a^{p+1}} + \frac{\psi^{(p)}(a+1)}{1-z} + \frac{(-1)^p\, p!\, z^2}{1-z}\, \Phi(z, p+1, a+1) \tag{31}$$
$$\left(|z| < 1;\ a \neq \mathbb{Z}_0^- \setminus \{0\}\right).$$

Proof. Denoting, for convenience, the left-hand side of the summation formula (31) by $\Xi(z)$, it is not difficult to find from 1.2(53) and 2.2(21) that

$$\begin{aligned}\Xi(z) &= (-1)^{p+1}\int_0^\infty \frac{t^p\, e^{-(a-1)t}}{e^t-1}\left(\sum_{k=1}^\infty z^k\, e^{-kt}\right)dt \\ &= -\int_0^1 \frac{\tau^{a-1}(\log\tau)^p}{(1-\tau)(1-z\tau)}\,d\tau \qquad (\Re(a)>0)\,.\end{aligned}$$

Upon evaluating this last integral, and waiving the restriction on the parameter a by appealing to the principle of analytic continuation, we complete the proof of Proposition 3.6.

Next we turn to a family of *linear* harmonic sums:

$$\mathcal{S}_{p,q} := \sum_{n=1}^\infty \frac{H_n^{(p)}}{n^q}\,, \tag{32}$$

which were discussed extensively by Flajolet and Salvy (1998). By applying (30), 2.2(21), and 2.3(30), it is not difficult to prove [see Adamchik and Srivastava (1998, p. 140, Proposition 6)]

Proposition 3.7. *Let $\mathcal{S}_{p,q}$ be defined by* (32). *Then*

$$\mathcal{S}_{p,q} = \zeta(p)\zeta(q) + \frac{(-1)^p}{(p-1)!}\int_0^1 (\log t)^{p-1}\,\mathrm{Li}_q(t)\,\frac{dt}{1-t} \tag{33}$$

and

$$\mathcal{S}_{p,q} = \zeta(p+q) - \frac{(-1)^q}{(q-1)!}\int_0^1 (\log t)^{q-1}\,\mathrm{Li}_p(t)\,\frac{dt}{1-t}. \tag{34}$$

Remark 3. In view of the following symmetry relation in Flajolet and Salvy (1998):

$$\mathcal{S}_{p,q} + \mathcal{S}_{q,p} = \zeta(p)\zeta(q) + \zeta(p+q), \tag{35}$$

the integral representations (33) and (34) are essentially the same. Although, in general, the integrals occurring in (33) and (34) cannot be evaluated in closed forms, many interesting particular cases of linear harmonic sums would follow from Proposition 3.7.

Series involving Polylogarithm functions

We shall derive, among other results, an integral representation for the Khintchine constant K_0 which arises in the measure theory of continued fractions. Every positive irrational number μ can indeed be written uniquely as a simple continued fraction as follows:

$$\mu = a_0 + \frac{a_1}{a_2+}\,\frac{a_3}{a_4+}\cdots\frac{a_n}{a_{n+1}+}\cdots\,, \tag{36}$$

that is, with a_0 a non-negative integer and with all other a_j $(j = 1, 2, 3, \cdots)$ positive integers. The Gauss-Kuz'min distribution [*cf.*, *e.g.*, Khintchine (1964)] predicts that the density of occurrence of some chosen positive integer k in the continued fraction (36) of a random real number is given by

$$\text{Prob}\{a_n = k\} = -\log_2\left(1 - \frac{1}{(k+1)^2}\right). \tag{37}$$

And, making use of the Gauss-Kuz'min distribution involving (37), Khintchine (1964) showed that, for almost all irrational numbers, the limiting geometric mean of the positive integer elements a_j $(j = 1, 2, 3, \cdots)$ of the relevant continued fraction exists and equals

$$\begin{aligned} K_0 &:= \prod_{k=1}^{\infty}\left\{1 + \frac{1}{k(k+2)}\right\}^{\log_2 k} \\ &= \prod_{k=1}^{\infty}\left\{k^{\log_2\left(1+\frac{1}{k(k+2)}\right)}\right\}. \end{aligned} \tag{38}$$

An interesting explicit representation of the Khintchine constant K_0 in terms of Polylogarithm functions was proven recently by Bailey *et al.* (1997, p. 422):

$$\log(K_0)\log 2 = (\log 2)^2 + \text{Li}_2\left(-\frac{1}{2}\right) + \frac{1}{2}\sum_{n=2}^{\infty}(-1)^n\,\text{Li}_2\left(\frac{4}{n^2}\right). \tag{39}$$

If we set

$$\mathcal{L}(n) := \sum_{n=2}^{\infty}(-1)^n\,\text{Li}_2\left(\frac{4}{n^2}\right), \tag{40}$$

replace the Dilogarithm function by its series representation given by 2.4(3), change the order of summation, and evaluate the inner sum, we shall obtain

$$\mathcal{L}(n) = 2\sum_{k=1}^{\infty}\frac{\zeta(2k)}{k^2} - \sum_{k=1}^{\infty}\{\zeta(2k) - 1\}\frac{4^k}{k^2}. \tag{41}$$

It seems very unlikely that the sums occurring in (41) can be evaluated in terms of well-known functions. Nevertheless, by noting that

$$\sum_{k=1}^{\infty}\frac{t^k}{k^2}\zeta(2k) = \int_0^1\frac{dt}{t}\sum_{k=1}^{\infty}\frac{t^k}{k}\zeta(2k), \tag{42}$$

and evaluating the inner sum by the method illustrated in the preceding sections, we find that

$$\sum_{k=1}^{\infty}\frac{t^k}{k^2}\zeta(2k) = \log\left(\pi\sqrt{t}\,\csc(\pi\sqrt{t})\right). \tag{43}$$

Combining (41), (42), and (43), we have

$$\mathcal{L}(n) := \sum_{n=2}^{\infty}(-1)^n\,\text{Li}_2\left(\frac{4}{n^2}\right) = \int_0^1\frac{dt}{t}\log\left(\frac{\pi\sqrt{t}\,\cot(\pi\sqrt{t})}{1-4t}\right), \tag{44}$$

which leads us immediately to the following integral representation for the Khintchine constant K_0 [see Adamchik and Srivastava (1998, p. 141, Eq. (4.10))]:

$$\log(K_0)\log 2 = \frac{\pi^2}{12} + \frac{(\log 2)^2}{2} + \int_0^{\pi}\log\left(t|\cot t|\right)\frac{dt}{t}. \tag{45}$$

Other sums involving the Polylogarithm function, which were also evaluated by Adamchik and Srivastava (1998), are given below.

$$(46)\qquad \sum_{k=1}^{\infty}\left(-\frac{1}{2}\right)^{k}\operatorname{Li}_k(z^2)=1-z^{-1}\operatorname{arctanh} z;$$

$$(47)\qquad \sum_{k=1}^{\infty}\left(\frac{1}{2}\right)^{k}\operatorname{Li}_k(z^2)=\frac{1}{2}\,z\log\left(\frac{1+z}{1-z}\right).$$

Since $\operatorname{arctanh}\left(\frac{1}{2}\right)=\log 3$, both (46) and (47) can be expressed in terms of $\log 3$ when $z=\frac{1}{2}$.

3.4 Use of Multiple Gamma Functions

We provide a rather extensive list of evaluations of series involving the Zeta functions by making use of the Gamma and multiple Gamma functions. Although our list in this section includes some of the series identities considered in previous sections or earlier works, we choose to present here all of the identities evaluated by the use of multiple Gamma functions for the sake of completeness.

Evaluation by using the Gamma function

Upon setting $n=0$ in 3.2(35) and applying 2.2(17), we obtain the familiar result [*cf.* Whittaker and Watson (1963, p. 276), Hansen (1975, p. 358, Entry (54.11.1), and Srivastava (1988a, p. 18)]:

$$(1)\qquad \sum_{k=2}^{\infty}\zeta(k,a)\frac{t^k}{k}=\log\Gamma(a-t)-\log\Gamma(a)+t\psi(a)\quad(|t|<|a|),$$

which, upon replacing t by $-t$, yields

$$(2)\qquad \sum_{k=2}^{\infty}(-1)^k\zeta(k,a)\frac{t^k}{k}=\log\Gamma(a+t)-\log\Gamma(a)-t\psi(a)\quad(|t|<|a|).$$

By adding and subtracting, we find from (1) and (2) that

$$(3)\qquad \sum_{k=1}^{\infty}\zeta(2k,a)\frac{t^{2k}}{k}=\log\Gamma(a+t)+\log\Gamma(a-t)-2\log\Gamma(a)\quad(|t|<|a|);$$

$$(4)\qquad \sum_{k=1}^{\infty}\zeta(2k+1,a)\frac{t^{2k+1}}{2k+1}=\frac{1}{2}\left\{\log\Gamma(a-t)-\log\Gamma(a+t)\right\}+t\psi(a)\quad(|t|<|a|).$$

Differentiating both sides of (1), (2), (3), and (4) with respect to t, we have

$$(5)\qquad \sum_{k=2}^{\infty}\zeta(k,a)t^{k-1}=-\psi(a-t)+\psi(a)\quad(|t|<|a|);$$

$$(6)\qquad \sum_{k=2}^{\infty}(-1)^k\zeta(k,a)t^{k-1}=\psi(a+t)-\psi(a)\quad (|t|<|a|);$$

$$(7)\qquad \sum_{k=1}^{\infty}\zeta(2k,a)t^{2k-1}=\frac{1}{2}\{\psi(a+t)-\psi(a-t)\}\quad (|t|<|a|);$$

$$(8)\qquad \sum_{k=1}^{\infty}\zeta(2k+1,a)t^{2k}=-\frac{1}{2}\{\psi(a+t)+\psi(a-t)\}+\psi(a)\quad (|t|<|a|).$$

Setting $a=1$ in (1) through (8), and using 1.2(4) and 2.3(2), we obtain

$$(9)\qquad \sum_{k=2}^{\infty}\zeta(k)\frac{t^k}{k}=\log\Gamma(1-t)-\gamma t\quad (|t|<1);$$

$$(10)\qquad \sum_{k=2}^{\infty}(-1)^k\zeta(k)\frac{t^k}{k}=\log\Gamma(1+t)+\gamma t\quad (|t|<1);$$

$$(11)\qquad \sum_{k=1}^{\infty}\zeta(2k)\frac{t^{2k}}{k}=\log\Gamma(1+t)+\log\Gamma(1-t)\quad (|t|<1);$$

$$(12)\qquad \sum_{k=1}^{\infty}\zeta(2k+1)\frac{t^{2k+1}}{2k+1}=\frac{1}{2}\{\log\Gamma(1-t)-\log\Gamma(1+t)\}-\gamma t\quad (|t|<1);$$

$$(13)\qquad \sum_{k=2}^{\infty}\zeta(k)t^{k-1}=-\psi(1-t)-\gamma\quad (|t|<1);$$

$$(14)\qquad \sum_{k=2}^{\infty}(-1)^k\zeta(k)t^{k-1}=\psi(1+t)+\gamma\quad (|t|<1);$$

$$(15)\qquad \sum_{k=1}^{\infty}\zeta(2k)t^{2k-1}=\frac{1}{2}\{\psi(1+t)-\psi(1-t)\}\quad (|t|<1);$$

$$(16)\qquad \sum_{k=1}^{\infty}\zeta(2k+1)t^{2k}=-\frac{1}{2}\{\psi(1+t)+\psi(1-t)\}-\gamma\quad (|t|<1).$$

Equations (11) and (15) can, in view of 1.1(12) and 1.2(9), be written in their respective equivalent forms:

$$\sum_{k=1}^{\infty}\zeta(2k)\frac{t^{2k}}{k}=\log\left(\frac{\pi t}{\sin\pi t}\right)\qquad(|t|<1); \tag{17}$$

$$\sum_{k=1}^{\infty}\zeta(2k)t^{2k-1}=-\frac{\pi}{2}\cot(\pi t)+\frac{1}{2t}\qquad(|t|<1). \tag{18}$$

If we use some elementary trigonometric identities, 1.2(8), and some special values of the ψ-function given in Section 1.2, we obtain the following trigonometric evaluations:

$$\sin\left(\frac{\pi}{8}\right)=\sin\left(\frac{7\pi}{8}\right)=\frac{1}{2}\sqrt{2-\sqrt{2}}; \tag{19}$$

$$\sin\left(\frac{3\pi}{8}\right)=\sin\left(\frac{5\pi}{8}\right)=\frac{1}{2}\sqrt{2+\sqrt{2}}; \tag{20}$$

$$\cot\left(\frac{\pi}{5}\right)=\sqrt{1+\frac{2}{5}\sqrt{5}}-\frac{1}{4\pi}\sqrt{5}\log\left(\frac{3\sqrt{5}-1}{4}\right); \tag{21}$$

$$\begin{aligned}\sin\left(\frac{\pi}{5}\right)=&\left\{2+\frac{2}{5}\sqrt{5}+\frac{5}{16\pi^2}\left[\log\left(\frac{3\sqrt{5}-1}{4}\right)\right]^2\right.\\&\left.-\frac{1}{2\pi}\sqrt{5+2\sqrt{5}}\log\left(\frac{3\sqrt{5}-1}{4}\right)\right\}^{-\frac{1}{2}}.\end{aligned} \tag{22}$$

Taking the limit as $t\to 1$ on both sides of (10), we have the well-known result:

$$\sum_{k=2}^{\infty}(-1)^k\frac{\zeta(k)}{k}=\gamma. \tag{23}$$

For suitable special values of the argument t, we can deduce the following series identities from (9) through (18) by making use of such evaluations as (19) through (22) as well as the ψ-function values listed in Section 1.2.

$$\sum_{k=2}^{\infty}\frac{\zeta(k)}{k\cdot 2^k}=\frac{1}{2}\log\pi-\frac{\gamma}{2}; \tag{24}$$

$$\sum_{k=2}^{\infty}(-1)^k\frac{\zeta(k)}{k\cdot 2^k}=\frac{\gamma}{2}+\frac{1}{2}\log\pi-\log 2; \tag{25}$$

(26) $$\sum_{k=1}^{\infty} \frac{\zeta(2k)}{k \cdot 2^{2k}} = \log\left(\frac{\pi}{2}\right);$$

(27) $$\sum_{k=1}^{\infty} \frac{\zeta(2k+1)}{(2k+1)2^{2k}} = \log 2 - \gamma;$$

(28) $$\sum_{k=1}^{\infty} \frac{\zeta(2k)}{k \cdot 3^{2k}} = \log\left(\frac{2\pi}{3\sqrt{3}}\right);$$

(29) $$\sum_{k=1}^{\infty} \frac{\zeta(2k)}{k}\left(\frac{2}{3}\right)^{2k} = \log\left(\frac{4\pi}{3\sqrt{3}}\right).$$

(30) $$\sum_{k=1}^{\infty} \frac{\zeta(2k)}{k \cdot 2^{4k}} = \log\left(\frac{\pi}{2\sqrt{2}}\right);$$

(31) $$\sum_{k=1}^{\infty} \frac{\zeta(2k)}{k}\left(\frac{3}{4}\right)^{2k} = \log\left(\frac{3\pi}{2\sqrt{2}}\right);$$

(32) $$\sum_{k=1}^{\infty} \frac{\zeta(2k)}{k \cdot 5^{2k}} = \frac{1}{2}\log\left\{\frac{\pi^2}{25}\left(2+\frac{2}{5}\sqrt{5}\right) + \frac{1}{80}\left[\log\left(\frac{3\sqrt{5}-1}{4}\right)\right]^2 - \frac{\pi}{50}\sqrt{5+2\sqrt{5}}\,\log\left(\frac{3\sqrt{5}-1}{4}\right)\right\};$$

(33) $$\sum_{k=1}^{\infty} \frac{\zeta(2k)}{k \cdot 6^{2k}} = \log\left(\frac{\pi}{3}\right);$$

(34) $$\sum_{k=1}^{\infty} \frac{\zeta(2k)}{k}\left(\frac{5}{6}\right)^{2k} = \log\left(\frac{5\pi}{3}\right);$$

(35) $$\sum_{k=1}^{\infty} \frac{\zeta(2k)}{k \cdot 2^{6k}} = \log\left(\frac{\pi}{4\sqrt{2-\sqrt{2}}}\right);$$

(36) $$\sum_{k=1}^{\infty} \frac{\zeta(2k)}{k}\left(\frac{3}{8}\right)^{2k} = \log\left(\frac{3\pi}{4\sqrt{2+\sqrt{2}}}\right);$$

(37) $$\sum_{k=1}^{\infty} \frac{\zeta(2k)}{k} \left(\frac{5}{8}\right)^{2k} = \log\left(\frac{5\pi}{4\sqrt{2+\sqrt{2}}}\right);$$

(38) $$\sum_{k=1}^{\infty} \frac{\zeta(2k)}{k} \left(\frac{7}{8}\right)^{2k} = \log\left[\frac{7\pi}{8}\left(\sqrt{2+\sqrt{2}}+\sqrt{2-\sqrt{2}}\right)\right];$$

(39) $$\sum_{k=1}^{\infty} \frac{\zeta(2k)}{k \cdot 12^{2k}} = \log\left(\frac{\pi(\sqrt{3}+1)}{6\sqrt{2}}\right);$$

(40) $$\sum_{k=1}^{\infty} \frac{\zeta(2k)}{k} \left(\frac{5}{12}\right)^{2k} = \log\left(\frac{5\pi(\sqrt{3}-1)}{6\sqrt{2}}\right);$$

(41) $$\sum_{k=1}^{\infty} \frac{\zeta(2k)}{k} \left(\frac{7}{12}\right)^{2k} = \log\left(\frac{7\pi(\sqrt{3}-1)}{6\sqrt{2}}\right);$$

(42) $$\sum_{k=1}^{\infty} \frac{\zeta(2k)}{k} \left(\frac{11}{12}\right)^{2k} = \log\left(\frac{11\pi(\sqrt{3}+1)}{6\sqrt{2}}\right);$$

(43) $$\sum_{k=2}^{\infty} (-1)^k \frac{\zeta(k)}{2^k} = 1 - \log 2;$$

(44) $$\sum_{k=2}^{\infty} \frac{\zeta(k)}{2^k} = \log 2;$$

(45) $$\sum_{k=1}^{\infty} \frac{\zeta(2k)}{2^{2k-1}} = 1;$$

(46) $$\sum_{k=1}^{\infty} \frac{\zeta(2k+1)}{2^{2k}} = 2\log 2 - 1.$$

(47) $$\sum_{k=2}^{\infty} (-1)^k \frac{\zeta(k)}{3^k} = 1 - \frac{1}{18}\pi\sqrt{3} - \frac{1}{2}\log 3;$$

(48)
$$\sum_{k=2}^{\infty} \frac{\zeta(k)}{3^k} = \frac{1}{2}\log 3 - \frac{1}{18}\pi\sqrt{3};$$

(49)
$$\sum_{k=1}^{\infty} \frac{\zeta(2k)}{3^{2k}} = \frac{1}{2} - \frac{1}{18}\pi\sqrt{3};$$

(50)
$$\sum_{k=1}^{\infty} \frac{\zeta(2k+1)}{3^{2k}} = \frac{3}{2}(\log 3 - 1);$$

(51)
$$\sum_{k=2}^{\infty} (-1)^k \zeta(k) \left(\frac{2}{3}\right)^k = 1 + \frac{1}{9}\pi\sqrt{3} - \log 3;$$

(52)
$$\sum_{k=2}^{\infty} \zeta(k) \left(\frac{2}{3}\right)^k = \frac{1}{9}\pi\sqrt{3} + \log 3;$$

(53)
$$\sum_{k=1}^{\infty} \zeta(2k) \left(\frac{2}{3}\right)^{2k} = \frac{1}{2} + \frac{1}{9}\pi\sqrt{3};$$

(54)
$$\sum_{k=1}^{\infty} \zeta(2k+1) \left(\frac{2}{3}\right)^{2k} = \frac{3}{2}\log 3 - \frac{3}{4};$$

(55)
$$\sum_{k=2}^{\infty} (-1)^k \frac{\zeta(k)}{2^{2k}} = 1 - \frac{1}{8}\pi - \frac{3}{4}\log 2;$$

(56)
$$\sum_{k=2}^{\infty} \frac{\zeta(k)}{2^{2k}} = \frac{3}{4}\log 2 - \frac{1}{8}\pi;$$

(57)
$$\sum_{k=1}^{\infty} \frac{\zeta(2k)}{2^{4k}} = \frac{1}{2} - \frac{1}{8}\pi;$$

(58)
$$\sum_{k=1}^{\infty} \frac{\zeta(2k+1)}{2^{4k}} = 3\log 2 - 2;$$

(59)
$$\sum_{k=2}^{\infty}(-1)^k\zeta(k)\left(\frac{3}{4}\right)^k = 1+\frac{3}{8}\pi-\frac{9}{4}\log 2;$$

(60)
$$\sum_{k=2}^{\infty}\zeta(k)\left(\frac{3}{4}\right)^k = \frac{3}{8}\pi+\frac{9}{4}\log 2;$$

(61)
$$\sum_{k=1}^{\infty}\zeta(2k)\left(\frac{3}{4}\right)^{2k} = \frac{1}{2}+\frac{3}{8}\pi;$$

(62)
$$\sum_{k=1}^{\infty}\zeta(2k+1)\left(\frac{3}{4}\right)^{2k} = 3\log 2-\frac{2}{3};$$

(63)
$$\sum_{k=2}^{\infty}(-1)^k\frac{\zeta(k)}{5^k} = 1-\frac{1}{10}\pi\sqrt{1+\frac{2}{5}\sqrt{5}}-\frac{1}{4}\log 5-\frac{1}{20}\sqrt{5}\log\left(\frac{3+\sqrt{5}}{2}\right);$$

(64)
$$\sum_{k=2}^{\infty}\frac{\zeta(k)}{5^k} = -\frac{1}{10}\pi\sqrt{1+\frac{2}{5}\sqrt{5}}+\frac{1}{4}\log 5+\frac{1}{20}\sqrt{5}\log\left(\frac{3+2\sqrt{5}}{2}\right);$$

(65)
$$\sum_{k=1}^{\infty}\frac{\zeta(2k)}{5^{2k}} = \frac{1}{2}-\frac{1}{10}\pi\sqrt{1+\frac{2}{5}\sqrt{5}}+\frac{1}{40}\sqrt{5}\log\left(\frac{3\sqrt{5}-1}{4}\right);$$

(66)
$$\sum_{k=1}^{\infty}\frac{\zeta(2k+1)}{5^{2k}} = -\frac{5}{2}+\frac{5}{4}\log 5+\frac{1}{8}\sqrt{5}\log\left(\frac{19+9\sqrt{5}}{4}\right).$$

(67)
$$\sum_{k=2}^{\infty}(-1)^k\zeta(k)\left(\frac{2}{5}\right)^k = 1-\frac{1}{5}\pi\sqrt{1-\frac{2}{5}\sqrt{5}}-\frac{1}{2}\log 5+\frac{1}{10}\sqrt{5}\log\left(\frac{3+\sqrt{5}}{2}\right);$$

(68)
$$\sum_{k=2}^{\infty}\zeta(k)\left(\frac{2}{5}\right)^k = -\frac{1}{5}\pi\sqrt{1-\frac{2}{5}\sqrt{5}}+\frac{1}{2}\log 5-\frac{1}{10}\sqrt{5}\log\left(\frac{3+2\sqrt{5}}{2}\right);$$

(69)
$$\sum_{k=1}^{\infty}\zeta(2k)\left(\frac{2}{5}\right)^{2k} = \frac{1}{2}-\frac{1}{5}\pi\sqrt{1-\frac{2}{5}\sqrt{5}}+\frac{1}{20}\sqrt{5}\log\left(\frac{1+3\sqrt{5}}{11}\right);$$

$$\text{(70)}\qquad \sum_{k=1}^{\infty}\zeta(2k+1)\left(\frac{2}{5}\right)^{2k}=-\frac{5}{4}+\frac{5}{4}\log 5-\frac{1}{8}\sqrt{5}\log\left(\frac{19+9\sqrt{5}}{4}\right);$$

$$\text{(71)}\qquad \sum_{k=2}^{\infty}(-1)^k\zeta(k)\left(\frac{3}{5}\right)^{k}=1+\frac{3}{10}\pi\sqrt{1-\frac{2}{5}\sqrt{5}}-\frac{3}{4}\log 5+\frac{3}{20}\sqrt{5}\log\left(\frac{3+2\sqrt{5}}{2}\right);$$

$$\text{(72)}\qquad \sum_{k=2}^{\infty}\zeta(k)\left(\frac{3}{5}\right)^{k}=\frac{3}{10}\pi\sqrt{1-\frac{2}{5}\sqrt{5}}+\frac{3}{4}\log 5-\frac{3}{20}\sqrt{5}\log\left(\frac{3+\sqrt{5}}{2}\right);$$

$$\text{(73)}\qquad \sum_{k=1}^{\infty}\zeta(2k)\left(\frac{3}{5}\right)^{2k}=\frac{1}{2}+\frac{3}{10}\pi\sqrt{1-\frac{2}{5}\sqrt{5}}+\frac{3}{40}\sqrt{5}\log\left(\frac{3\sqrt{5}-1}{4}\right);$$

$$\text{(74)}\qquad \sum_{k=1}^{\infty}\zeta(2k+1)\left(\frac{3}{5}\right)^{2k}=-\frac{5}{6}+\frac{5}{4}\log 5-\frac{1}{8}\sqrt{5}\log\left(\frac{19+9\sqrt{5}}{4}\right);$$

$$\text{(75)}\qquad \sum_{k=2}^{\infty}(-1)^k\zeta(k)\left(\frac{4}{5}\right)^{k}=1+\frac{2}{5}\pi\sqrt{1+\frac{2}{5}\sqrt{5}}-\log 5-\frac{1}{5}\sqrt{5}\log\left(\frac{3+2\sqrt{5}}{2}\right);$$

$$\text{(76)}\qquad \sum_{k=2}^{\infty}\zeta(k)\left(\frac{4}{5}\right)^{k}=\frac{2}{5}\pi\sqrt{1+\frac{2}{5}\sqrt{5}}+\log 5+\frac{1}{5}\sqrt{5}\log\left(\frac{3+\sqrt{5}}{2}\right);$$

$$\text{(77)}\qquad \sum_{k=1}^{\infty}\zeta(2k)\left(\frac{4}{5}\right)^{2k}=\frac{1}{2}+\frac{2}{5}\pi\sqrt{1+\frac{2}{5}\sqrt{5}}+\frac{1}{10}\sqrt{5}\log\left(\frac{1+3\sqrt{5}}{11}\right);$$

$$\text{(78)}\qquad \sum_{k=1}^{\infty}\zeta(2k+1)\left(\frac{4}{5}\right)^{2k}=-\frac{5}{8}+\frac{5}{4}\log 5+\frac{1}{8}\sqrt{5}\log\left(\frac{19+9\sqrt{5}}{4}\right);$$

$$\text{(79)}\qquad \sum_{k=2}^{\infty}(-1)^k\frac{\zeta(k)}{6^k}=1-\frac{1}{12}\pi\sqrt{3}-\frac{1}{4}\log 3-\frac{1}{3}\log 2;$$

$$\text{(80)}\qquad \sum_{k=2}^{\infty}\frac{\zeta(k)}{6^k}=-\frac{1}{12}\pi\sqrt{3}+\frac{1}{4}\log 3+\frac{1}{3}\log 2;$$

(81) $$\sum_{k=1}^{\infty} \frac{\zeta(2k)}{6^{2k}} = \frac{1}{2} - \frac{1}{12}\pi\sqrt{3};$$

(82) $$\sum_{k=1}^{\infty} \frac{\zeta(2k+1)}{6^{2k}} = -3 + \frac{3}{2}\log 3 + 2\log 2;$$

(83) $$\sum_{k=2}^{\infty} (-1)^k \zeta(k) \left(\frac{5}{6}\right)^k = 1 + \frac{5}{12}\pi\sqrt{3} - \frac{5}{4}\log 3 - \frac{5}{3}\log 2;$$

(84) $$\sum_{k=2}^{\infty} \zeta(k) \left(\frac{5}{6}\right)^k = \frac{5}{12}\pi\sqrt{3} + \frac{5}{4}\log 3 + \frac{5}{3}\log 2;$$

(85) $$\sum_{k=1}^{\infty} \zeta(2k) \left(\frac{5}{6}\right)^{2k} = \frac{1}{2} + \frac{5}{12}\pi\sqrt{3};$$

(86) $$\sum_{k=1}^{\infty} \zeta(2k+1) \left(\frac{5}{6}\right)^{2k} = -\frac{3}{5} + \frac{3}{2}\log 3 + 2\log 2;$$

(87) $$\sum_{k=2}^{\infty} (-1)^k \frac{\zeta(k)}{2^{3k}} = 1 - \frac{1}{16}\pi(\sqrt{2}+1) - \frac{1}{2}\log 2 + \frac{\sqrt{2}}{16}\log(3-2\sqrt{2});$$

(88) $$\sum_{k=2}^{\infty} \frac{\zeta(k)}{2^{3k}} = -\frac{1}{16}\pi(\sqrt{2}+1) + \frac{1}{2}\log 2 - \frac{\sqrt{2}}{16}\log(3-2\sqrt{2});$$

(89) $$\sum_{k=1}^{\infty} \frac{\zeta(2k)}{2^{6k}} = \frac{1}{2} - \frac{1}{16}\pi(\sqrt{2}+1);$$

(90) $$\sum_{k=1}^{\infty} \frac{\zeta(2k+1)}{2^{6k}} = -4 + 4\log 2 - \frac{\sqrt{2}}{2}\log(3-2\sqrt{2});$$

(91) $$\sum_{k=2}^{\infty} (-1)^k \zeta(k) \left(\frac{3}{8}\right)^k = 1 - \frac{3}{16}\pi(\sqrt{2}-1) - \frac{3}{2}\log 2 - \frac{3\sqrt{2}}{16}\log(3-2\sqrt{2});$$

$$(92)\qquad \sum_{k=2}^{\infty} \zeta(k) \left(\frac{3}{8}\right)^k = -\frac{3}{16}\pi(\sqrt{2}-1) + \frac{3}{2}\log 2 + \frac{3\sqrt{2}}{16}\log(3-2\sqrt{2});$$

$$(93)\qquad \sum_{k=1}^{\infty} \zeta(2k) \left(\frac{3}{8}\right)^{2k} = \frac{1}{2} - \frac{3}{16}\pi(\sqrt{2}-1);$$

$$(94)\qquad \sum_{k=1}^{\infty} \zeta(2k+1) \left(\frac{3}{8}\right)^{2k} = -\frac{4}{3} + 4\log 2 + \frac{\sqrt{2}}{2}\log(3-2\sqrt{2});$$

$$(95)\qquad \sum_{k=2}^{\infty} (-1)^k \zeta(k) \left(\frac{5}{8}\right)^k = 1 + \frac{5}{16}\pi(\sqrt{2}-1) - \frac{5}{2}\log 2 - \frac{5\sqrt{2}}{16}\log(3-2\sqrt{2});$$

$$(96)\qquad \sum_{k=2}^{\infty} \zeta(k) \left(\frac{5}{8}\right)^k = \frac{5}{16}\pi(\sqrt{2}-1) + \frac{5}{2}\log 2 + \frac{5\sqrt{2}}{16}\log(3-2\sqrt{2});$$

$$(97)\qquad \sum_{k=1}^{\infty} \zeta(2k) \left(\frac{5}{8}\right)^{2k} = \frac{1}{2} + \frac{5}{16}\pi(\sqrt{2}-1);$$

$$(98)\qquad \sum_{k=1}^{\infty} \zeta(2k+1) \left(\frac{5}{8}\right)^{2k} = -\frac{4}{5} + 4\log 2 + \frac{\sqrt{2}}{2}\log(3-2\sqrt{2});$$

$$(99)\qquad \sum_{k=2}^{\infty} (-1)^k \zeta(k) \left(\frac{7}{8}\right)^k = 1 + \frac{7}{16}\pi(\sqrt{2}+1) - \frac{7}{2}\log 2 + \frac{7\sqrt{2}}{16}\log(3-2\sqrt{2});$$

$$(100)\qquad \sum_{k=2}^{\infty} \zeta(k) \left(\frac{7}{8}\right)^k = \frac{7}{16}\pi(\sqrt{2}+1) + \frac{7}{2}\log 2 - \frac{7\sqrt{2}}{16}\log(3-2\sqrt{2});$$

$$(101)\qquad \sum_{k=1}^{\infty} \zeta(2k) \left(\frac{7}{8}\right)^{2k} = \frac{1}{2} + \frac{7}{16}\pi(\sqrt{2}+1);$$

$$(102)\qquad \sum_{k=1}^{\infty} \zeta(2k+1) \left(\frac{7}{8}\right)^{2k} = -\frac{4}{7} + 4\log 2 - \frac{\sqrt{2}}{2}\log(3-2\sqrt{2});$$

(103)
$$\sum_{k=2}^{\infty}(-1)^k\frac{\zeta(k)}{10^k} = 1-\frac{\pi}{20}\left(2\sqrt{1+\frac{2}{5}\sqrt{5}}+\sqrt{1-\frac{2}{5}\sqrt{5}}\right)-\frac{1}{8}\log 5 - \frac{1}{40}\sqrt{5}\,\log\left(\frac{51+23\sqrt{5}}{4}\right)-\frac{1}{5}\log 2;$$

(104)
$$\sum_{k=2}^{\infty}\frac{\zeta(k)}{10^k} = \frac{\pi}{20}\left(2\sqrt{1+\frac{2}{5}\sqrt{5}}+\sqrt{1-\frac{2}{5}\sqrt{5}}\right)+\frac{1}{8}\log 5 + \frac{1}{40}\sqrt{5}\,\log\left(\frac{51+23\sqrt{5}}{4}\right)+\frac{1}{5}\log 2;$$

(105)
$$\sum_{k=1}^{\infty}\frac{\zeta(2k)}{10^{2k}} = \frac{1}{2}-\frac{\pi}{20}\left(2\sqrt{1+\frac{2}{5}\sqrt{5}}+\sqrt{1-\frac{2}{5}\sqrt{5}}\right) + \frac{1}{80}\sqrt{5}\,\log\left(\frac{3\sqrt{5}-1}{4}\right);$$

(106)
$$\sum_{k=1}^{\infty}\frac{\zeta(2k+1)}{10^{2k}} = -5+\frac{5}{4}\log 5+\frac{1}{8}\sqrt{5}\,\log\left(\frac{3743+1674\sqrt{5}}{8}\right)+2\log 2;$$

(107)
$$\sum_{k=2}^{\infty}(-1)^k\zeta(k)\left(\frac{3}{10}\right)^k = 1+\frac{3\pi}{20}\left(2\sqrt{1-\frac{2}{5}\sqrt{5}}-\sqrt{1+\frac{2}{5}\sqrt{5}}\right)-\frac{3}{8}\log 5 + \frac{9}{40}\sqrt{5}\,\log\left(\frac{3+2\sqrt{5}}{2}\right)-\frac{3}{5}\log 2;$$

(108)
$$\sum_{k=2}^{\infty}\zeta(k)\left(\frac{3}{10}\right)^k = -\frac{3\pi}{20}\left(\sqrt{1+\frac{2}{5}\sqrt{5}}-2\sqrt{1-\frac{2}{5}\sqrt{5}}\right)+\frac{3}{8}\log 5 - \frac{9}{40}\sqrt{5}\,\log\left(\frac{3+\sqrt{5}}{2}\right)+\frac{3}{5}\log 2;$$

(109)
$$\sum_{k=1}^{\infty}\zeta(2k)\left(\frac{3}{10}\right)^{2k} = \frac{1}{2}+\frac{3\pi}{20}\left(2\sqrt{1-\frac{2}{5}\sqrt{5}}-\sqrt{1+\frac{2}{5}\sqrt{5}}\right)+\frac{9}{80}\sqrt{5}\,\log\left(\frac{3\sqrt{5}-1}{4}\right);$$

(110) $$\sum_{k=1}^{\infty}\zeta(2k+1)\left(\frac{3}{10}\right)^{2k} = -\frac{5}{3}+\frac{5}{4}\log 5-\frac{3}{8}\sqrt{5}\,\log\left(\frac{19+9\sqrt{5}}{4}\right)+2\log 2;$$

(111) $$\begin{aligned}\sum_{k=2}^{\infty}(-1)^k\zeta(k)\left(\frac{7}{10}\right)^{k} =&1+\frac{7\pi}{20}\left(\sqrt{1+\frac{2}{5}\sqrt{5}}-2\sqrt{1-\frac{2}{5}\sqrt{5}}\right)-\frac{7}{8}\log 5\\ &+\frac{21}{40}\sqrt{5}\,\log\left(\frac{3+\sqrt{5}}{2}\right)-\frac{7}{5}\log 2;\end{aligned}$$

(112) $$\begin{aligned}\sum_{k=2}^{\infty}\zeta(k)\left(\frac{7}{10}\right)^{k} =&-\frac{7\pi}{20}\left(2\sqrt{1-\frac{2}{5}\sqrt{5}}-\sqrt{1+\frac{2}{5}\sqrt{5}}\right)+\frac{7}{8}\log 5\\ &-\frac{21}{40}\sqrt{5}\,\log\left(\frac{3+2\sqrt{5}}{2}\right)+\frac{7}{5}\log 2;\end{aligned}$$

(113) $$\begin{aligned}&\sum_{k=1}^{\infty}\zeta(2k)\left(\frac{7}{10}\right)^{2k}\\ &=\frac{1}{2}-\frac{7\pi}{20}\left(2\sqrt{1-\frac{2}{5}\sqrt{5}}-\sqrt{1+\frac{2}{5}\sqrt{5}}\right)-\frac{21}{80}\sqrt{5}\,\log\left(\frac{3\sqrt{5}-1}{4}\right);\end{aligned}$$

(114) $$\begin{aligned}&\sum_{k=1}^{\infty}\zeta(2k+1)\left(\frac{7}{10}\right)^{2k}\\ &=-\frac{5}{7}+\frac{5}{4}\log 5-\frac{3}{8}\sqrt{5}\log\left(\frac{19+9\sqrt{5}}{4}\right)+2\log 2;\end{aligned}$$

(115) $$\begin{aligned}\sum_{k=2}^{\infty}(-1)^k\zeta(k)\left(\frac{9}{10}\right)^{k} =&1+\frac{9\pi}{20}\left(\sqrt{1-\frac{2}{5}\sqrt{5}}+2\sqrt{1+\frac{2}{5}\sqrt{5}}\right)-\frac{9}{8}\log 5\\ &-\frac{9}{40}\sqrt{5}\,\log\left(\frac{147+65\sqrt{5}}{8}\right)-\frac{9}{5}\log 2;\end{aligned}$$

(116) $$\begin{aligned}\sum_{k=2}^{\infty}\zeta(k)\left(\frac{9}{10}\right)^{k} =&\frac{9\pi}{20}\left(2\sqrt{1+\frac{2}{5}\sqrt{5}}+\sqrt{1-\frac{2}{5}\sqrt{5}}\right)+\frac{9}{8}\log 5\\ &+\frac{9}{40}\sqrt{5}\,\log\left(\frac{51+23\sqrt{5}}{4}\right)+\frac{9}{5}\log 2;\end{aligned}$$

(117)
$$\sum_{k=1}^{\infty} \zeta(2k) \left(\frac{9}{10}\right)^{2k} = \frac{1}{2} + \frac{9\pi}{20} \left(2\sqrt{1+\frac{2}{5}\sqrt{5}} + \sqrt{1-\frac{2}{5}\sqrt{5}}\right) - \frac{9}{80}\sqrt{5}\, \log\left(\frac{3\sqrt{5}-1}{4}\right);$$

(118)
$$\sum_{k=1}^{\infty} \zeta(2k+1) \left(\frac{9}{10}\right)^{2k} = -\frac{5}{9} + \frac{5}{4}\log 5 + \frac{1}{8}\sqrt{5}\log\left(\frac{3743+1674\sqrt{5}}{8}\right) + 2\log 2;$$

(119)
$$\sum_{k=2}^{\infty} (-1)^k \frac{\zeta(k)}{12^k} = 1 - \frac{\pi}{24}\left(2+\sqrt{3}\right) + \frac{\sqrt{3}}{24}\log(7-4\sqrt{3}) - \frac{1}{8}\log 3 + \frac{1}{4}\log 2;$$

(120)
$$\sum_{k=2}^{\infty} \frac{\zeta(k)}{12^k} = -\frac{\pi}{24}\left(2+\sqrt{3}\right) - \frac{\sqrt{3}}{24}\log(7-4\sqrt{3}) + \frac{1}{8}\log 3 + \frac{1}{4}\log 2;$$

(121)
$$\sum_{k=1}^{\infty} \frac{\zeta(2k)}{12^{2k}} = \frac{1}{2} - \frac{\pi}{24}\left(2+\sqrt{3}\right);$$

(122)
$$\sum_{k=1}^{\infty} \frac{\zeta(2k+1)}{12^{2k}} = -6 - \frac{1}{2}\sqrt{3}\log(7-4\sqrt{3}) + \frac{3}{2}\log 3 + 3\log 2;$$

(123)
$$\sum_{k=2}^{\infty} (-1)^k \zeta(k) \left(\frac{5}{12}\right)^{k} = 1 - \frac{5\pi}{24}\left(2-\sqrt{3}\right) - \frac{5\sqrt{3}}{24}\log(7-4\sqrt{3}) - \frac{5}{8}\log 3 - \frac{5}{4}\log 2;$$

(124)
$$\sum_{k=2}^{\infty} \zeta(k) \left(\frac{5}{12}\right)^{k} = -\frac{5\pi}{24}\left(2-\sqrt{3}\right) + \frac{5\sqrt{3}}{24}\log(7-4\sqrt{3}) + \frac{5}{8}\log 3 + \frac{5}{4}\log 2;$$

(125)
$$\sum_{k=1}^{\infty} \zeta(2k) \left(\frac{5}{12}\right)^{2k} = \frac{1}{2} - \frac{5\pi}{24}\left(2-\sqrt{3}\right);$$

(126) $$\sum_{k=1}^{\infty} \zeta(2k+1)\left(\frac{5}{12}\right)^{2k} = -\frac{6}{5} + \frac{\sqrt{3}}{2}\log(7-4\sqrt{3}) + \frac{3}{2}\log 3 + 3\log 2;$$

(127) $$\sum_{k=2}^{\infty}(-1)^k \zeta(k)\left(\frac{7}{12}\right)^{k} = 1 + \frac{7\pi}{24}\left(2-\sqrt{3}\right) - \frac{7\sqrt{3}}{24}\log(7-4\sqrt{3}) - \frac{7}{8}\log 3 - \frac{7}{4}\log 2;$$

(128) $$\sum_{k=2}^{\infty} \zeta(k)\left(\frac{7}{12}\right)^{k} = \frac{7\pi}{24}\left(2-\sqrt{3}\right) + \frac{7\sqrt{3}}{24}\log(7-4\sqrt{3}) + \frac{7}{8}\log 3 + \frac{7}{4}\log 2;$$

(129) $$\sum_{k=1}^{\infty} \zeta(2k)\left(\frac{7}{12}\right)^{2k} = \frac{1}{2} + \frac{7\pi}{24}\left(2-\sqrt{3}\right);$$

(130) $$\sum_{k=1}^{\infty} \zeta(2k+1)\left(\frac{7}{12}\right)^{2k} = -\frac{6}{7} + \frac{\sqrt{3}}{2}\log(7-4\sqrt{3}) + \frac{3}{2}\log 3 + 3\log 2;$$

(131) $$\sum_{k=2}^{\infty}(-1)^k \zeta(k)\left(\frac{11}{12}\right)^{k} = 1 + \frac{11\pi}{24}\left(2+\sqrt{3}\right) + \frac{11\sqrt{3}}{24}\log(7-4\sqrt{3}) - \frac{11}{8}\log 3 - \frac{11}{4}\log 2;$$

(132) $$\sum_{k=2}^{\infty} \zeta(k)\left(\frac{11}{12}\right)^{k} = \frac{11\pi}{24}\left(2+\sqrt{3}\right) - \frac{11\sqrt{3}}{24}\log(7-4\sqrt{3}) + \frac{11}{8}\log 3 + \frac{11}{4}\log 2;$$

(133) $$\sum_{k=1}^{\infty} \zeta(2k)\left(\frac{11}{12}\right)^{2k} = \frac{1}{2} + \frac{11\pi}{24}\left(2+\sqrt{3}\right);$$

(134) $$\sum_{k=1}^{\infty} \zeta(2k+1)\left(\frac{11}{12}\right)^{2k} = -\frac{6}{11} - \frac{\sqrt{3}}{2}\log(7-4\sqrt{3}) + \frac{3}{2}\log 3 + 3\log 2.$$

In view of the various identities given in previous sections and the special case of 2.3(9) when $n = 1$, by setting $a = 2$ in (1) through (8), we obtain the following additional series identities:

(135) $$\sum_{k=2}^{\infty} \{\zeta(k) - 1\} \frac{t^k}{k} = \log \Gamma(2-t) + (1-\gamma)t \quad (|t| < 2);$$

(136) $$\sum_{k=2}^{\infty} (-1)^k \{\zeta(k) - 1\} \frac{t^k}{k} = \log \Gamma(2+t) + (\gamma - 1)t \quad (|t| < 2);$$

(137) $$\sum_{k=1}^{\infty} \{\zeta(2k) - 1\} \frac{t^{2k}}{k} = \log \Gamma(2+t) + \log \Gamma(2-t) \quad (|t| < 2);$$

(138) $$\sum_{k=1}^{\infty} \{\zeta(2k+1) - 1\} \frac{t^{2k+1}}{2k+1} = \frac{1}{2} \{\log \Gamma(2-t) - \log \Gamma(2+t)\} + (1-\gamma)t \quad (|t| < 2);$$

(139) $$\sum_{k=2}^{\infty} \{\zeta(k) - 1\} t^{k-1} = -\psi(2-t) + 1 - \gamma \quad (|t| < 2);$$

(140) $$\sum_{k=2}^{\infty} (-1)^k \{\zeta(k) - 1\} t^{k-1} = \psi(2+t) + \gamma - 1 \quad (|t| < 1);$$

(141) $$\sum_{k=1}^{\infty} \{\zeta(2k) - 1\} t^{2k-1} = \frac{1}{2} \{\psi(2+t) - \psi(2-t)\} \quad (|t| < 2);$$

(142) $$\sum_{k=1}^{\infty} \{\zeta(2k+1) - 1\} t^{2k} = -\frac{1}{2} \{\psi(2+t) + \psi(2-t)\} + 1 - \gamma \quad (|t| < 2);$$

(143) $$\sum_{k=1}^{\infty} \{\zeta(2k) - 1\} \frac{t^{2k}}{k} = \log \left(\frac{\pi t (1-t^2)}{\sin \pi t} \right) \quad (|t| < 2);$$

(144) $$\sum_{k=1}^{\infty} \{\zeta(2k) - 1\} t^{2k-1} = -\frac{\pi}{2} \cot(\pi t) + \frac{3t^2 - 1}{2t(t^2-1)} \quad (|t| < 2).$$

Similarly as above, setting various suitable arguments in (135) through (144), we readily obtain the following evaluations. For instance, by taking the limit in (136) as $t \to 2$, we obtain

$$\sum_{k=2}^{\infty}(-1)^k\{\zeta(k)-1\}\frac{2^k}{k} = 2\gamma - 2 + \log 6; \tag{145}$$

$$\sum_{k=2}^{\infty}(-1)^k\frac{\zeta(k)-1}{k\cdot 2^k} = \frac{\gamma}{2} - \frac{1}{2} + \log\left(\frac{3\sqrt{\pi}}{4}\right); \tag{146}$$

$$\sum_{k=2}^{\infty}\frac{\zeta(k)-1}{k\cdot 2^k} = \frac{1}{2} - \frac{\gamma}{2} + \log\left(\frac{\sqrt{\pi}}{2}\right); \tag{147}$$

$$\sum_{k=1}^{\infty}\frac{\zeta(2k)-1}{k\cdot 2^{2k}} = \log\left(\frac{3\pi}{8}\right); \tag{148}$$

$$\sum_{k=1}^{\infty}\frac{\zeta(2k+1)-1}{(2k+1)2^{2k}} = 1 - \gamma + \log\left(\frac{2}{3}\right); \tag{149}$$

$$\sum_{k=2}^{\infty}(-1)^k\frac{\zeta(k)-1}{k} = \gamma - 1 + \log 2; \tag{150}$$

$$\sum_{k=2}^{\infty}\frac{\zeta(k)-1}{k} = 1 - \gamma; \tag{151}$$

$$\sum_{k=1}^{\infty}\frac{\zeta(2k)-1}{k} = \log 2; \tag{152}$$

$$\sum_{k=1}^{\infty}\frac{\zeta(2k+1)-1}{2k+1} = 1 - \gamma - \frac{1}{2}\log 2; \tag{153}$$

$$\sum_{k=2}^{\infty}(-1)^k\frac{\zeta(k)-1}{k}\left(\frac{3}{2}\right)^k = \frac{3\gamma}{2} - \frac{3}{2} + \log\left(\frac{15\sqrt{\pi}}{8}\right); \tag{154}$$

(155)
$$\sum_{k=2}^{\infty} \frac{\zeta(k)-1}{k}\left(\frac{3}{2}\right)^{k} = \frac{3}{2} - \frac{3\gamma}{2} + \frac{1}{2}\log\pi;$$

(156)
$$\sum_{k=1}^{\infty} \frac{\zeta(2k)-1}{k}\left(\frac{3}{2}\right)^{2k} = \log\left(\frac{15\pi}{8}\right);$$

(157)
$$\sum_{k=1}^{\infty} \frac{\zeta(2k+1)-1}{2k+1}\left(\frac{3}{2}\right)^{2k} = 1 - \gamma + \frac{1}{3}\log\left(\frac{8}{15}\right);$$

(158)
$$\sum_{k=1}^{\infty} \frac{\zeta(2k)-1}{k\cdot 3^{2k}} = \log\left(\frac{16\pi}{27\sqrt{3}}\right);$$

(159)
$$\sum_{k=1}^{\infty} \frac{\zeta(2k)-1}{k}\left(\frac{2}{3}\right)^{2k} = \log\left(\frac{20\pi}{27\sqrt{3}}\right);$$

(160)
$$\sum_{k=1}^{\infty} \frac{\zeta(2k)-1}{k}\left(\frac{4}{3}\right)^{2k} = \log\left(\frac{56\pi}{27\sqrt{3}}\right);$$

(161)
$$\sum_{k=1}^{\infty} \frac{\zeta(2k)-1}{k}\left(\frac{5}{3}\right)^{2k} = \log\left(\frac{160\pi}{27\sqrt{3}}\right);$$

(162)
$$\sum_{k=1}^{\infty} \frac{\zeta(2k)-1}{k\cdot 2^{4k}} = \log\left(\frac{15\pi}{32\sqrt{2}}\right);$$

(163)
$$\sum_{k=1}^{\infty} \frac{\zeta(2k)-1}{k}\left(\frac{3}{4}\right)^{2k} = \log\left(\frac{21\pi}{32\sqrt{2}}\right);$$

(164)
$$\sum_{k=1}^{\infty} \frac{\zeta(2k)-1}{k}\left(\frac{5}{4}\right)^{2k} = \log\left(\frac{45\pi}{32\sqrt{2}}\right);$$

(165)
$$\sum_{k=1}^{\infty} \frac{\zeta(2k)-1}{k}\left(\frac{7}{4}\right)^{2k} = \log\left(\frac{231\pi}{32\sqrt{2}}\right);$$

(166)
$$\sum_{k=1}^{\infty}\frac{\zeta(2k)-1}{k\cdot 5^{2k}}=\log\frac{24\pi}{125}+\frac{1}{2}\log\left\{2+\frac{2}{5}\sqrt{5}+\frac{5}{16\pi^2}\left[\log\left(\frac{3\sqrt{5}-1}{4}\right)\right]^2 -\frac{1}{2\pi}\sqrt{5+2\sqrt{5}}\log\left(\frac{3\sqrt{5}-1}{4}\right)\right\};$$

(167)
$$\sum_{k=1}^{\infty}\frac{\zeta(2k)-1}{k\cdot 6^{2k}}=\log\left(\frac{35\pi}{108}\right);$$

(168)
$$\sum_{k=1}^{\infty}\frac{\zeta(2k)-1}{k}\left(\frac{5}{6}\right)^{2k}=\log\left(\frac{55\pi}{108}\right);$$

(169)
$$\sum_{k=1}^{\infty}\frac{\zeta(2k)-1}{k}\left(\frac{7}{6}\right)^{2k}=\log\left(\frac{91\pi}{108}\right);$$

(170)
$$\sum_{k=1}^{\infty}\frac{\zeta(2k)-1}{k}\left(\frac{11}{6}\right)^{2k}=\log\left(\frac{935\pi}{108}\right);$$

(171)
$$\sum_{k=1}^{\infty}\frac{\zeta(2k)-1}{k\cdot 2^{6k}}=\log\left(\frac{63\pi}{256\sqrt{2-\sqrt{2}}}\right);$$

(172)
$$\sum_{k=1}^{\infty}\frac{\zeta(2k)-1}{k}\left(\frac{3}{8}\right)^{2k}=\log\left(\frac{165\pi}{256\sqrt{2+\sqrt{2}}}\right);$$

(173)
$$\sum_{k=1}^{\infty}\frac{\zeta(2k)-1}{k}\left(\frac{5}{8}\right)^{2k}=\log\left(\frac{195\pi}{256\sqrt{2+\sqrt{2}}}\right);$$

(174)
$$\sum_{k=1}^{\infty}\frac{\zeta(2k)-1}{k}\left(\frac{7}{8}\right)^{2k}=\log\left(\frac{105\pi}{256\sqrt{2-\sqrt{2}}}\right);$$

(175)
$$\sum_{k=1}^{\infty}\frac{\zeta(2k)-1}{k}\left(\frac{9}{8}\right)^{2k}=\log\left(\frac{153\pi}{256\sqrt{2-\sqrt{2}}}\right);$$

(176) $$\sum_{k=1}^{\infty} \frac{\zeta(2k)-1}{k}\left(\frac{11}{8}\right)^{2k} = \log\left(\frac{627\pi}{256\sqrt{2+\sqrt{2}}}\right);$$

(177) $$\sum_{k=1}^{\infty} \frac{\zeta(2k)-1}{k}\left(\frac{13}{8}\right)^{2k} = \log\left(\frac{1365\pi}{256\sqrt{2+\sqrt{2}}}\right);$$

(178) $$\sum_{k=1}^{\infty} \frac{\zeta(2k)-1}{k}\left(\frac{15}{8}\right)^{2k} = \log\left(\frac{2415\pi}{256\sqrt{2-\sqrt{2}}}\right);$$

(179) $$\sum_{k=1}^{\infty} \frac{\zeta(2k)-1}{k\cdot 12^{2k}} = \log\left(\frac{143\pi(\sqrt{3}+1)}{864\sqrt{2}}\right);$$

(180) $$\sum_{k=1}^{\infty} \frac{\zeta(2k)-1}{k}\left(\frac{5}{12}\right)^{2k} = \log\left(\frac{595\pi(\sqrt{3}-1)}{864\sqrt{2}}\right);$$

(181) $$\sum_{k=1}^{\infty} \frac{\zeta(2k)-1}{k}\left(\frac{7}{12}\right)^{2k} = \log\left(\frac{665\pi(\sqrt{3}-1)}{864\sqrt{2}}\right);$$

(182) $$\sum_{k=1}^{\infty} \frac{\zeta(2k)-1}{k}\left(\frac{11}{12}\right)^{2k} = \log\left(\frac{253\pi(\sqrt{3}+1)}{864\sqrt{2}}\right);$$

(183) $$\sum_{k=1}^{\infty} \frac{\zeta(2k)-1}{k}\left(\frac{13}{12}\right)^{2k} = \log\left(\frac{325\pi(\sqrt{3}+1)}{864\sqrt{2}}\right);$$

(184) $$\sum_{k=1}^{\infty} \frac{\zeta(2k)-1}{k}\left(\frac{17}{12}\right)^{2k} = \log\left(\frac{2465\pi(\sqrt{3}-1)}{864\sqrt{2}}\right);$$

(185) $$\sum_{k=1}^{\infty} \frac{\zeta(2k)-1}{k}\left(\frac{19}{12}\right)^{2k} = \log\left(\frac{4123\pi(\sqrt{3}-1)}{864\sqrt{2}}\right);$$

(186) $$\sum_{k=1}^{\infty} \frac{\zeta(2k)-1}{k} \left(\frac{23}{12}\right)^{2k} = \log\left(\frac{8855\pi(\sqrt{3}+1)}{864\sqrt{2}}\right);$$

(187) $$\sum_{k=2}^{\infty} (-1)^k \frac{\zeta(k)-1}{2^k} = \frac{5}{6} - \log 2;$$

(188) $$\sum_{k=2}^{\infty} \frac{\zeta(k)-1}{2^k} = \log 2 - \frac{1}{2};$$

(189) $$\sum_{k=1}^{\infty} \frac{\zeta(2k)-1}{2^{2k}} = \frac{1}{6};$$

(190) $$\sum_{k=1}^{\infty} \frac{\zeta(2k+1)-1}{2^{2k}} = -\frac{4}{3} + 2\log 2;$$

(191) $$\sum_{k=2}^{\infty} (-1)^k \{\zeta(k)-1\} = \frac{1}{2};$$

(192) $$\sum_{k=2}^{\infty} \{\zeta(k)-1\} = 1;$$

(193) $$\sum_{k=1}^{\infty} \{\zeta(2k)-1\} = \frac{3}{4};$$

(194) $$\sum_{k=1}^{\infty} \{\zeta(2k+1)-1\} = \frac{1}{4};$$

(195) $$\sum_{k=2}^{\infty} (-1)^k \{\zeta(k)-1\} \left(\frac{3}{2}\right)^k = \frac{31}{10} - 3\log 2;$$

(196) $$\sum_{k=2}^{\infty} \{\zeta(k)-1\} \left(\frac{3}{2}\right)^k = \frac{3}{2} + 3\log 2;$$

(197) $$\sum_{k=1}^{\infty}\{\zeta(2k)-1\}\left(\frac{3}{2}\right)^{2k} = \frac{23}{10};$$

(198) $$\sum_{k=1}^{\infty}\{\zeta(2k+1)-1\}\left(\frac{3}{2}\right)^{2k} = -\frac{8}{15}+2\log 2;$$

(199) $$\sum_{k=2}^{\infty}(-1)^k\frac{\zeta(k)-1}{3^k} = \frac{11}{12}-\frac{\pi}{18}\sqrt{3}-\frac{1}{2}\log 3;$$

(200) $$\sum_{k=2}^{\infty}\frac{\zeta(k)-1}{3^k} = -\frac{1}{6}-\frac{\pi}{18}\sqrt{3}+\frac{1}{2}\log 3;$$

(201) $$\sum_{k=1}^{\infty}\frac{\zeta(2k)-1}{3^{2k}} = \frac{3}{8}-\frac{1}{18}\pi\sqrt{3};$$

(202) $$\sum_{k=1}^{\infty}\frac{\zeta(2k+1)-1}{3^{2k}} = -\frac{13}{8}+\frac{3}{2}\log 3;$$

(203) $$\sum_{k=2}^{\infty}(-1)^k\{\zeta(k)-1\}\left(\frac{2}{3}\right)^{k} = \frac{11}{15}+\frac{1}{9}\pi\sqrt{3}-\log 3;$$

(204) $$\sum_{k=2}^{\infty}\{\zeta(k)-1\}\left(\frac{2}{3}\right)^{k} = -\frac{4}{3}+\frac{1}{9}\pi\sqrt{3}+\log 3;$$

(205) $$\sum_{k=1}^{\infty}\{\zeta(2k)-1\}\left(\frac{2}{3}\right)^{2k} = -\frac{3}{10}+\frac{1}{9}\pi\sqrt{3};$$

(206) $$\sum_{k=1}^{\infty}\{\zeta(2k+1)-1\}\left(\frac{2}{3}\right)^{2k} = -\frac{31}{20}+\frac{3}{2}\log 3;$$

(207) $$\sum_{k=2}^{\infty}(-1)^k\{\zeta(k)-1\}\left(\frac{4}{3}\right)^{k} = \frac{89}{21}-\frac{2}{9}\pi\sqrt{3}-2\log 3;$$

(208) $$\sum_{k=2}^{\infty}\{\zeta(k)-1\}\left(\frac{4}{3}\right)^{k}=\frac{4}{3}-\frac{2}{9}\pi\sqrt{3}+2\log 3;$$

(209) $$\sum_{k=1}^{\infty}\{\zeta(2k)-1\}\left(\frac{4}{3}\right)^{2k}=\frac{39}{14}-\frac{2}{9}\pi\sqrt{3};$$

(210) $$\sum_{k=1}^{\infty}\{\zeta(2k+1)-1\}\left(\frac{4}{3}\right)^{2k}=-\frac{61}{56}+\frac{3}{2}\log 3;$$

(211) $$\sum_{k=2}^{\infty}(-1)^{k}\{\zeta(k)-1\}\left(\frac{5}{3}\right)^{k}=\frac{59}{24}+\frac{5}{18}\pi\sqrt{3}-\frac{5}{2}\log 3;$$

(212) $$\sum_{k=2}^{\infty}\{\zeta(k)-1\}\left(\frac{5}{3}\right)^{k}=\frac{5}{3}+\frac{5}{18}\pi\sqrt{3}+\frac{5}{2}\log 3;$$

(213) $$\sum_{k=1}^{\infty}\{\zeta(2k)-1\}\left(\frac{5}{3}\right)^{2k}=\frac{33}{16}+\frac{5}{18}\pi\sqrt{3};$$

(214) $$\sum_{k=1}^{\infty}\{\zeta(2k+1)-1\}\left(\frac{5}{3}\right)^{2k}=-\frac{19}{80}+\frac{3}{2}\log 3;$$

(215) $$\sum_{k=2}^{\infty}(-1)^{k}\frac{\zeta(k)-1}{2^{2k}}=\frac{19}{20}-\frac{1}{8}\pi-\frac{3}{4}\log 2;$$

(216) $$\sum_{k=2}^{\infty}\frac{\zeta(k)-1}{2^{2k}}=-\frac{1}{12}-\frac{1}{8}\pi+\frac{3}{4}\log 2;$$

(217) $$\sum_{k=1}^{\infty}\frac{\zeta(2k)-1}{2^{4k}}=\frac{13}{30}-\frac{1}{8}\pi;$$

(218) $$\sum_{k=1}^{\infty}\frac{\zeta(2k+1)-1}{2^{4k}}=-\frac{31}{15}+3\log 2;$$

$$(219)\qquad \sum_{k=2}^{\infty}(-1)^k\{\zeta(k)-1\}\left(\frac{3}{4}\right)^k = \frac{19}{28}+\frac{3}{8}\pi-\frac{9}{4}\log 2;$$

$$(220)\qquad \sum_{k=2}^{\infty}\{\zeta(k)-1\}\left(\frac{3}{4}\right)^k = -\frac{9}{4}+\frac{3}{8}\pi+\frac{9}{4}\log 2;$$

$$(221)\qquad \sum_{k=1}^{\infty}\{\zeta(2k)-1\}\left(\frac{3}{4}\right)^{2k} = -\frac{11}{14}+\frac{3}{8}\pi;$$

$$(222)\qquad \sum_{k=1}^{\infty}\{\zeta(2k+1)-1\}\left(\frac{3}{4}\right)^{2k} = -\frac{41}{21}+3\log 2;$$

$$(223)\qquad \sum_{k=2}^{\infty}(-1)^k\{\zeta(k)-1\}\left(\frac{5}{4}\right)^k = \frac{191}{36}-\frac{5}{8}\pi-\frac{15}{4}\log 2;$$

$$(224)\qquad \sum_{k=2}^{\infty}\{\zeta(k)-1\}\left(\frac{5}{4}\right)^k = \frac{5}{4}-\frac{5}{8}\pi+\frac{15}{4}\log 2;$$

$$(225)\qquad \sum_{k=1}^{\infty}\{\zeta(2k)-1\}\left(\frac{5}{4}\right)^{2k} = \frac{59}{18}-\frac{5}{8}\pi;$$

$$(226)\qquad \sum_{k=1}^{\infty}\{\zeta(2k+1)-1\}\left(\frac{5}{4}\right)^{2k} = -\frac{73}{45}+3\log 2;$$

$$(227)\qquad \sum_{k=2}^{\infty}(-1)^k\{\zeta(k)-1\}\left(\frac{7}{4}\right)^k = \frac{293}{132}+\frac{7}{8}\pi-\frac{21}{4}\log 2;$$

$$(228)\qquad \sum_{k=2}^{\infty}\{\zeta(k)-1\}\left(\frac{7}{4}\right)^k = \frac{7}{4}+\frac{7}{8}\pi+\frac{21}{4}\log 2;$$

$$(229)\qquad \sum_{k=1}^{\infty}\{\zeta(2k)-1\}\left(\frac{7}{4}\right)^{2k} = \frac{131}{66}+\frac{7}{8}\pi;$$

(230) $$\sum_{k=1}^{\infty}\{\zeta(2k+1)-1\}\left(\frac{7}{4}\right)^{2k}=-\frac{31}{231}+3\log 2;$$

(231) $$\sum_{k=2}^{\infty}(-1)^k\frac{\zeta(k)-1}{5^k}=\frac{29}{30}-\frac{1}{10}\pi\sqrt{1+\frac{2}{5}\sqrt{5}}-\frac{1}{4}\log 5-\frac{1}{20}\sqrt{5}\log\left(\frac{3+\sqrt{5}}{2}\right);$$

(232) $$\sum_{k=2}^{\infty}\frac{\zeta(k)-1}{5^k}=-\frac{1}{20}-\frac{1}{10}\pi\sqrt{1+\frac{2}{5}\sqrt{5}}+\frac{1}{4}\log 5+\frac{1}{20}\sqrt{5}\log\left(\frac{3+2\sqrt{5}}{2}\right);$$

(233) $$\sum_{k=1}^{\infty}\frac{\zeta(2k)-1}{5^{2k}}=\frac{11}{24}-\frac{1}{10}\pi\sqrt{1+\frac{2}{5}\sqrt{5}}+\frac{1}{40}\sqrt{5}\log\left(\frac{3\sqrt{5}-1}{4}\right);$$

(234) $$\sum_{k=1}^{\infty}\frac{\zeta(2k+1)-1}{5^{2k}}=-\frac{61}{24}+\frac{5}{4}\log 5+\frac{1}{8}\sqrt{5}\log\left(\frac{19+9\sqrt{5}}{4}\right);$$

(235) $$\begin{aligned}&\sum_{k=2}^{\infty}(-1)^k\{\zeta(k)-1\}\left(\frac{2}{5}\right)^{k}\\&\quad=\frac{31}{35}-\frac{1}{5}\pi\sqrt{1-\frac{2}{5}\sqrt{5}}-\frac{1}{2}\log 5+\frac{1}{10}\sqrt{5}\log\left(\frac{3+\sqrt{5}}{2}\right);\end{aligned}$$

(236) $$\begin{aligned}&\sum_{k=2}^{\infty}\{\zeta(k)-1\}\left(\frac{2}{5}\right)^{k}\\&\quad=-\frac{4}{15}-\frac{1}{5}\pi\sqrt{1-\frac{2}{5}\sqrt{5}}+\frac{1}{2}\log 5-\frac{1}{10}\sqrt{5}\log\left(\frac{3+2\sqrt{5}}{2}\right);\end{aligned}$$

(237) $$\sum_{k=1}^{\infty}\{\zeta(2k)-1\}\left(\frac{2}{5}\right)^{2k}=\frac{13}{42}-\frac{1}{5}\pi\sqrt{1-\frac{2}{5}\sqrt{5}}+\frac{1}{20}\sqrt{5}\log\left(\frac{3\sqrt{5}+1}{11}\right);$$

(238) $$\sum_{k=1}^{\infty}\{\zeta(2k+1)-1\}\left(\frac{2}{5}\right)^{2k}=-\frac{121}{84}+\frac{5}{4}\log 5-\frac{1}{8}\sqrt{5}\log\left(\frac{19+9\sqrt{5}}{4}\right);$$

(239)
$$\sum_{k=2}^{\infty}(-1)^k\{\zeta(k)-1\}\left(\frac{3}{5}\right)^k = \frac{31}{40}+\frac{3}{10}\pi\sqrt{1-\frac{2}{5}\sqrt{5}}-\frac{3}{4}\log 5+\frac{3}{20}\sqrt{5}\log\left(\frac{3+2\sqrt{5}}{2}\right);$$

(240)
$$\sum_{k=2}^{\infty}\{\zeta(k)-1\}\left(\frac{3}{5}\right)^k = -\frac{9}{10}+\frac{3}{10}\pi\sqrt{1-\frac{2}{5}\sqrt{5}}+\frac{3}{4}\log 5-\frac{3}{20}\sqrt{5}\log\left(\frac{3+\sqrt{5}}{2}\right);$$

(241)
$$\sum_{k=1}^{\infty}\{\zeta(2k)-1\}\left(\frac{3}{5}\right)^{2k} = -\frac{1}{16}+\frac{3}{10}\pi\sqrt{1-\frac{2}{5}\sqrt{5}}+\frac{3}{40}\sqrt{5}\log\left(\frac{3\sqrt{5}-1}{4}\right);$$

(242)
$$\sum_{k=1}^{\infty}\{\zeta(2k+1)-1\}\left(\frac{3}{5}\right)^{2k} = -\frac{67}{48}+\frac{5}{4}\log 5-\frac{1}{8}\sqrt{5}\log\left(\frac{19+9\sqrt{5}}{4}\right);$$

(243)
$$\sum_{k=2}^{\infty}(-1)^k\{\zeta(k)-1\}\left(\frac{4}{5}\right)^k = \frac{29}{45}+\frac{2}{5}\pi\sqrt{1+\frac{2}{5}\sqrt{5}}-\log 5-\frac{1}{5}\sqrt{5}\log\left(\frac{3+2\sqrt{5}}{2}\right);$$

(244)
$$\sum_{k=2}^{\infty}\{\zeta(k)-1\}\left(\frac{4}{5}\right)^k = -\frac{16}{5}+\frac{2}{5}\pi\sqrt{1+\frac{2}{5}\sqrt{5}}+\log 5+\frac{1}{5}\sqrt{5}\log\left(\frac{3+\sqrt{5}}{2}\right);$$

(245)
$$\sum_{k=1}^{\infty}\{\zeta(2k)-1\}\left(\frac{4}{5}\right)^{2k} = -\frac{23}{18}+\frac{2}{5}\pi\sqrt{1+\frac{2}{5}\sqrt{5}}+\frac{1}{10}\sqrt{5}\log\left(\frac{1+3\sqrt{5}}{11}\right);$$

(246)
$$\sum_{k=1}^{\infty}\{\zeta(2k+1)-1\}\left(\frac{4}{5}\right)^{2k} = -\frac{173}{72}+\frac{5}{4}\log 5+\frac{1}{8}\sqrt{5}\log\left(\frac{19+9\sqrt{5}}{4}\right);$$

$$\sum_{k=2}^{\infty}(-1)^k\{\zeta(k)-1\}\left(\frac{6}{5}\right)^k = \frac{349}{55}-\frac{3}{5}\pi\sqrt{1+\frac{2}{5}\sqrt{5}}-\frac{3}{2}\log 5-\frac{3}{10}\sqrt{5}\log\left(\frac{3+\sqrt{5}}{2}\right); \tag{247}$$

$$\sum_{k=2}^{\infty}\{\zeta(k)-1\}\left(\frac{6}{5}\right)^k = \frac{6}{5}-\frac{3}{5}\pi\sqrt{1+\frac{2}{5}\sqrt{5}}+\frac{3}{2}\log 5+\frac{3}{10}\sqrt{5}\log\left(\frac{3+2\sqrt{5}}{2}\right); \tag{248}$$

$$\sum_{k=1}^{\infty}\{\zeta(2k)-1\}\left(\frac{6}{5}\right)^{2k} = \frac{83}{22}-\frac{3}{5}\pi\sqrt{1+\frac{2}{5}\sqrt{5}}+\frac{3}{20}\sqrt{5}\log\left(\frac{3\sqrt{5}-1}{4}\right); \tag{249}$$

$$\sum_{k=1}^{\infty}\{\zeta(2k+1)-1\}\left(\frac{6}{5}\right)^{2k} = -\frac{283}{132}+\frac{5}{4}\log 5+\frac{1}{8}\sqrt{5}\log\left(\frac{19+9\sqrt{5}}{4}\right); \tag{250}$$

$$\sum_{k=2}^{\infty}(-1)^k\{\zeta(k)-1\}\left(\frac{7}{5}\right)^k = \frac{221}{60}-\frac{7}{10}\pi\sqrt{1-\frac{2}{5}\sqrt{5}}-\frac{7}{4}\log 5+\frac{7}{20}\sqrt{5}\log\left(\frac{3+\sqrt{5}}{2}\right); \tag{251}$$

$$\sum_{k=2}^{\infty}\{\zeta(k)-1\}\left(\frac{7}{5}\right)^k = \frac{7}{5}-\frac{7}{10}\pi\sqrt{1-\frac{2}{5}\sqrt{5}}+\frac{7}{4}\log 5-\frac{7}{20}\sqrt{5}\log\left(\frac{3+2\sqrt{5}}{2}\right); \tag{252}$$

$$\sum_{k=1}^{\infty}\{\zeta(2k)-1\}\left(\frac{7}{5}\right)^{2k} = \frac{61}{24}-\frac{7}{10}\pi\sqrt{1-\frac{2}{5}\sqrt{5}}+\frac{7}{40}\sqrt{5}\log\left(\frac{1+3\sqrt{5}}{11}\right); \tag{253}$$

$$\sum_{k=1}^{\infty}\{\zeta(2k+1)-1\}\left(\frac{7}{5}\right)^{2k} = -\frac{137}{168}+\frac{5}{4}\log 5-\frac{1}{8}\sqrt{5}\log\left(\frac{19+9\sqrt{5}}{4}\right); \tag{254}$$

(255)
$$\sum_{k=2}^{\infty}(-1)^k\{\zeta(k)-1\}\left(\frac{8}{5}\right)^k = \frac{523}{195}+\frac{4}{5}\pi\sqrt{1-\frac{2}{5}\sqrt{5}}-2\log 5+\frac{2}{5}\sqrt{5}\,\log\left(\frac{3+2\sqrt{5}}{2}\right);$$

(256)
$$\sum_{k=2}^{\infty}\{\zeta(k)-1\}\left(\frac{8}{5}\right)^k = \frac{8}{5}+\frac{4}{5}\pi\sqrt{1-\frac{2}{5}\sqrt{5}}+2\log 5-\frac{2}{5}\sqrt{5}\,\log\left(\frac{3+\sqrt{5}}{2}\right);$$

(257)
$$\sum_{k=1}^{\infty}\{\zeta(2k)-1\}\left(\frac{8}{5}\right)^{2k} = \frac{835}{546}+\frac{4}{7}\pi\sqrt{1-\frac{2}{5}\sqrt{5}}+\frac{1}{7}\sqrt{5}\,\log\left(\frac{3\sqrt{5}-1}{4}\right);$$

(258)
$$\sum_{k=1}^{\infty}\{\zeta(2k+1)-1\}\left(\frac{8}{5}\right)^{2k} = -\frac{211}{624}+\frac{5}{4}\log 5-\frac{1}{8}\sqrt{5}\,\log\left(\frac{19+9\sqrt{5}}{4}\right);$$

(259)
$$\sum_{k=2}^{\infty}(-1)^k\{\zeta(k)-1\}\left(\frac{9}{5}\right)^k = \frac{293}{140}+\frac{9}{10}\pi\sqrt{1+\frac{2}{5}\sqrt{5}}-\frac{9}{4}\log 5-\frac{9}{20}\sqrt{5}\,\log\left(\frac{3+2\sqrt{5}}{2}\right);$$

(260)
$$\sum_{k=2}^{\infty}\{\zeta(k)-1\}\left(\frac{9}{5}\right)^k = \frac{9}{5}+\frac{9}{10}\pi\sqrt{1+\frac{2}{5}\sqrt{5}}+\frac{9}{4}\log 5+\frac{9}{20}\sqrt{5}\,\log\left(\frac{3+\sqrt{5}}{2}\right);$$

(261)
$$\sum_{k=1}^{\infty}\{\zeta(2k)-1\}\left(\frac{9}{5}\right)^{2k} = \frac{109}{56}+\frac{9}{10}\pi\sqrt{1+\frac{2}{5}\sqrt{5}}+\frac{9}{40}\sqrt{5}\,\log\left(\frac{1+3\sqrt{5}}{11}\right);$$

(262)
$$\sum_{k=1}^{\infty}\{\zeta(2k+1)-1\}\left(\frac{9}{5}\right)^{2k} = -\frac{41}{504}+\frac{5}{4}\log 5+\frac{1}{8}\sqrt{5}\,\log\left(\frac{19+9\sqrt{5}}{}\right);$$

(263) $$\sum_{k=2}^{\infty}(-1)^k\frac{\zeta(k)-1}{6^k}=\frac{41}{42}-\frac{1}{12}\pi\sqrt{3}-\frac{1}{4}\log 3-\frac{1}{3}\log 2;$$

(264) $$\sum_{k=2}^{\infty}\frac{\zeta(k)-1}{6^k}=-\frac{1}{30}-\frac{1}{12}\pi\sqrt{3}+\frac{1}{4}\log 3+\frac{1}{3}\log 2;$$

(265) $$\sum_{k=1}^{\infty}\frac{\zeta(2k)-1}{6^{2k}}=\frac{33}{70}-\frac{1}{12}\pi\sqrt{3};$$

(266) $$\sum_{k=1}^{\infty}\frac{\zeta(2k+1)-1}{6^{2k}}=-\frac{106}{35}+\frac{3}{2}\log 3+2\log 2;$$

(267) $$\sum_{k=2}^{\infty}(-1)^k\{\zeta(k)-1\}\left(\frac{5}{6}\right)^k=\frac{41}{66}+\frac{5}{12}\pi\sqrt{3}-\frac{5}{4}\log 3-\frac{5}{3}\log 2;$$

(268) $$\sum_{k=2}^{\infty}\{\zeta(k)-1\}\left(\frac{5}{6}\right)^k=-\frac{25}{6}+\frac{5}{12}\pi\sqrt{3}+\frac{5}{4}\log 3+\frac{5}{3}\log 2;$$

(269) $$\sum_{k=1}^{\infty}\{\zeta(2k)-1\}\left(\frac{5}{6}\right)^{2k}=-\frac{107}{66}+\frac{5}{12}\pi\sqrt{3};$$

(270) $$\sum_{k=1}^{\infty}\{\zeta(2k+1)-1\}\left(\frac{5}{6}\right)^{2k}=-\frac{158}{55}+\frac{3}{2}\log 3+2\log 2;$$

(271) $$\sum_{k=2}^{\infty}(-1)^k\{\zeta(k)-1\}\left(\frac{7}{6}\right)^k=\frac{575}{78}-\frac{7}{12}\pi\sqrt{3}-\frac{7}{4}\log 3-\frac{7}{3}\log 2;$$

(272) $$\sum_{k=2}^{\infty}\{\zeta(k)-1\}\left(\frac{7}{6}\right)^k=\frac{7}{6}-\frac{7}{12}\pi\sqrt{3}+\frac{7}{4}\log 3+\frac{7}{3}\log 2;$$

(273) $$\sum_{k=1}^{\infty}\{\zeta(2k)-1\}\left(\frac{7}{6}\right)^{2k}=\frac{111}{26}-\frac{7}{12}\pi\sqrt{3};$$

(274) $$\sum_{k=1}^{\infty}\{\zeta(2k+1)-1\}\left(\frac{7}{6}\right)^{2k}=-\frac{242}{91}+\frac{3}{2}\log 3+2\log 2;$$

(275) $$\sum_{k=2}^{\infty}(-1)^k\{\zeta(k)-1\}\left(\frac{11}{6}\right)^{k}=\frac{11297}{5610}+\frac{11}{12}\pi\sqrt{3}-\frac{11}{4}\log 3-\frac{11}{3}\log 2;$$

(276) $$\sum_{k=2}^{\infty}\{\zeta(k)-1\}\left(\frac{11}{6}\right)^{k}=\frac{11}{6}+\frac{11}{12}\pi\sqrt{3}+\frac{11}{4}\log 3+\frac{11}{3}\log 2;$$

(277) $$\sum_{k=1}^{\infty}\{\zeta(2k)-1\}\left(\frac{11}{6}\right)^{2k}=\frac{3597}{1870}+\frac{11}{12}\pi\sqrt{3};$$

(278) $$\sum_{k=1}^{\infty}\{\zeta(2k+1)-1\}\left(\frac{11}{6}\right)^{2k}=-\frac{46}{935}+\frac{3}{2}\log 3+2\log 2;$$

(279) $$\sum_{k=2}^{\infty}(-1)^k\frac{\zeta(k)-1}{2^{3k}}=\frac{71}{72}-\frac{1}{16}\pi(\sqrt{2}+1)-\frac{1}{2}\log 2+\frac{\sqrt{2}}{16}\log(3-2\sqrt{2});$$

(280) $$\sum_{k=2}^{\infty}\frac{\zeta(k)-1}{2^{3k}}=-\frac{1}{56}-\frac{1}{16}\pi(\sqrt{2}+1)+\frac{1}{2}\log 2-\frac{\sqrt{2}}{16}\log(3-2\sqrt{2});$$

(281) $$\sum_{k=1}^{\infty}\frac{\zeta(2k)-1}{2^{6k}}=\frac{61}{126}-\frac{1}{16}\pi(\sqrt{2}+1);$$

(282) $$\sum_{k=1}^{\infty}\frac{\zeta(2k+1)-1}{2^{6k}}=-\frac{253}{63}+4\log 2-\frac{\sqrt{2}}{2}\log(3-2\sqrt{2});$$

(283) $$\sum_{k=2}^{\infty}(-1)^k\{\zeta(k)-1\}\left(\frac{3}{8}\right)^{k}$$
$$=\frac{79}{88}-\frac{3}{16}\pi(\sqrt{2}-1)-\frac{3}{2}\log 2-\frac{3\sqrt{2}}{16}\log(3-2\sqrt{2});$$

$$\sum_{k=2}^{\infty}\{\zeta(k)-1\}\left(\frac{3}{8}\right)^{k} = -\frac{9}{40}-\frac{3}{16}\pi(\sqrt{2}-1)+\frac{3}{2}\log 2+\frac{3}{16}\sqrt{2}\log(3-2\sqrt{2}); \tag{284}$$

$$\sum_{k=1}^{\infty}\{\zeta(2k)-1\}\left(\frac{3}{8}\right)^{2k} = \frac{37}{110}-\frac{3}{16}\pi(\sqrt{2}-1); \tag{285}$$

$$\sum_{k=1}^{\infty}\{\zeta(2k+1)-1\}\left(\frac{3}{8}\right)^{2k} = -\frac{247}{165}+4\log 2+\frac{\sqrt{2}}{2}\log(3-2\sqrt{2}); \tag{286}$$

$$\sum_{k=2}^{\infty}(-1)^k\{\zeta(k)-1\}\left(\frac{5}{8}\right)^{k} = \frac{79}{104}+\frac{5}{16}\pi(\sqrt{2}-1)-\frac{5}{2}\log 2-\frac{5\sqrt{2}}{16}\log(3-2\sqrt{2}); \tag{287}$$

$$\sum_{k=2}^{\infty}\{\zeta(k)-1\}\left(\frac{5}{8}\right)^{k} = -\frac{25}{24}+\frac{5}{16}\pi(\sqrt{2}-1)+\frac{5}{2}\log 2+\frac{5\sqrt{2}}{16}\log(3-2\sqrt{2}); \tag{288}$$

$$\sum_{k=1}^{\infty}\{\zeta(2k)-1\}\left(\frac{5}{8}\right)^{2k} = -\frac{55}{78}+\frac{5}{16}\pi(\sqrt{2}-1); \tag{289}$$

$$\sum_{k=1}^{\infty}\{\zeta(2k+1)-1\}\left(\frac{5}{8}\right)^{2k} = -\frac{281}{195}+4\log 2+\frac{\sqrt{2}}{2}\log(3-2\sqrt{2}); \tag{290}$$

$$\sum_{k=2}^{\infty}(-1)^k\{\zeta(k)-1\}\left(\frac{7}{8}\right)^{k} = \frac{71}{120}+\frac{7}{16}\pi(\sqrt{2}+1)-\frac{7}{2}\log 2+\frac{7\sqrt{2}}{16}\log(3-2\sqrt{2}); \tag{291}$$

$$\sum_{k=2}^{\infty}\{\zeta(k)-1\}\left(\frac{7}{8}\right)^{k} = -\frac{49}{8}+\frac{7}{16}\pi(\sqrt{2}+1)+\frac{7}{2}\log 2-\frac{7\sqrt{2}}{16}\log(3-2\sqrt{2}); \tag{292}$$

(293) $$\sum_{k=1}^{\infty}\{\zeta(2k)-1\}\left(\frac{7}{8}\right)^{2k}=-\frac{83}{30}+\frac{7}{16}\pi(\sqrt{2}+1);$$

(294) $$\sum_{k=1}^{\infty}\{\zeta(2k+1)-1\}\left(\frac{7}{8}\right)^{2k}=-\frac{403}{105}+4\log 2-\frac{\sqrt{2}}{2}\log(3-2\sqrt{2});$$

(295) $$\sum_{k=2}^{\infty}(-1)^k\{\zeta(k)-1\}\left(\frac{9}{8}\right)^{k}$$
$$=\frac{1279}{136}-\frac{9}{16}\pi(\sqrt{2}+1)-\frac{9}{2}\log 2+\frac{9\sqrt{2}}{16}\log(3-2\sqrt{2});$$

(296) $$\sum_{k=2}^{\infty}\{\zeta(k)-1\}\left(\frac{9}{8}\right)^{k}$$
$$=\frac{9}{8}-\frac{9}{16}\pi(\sqrt{2}+1)+\frac{9}{2}\log 2-\frac{9\sqrt{2}}{16}\log(3-2\sqrt{2});$$

(297) $$\sum_{k=1}^{\infty}\{\zeta(2k)-1\}\left(\frac{9}{8}\right)^{2k}=\frac{179}{34}-\frac{9}{16}\pi(\sqrt{2}+1);$$

(298) $$\sum_{k=1}^{\infty}\{\zeta(2k+1)-1\}\left(\frac{9}{8}\right)^{2k}=-\frac{563}{153}+4\log 2-\frac{\sqrt{2}}{2}\log(3-2\sqrt{2});$$

(299) $$\sum_{k=2}^{\infty}(-1)^k\{\zeta(k)-1\}\left(\frac{11}{8}\right)^{k}$$
$$=\frac{1765}{456}-\frac{11}{16}\pi(\sqrt{2}-1)-\frac{11}{2}\log 2-\frac{11\sqrt{2}}{16}\log(3-2\sqrt{2});$$

(300) $$\sum_{k=2}^{\infty}\{\zeta(k)-1\}\left(\frac{11}{8}\right)^{k}$$
$$=\frac{11}{8}-\frac{11}{16}\pi(\sqrt{2}-1)+\frac{11}{2}\log 2+\frac{11\sqrt{2}}{16}\log(3-2\sqrt{2});$$

(301) $$\sum_{k=1}^{\infty}\{\zeta(2k)-1\}\left(\frac{11}{8}\right)^{2k}=\frac{299}{114}-\frac{11}{16}\pi(\sqrt{2}-1);$$

(302) $$\sum_{k=1}^{\infty}\{\zeta(2k+1)-1\}\left(\frac{11}{8}\right)^{2k}=-\frac{569}{627}+4\log 2+\frac{\sqrt{2}}{2}\log(3-2\sqrt{2});$$

(303) $$\sum_{k=2}^{\infty}(-1)^k\{\zeta(k)-1\}\left(\frac{13}{8}\right)^{k}=\frac{2179}{840}+\frac{13}{16}\pi(\sqrt{2}-1)-\frac{13}{2}\log 2-\frac{13\sqrt{2}}{16}\log(3-2\sqrt{2});$$

(304) $$\sum_{k=2}^{\infty}\{\zeta(k)-1\}\left(\frac{13}{8}\right)^{k}=\frac{13}{8}+\frac{13}{16}\pi(\sqrt{2}-1)+\frac{13}{2}\log 2+\frac{13\sqrt{2}}{16}\log(3-2\sqrt{2});$$

(305) $$\sum_{k=1}^{\infty}\{\zeta(2k)-1\}\left(\frac{13}{8}\right)^{2k}=\frac{443}{210}+\frac{13}{16}\pi(\sqrt{2}-1);$$

(306) $$\sum_{k=1}^{\infty}\{\zeta(2k+1)-1\}\left(\frac{13}{8}\right)^{2k}=-\frac{407}{1365}+4\log 2+\frac{\sqrt{2}}{2}\log(3-2\sqrt{2});$$

(307) $$\sum_{k=2}^{\infty}(-1)^k\{\zeta(k)-1\}\left(\frac{15}{8}\right)^{k}=\frac{2473}{1288}+\frac{15}{16}\pi(\sqrt{2}+1)-\frac{15}{2}\log 2+\frac{15\sqrt{2}}{16}\log(3-2\sqrt{2});$$

(308) $$\sum_{k=2}^{\infty}\{\zeta(k)-1\}\left(\frac{15}{8}\right)^{k}=\frac{15}{8}+\frac{15}{16}\pi(\sqrt{2}+1)+\frac{15}{2}\log 2-\frac{15\sqrt{2}}{16}\log(3-2\sqrt{2});$$

(309) $$\sum_{k=1}^{\infty}\{\zeta(2k)-1\}\left(\frac{15}{8}\right)^{2k}=\frac{611}{322}+\frac{15}{16}\pi(\sqrt{2}+1);$$

(310) $$\sum_{k=1}^{\infty}\{\zeta(2k+1)-1\}\left(\frac{15}{8}\right)^{2k}=-\frac{29}{2415}+4\log 2-\frac{\sqrt{2}}{2}\log(3-2\sqrt{2});$$

(311)
$$\sum_{k=2}^{\infty}(-1)^k\frac{\zeta(k)-1}{10^k}=\frac{109}{110}-\frac{\pi}{20}\left(2\sqrt{1+\frac{2}{5}\sqrt{5}}+\sqrt{1-\frac{2}{5}\sqrt{5}}\right)-\frac{1}{8}\log 5-\frac{1}{40}\sqrt{5}\log\left(\frac{102+46\sqrt{5}}{8}\right)-\frac{1}{5}\log 2;$$

(312)
$$\sum_{k=2}^{\infty}\frac{\zeta(k)-1}{10^k}=-\frac{1}{90}-\frac{\pi}{20}\left(2\sqrt{1+\frac{2}{5}\sqrt{5}}+\sqrt{1-\frac{2}{5}\sqrt{5}}\right)+\frac{1}{8}\log 5+\frac{1}{40}\sqrt{5}\log\left(\frac{147+65\sqrt{5}}{8}\right)+\frac{1}{5}\log 2;$$

(313)
$$\sum_{k=1}^{\infty}\frac{\zeta(2k)-1}{10^{2k}}=\frac{97}{198}-\frac{\pi}{20}\left(2\sqrt{1+\frac{2}{5}\sqrt{5}}+\sqrt{1-\frac{2}{5}\sqrt{5}}\right)+\frac{1}{80}\sqrt{5}\log\left(\frac{3\sqrt{5}-1}{4}\right);$$

(314)
$$\sum_{k=1}^{\infty}\frac{\zeta(2k+1)-1}{10^{2k}}=-\frac{496}{99}+\frac{5}{4}\log 5+\frac{1}{8}\sqrt{5}\log\left(\frac{3743+1674\sqrt{5}}{8}\right)+2\log 2;$$

(315)
$$\sum_{k=2}^{\infty}(-1)^k\{\zeta(k)-1\}\left(\frac{3}{10}\right)^k=\frac{363}{390}+\frac{3\pi}{20}\left(2\sqrt{1-\frac{2}{5}\sqrt{5}}-\sqrt{1+\frac{2}{5}\sqrt{5}}\right)-\frac{3}{8}\log 5+\frac{9}{40}\sqrt{5}\log\left(\frac{3+2\sqrt{5}}{2}\right)-\frac{3}{5}\log 2;$$

(316)
$$\sum_{k=2}^{\infty}\{\zeta(k)-1\}\left(\frac{3}{10}\right)^k=-\frac{9}{70}+\frac{3\pi}{20}\left(2\sqrt{1-\frac{2}{5}\sqrt{5}}-\sqrt{1+\frac{2}{5}\sqrt{5}}\right)+\frac{3}{8}\log 5-\frac{9}{40}\sqrt{5}\log\left(\frac{3+\sqrt{5}}{2}\right)+\frac{3}{5}\log 2;$$

(317)
$$\sum_{k=1}^{\infty}\{\zeta(2k)-1\}\left(\frac{3}{10}\right)^{2k}=\frac{73}{182}+\frac{3\pi}{20}\left(2\sqrt{1-\frac{2}{5}\sqrt{5}}-\sqrt{1+\frac{2}{5}\sqrt{5}}\right)+\frac{9}{80}\sqrt{5}\log\left(\frac{3\sqrt{5}-1}{4}\right);$$

(318)
$$\sum_{k=1}^{\infty}\{\zeta(2k+1)-1\}\left(\frac{3}{10}\right)^{2k} = -\frac{482}{273}+\frac{5}{4}\log 5-\frac{3}{8}\sqrt{5}\log\left(\frac{19+9\sqrt{5}}{4}\right)+2\log 2;$$

(319)
$$\sum_{k=2}^{\infty}(-1)^k\{\zeta(k)-1\}\left(\frac{7}{10}\right)^{k} = \frac{121}{170}+\frac{7\pi}{20}\left(\sqrt{1+\frac{2}{5}\sqrt{5}}-2\sqrt{1-\frac{2}{5}\sqrt{5}}\right) -\frac{7}{8}\log 5+\frac{21}{40}\sqrt{5}\log\left(\frac{3+\sqrt{5}}{2}\right)-\frac{7}{5}\log 2;$$

(320)
$$\sum_{k=2}^{\infty}\{\zeta(k)-1\}\left(\frac{7}{10}\right)^{k} = -\frac{49}{30}-\frac{7\pi}{20}\left(2\sqrt{1-\frac{2}{5}\sqrt{5}}-\sqrt{1+\frac{2}{5}\sqrt{5}}\right) +\frac{7}{8}\log 5-\frac{21}{40}\sqrt{5}\log\left(\frac{3+2\sqrt{5}}{2}\right)+\frac{7}{5}\log 2;$$

(321)
$$\sum_{k=1}^{\infty}\{\zeta(2k)-1\}\left(\frac{7}{10}\right)^{2k} = -\frac{329}{714}+\frac{7\pi}{20}\left(2\sqrt{1-\frac{2}{5}\sqrt{5}}-\sqrt{1+\frac{2}{5}\sqrt{5}}\right)+\frac{21}{80}\sqrt{5}\log(3\sqrt{5}+1);$$

(322)
$$\sum_{k=1}^{\infty}\{\zeta(2k+1)-1\}\left(\frac{7}{10}\right)^{2k} = -\frac{598}{357}+\frac{5}{4}\log 5-\frac{3}{8}\sqrt{5}\log\left(\frac{19+9\sqrt{5}}{4}\right)+2\log 2;$$

(323)
$$\sum_{k=2}^{\infty}(-1)^k\{\zeta(k)-1\}\left(\frac{9}{10}\right)^{k} = \frac{109}{190}+\frac{9\pi}{20}\left(2\sqrt{1+\frac{2}{5}\sqrt{5}}+\sqrt{1-\frac{2}{5}\sqrt{5}}\right) -\frac{9}{8}\log 5-\frac{9}{40}\sqrt{5}\log\left(\frac{147+65\sqrt{5}}{8}\right)-\frac{9}{5}\log 2;$$

$$(324)\quad \sum_{k=2}^{\infty}\{\zeta(k)-1\}\left(\frac{9}{10}\right)^{k} = -\frac{81}{10}+\frac{9\pi}{20}\left(2\sqrt{1+\frac{2}{5}\sqrt{5}}+\sqrt{1-\frac{2}{5}\sqrt{5}}\right) + \frac{9}{8}\log 5+\frac{9}{40}\sqrt{5}\,\log\left(\frac{102+46\sqrt{5}}{8}\right)+\frac{9}{5}\log 2;$$

$$(325)\quad \sum_{k=1}^{\infty}\{\zeta(2k)-1\}\left(\frac{9}{10}\right)^{2k} = -\frac{143}{38}+\frac{9\pi}{20}\left(2\sqrt{1+\frac{2}{5}\sqrt{5}}+\sqrt{1-\frac{2}{5}\sqrt{5}}\right) + \frac{9}{80}\sqrt{5}\,\log\left(\frac{1+3\sqrt{5}}{11}\right);$$

$$(326)\quad \sum_{k=1}^{\infty}\{\zeta(2k+1)-1\}\left(\frac{9}{10}\right)^{2k} = -\frac{824}{171}+\frac{5}{4}\log 5+\frac{1}{8}\sqrt{5}\,\log\left(\frac{3743+1674\sqrt{5}}{8}\right)+2\log 2;$$

$$(327)\quad \sum_{k=2}^{\infty}(-1)^{k}\{\zeta(k)-1\}\left(\frac{11}{10}\right)^{k} = \frac{2399}{210}-\frac{11\pi}{20}\left(2\sqrt{1+\frac{2}{5}\sqrt{5}}+\sqrt{1-\frac{2}{5}\sqrt{5}}\right) - \frac{11}{8}\log 5-\frac{11}{40}\sqrt{5}\,\log\left(\frac{102+46\sqrt{5}}{8}\right)-\frac{11}{5}\log 2;$$

$$(328)\quad \sum_{k=2}^{\infty}\{\zeta(k)-1\}\left(\frac{11}{10}\right)^{k} = \frac{11}{10}-\frac{11\pi}{20}\left(2\sqrt{1+\frac{2}{5}\sqrt{5}}+\sqrt{1-\frac{2}{5}\sqrt{5}}\right) + \frac{11}{8}\log 5+\frac{11}{40}\sqrt{5}\,\log\left(\frac{147+65\sqrt{5}}{8}\right)+\frac{11}{5}\log 2;$$

$$(329)\quad \sum_{k=1}^{\infty}\{\zeta(2k)-1\}\left(\frac{11}{10}\right)^{2k} = \frac{263}{42}-\frac{11\pi}{20}\left(2\sqrt{1+\frac{2}{5}\sqrt{5}}+\sqrt{1-\frac{2}{5}\sqrt{5}}\right) + \frac{11}{80}\sqrt{5}\,\log\left(\frac{3\sqrt{5}-1}{4}\right);$$

$$(330)\quad \sum_{k=1}^{\infty}\{\zeta(2k+1)-1\}\left(\frac{11}{10}\right)^{2k} = -\frac{1084}{231}+\frac{5}{4}\log 5+\frac{1}{8}\sqrt{5}\,\log\left(\frac{3743+1674\sqrt{5}}{8}\right)+2\log 2.$$

(331)
$$\sum_{k=2}^{\infty}(-1)^k\{\zeta(k)-1\}\left(\frac{13}{10}\right)^k=\frac{3173}{690}+\frac{13\pi}{20}\left(2\sqrt{1-\frac{2}{5}\sqrt{5}}-\sqrt{1+\frac{2}{5}\sqrt{5}}\right)-\frac{13}{8}\log 5+\frac{39}{40}\sqrt{5}\log\left(\frac{3+2\sqrt{5}}{2}\right)-\frac{13}{5}\log 2;$$

(332)
$$\sum_{k=2}^{\infty}\{\zeta(k)-1\}\left(\frac{13}{10}\right)^k=\frac{13}{10}+\frac{13\pi}{20}\left(2\sqrt{1-\frac{2}{5}\sqrt{5}}+\sqrt{1+\frac{2}{5}\sqrt{5}}\right)+\frac{13}{8}\log 5-\frac{39}{40}\sqrt{5}\log\left(\frac{3+\sqrt{5}}{2}\right)+\frac{13}{5}\log 2;$$

(333)
$$\sum_{k=1}^{\infty}\{\zeta(2k)-1\}\left(\frac{13}{10}\right)^{2k}=\frac{407}{138}+\frac{13\pi}{20}\left(2\sqrt{1-\frac{2}{5}\sqrt{5}}-\sqrt{1+\frac{2}{5}\sqrt{5}}\right)+\frac{39}{80}\sqrt{5}\log\left(\frac{3\sqrt{5}-1}{4}\right);$$

(334)
$$\sum_{k=1}^{\infty}\{\zeta(2k+1)-1\}\left(\frac{13}{10}\right)^{2k}=-\frac{1138}{897}+\frac{5}{4}\log 5-\frac{3}{8}\sqrt{5}\log\left(\frac{19+9\sqrt{5}}{4}\right)+2\log 2;$$

(335)
$$\sum_{k=2}^{\infty}(-1)^k\{\zeta(k)-1\}\left(\frac{17}{10}\right)^k=\frac{4457}{1890}+\frac{17\pi}{20}\left(\sqrt{1+\frac{2}{5}\sqrt{5}}-2\sqrt{1-\frac{2}{5}\sqrt{5}}\right)-\frac{17}{8}\log 5+\frac{51}{40}\sqrt{5}\log\left(\frac{3+\sqrt{5}}{2}\right)-\frac{17}{5}\log 2;$$

(336)
$$\sum_{k=2}^{\infty}\{\zeta(k)-1\}\left(\frac{17}{10}\right)^k=\frac{17}{10}-\frac{17\pi}{20}\left(2\sqrt{1-\frac{2}{5}\sqrt{5}}-\sqrt{1+\frac{2}{5}\sqrt{5}}\right)+\frac{17}{8}\log 5-\frac{51}{40}\sqrt{5}\log\left(\frac{3+2\sqrt{5}}{2}\right)+\frac{17}{5}\log 2;$$

(337)
$$\sum_{k=1}^{\infty}\{\zeta(2k)-1\}\left(\frac{17}{10}\right)^{2k}=\frac{767}{378}+\frac{17\pi}{20}\left(\sqrt{1+\frac{2}{5}\sqrt{5}}-2\sqrt{1-\frac{2}{5}\sqrt{5}}\right)+\frac{51}{80}\sqrt{5}\log(3\sqrt{5}+1);$$

(338)
$$\sum_{k=1}^{\infty}\{\zeta(2k+1)-1\}\left(\frac{17}{10}\right)^{2k} = -\frac{622}{3213}+\frac{5}{4}\log 5-\frac{3}{8}\sqrt{5}\,\log\left(\frac{19+9\sqrt{5}}{4}\right)+2\log 2;$$

(339)
$$\sum_{k=2}^{\infty}(-1)^k\{\zeta(k)-1\}\left(\frac{19}{10}\right)^{k} = \frac{4871}{261}+\frac{19\pi}{20}\left(2\sqrt{1+\frac{2}{5}\sqrt{5}}+\sqrt{1-\frac{2}{5}\sqrt{5}}\right) -\frac{19}{8}\log 5-\frac{19}{40}\sqrt{5}\,\log\left(\frac{147+65\sqrt{5}}{8}\right)-\frac{19}{5}\log 2;$$

(340)
$$\sum_{k=2}^{\infty}\{\zeta(k)-1\}\left(\frac{19}{10}\right)^{k} = \frac{19}{10}+\frac{19\pi}{20}\left(2\sqrt{1+\frac{2}{5}\sqrt{5}}+\sqrt{1+\frac{2}{5}\sqrt{5}}\right) +\frac{19}{8}\log 5+\frac{19}{40}\sqrt{5}\,\log\left(\frac{102+46\sqrt{5}}{8}\right)+\frac{19}{5}\log 2;$$

(341)
$$\sum_{k=1}^{\infty}\{\zeta(2k)-1\}\left(\frac{19}{10}\right)^{2k} = \frac{983}{522}+\frac{19\pi}{20}\left(2\sqrt{1+\frac{2}{5}\sqrt{5}}+\sqrt{1-\frac{2}{5}\sqrt{5}}\right) +\frac{19}{80}\sqrt{5}\,\log\left(\frac{3\sqrt{5}+1}{11}\right);$$

(342)
$$\sum_{k=1}^{\infty}\{\zeta(2k+1)-1\}\left(\frac{19}{10}\right)^{2k} = \frac{44}{4959}+\frac{5}{4}\log 5+\frac{1}{8}\sqrt{5}\,\log\left(\frac{3743+1674\sqrt{5}}{8}\right)+2\log 2;$$

(343)
$$\sum_{k=2}^{\infty}(-1)^k\frac{\zeta(k)-1}{12^k} = \frac{155}{156}-\frac{\pi}{24}(2+\sqrt{3})+\frac{\sqrt{3}}{24}\log(7-4\sqrt{3})-\frac{1}{8}\log 3-\frac{1}{4}\log 2;$$

(344)
$$\sum_{k=2}^{\infty}\frac{\zeta(k)-1}{12^k} = -\frac{1}{132}-\frac{\pi}{24}(2+\sqrt{3})-\frac{\sqrt{3}}{24}\log(7-4\sqrt{3})+\frac{1}{8}\log 3+\frac{1}{4}\log 2;$$

(345) $$\sum_{k=1}^{\infty} \frac{\zeta(2k)-1}{12^{2k}} = \frac{141}{286} - \frac{\pi}{24}(2+\sqrt{3});$$

(346) $$\sum_{k=1}^{\infty} \frac{\zeta(2k+1)-1}{12^{2k}} = -\frac{859}{143} - \frac{\sqrt{3}}{2}\log(7-4\sqrt{3}) + \frac{3}{2}\log 3 + 3\log 2;$$

(347) $$\sum_{k=2}^{\infty}(-1)^k\{\zeta(k)-1\}\left(\frac{5}{12}\right)^k = \frac{179}{204} - \frac{5\pi}{24}(2-\sqrt{3}) - \frac{5\sqrt{3}}{24}\log(7-4\sqrt{3}) - \frac{5}{8}\log 3 - \frac{5}{4}\log 2;$$

(348) $$\sum_{k=2}^{\infty}\{\zeta(k)-1\}\left(\frac{5}{12}\right)^k = -\frac{25}{84} - \frac{5\pi}{24}(2-\sqrt{3}) + \frac{5\sqrt{3}}{24}\log(7-4\sqrt{3}) + \frac{5}{8}\log 3 + \frac{5}{4}\log 2;$$

(349) $$\sum_{k=1}^{\infty}\{\zeta(2k)-1\}\left(\frac{5}{12}\right)^{2k} = \frac{69}{238} - \frac{5\pi}{24}(2-\sqrt{3});$$

(350) $$\sum_{k=1}^{\infty}\{\zeta(2k+1)-1\}\left(\frac{5}{12}\right)^{2k} = -\frac{839}{595} + \frac{\sqrt{3}}{2}\log(7-4\sqrt{3}) + \frac{3}{2}\log 3 + 3\log 2;$$

(351) $$\sum_{k=2}^{\infty}(-1)^k\{\zeta(k)-1\}\left(\frac{7}{12}\right)^k = \frac{179}{228} + \frac{7\pi}{24}(2-\sqrt{3}) - \frac{7\sqrt{3}}{24}\log(7-4\sqrt{3}) - \frac{7}{8}\log 3 - \frac{7}{4}\log 2;$$

(352) $$\sum_{k=2}^{\infty}\{\zeta(k)-1\}\left(\frac{7}{12}\right)^k = -\frac{49}{60} + \frac{7\pi}{24}(2-\sqrt{3}) + \frac{7\sqrt{3}}{24}\log(7-4\sqrt{3}) + \frac{7}{8}\log 3 + \frac{7}{4}\log 2;$$

(353) $$\sum_{k=1}^{\infty}\{\zeta(2k)-1\}\left(\frac{7}{12}\right)^{2k} = -\frac{3}{190} + \frac{7\pi}{24}(2-\sqrt{3});$$

(354) $$\sum_{k=1}^{\infty}\{\zeta(2k+1)-1\}\left(\frac{7}{12}\right)^{2k} = -\frac{913}{665}+\frac{\sqrt{3}}{2}\log(7-4\sqrt{3})+\frac{3}{2}\log 3+3\log 2;$$

(355) $$\sum_{k=2}^{\infty}(-1)^k\{\zeta(k)-1\}\left(\frac{11}{12}\right)^{k} = \frac{155}{276}+\frac{11\pi}{24}(2+\sqrt{3})+\frac{11\sqrt{3}}{24}\log(7-4\sqrt{3})-\frac{11}{8}\log 3-\frac{11}{4}\log 2;$$

(356) $$\sum_{k=2}^{\infty}\{\zeta(k)-1\}\left(\frac{11}{12}\right)^{k} = -\frac{121}{12}+\frac{11\pi}{24}(2+\sqrt{3})-\frac{11\sqrt{3}}{24}\log(7-4\sqrt{3})+\frac{11}{8}\log 3+\frac{11}{4}\log 2;$$

(357) $$\sum_{k=1}^{\infty}\{\zeta(2k)-1\}\left(\frac{11}{12}\right)^{2k} = -\frac{219}{46}+\frac{11\pi}{24}(2+\sqrt{3});$$

(358) $$\sum_{k=1}^{\infty}\{\zeta(2k+1)-1\}\left(\frac{11}{12}\right)^{2k} = -\frac{1469}{253}-\frac{\sqrt{3}}{2}\log(7-4\sqrt{3})+\frac{3}{2}\log 3+3\log 2;$$

(359) $$\sum_{k=2}^{\infty}(-1)^k\{\zeta(k)-1\}\left(\frac{13}{12}\right)^{k} = \frac{4031}{300}-\frac{13\pi}{24}(2+\sqrt{3})+\frac{13\sqrt{3}}{24}\log(7-4\sqrt{3})-\frac{13}{8}\log 3-\frac{13}{4}\log 2;$$

(360) $$\sum_{k=2}^{\infty}\{\zeta(k)-1\}\left(\frac{13}{12}\right)^{k} = \frac{13}{12}-\frac{13\pi}{24}(2+\sqrt{3})-\frac{13\sqrt{3}}{24}\log(7-4\sqrt{3})+\frac{13}{8}\log 3+\frac{13}{4}\log 2;$$

(361) $$\sum_{k=1}^{\infty}\{\zeta(2k)-1\}\left(\frac{13}{12}\right)^{2k} = \frac{363}{50}-\frac{13\pi}{24}(2+\sqrt{3});$$

$$(362)\quad \sum_{k=1}^{\infty}\{\zeta(2k+1)-1\}\left(\frac{13}{12}\right)^{2k}=-\frac{1853}{325}-\frac{\sqrt{3}}{2}\log(7-4\sqrt{3})+\frac{3}{2}\log 3+3\log 2;$$

$$(363)\quad \sum_{k=2}^{\infty}(-1)^k\{\zeta(k)-1\}\left(\frac{17}{12}\right)^{k}=\frac{6211}{1740}-\frac{17\pi}{24}(2-\sqrt{3})-\frac{17\sqrt{3}}{24}\log(7-4\sqrt{3})-\frac{17}{8}\log 3-\frac{17}{4}\log 2;$$

$$(364)\quad \sum_{k=2}^{\infty}\{\zeta(k)-1\}\left(\frac{17}{12}\right)^{k}=\frac{17}{12}-\frac{17\pi}{24}(2-\sqrt{3})+\frac{17\sqrt{3}}{24}\log(7-4\sqrt{3})+\frac{17}{8}\log 3+\frac{17}{4}\log 2;$$

$$(365)\quad \sum_{k=1}^{\infty}\{\zeta(2k)-1\}\left(\frac{17}{12}\right)^{2k}=\frac{723}{290}-\frac{17\pi}{24}(2-\sqrt{3});$$

$$(366)\quad \sum_{k=1}^{\infty}\{\zeta(2k+1)-1\}\left(\frac{17}{12}\right)^{2k}=-\frac{1873}{2465}+\frac{\sqrt{3}}{2}\log(7-4\sqrt{3})+\frac{3}{2}\log 3+3\log 2;$$

$$(367)\quad \sum_{k=2}^{\infty}(-1)^k\{\zeta(k)-1\}\left(\frac{19}{12}\right)^{k}=\frac{7145}{2604}+\frac{19\pi}{24}(2-\sqrt{3})-\frac{19\sqrt{3}}{24}\log(7-4\sqrt{3})-\frac{19}{8}\log 3-\frac{19}{4}\log 2;$$

$$(368)\quad \sum_{k=2}^{\infty}\{\zeta(k)-1\}\left(\frac{19}{12}\right)^{k}=\frac{19}{12}+\frac{19\pi}{24}(2-\sqrt{3})+\frac{19\sqrt{3}}{24}\log(7-4\sqrt{3})+\frac{19}{8}\log 3+\frac{19}{4}\log 2;$$

$$(369)\quad \sum_{k=1}^{\infty}\{\zeta(2k)-1\}\left(\frac{19}{12}\right)^{2k}=\frac{939}{434}+\frac{19\pi}{24}(2-\sqrt{3});$$

$$(370)\quad \sum_{k=1}^{\infty}\{\zeta(2k+1)-1\}\left(\frac{19}{12}\right)^{2k}=-\frac{1511}{4123}+\frac{\sqrt{3}}{2}\log(7-4\sqrt{3})+\frac{3}{2}\log 3+3\log 2;$$

(371)
$$\sum_{k=2}^{\infty}(-1)^k\{\zeta(k)-1\}\left(\frac{23}{12}\right)^k = \frac{8461}{4620}+\frac{23\pi}{24}(2+\sqrt{3})+\frac{23\sqrt{3}}{24}\log(7-4\sqrt{3})-\frac{23}{8}\log 3-\frac{23}{4}\log 2;$$

(372)
$$\sum_{k=2}^{\infty}\{\zeta(k)-1\}\left(\frac{23}{12}\right)^k = \frac{23}{12}+\frac{23\pi}{24}(2+\sqrt{3})-\frac{23\sqrt{3}}{24}\log(7-4\sqrt{3})+\frac{23}{8}\log 3+\frac{23}{4}\log 2;$$

(373)
$$\sum_{k=1}^{\infty}\{\zeta(2k)-1\}\left(\frac{23}{12}\right)^{2k} = \frac{1443}{770}+\frac{23\pi}{24}(2+\sqrt{3});$$

(374)
$$\sum_{k=1}^{\infty}\{\zeta(2k+1)-1\}\left(\frac{23}{12}\right)^{2k} = \frac{197}{8855}-\frac{\sqrt{3}}{2}\log(7-4\sqrt{3})+\frac{3}{2}\log 3+3\log 2.$$

Setting $a=\frac{1}{2}$ in (1) through (8), and using 1.1(14), 2.3(2), and appropriate special values of the ψ-function listed in Section 1.2, we obtain

(375)
$$\sum_{k=2}^{\infty}(2^k-1)\zeta(k)\frac{t^k}{k} = \log\Gamma\left(\frac{1}{2}-t\right)-(\gamma+2\log 2)t-\frac{1}{2}\log\pi \quad \left(|t|<\frac{1}{2}\right);$$

(376)
$$\sum_{k=2}^{\infty}(-1)^k(2^k-1)\zeta(k)\frac{t^k}{k} = \log\Gamma\left(\frac{1}{2}+t\right)+(\gamma+2\log 2)t-\frac{1}{2}\log\pi \quad \left(|t|<\frac{1}{2}\right);$$

(377)
$$\sum_{k=1}^{\infty}\left(2^{2k}-1\right)\zeta(2k)\frac{t^{2k}}{k} = \log\Gamma\left(\frac{1}{2}+t\right)+\log\Gamma\left(\frac{1}{2}-t\right)-\log\pi \quad \left(|t|<\frac{1}{2}\right);$$

(378)
$$\sum_{k=1}^{\infty}\left(2^{2k+1}-1\right)\zeta(2k+1)\frac{t^{2k+1}}{2k+1} = \frac{1}{2}\left\{\log\Gamma\left(\frac{1}{2}-t\right)-\log\Gamma\left(\frac{1}{2}+t\right)\right\}-(\gamma+2\log 2)t \quad \left(|t|<\frac{1}{2}\right);$$

(379)
$$\sum_{k=2}^{\infty}(2^k-1)\zeta(k)t^{k-1} = -\psi\left(\frac{1}{2}-t\right)-\gamma-2\log 2 \quad \left(|t|<\frac{1}{2}\right);$$

$$\sum_{k=2}^{\infty}(-1)^k(2^k-1)\zeta(k)t^{k-1} = \psi\left(\frac{1}{2}+t\right)+\gamma+2\log 2 \quad \left(|t|<\frac{1}{2}\right); \tag{380}$$

$$\sum_{k=1}^{\infty}(2^{2k}-1)\,\zeta(2k)t^{2k-1} = \frac{1}{2}\left\{\psi\left(\frac{1}{2}+t\right)-\psi\left(\frac{1}{2}-t\right)\right\} \quad \left(|t|<\frac{1}{2}\right); \tag{381}$$

$$\begin{aligned}&\sum_{k=1}^{\infty}(2^{2k+1}-1)\,\zeta(2k+1)t^{2k}\\ &\quad = -\frac{1}{2}\left\{\psi\left(\frac{1}{2}+t\right)+\psi\left(\frac{1}{2}-t\right)\right\}-\gamma-2\log 2 \quad \left(|t|<\frac{1}{2}\right).\end{aligned} \tag{382}$$

For $z = t+\frac{1}{2}$, 1.1(12) yields

$$\Gamma\left(\frac{1}{2}+t\right)\Gamma\left(\frac{1}{2}-t\right) = \frac{\pi}{\cos(\pi t)}, \tag{383}$$

which can be used to rewrite the series identity (377) in its equivalent form:

$$\sum_{k=1}^{\infty}(2^{2k}-1)\,\zeta(2k)\frac{t^{2k}}{k} = -\log\cos(\pi t) \quad \left(|t|<\frac{1}{2}\right). \tag{384}$$

Furthermore, in view of 1.2(11), the series identity (381) can be rewritten in its equivalent form:

$$\sum_{k=1}^{\infty}(2^{2k}-1)\,\zeta(2k)t^{2k-1} = \frac{\pi}{2}\tan(\pi t) \quad (|t|<\frac{1}{2}). \tag{385}$$

For suitable special values of the argument t, we can easily deduce the following series identities from (375) through (382), (384), and (385):

$$\sum_{k=2}^{\infty}(-1)^k\left(1-\frac{1}{2^k}\right)\frac{\zeta(k)}{k} = \frac{1}{2}\gamma-\frac{1}{2}\log\pi+\log 2; \tag{386}$$

$$\sum_{k=1}^{\infty}\frac{2^{2k}-1}{k\cdot 3^{2k}}\zeta(2k) = \log 2; \tag{387}$$

$$\sum_{k=1}^{\infty}\frac{2^{2k}-1}{k\cdot 4^{2k}}\zeta(2k) = \frac{1}{2}\log 2; \tag{388}$$

(389) $$\sum_{k=1}^{\infty} \frac{2^{2k}-1}{k \cdot 6^{2k}} \zeta(2k) = \log 2 - \frac{1}{2}\log 3;$$

(390) $$\sum_{k=1}^{\infty} \frac{2^{2k}-1}{k \cdot 8^{2k}} \zeta(2k) = \frac{1}{2}\log(4-2\sqrt{2});$$

(391) $$\sum_{k=1}^{\infty} \frac{2^{2k}-1}{k} \zeta(2k) \left(\frac{3}{8}\right)^{2k} = \frac{1}{2}\log(4+2\sqrt{2});$$

(392) $$\sum_{k=1}^{\infty} \frac{2^{2k}-1}{k \cdot 12^{2k}} \zeta(2k) = \log(\sqrt{6}-\sqrt{2});$$

(393) $$\sum_{k=1}^{\infty} \frac{2^{2k}-1}{k} \zeta(2k) \left(\frac{5}{12}\right)^{2k} = \log(\sqrt{6}+\sqrt{2});$$

(394) $$\sum_{k=2}^{\infty} (-1)^k \left(1 - \frac{1}{2^k}\right) \zeta(k) = \log 2;$$

(395) $$\sum_{k=2}^{\infty} (-1)^k \frac{2^k-1}{3^k} \zeta(k) = \frac{1}{6}\pi\sqrt{3} - \frac{1}{2}\log 3;$$

(396) $$\sum_{k=2}^{\infty} \frac{2^k-1}{3^k} \zeta(k) = \frac{1}{6}\pi\sqrt{3} + \frac{1}{2}\log 3;$$

(397) $$\sum_{k=1}^{\infty} \frac{2^{2k}-1}{3^{2k}} \zeta(2k) = \frac{1}{6}\pi\sqrt{3};$$

(398) $$\sum_{k=1}^{\infty} \frac{2^{2k+1}-1}{3^{2k+1}} \zeta(2k+1) = \frac{1}{2}\log 3;$$

(399) $$\sum_{k=2}^{\infty} (-1)^k \frac{2^k-1}{4^k} \zeta(k) = \frac{1}{8}\pi - \frac{1}{4}\log 2;$$

(400) $$\sum_{k=2}^{\infty}\frac{2^k-1}{4^k}\zeta(k)=\frac{1}{8}\pi+\frac{1}{4}\log 2;$$

(401) $$\sum_{k=1}^{\infty}\frac{2^{2k}-1}{4^{2k}}\zeta(2k)=\frac{1}{8}\pi;$$

(402) $$\sum_{k=1}^{\infty}\frac{2^{2k+1}-1}{4^{2k}}\zeta(2k+1)=\log 2;$$

(403) $$\sum_{k=2}^{\infty}(-1)^k\frac{2^k-1}{5^k}\zeta(k) = \frac{\pi}{10}\left(\sqrt{1+\frac{2}{5}\sqrt{5}}-2\sqrt{1-\frac{2}{5}\sqrt{5}}\right)-\frac{1}{4}\log 5+\frac{3}{20}\sqrt{5}\log\left(\frac{3+\sqrt{5}}{2}\right);$$

(404) $$\sum_{k=2}^{\infty}\frac{2^k-1}{5^k}\zeta(k) = \frac{\pi}{10}\left(\sqrt{1+\frac{2}{5}\sqrt{5}}-2\sqrt{1-\frac{2}{5}\sqrt{5}}\right)+\frac{1}{4}\log 5-\frac{3}{20}\sqrt{5}\log\left(\frac{3+2\sqrt{5}}{2}\right);$$

(405) $$\sum_{k=1}^{\infty}\frac{2^{2k}-1}{5^{2k}}\zeta(2k) = \frac{\pi}{10}\left(\sqrt{1+\frac{2}{5}\sqrt{5}}-2\sqrt{1-\frac{2}{5}\sqrt{5}}\right)+\frac{3}{40}\sqrt{5}\log(3\sqrt{5}+1);$$

(406) $$\sum_{k=1}^{\infty}\frac{2^{2k+1}-1}{5^{2k+1}}\zeta(2k+1)=\frac{1}{4}\log 5-\frac{3}{40}\sqrt{5}\log\left(\frac{19+9\sqrt{5}}{4}\right);$$

(407) $$\sum_{k=2}^{\infty}(-1)^k\left(2^k-1\right)\zeta(k)\left(\frac{2}{5}\right)^k=\frac{\pi}{5}\left(2\sqrt{1+\frac{2}{5}\sqrt{5}}+\sqrt{1-\frac{2}{5}\sqrt{5}}\right) -\frac{1}{2}\log 5-\frac{1}{10}\sqrt{5}\log\left(\frac{147+65\sqrt{5}}{8}\right);$$

(408)
$$\sum_{k=2}^{\infty}\left(2^k-1\right)\zeta(k)\left(\frac{2}{5}\right)^k=\frac{\pi}{5}\left(2\sqrt{1+\frac{2}{5}\sqrt{5}}+\sqrt{1-\frac{2}{5}\sqrt{5}}\right)+\frac{1}{2}\log 5+\frac{1}{10}\sqrt{5}\log\left(\frac{102+46\sqrt{5}}{8}\right);$$

(409)
$$\sum_{k=1}^{\infty}\left(2^{2k}-1\right)\zeta(2k)\left(\frac{2}{5}\right)^{2k}=\frac{\pi}{5}\left(2\sqrt{1+\frac{2}{5}\sqrt{5}}+\sqrt{1-\frac{2}{5}\sqrt{5}}\right)+\frac{1}{20}\sqrt{5}\log\left(\frac{1+3\sqrt{5}}{11}\right);$$

(410)
$$\sum_{k=1}^{\infty}\left(2^{2k+1}-1\right)\zeta(2k+1)\left(\frac{2}{5}\right)^{2k}=\frac{5}{4}\log 5+\frac{1}{8}\sqrt{5}\log\left(\frac{3743+1674\sqrt{5}}{8}\right);$$

(411)
$$\sum_{k=2}^{\infty}(-1)^k\frac{2^k-1}{6^k}\zeta(k)=\frac{1}{36}\pi\sqrt{3}-\frac{1}{4}\log 3+\frac{1}{3}\log 2;$$

(412)
$$\sum_{k=2}^{\infty}\frac{2^k-1}{6^k}\zeta(k)=\frac{1}{36}\pi\sqrt{3}+\frac{1}{4}\log 3-\frac{1}{3}\log 2;$$

(413)
$$\sum_{k=1}^{\infty}\frac{2^{2k}-1}{6^{2k}}\zeta(2k)=\frac{1}{36}\pi\sqrt{3};$$

(414)
$$\sum_{k=1}^{\infty}\frac{2^{2k+1}-1}{6^{2k+1}}\zeta(2k+1)=\frac{1}{4}\log 3-\frac{1}{3}\log 2;$$

(415)
$$\sum_{k=2}^{\infty}(-1)^k\frac{2^k-1}{8^k}\zeta(k)=\frac{1}{16}\pi(\sqrt{2}-1)-\frac{\sqrt{2}}{16}\log(3-2\sqrt{2})-\frac{1}{4}\log 2;$$

(416)
$$\sum_{k=2}^{\infty}\frac{2^k-1}{8^k}\zeta(k)=\frac{1}{16}\pi(\sqrt{2}-1)+\frac{\sqrt{2}}{16}\log(3-2\sqrt{2})+\frac{1}{4}\log 2;$$

(417) $$\sum_{k=1}^{\infty} \frac{2^{2k}-1}{8^{2k}}\zeta(2k) = \frac{1}{16}\pi(\sqrt{2}-1);$$

(418) $$\sum_{k=1}^{\infty} \frac{2^{2k+1}-1}{8^{2k+1}}\zeta(2k+1) = \frac{1}{4}\log 2 + \frac{\sqrt{2}}{16}\log(3-2\sqrt{2});$$

(419) $$\sum_{k=2}^{\infty}(-1)^k\left(2^k-1\right)\zeta(k)\left(\frac{3}{8}\right)^k = \frac{3\pi}{16}(\sqrt{2}+1) - \frac{3}{4}\log 2 + \frac{3\sqrt{2}}{16}\log(3-2\sqrt{2});$$

(420) $$\sum_{k=2}^{\infty}\left(2^k-1\right)\zeta(k)\left(\frac{3}{8}\right)^k = \frac{3\pi}{16}(\sqrt{2}+1) + \frac{3}{4}\log 2 - \frac{3\sqrt{2}}{16}\log(3-2\sqrt{2});$$

(421) $$\sum_{k=1}^{\infty}\left(2^{2k}-1\right)\zeta(2k)\left(\frac{3}{8}\right)^{2k} = \frac{3\pi}{16}(\sqrt{2}+1);$$

(422) $$\sum_{k=1}^{\infty}\left(2^{2k+1}-1\right)\zeta(2k+1)\left(\frac{3}{8}\right)^{2k} = 2\log 2 - \frac{\sqrt{2}}{2}\log(3-2\sqrt{2});$$

(423) $$\begin{aligned}\sum_{k=2}^{\infty}(-1)^k\frac{2^k-1}{10^k}\zeta(k) =& \frac{\pi}{20}\sqrt{1-\frac{2}{5}\sqrt{5}} + \frac{1}{40}\sqrt{5}\log\left(\frac{3+2\sqrt{5}}{2}\right)\\ &- \frac{1}{8}\log 5 + \frac{1}{5}\log 2;\end{aligned}$$

(424) $$\begin{aligned}\sum_{k=2}^{\infty}\frac{2^k-1}{10^k}\zeta(k) =& \frac{\pi}{20}\sqrt{1-\frac{2}{5}\sqrt{5}} - \frac{1}{40}\sqrt{5}\log\left(\frac{3+2\sqrt{5}}{2}\right)\\ &+ \frac{1}{8}\log 5 - \frac{1}{5}\log 2;\end{aligned}$$

(425) $$\sum_{k=1}^{\infty}\frac{2^{2k}-1}{10^{2k}}\zeta(2k) = \frac{\pi}{20}\sqrt{1-\frac{2}{5}\sqrt{5}} + \frac{1}{80}\sqrt{5}\log\left(\frac{3\sqrt{5}-1}{4}\right);$$

(426) $$\sum_{k=1}^{\infty}\frac{2^{2k+1}-1}{10^{2k+1}}\zeta(2k+1) = \frac{1}{8}\log 5 - \frac{1}{5}\log 2 - \frac{1}{80}\sqrt{5}\log\left(\frac{19+9\sqrt{5}}{4}\right);$$

(427) $$\sum_{k=2}^{\infty}(-1)^k\left(2^k-1\right)\zeta(k)\left(\frac{3}{10}\right)^k = \frac{3\pi}{20}\sqrt{1+\frac{2}{5}\sqrt{5}}-\frac{3}{40}\sqrt{5}\log\left(\frac{3+2\sqrt{5}}{2}\right) - \frac{3}{8}\log 5+\frac{3}{5}\log 2;$$

(428) $$\sum_{k=2}^{\infty}\left(2^k-1\right)\zeta(k)\left(\frac{3}{10}\right)^k = \frac{3\pi}{20}\sqrt{1+\frac{2}{5}\sqrt{5}}+\frac{3}{40}\sqrt{5}\log\left(\frac{3+\sqrt{5}}{2}\right) + \frac{3}{8}\log 5-\frac{3}{5}\log 2;$$

(429) $$\sum_{k=1}^{\infty}\left(2^{2k}-1\right)\zeta(2k)\left(\frac{3}{10}\right)^{2k} = \frac{3\pi}{20}\sqrt{1+\frac{2}{5}\sqrt{5}}+\frac{3}{80}\sqrt{5}\log\left(\frac{1+3\sqrt{5}}{11}\right);$$

(430)
$$\sum_{k=1}^{\infty}\left(2^{2k+1}-1\right)\zeta(2k+1)\left(\frac{3}{10}\right)^{2k} = \frac{5}{4}\log 5-2\log 2+\frac{1}{8}\sqrt{5}\log\left(\frac{19+9\sqrt{5}}{4}\right);$$

(431) $$\sum_{k=2}^{\infty}(-1)^k\frac{2^k-1}{12^k}\zeta(k) = \frac{\pi}{24}(2-\sqrt{3})-\frac{\sqrt{3}}{24}\log(7-4\sqrt{3})-\frac{1}{8}\log 3-\frac{1}{12}\log 2;$$

(432) $$\sum_{k=2}^{\infty}\frac{2^k-1}{12^k}\zeta(k) = \frac{\pi}{24}(2-\sqrt{3})+\frac{\sqrt{3}}{24}\log(7-4\sqrt{3})+\frac{1}{8}\log 3+\frac{1}{12}\log 2;$$

(433) $$\sum_{k=1}^{\infty}\frac{2^{2k}-1}{12^{2k}}\zeta(2k) = \frac{1}{24}\pi(2-\sqrt{3});$$

(434) $$\sum_{k=1}^{\infty}\frac{2^{2k+1}-1}{12^{2k+1}}\zeta(2k+1) = \frac{\sqrt{3}}{24}\log(7-4\sqrt{3})+\frac{1}{8}\log 3+\frac{1}{12}\log 2;$$

(435) $$\sum_{k=2}^{\infty}(-1)^k\left(2^k-1\right)\zeta(k)\left(\frac{5}{12}\right)^k = \frac{5\pi}{24}(2+\sqrt{3})+\frac{5\sqrt{3}}{24}\log(7-4\sqrt{3}) - \frac{5}{8}\log 3-\frac{5}{12}\log 2;$$

$$\sum_{k=2}^{\infty}(2^k-1)\,\zeta(k)\left(\frac{5}{12}\right)^k = \frac{5\pi}{24}(2+\sqrt{3}) - \frac{5\sqrt{3}}{24}\log(7-4\sqrt{3}) + \frac{5}{8}\log 3 + \frac{5}{12}\log 2; \tag{436}$$

$$\sum_{k=1}^{\infty}(2^{2k}-1)\,\zeta(2k)\left(\frac{5}{12}\right)^{2k} = \frac{5}{24}\pi(2+\sqrt{3}); \tag{437}$$

$$\sum_{k=1}^{\infty}(2^{2k+1}-1)\,\zeta(2k+1)\left(\frac{5}{12}\right)^{2k} = -\frac{\sqrt{3}}{2}\log(7-4\sqrt{3}) + \frac{3}{2}\log 3 + \log 2. \tag{438}$$

Evaluation by using the double and triple Gamma functions

We first derive some useful integral expressions for $\log\Gamma(z+a)$, $\log G(z+a)$, and $\log\Gamma_3(z+a)$. Setting $z = t+a-1$ in 1.1(2) and 1.3(3), and taking the logarithmic derivatives of the resulting equations, we obtain

$$\sum_{k=1}^{\infty}\left(\frac{1}{t+a-1+k} - \frac{1}{k}\right) = -\frac{\Gamma'(t+a)}{\Gamma(t+a)} - \gamma \tag{439}$$

and

$$\sum_{k=1}^{\infty}\left(\frac{k}{t+a-1+k} - 1 + \frac{t+a-1}{k}\right) = \frac{G'(t+a)}{G(t+a)} - \frac{1}{2}\log(2\pi) + \frac{1}{2} + (1+\gamma)(t+a-1), \tag{440}$$

respectively.

Now we set $z = t+a-1$ in 1.3(90) and take the logarithmic derivative of the resulting equation. We thus find that

(441)

$$\begin{aligned}\frac{\Gamma_3'(t+a)}{\Gamma_3(t+a)} &= \left(\frac{3}{8} - \frac{1}{4}\log(2\pi) - \log A\right)\\ &+ \frac{1}{2}\left(\gamma + \log(2\pi) + \frac{1}{2}\right)(t+a-1) - \frac{1}{2}\left(\gamma + \frac{\pi^2}{6} + \frac{3}{2}\right)(t+a-1)^2\\ &+ \frac{t+a-1}{2}\left[\sum_{k=1}^{\infty}\left(\frac{k}{t+a-1+k} - 1 + \frac{t+a-1}{k}\right)\right.\\ &\left. + \sum_{k=1}^{\infty}\left(\frac{1}{t+a-1+k} - \frac{1}{k}\right) + \frac{\pi^2}{6}(t+a-1)\right].\end{aligned}$$

A combination of (439) and (440) yields

$$\frac{G'(t+a)}{G(t+a)} = \frac{1}{2} + \frac{1}{2}\log(2\pi) - a - t + (t+a-1)\frac{\Gamma'(t+a)}{\Gamma(t+a)}. \tag{442}$$

Equation (441), by virtue of (439) and (440), becomes

$$\begin{aligned}\frac{\Gamma_3'(t+a)}{\Gamma_3(t+a)} &= \left(\frac{3}{8} - \frac{1}{4}\log(2\pi) - \log A\right) \\ &\quad + \left(\frac{1}{2} + \frac{1}{4}\log(2\pi)\right)(t+a-1) - \frac{(t+a-1)^2}{4} \\ &\quad + \frac{1}{2}(t+a-1)\frac{G'(t+a)}{G(t+a)} - \frac{1}{2}(t+a-1)\frac{\Gamma'(t+a)}{\Gamma(t+a)}.\end{aligned} \tag{443}$$

Integrating both sides of (442) from $t = 0$ to $t = z$, we have

$$\begin{aligned}&\int_0^z \log\Gamma(t+a)dt \\ &= \left(\frac{1}{2} + \frac{1}{2}\log(2\pi) - a\right)z - \frac{z^2}{2} + (z+a-1)\log\Gamma(z+a) \\ &\quad - \log G(z+a) + (1-a)\log\Gamma(a) + \log G(a),\end{aligned} \tag{444}$$

which is due to Barnes (1899b, p. 283) and, in the special case when $a = 1$, reduces at once to Alexeiewsky's theorem 1.3(42).

Integrating both sides of (443) with respect to t from $t = 0$ to $t = z$ with the aid of Barnes's integral formula (444), we obtain

$$\begin{aligned}\int_0^z \log G(t+a)dt &= \left[\frac{1}{2}(a-1)\log(2\pi) - 2\log A - \frac{a^2}{2} + a - \frac{1}{4}\right]z \\ &\quad + \frac{1}{4}\left[\log(2\pi) + 2 - 2a\right]z^2 - \frac{1}{6}z^3 \\ &\quad + (z+a-2)\log G(z+a) - 2\log\Gamma_3(z+a) \\ &\quad + (2-a)\log G(a) + 2\log\Gamma_3(a).\end{aligned} \tag{445}$$

In addition to 1.3(73) and 1.3(77), we can also evaluate each of the following integrals by direct use of the triple Gamma function Γ_3 in 1.3(90):

$$\int_0^1 \log\Gamma_3(t+1)dt = -\frac{1}{24}\log(2\pi) + \frac{3}{2}\log B \tag{446}$$

and

$$\begin{aligned}\int_0^{\frac{1}{2}} \log\Gamma_3(t+1)dt &= -\frac{1}{256} - \frac{29}{1920}\log 2 - \frac{1}{48}\log\pi \\ &\quad + \frac{1}{16}\log A + \frac{3}{4}\log B + \frac{15}{16}\log C.\end{aligned} \tag{447}$$

For example, in order to evaluate the integral in (447), we take the logarithms on both sides of the equation 1.3(90) and integrate the resulting equation from $t = 0$ to $t = \frac{1}{2}$. We then obtain

$$\text{(448)} \qquad \int_0^{\frac{1}{2}} \log \Gamma_3(t+1)dt = \frac{37}{768} + \frac{1}{128}\gamma - \frac{1}{48}\log(2\pi) - \frac{1}{8}\log A + \frac{1}{16}\lim_{n\to\infty} S_n,$$

where

$$\begin{aligned} S_n := \sum_{k=1}^{n} \Big[&- \{(2k+1)^3 \log(2k+1) + (2k)^3 \log(2k)\} \\ &+ \{(2k+1)\log(2k+1) + (2k)\log(2k)\} \\ &+ (16k^3 + 12k^2 + 2k)\log(2k) + (4k^2 + 5k) + \frac{5}{6} - \frac{1}{8k}\Big], \end{aligned}$$

which immediately yields

$$\begin{aligned} S_n = &- \sum_{k=1}^{2n+1} k^3 \log k + \sum_{k=1}^{2n+1} k \log k + 16 \sum_{k=1}^{n} k^3 \log k + 12 \sum_{k=1}^{n} k^2 \log k \\ &+ 2\sum_{k=1}^{n} k \log k + 16 \log 2 \sum_{k=1}^{n} k^3 + (4 + 12\log 2)\sum_{k=1}^{n} k^2 \\ &+ (5 + 2\log 2)\sum_{k=1}^{n} k - \frac{1}{8}\sum_{k=1}^{n}\frac{1}{k} + \frac{5}{6}n. \end{aligned}$$

Using 1.1(3), 1.3(2), 1.3(69), and 1.3(70), we have

$$\begin{aligned} S_n = &\frac{11}{120}\log 2 - \frac{1}{8}\gamma + 3\log A + 12\log B + 15\log C \\ &- \left(4n^4 + 12n^3 + 11n^2 + 3n - \frac{11}{120}\right)\log\left(1 + \frac{1}{2n}\right) \\ &+ 2n^3 + \frac{11}{2}n^2 + \frac{25}{6}n - \frac{13}{48} + O\left(\frac{1}{n}\right) \quad (n \to \infty), \end{aligned}$$

which, by means of the asymptotic formula 1.3(76), yields

$$\text{(449)} \qquad \lim_{n\to\infty} S_n = -\frac{5}{6} + \frac{11}{120}\log 2 - \frac{1}{8}\gamma + 3\log A + 12\log B + 15\log C.$$

Finally, by substituting (449) into (448), we obtain the desired formula (447).

We shall show that integrals of the forms:

$$\int_0^z t^k \psi(t+a)dt \ \ (k \in \mathbb{N})$$

can be expressed in terms of multiple Gamma functions.

First of all, integrating by parts with the aid of (444), we obtain

(450)
$$\int_0^z t\psi(t+a)dt = \frac{1}{2}[2a-1-\log(2\pi)]z + \frac{z^2}{2} + (1-a)\log\Gamma(z+a) + \log G(z+a) + (a-1)\log\Gamma(a) - \log G(a).$$

On the other hand, integrating by parts with the aid of (444) and (445), we obtain

(451)
$$\begin{aligned} 2\int_0^z t\log\Gamma(t+a)dt =& \left(\frac{1}{4} - \frac{1}{2}a + \frac{1}{2}a^2 - 2\log A\right)z + \left(\frac{1}{2}\log(2\pi) - \frac{a}{2} + \frac{1}{4}\right)z^2 \\ & - \frac{z^3}{2} + (z^2 - (a-1)^2)\log\Gamma(z+a) + (2a-3)\log G(z+a) \\ & - 2\log\Gamma_3(z+a) + (a-1)^2\log\Gamma(a) + (3-2a)\log G(a) + 2\log\Gamma_3(a). \end{aligned}$$

Next, integrating by parts with the aid of (451), we obtain

(452)
$$\begin{aligned} \int_0^z t^2\psi(t+a)dt =& \frac{a-1}{2}\{1-2a+\log(2\pi)\}z + \{1-\log(2\pi)\}\frac{z^2}{4} + \frac{z^3}{3} \\ & + (a-1)^2\log\Gamma(a+z) + (z-a+1)\log G(a+z) \\ & - (a-1)^2\log\Gamma(a) + (a-1)\log G(a) - \int_0^z \log G(t+a)\,dt \\ =& \left(-\frac{1}{4} + \frac{1}{2}a - \frac{1}{2}a^2 + 2\log A\right)z + \left(-\frac{1}{2}\log(2\pi) + \frac{a}{2} - \frac{1}{4}\right)z^2 \\ & + \frac{z^3}{2} + (a-1)^2\log\Gamma(z+a) + (3-2a)\log G(z+a) \\ & + 2\log\Gamma_3(z+a) - (a-1)^2\log\Gamma(a) + (2a-3)\log G(a) - 2\log\Gamma_3(a). \end{aligned}$$

Furthermore, integrating by parts with the aid of (445), we obtain

(453)
$$\begin{aligned} 2\int_0^z t\log G(t+a)dt =& (2-a)\left(-\frac{1}{4} + \frac{1}{2}(a-1)\log(2\pi) - 2\log A - \frac{a^2}{2} + a\right)z \\ & + \frac{1}{2}\left(\frac{7}{4} + \frac{1}{2}\log(2\pi) - 2\log A + \frac{a^2}{2} - 2a\right)z^2 \\ & + \frac{1}{6}(\log(2\pi) - a)z^3 - \frac{1}{8}z^4 + (z^2 - a^2 + 4a - 4)\log G(z+a) \\ & + 2(a-2-z)\log\Gamma_3(z+a) + (a-2)^2\log G(a) \\ & + 2(2-a)\log\Gamma_3(a) + 2\int_0^z \log\Gamma_3(t+a)dt. \end{aligned}$$

Finally, integrating by parts with the aid of (451) and (453), we obtain

(454)

$$\begin{aligned}
&3\int_0^z t^2 \log\Gamma(t+a)dt \\
&= \left[-a^2+\frac{3}{2}a-\frac{1}{4}+2(2a-3)\log A-\frac{1}{2}(a^2-3a+2)\log(2\pi)\right]z \\
&+\left[\frac{9}{8}-\frac{7}{4}a+\frac{3}{4}a^2+\frac{1}{4}(3-2a)\log(2\pi)-\log A\right]z^2 \\
&+\frac{1}{3}[\log(2\pi)-1]z^3-\frac{3}{8}z^4+\left\{z^3+(a-1)^3\right\}\log\Gamma(z+a) \\
&-(3a^2-9a+7)\log G(z+a)+2(2a-3-z)\log\Gamma_3(z+a)+(1-a)^3\log\Gamma(a) \\
&+(3a^2-9a+7)\log G(a)+2(3-2a)\log\Gamma_3(a)+2\int_0^z \log\Gamma_3(t+a)dt
\end{aligned}$$

and

(455)

$$\begin{aligned}
&\int_0^z t^3\psi(t+a)dt \\
&= \left[a^2-\frac{3}{2}a+\frac{1}{4}+2(3-2a)\log A+\frac{1}{2}(a^2-3a+2)\log(2\pi)\right]z \\
&+\left[-\frac{3}{4}a^2+\frac{7}{4}a-\frac{9}{8}+\frac{1}{4}(2a-3)\log(2\pi)+\log A\right]z^2 \\
&+\frac{1}{3}[1-\log(2\pi)]z^3+\frac{3}{8}z^4+(1-a)^3\log\Gamma(z+a) \\
&+(3a^2-9a+7)\log G(z+a)+2(z-2a+3)\log\Gamma_3(z+a)+(a-1)^3\log\Gamma(a) \\
&-(3a^2-9a+7)\log G(a)+2(2a-3)\log\Gamma_3(a)-2\int_0^z \log\Gamma_3(t+a)dt.
\end{aligned}$$

By taking logarithms on both sides of 1.3(3) and using the Maclaurin series of $\log(1+x)$, we are led to the following identity [*cf.* Srivastava (1988a, p. 13, Eq. (5.2))]:

$$(456)\quad \sum_{k=2}^{\infty}(-1)^k\frac{\zeta(k)}{k+1}z^{k+1}=[1-\log(2\pi)]\frac{z}{2}+(1+\gamma)\frac{z^2}{2}+\log G(z+1)\quad (|z|<1).$$

Letting $z\to 1$ in (456), in view of 1.3(6), we obtain

$$(457)\qquad \sum_{k=2}^{\infty}(-1)^k\frac{\zeta(k)}{k+1}=1+\frac{\gamma}{2}-\frac{1}{2}\log(2\pi),$$

which was proved by Suryanarayana (1974), and again by Singh and Verma (1983, p. 3, Section 4).

If we set $z = \frac{1}{2}$ in (456) and make use of 1.3(6) and 1.3(8), we get

$$\sum_{k=2}^{\infty} (-1)^k \frac{\zeta(k)}{k+1} 2^{-k} = 1 + \frac{\gamma}{4} - \frac{5}{12} \log 2 - 3 \log A. \tag{458}$$

Multiplying each member of (5) through (8) by t, integrating both sides of the resulting equations from $t = 0$ to $t = z$, and using (450), we obtain

$$\begin{aligned} \sum_{k=2}^{\infty} \zeta(k, a) \frac{z^{k+1}}{k+1} &= \frac{\psi(a)}{2} z^2 + z \log \Gamma(a - z) + \int_0^{-z} \log \Gamma(a + t)\, dt \\ &= [2a - 1 - \log(2\pi)] \frac{z}{2} + [\psi(a) - 1] \frac{z^2}{2} + (a - 1) \log \Gamma(a - z) \\ &\quad - \log G(a - z) + (1 - a) \log \Gamma(a) + \log G(a) \quad (|z| < |a|); \end{aligned} \tag{459}$$

$$\begin{aligned} \sum_{k=2}^{\infty} (-1)^k \zeta(k, a) \frac{z^{k+1}}{k+1} &= -\frac{\psi(a)}{2} z^2 + z \log \Gamma(a + z) - \int_0^{z} \log \Gamma(a + t)\, dt \\ &= [2a - 1 - \log(2\pi)] \frac{z}{2} + [1 - \psi(a)] \frac{z^2}{2} + (1 - a) \log \Gamma(a + z) \\ &\quad + \log G(a + z) + (a - 1) \log \Gamma(a) - \log G(a) \quad (|z| < |a|), \end{aligned} \tag{460}$$

which, in the special case when $a = 1$, corresponds to (456);

$$\begin{aligned} \sum_{k=1}^{\infty} \zeta(2k, a) \frac{z^{2k+1}}{2k+1} &= \frac{z}{2} \log \Gamma(a + z)\Gamma(a - z) - \frac{1}{2} \int_{-z}^{z} \log \Gamma(a + t)\, dt \\ &= \frac{1}{2} \left\{ [2a - 1 - \log(2\pi)] z + (1 - a) \log \frac{\Gamma(a + z)}{\Gamma(a - z)} + \log \frac{G(a + z)}{G(a - z)} \right\} \\ &\qquad (|z| < |a|); \end{aligned} \tag{461}$$

$$\begin{aligned} &\sum_{k=1}^{\infty} \zeta(2k + 1, a) \frac{z^{2k+2}}{k+1} \\ &= \psi(a) z^2 + z \log \frac{\Gamma(a - z)}{\Gamma(a + z)} + \int_0^{z} \log \Gamma(a + t)\, dt + \int_0^{-z} \log \Gamma(a + t)\, dt \\ &= [\psi(a) - 1] z^2 + (a - 1) \log \Gamma(a + z)\Gamma(a - z) - \log G(a + z) G(a - z) \\ &\quad + 2(1 - a) \log \Gamma(a) + 2 \log G(a) \quad (|z| < |a|). \end{aligned} \tag{462}$$

Taking $a = 1$ in (459), (461), and (462), and using 1.3(6), 2.3(2), and 1.2(4), we obtain

$$\sum_{k=2}^{\infty} \zeta(k) \frac{z^{k+1}}{k+1} = [1 - \log(2\pi)] \frac{z}{2} - \frac{1 + \gamma}{2} z^2 - \log G(1 - z) \quad (|z| < 1); \tag{463}$$

(464) $$\sum_{k=1}^{\infty} \zeta(2k) \frac{z^{2k+1}}{2k+1} = \frac{1}{2} \left\{ [1 - \log(2\pi)]z + \log \frac{G(1+z)}{G(1-z)} \right\} \quad (|z| < 1);$$

(465) $$\sum_{k=1}^{\infty} \zeta(2k+1) \frac{z^{2k+2}}{k+1} = -(1+\gamma)z^2 - \log G(1+z)G(1-z) \quad (|z| < 1).$$

Setting $z = \frac{1}{2}$ in (463), (464), and (465), and using 1.3(6) and 1.3(8), we obtain

(466) $$\sum_{k=2}^{\infty} \frac{\zeta(k)}{(k+1) \cdot 2^k} = -\frac{\gamma}{4} - \frac{7}{12} \log 2 + 3 \log A;$$

(467) $$\sum_{k=1}^{\infty} \frac{\zeta(2k)}{(2k+1) \cdot 2^{2k}} = \frac{1}{2} - \frac{1}{2} \log 2;$$

(468) $$\sum_{k=1}^{\infty} \frac{\zeta(2k+1)}{(k+1) \cdot 2^{2k}} = -2 - \gamma - \frac{1}{3} \log 2 + 12 \log A.$$

Taking $a = 2$ in (459) through (462), and using 1.3(6), 2.3(2), and 1.2(50), we obtain

(469) $$\sum_{k=2}^{\infty} \{\zeta(k) - 1\} \frac{z^{k+1}}{k+1} = [3 - \log(2\pi)] \frac{z}{2} - \frac{\gamma}{2} z^2 + \log \frac{\Gamma(2-z)}{G(2-z)} \quad (|z| < 2);$$

(470) $$\sum_{k=2}^{\infty} (-1)^k \{\zeta(k) - 1\} \frac{z^{k+1}}{k+1} = [3 - \log(2\pi)] \frac{z}{2} + \frac{\gamma}{2} z^2 + \log \frac{G(2+z)}{\Gamma(2+z)} \quad (|z| < 2);$$

(471)
$$\sum_{k=1}^{\infty} \{\zeta(2k) - 1\} \frac{z^{2k+1}}{2k+1} = \frac{1}{2} \left\{ [3 - \log(2\pi)]z + \log \frac{G(2+z)\Gamma(2-z)}{G(2-z)\Gamma(2+z)} \right\} \quad (|z| < 2);$$

(472) $$\sum_{k=1}^{\infty} \{\zeta(2k+1) - 1\} \frac{z^{2k+2}}{k+1} = -\gamma z^2 + \log \frac{\Gamma(2+z)\Gamma(2-z)}{G(2+z)G(2-z)} \quad (|z| < 2).$$

Taking the limit as $z \to 2$ on both sides of (470), and using 1.3(6), we obtain

(473) $$\sum_{k=2}^{\infty} (-1)^k \{\zeta(k) - 1\} \frac{2^{k+1}}{k+1} = 3 + 2\gamma - \log(6\pi).$$

Setting $z = 1$, $z = \frac{1}{2}$, and $z = \frac{3}{2}$ in (469) through (472), and using 1.3(6) and 1.3(8), we obtain

(474) $$\sum_{k=2}^{\infty} \frac{\zeta(k)-1}{k+1} = \frac{3}{2} - \frac{\gamma}{2} - \frac{1}{2}\log(2\pi);$$

(475) $$\sum_{k=2}^{\infty} \frac{(-1)^k}{k+1}\{\zeta(k)-1\} = \frac{3}{2} + \frac{\gamma}{2} - \frac{1}{2}\log(8\pi);$$

(476) $$\sum_{k=1}^{\infty} \frac{\zeta(2k)-1}{2k+1} = \frac{3}{2} - \frac{1}{2}\log(4\pi);$$

(477) $$\sum_{k=1}^{\infty} \frac{\zeta(2k+1)-1}{k+1} = -\gamma + \log 2;$$

(478) $$\sum_{k=2}^{\infty} \frac{\zeta(k)-1}{(k+1)\cdot 2^k} = \frac{5}{4} - \frac{\gamma}{4} + \log\left(2^{-\frac{31}{12}} \cdot A^3\right);$$

(479) $$\sum_{k=2}^{\infty} (-1)^k \frac{\zeta(k)-1}{(k+1)\cdot 2^k} = \frac{7}{4} + \frac{\gamma}{4} + \log\left(2^{\frac{19}{12}} \cdot 3^{-2} \cdot A^{-3}\right);$$

(480) $$\sum_{k=1}^{\infty} \frac{\zeta(2k)-1}{(2k+1)\cdot 2^{2k}} = \frac{3}{2} + \log\left(2^{-\frac{1}{2}} \cdot 3^{-1}\right);$$

(481) $$\sum_{k=1}^{\infty} \frac{\zeta(2k+1)-1}{(k+1)\cdot 2^{2k}} = -1 - \gamma + \log\left(2^{-\frac{25}{3}} \cdot 3^4 \cdot A^{12}\right);$$

(482) $$\sum_{k=2}^{\infty} \frac{\zeta(k)-1}{k+1}\left(\frac{3}{2}\right)^k = \frac{17}{12} - \frac{3\gamma}{4} + \log\left(2^{-\frac{19}{36}} \cdot A\right);$$

(483) $$\sum_{k=2}^{\infty} (-1)^k \frac{\zeta(k)-1}{k+1}\left(\frac{3}{2}\right)^k = \frac{19}{12} + \frac{3\gamma}{4} + \log\left(2^{-\frac{17}{36}} \cdot 5^{-\frac{2}{3}} \cdot A^{-1}\right);$$

$$\sum_{k=1}^{\infty} \frac{\zeta(2k)-1}{2k+1}\left(\frac{3}{2}\right)^{2k} = \frac{3}{2} + \log\left(2^{-\frac{1}{2}} \cdot 5^{-\frac{1}{3}}\right); \tag{484}$$

$$\sum_{k=1}^{\infty} \frac{\zeta(2k+1)-1}{k+1}\left(\frac{3}{2}\right)^{2k} = -\frac{1}{9} - \gamma + \log\left(2^{-\frac{1}{27}} \cdot 5^{\frac{4}{9}} \cdot A^{\frac{4}{3}}\right). \tag{485}$$

Taking $a = \frac{1}{2}$ in (459) through (462), and using 1.3(8), 2.3(2), and 1.2(51), we obtain

$$\begin{aligned} &\sum_{k=2}^{\infty} \left(2^k - 1\right)\zeta(k)\frac{z^{k+1}}{k+1} \\ &\quad = \frac{1}{8} - \frac{z}{2}\log(2\pi) - (1+\gamma+2\log 2)\frac{z^2}{2} - \frac{1}{2}\log\Gamma\left(\frac{1}{2} - z\right) \\ &\quad - \log G\left(\frac{1}{2} - z\right) + \log\left(2^{\frac{1}{24}} \cdot A^{-\frac{3}{2}}\right) \quad (|z| < \frac{1}{2}); \end{aligned} \tag{486}$$

$$\begin{aligned} &\sum_{k=2}^{\infty} (-1)^k \left(2^k - 1\right)\zeta(k)\frac{z^{k+1}}{k+1} \\ &\quad = -\frac{1}{8} - \frac{z}{2}\log(2\pi) + (1+\gamma+2\log 2)\frac{z^2}{2} + \frac{1}{2}\log\Gamma\left(\frac{1}{2} + z\right) \\ &\quad + \log G\left(\frac{1}{2} + z\right) + \log\left(2^{-\frac{1}{24}} \cdot A^{\frac{3}{2}}\right) \quad (|z| < \frac{1}{2}); \end{aligned} \tag{487}$$

$$\begin{aligned} &\sum_{k=1}^{\infty} \left(2^{2k} - 1\right)\zeta(2k)\frac{z^{2k+1}}{2k+1} \\ &\quad = -\frac{z}{2}\log(2\pi) + \frac{1}{4}\log\frac{\Gamma\left(\frac{1}{2}+z\right)}{\Gamma\left(\frac{1}{2}-z\right)} + \frac{1}{2}\log\frac{G\left(\frac{1}{2}+z\right)}{G\left(\frac{1}{2}-z\right)} \quad (|z| < \frac{1}{2}); \end{aligned} \tag{488}$$

$$\begin{aligned} &\sum_{k=1}^{\infty} \left(2^{2k+1} - 1\right)\zeta(2k+1)\frac{z^{2k+2}}{k+1} \\ &\quad = \frac{1}{4} - (1+\gamma+2\log 2)z^2 - \frac{1}{2}\log\Gamma\left(\frac{1}{2} + z\right)\Gamma\left(\frac{1}{2} - z\right) \\ &\quad - \log G\left(\frac{1}{2} + z\right)G\left(\frac{1}{2} - z\right) + \log\left(2^{\frac{1}{12}} \cdot A^{-3}\right) \quad (|z| < \frac{1}{2}). \end{aligned} \tag{489}$$

Taking the limit in (487) as $z \to \frac{1}{2}$, and using 1.3(6), we obtain

$$\sum_{k=2}^{\infty} (-1)^k \frac{2^k - 1}{(k+1)\cdot 2^k}\zeta(k) = \frac{\gamma}{4} + \log\left(2^{-\frac{1}{12}} \cdot \pi^{-\frac{1}{2}} \cdot A^3\right). \tag{490}$$

Multiplying each member of (5) through (8) by t^2, integrating both sides of the resulting equations from $t=0$ to $t=z$, and using (452), we obtain

(491)
$$\begin{aligned}&\sum_{k=2}^{\infty}\zeta(k,a)\frac{z^{k+2}}{k+2}\\&\quad=\frac{1-a}{2}\left[1-2a+\log(2\pi)\right]z+\left[1-\log(2\pi)\right]\frac{z^2}{4}+\left[\psi(a)-1\right]\frac{z^3}{3}\\&\qquad+(a-1)^2\log\Gamma(a-z)-(z+a-1)\log G(a-z)-(a-1)^2\log\Gamma(a)\\&\qquad+(a-1)\log G(a)-\int_0^{-z}\log G(t+a)\,dt\quad(|z|<|a|);\end{aligned}$$

(492)
$$\begin{aligned}&\sum_{k=2}^{\infty}(-1)^k\zeta(k,a)\frac{z^{k+2}}{k+2}\\&\quad=\frac{a-1}{2}\left[1-2a+\log(2\pi)\right]z+\left[1-\log(2\pi)\right]\frac{z^2}{4}+\left[1-\psi(a)\right]\frac{z^3}{3}\\&\qquad+(a-1)^2\log\Gamma(a+z)+(z-a+1)\log G(a+z)-(a-1)^2\log\Gamma(a)\\&\qquad+(a-1)\log G(a)-\int_0^{z}\log G(t+a)\,dt\quad(|z|<|a|);\end{aligned}$$

(493)
$$\begin{aligned}&\sum_{k=1}^{\infty}\zeta(2k,a)\frac{z^{2k+2}}{k+1}\\&\quad=2(a-1)\log G(a)-2(a-1)^2\log\Gamma(a)+\left[1-\log(2\pi)\right]\frac{z^2}{2}\\&\qquad+(a-1)^2\log\Gamma(a+z)\Gamma(a-z)+(z-a+1)\log G(a+z)\\&\qquad-(z+a-1)\log G(a-z)-\int_0^{z}\log G(t+a)\,dt\\&\qquad-\int_0^{-z}\log G(t+a)\,dt\quad(|z|<|a|);\end{aligned}$$

(494)
$$\begin{aligned}&\sum_{k=1}^{\infty}\zeta(2k+1,a)\frac{z^{2k+3}}{2k+3}\\&\quad=\frac{1-a}{2}[1-2a+\log(2\pi)]z+[\psi(a)-1]\frac{z^3}{3}-\frac{1}{2}(a-1)^2\log\frac{\Gamma(a+z)}{\Gamma(a-z)}\\&\qquad-\frac{1}{2}(z-a+1)\log G(a+z)-\frac{1}{2}(z+a-1)\log G(a-z)\\&\qquad+\frac{1}{2}\left\{\int_0^{z}\log G(t+a)\,dt-\int_0^{-z}\log G(t+a)\,dt\right\}\quad(|z|<|a|);\end{aligned}$$

If we use 1.3(6), 1.3(55), 1.3(56), 1.3(73), and 1.3(77), we readily find that

$$\int_0^{-1} \log \Gamma(t+1)\, dt = -\frac{1}{2} \log(2\pi); \tag{495}$$

$$\int_0^{-1} \log G(t+1)\, dt = -\frac{1}{12} + \frac{1}{4} \log(2\pi) + 2 \log A; \tag{496}$$

$$\int_0^{-\frac{1}{2}} \log \Gamma(t+1)\, dt = -\frac{7}{24} \log 2 - \frac{1}{4} \log \pi + \frac{3}{2} \log A; \tag{497}$$

$$\int_0^{-\frac{1}{2}} \log G(t+1)\, dt = -\frac{1}{24} + \frac{1}{12} \log 2 + \frac{1}{16} \log \pi + \frac{1}{4} \log A - \frac{7}{4} \log B. \tag{498}$$

Taking $a = 1$ in (491) through (494), and using 1.3(6), 2.3(2), and 1.2(4), we obtain

$$\begin{aligned} &\sum_{k=2}^{\infty} \zeta(k) \frac{z^{k+2}}{k+2} \\ &\quad = [1 - \log(2\pi)] \frac{z^2}{4} - (1+\gamma) \frac{z^3}{3} - z \log G(1-z) \\ &\quad - \int_0^{-z} \log G(t+1)\, dt \quad (|z| < 1); \end{aligned} \tag{499}$$

$$\begin{aligned} &\sum_{k=2}^{\infty} (-1)^k \zeta(k) \frac{z^{k+2}}{k+2} \\ &\quad = [1 - \log(2\pi)] \frac{z^2}{4} + (1+\gamma) \frac{z^3}{3} + z \log G(1+z) \\ &\quad - \int_0^{z} \log G(t+1)\, dt \quad (|z| < 1); \end{aligned} \tag{500}$$

$$\begin{aligned} &\sum_{k=1}^{\infty} \zeta(2k) \frac{z^{2k+2}}{k+1} \\ &\quad = [1 - \log(2\pi)] \frac{z^2}{2} + z \log \frac{G(1+z)}{G(1-z)} \\ &\quad - \int_0^{z} \log G(t+1)\, dt - \int_0^{-z} \log G(t+1)\, dt \quad (|z| < 1); \end{aligned} \tag{501}$$

$$\begin{aligned}\sum_{k=1}^{\infty}&\zeta(2k+1)\frac{z^{2k+3}}{2k+3}\\ &=-(1+\gamma)\frac{z^3}{3}-\frac{1}{2}z\log G(1+z)G(1-z)\\ &+\frac{1}{2}\left\{\int_0^z \log G(t+1)\,dt-\int_0^{-z}\log G(t+1)\,dt\right\}\qquad(|z|<1).\end{aligned}\tag{502}$$

Taking the limit in (500) as $z\to 1$, and using 1.3(6) and 1.3(73), we obtain

$$\sum_{k=2}^{\infty}(-1)^k\frac{\zeta(k)}{k+2}=\frac{1}{2}+\frac{\gamma}{3}+\log\left(2^{-\frac{1}{2}}\cdot\pi^{-\frac{1}{2}}\cdot A^2\right).\tag{503}$$

If we set $z=\frac{1}{2}$ in (499) through (502), by virtue of 1.3(6), 1.3(8), 1.3(77), and (498), we get

$$\sum_{k=2}^{\infty}\frac{\zeta(k)}{(k+2)\cdot 2^k}=-\frac{\gamma}{6}+\log\left(2^{-\frac{2}{3}}\cdot A^2\cdot B^7\right);\tag{504}$$

$$\sum_{k=2}^{\infty}(-1)^k\frac{\zeta(k)}{(k+2)\cdot 2^k}=\frac{1}{2}+\frac{\gamma}{6}+\log\left(2^{-\frac{1}{3}}\cdot A^{-2}\cdot B^7\right);\tag{505}$$

$$\sum_{k=1}^{\infty}\frac{\zeta(2k)}{(k+1)\cdot 2^{2k}}=\frac{1}{2}+\log\left(2^{-1}\cdot B^{14}\right);\tag{506}$$

$$\sum_{k=1}^{\infty}\frac{\zeta(2k+1)}{(2k+3)\cdot 2^{2k}}=-\frac{1}{2}-\frac{\gamma}{3}+\log\left(2^{-\frac{1}{3}}\cdot A^4\right).\tag{507}$$

Now we recall the following integral formula [*cf.* Choi *et al.* (1995, p. 386)]:

$$\begin{aligned}\int_0^1&\log G(t+a)\,dt\\ &=\frac{a(1-a)}{2}+\frac{a}{2}\log(2\pi)+a\log\Gamma(a)-\log G(a+1)\\ &+\log\left(2^{-\frac{1}{4}}\cdot\pi^{-\frac{1}{4}}\cdot e^{\frac{1}{12}}\cdot A^{-2}\right).\end{aligned}\tag{508}$$

In view of 1.3(6), (508) with $a=2$ yields

$$\int_0^1\log G(t+2)\,dt=-\frac{11}{12}+\frac{3}{4}\log(2\pi)-2\log A.\tag{509}$$

By setting $a = 2$ in (491) through (494), and using 1.3(6), 2.3(2), and 1.2(50), we find that

(510)
$$\begin{aligned}\sum_{k=2}^{\infty}\{\zeta(k)-1\}\frac{z^{k+2}}{k+2} &= [3-\log(2\pi)]\,\frac{z}{2}+[1-\log(2\pi)]\,\frac{z^2}{4}-\frac{\gamma}{3}z^3+\log\Gamma(2-z)\\ &\quad-(z+1)\log G(2-z)-\int_0^{-z}\log G(t+2)\,dt \quad (|z|<2);\end{aligned}$$

(511)
$$\begin{aligned}\sum_{k=2}^{\infty}(-1)^k\{\zeta(k)-1\}\frac{z^{k+2}}{k+2} &= [\log(2\pi)-3]\,\frac{z}{2}+[1-\log(2\pi)]\,\frac{z^2}{4}+\frac{\gamma}{3}z^3+\log\Gamma(2+z)\\ &\quad+(z-1)\log G(2+z)-\int_0^{z}\log G(t+2)\,dt \quad (|z|<2);\end{aligned}$$

(512)
$$\begin{aligned}\sum_{k=1}^{\infty}\{\zeta(2k)-1\}\frac{z^{2k+2}}{k+1} &= [1-\log(2\pi)]\,\frac{z^2}{2}+\log\Gamma(2+z)\Gamma(2-z)+(z-1)\log G(2+z)\\ &\quad-(z+1)\log G(2-z)-\int_0^{z}\log G(t+2)\,dt\\ &\quad-\int_0^{-z}\log G(t+2)\,dt \quad (|z|<2);\end{aligned}$$

(513)
$$\begin{aligned}\sum_{k=1}^{\infty}\{\zeta(2k+1)-1\}\frac{z^{2k+3}}{2k+3} &= [3-\log(2\pi)]\,\frac{z}{2}-\frac{\gamma}{3}z^3-\frac{1}{2}\log\frac{\Gamma(2+z)}{\Gamma(2-z)}\\ &\quad-\frac{1}{2}(z-1)\log G(2+z)-\frac{1}{2}(z+1)\log G(2-z)\\ &\quad+\frac{1}{2}\left\{\int_0^{z}\log G(t+2)\,dt-\int_0^{-z}\log G(t+2)\,dt\right\} \quad (|z|<2).\end{aligned}$$

Taking the limit in (511) as $z \to 2$, and using 1.3(6), 1.3(55), and (509), we obtain

(514)
$$\sum_{k=2}^{\infty}(-1)^k(\zeta(k)-1)\frac{2^k}{k+2}=\frac{11}{24}+\frac{2\gamma}{3}+\log\left(3^{\frac{1}{4}}\cdot\pi^{-\frac{1}{2}}\cdot A\right).$$

If use is made of 1.3(6), 1.3(56), 1.3(73), 1.3(77), (498), and (509), the following integral formulas are readily obtained:

(515)
$$\begin{aligned}\int_0^{\frac{3}{2}} \log G(t+2)\,dt &= \int_0^1 \log G(t+2)\,dt + \int_0^{\frac{1}{2}} \log G(t+2)\,dt + \int_0^{\frac{1}{2}} \log\Gamma(t+2)\,dt \\ &= -\frac{19}{8} - \frac{31}{24}\log 2 + \frac{3}{2}\log 3 + \frac{21}{16}\log\pi + \frac{3}{4}\log A - \frac{7}{4}\log B;\end{aligned}$$

(516)
$$\begin{aligned}\int_0^{-\frac{3}{2}} \log G(t+2)\,dt &= -\int_0^1 \log G(t+1)\,dt + \int_0^{-\frac{1}{2}} \log G(t+1)\,dt \\ &= -\frac{1}{8} - \frac{1}{6}\log 2 - \frac{3}{16}\log\pi + \frac{9}{4}\log A - \frac{7}{4}\log B.\end{aligned}$$

If we set $z = 1$, $z = \frac{1}{2}$, and $z = \frac{3}{2}$ in (510) through (513), in view of 1.3(6), 1.3(8), 1.3(56), 1.3(73), 1.3(77), (497), (498), (509), and (515), we get

(517)
$$\sum_{k=2}^{\infty} \frac{\zeta(k)-1}{k+2} = \frac{11}{6} - \frac{\gamma}{3} + \log\left(2^{-\frac{1}{2}} \cdot \pi^{-\frac{1}{2}} \cdot A^{-2}\right);$$

(518)
$$\sum_{k=2}^{\infty} (-1)^k \frac{\zeta(k)-1}{k+2} = -\frac{1}{3} + \frac{\gamma}{3} + \log\left(2^{\frac{1}{2}} \cdot \pi^{-\frac{1}{2}} \cdot A^{2}\right);$$

(519)
$$\sum_{k=1}^{\infty} \frac{\zeta(2k)-1}{k+1} = \frac{3}{2} - \log\pi;$$

(520)
$$\sum_{k=1}^{\infty} \frac{\zeta(2k+1)-1}{2k+3} = \frac{13}{12} - \frac{\gamma}{3} + \log\left(2^{-\frac{1}{2}} \cdot A^{-2}\right);$$

(521)
$$\sum_{k=2}^{\infty} \frac{\zeta(k)-1}{(k+2)\cdot 2^k} = \frac{8}{3} - \frac{\gamma}{6} + \log\left(2^{-\frac{14}{3}} \cdot A^{2} \cdot B^{7}\right);$$

(522)
$$\sum_{k=2}^{\infty} (-1)^k \frac{\zeta(k)-1}{(k+2)\cdot 2^k} = -\frac{7}{6} + \frac{\gamma}{6} + \log\left(2^{-\frac{13}{3}} \cdot 3^{4} \cdot A^{-2} \cdot B^{7}\right);$$

(523) $$\sum_{k=1}^{\infty} \frac{\zeta(2k)-1}{(k+1)\cdot 2^{2k}} = \frac{3}{2} + \log\left(2^{-9}\cdot 3^{4}\cdot B^{14}\right);$$

(524) $$\sum_{k=1}^{\infty} \frac{\zeta(2k+1)-1}{(2k+3)\cdot 2^{2k}} = \frac{23}{6} - \frac{\gamma}{3} + \log\left(2^{\frac{1}{2}}\cdot 3^{-4}\cdot A^{4}\right);$$

(525) $$\sum_{k=2}^{\infty} \frac{\zeta(k)-1}{k+2}\left(\frac{3}{2}\right)^{k} = \frac{7}{6} - \frac{\gamma}{2} + \log\left(2^{-\frac{5}{9}}\cdot A^{\frac{2}{3}}\cdot B^{\frac{7}{9}}\right);$$

(526) $$\sum_{k=2}^{\infty} (-1)^k \frac{\zeta(k)-1}{k+2}\left(\frac{3}{2}\right)^{k} = \frac{1}{3} + \frac{\gamma}{2} + \log\left(2^{-\frac{4}{3}}\cdot 5^{\frac{4}{9}}\cdot A^{-\frac{2}{3}}\cdot B^{\frac{7}{9}}\right);$$

(527) $$\sum_{k=1}^{\infty} \frac{\zeta(2k)-1}{k+1}\left(\frac{3}{2}\right)^{2k} = \frac{3}{2} + \log\left(2^{-\frac{17}{9}}\cdot 5^{\frac{4}{9}}\cdot B^{\frac{14}{9}}\right);$$

(528) $$\sum_{k=1}^{\infty} \frac{\zeta(2k+1)-1}{2k+3}\left(\frac{3}{2}\right)^{2k} = \frac{5}{18} - \frac{\gamma}{3} + \log\left(2^{\frac{7}{27}}\cdot 5^{-\frac{4}{27}}\cdot A^{\frac{4}{9}}\right).$$

Taking $a = \frac{1}{2}$ in (491) through (494), and using 1.3(8), 2.3(2) and 1.2(51), we obtain

(529) $$\begin{aligned}&\sum_{k=2}^{\infty} \left(2^k-1\right)\zeta(k)\frac{z^{k+2}}{k+2}\\ &\quad= \log\left(2^{-\frac{1}{48}}\cdot e^{-\frac{1}{16}}\cdot A^{\frac{3}{4}}\right) + \frac{z}{4}\log(2\pi) + [1-\log(2\pi)]\frac{z^2}{4}\\ &\qquad - (1+\gamma+2\log 2)\frac{z^3}{3} + \frac{1}{4}\log\Gamma\left(\frac{1}{2}-z\right) + \left(\frac{1}{2}-z\right)\log G\left(\frac{1}{2}-z\right)\\ &\qquad - \int_0^{-z}\log G\left(t+\frac{1}{2}\right)dt \quad (|z|<\frac{1}{2});\end{aligned}$$

(530) $$\begin{aligned}&\sum_{k=2}^{\infty} (-1)^k\left(2^k-1\right)\zeta(k)\frac{z^{k+2}}{k+2}\\ &\quad= \log\left(2^{-\frac{1}{48}}\cdot e^{-\frac{1}{16}}\cdot A^{\frac{3}{4}}\right) - \frac{z}{4}\log(2\pi) + [1-\log(2\pi)]\frac{z^2}{4}\\ &\qquad + (1+\gamma+2\log 2)\frac{z^3}{3} + \frac{1}{4}\log\Gamma\left(\frac{1}{2}+z\right) + \left(\frac{1}{2}+z\right)\log G\left(\frac{1}{2}+z\right)\\ &\qquad - \int_0^{z}\log G\left(t+\frac{1}{2}\right)dt \quad (|z|<\frac{1}{2});\end{aligned}$$

(531)

$$\begin{aligned}&\sum_{k=1}^{\infty}\left(2^{2k}-1\right)\zeta(2k)\frac{z^{2k+2}}{k+1}\\&\quad=\log\left(2^{-\frac{1}{24}}\cdot e^{-\frac{1}{8}}\cdot A^{\frac{3}{2}}\right)+[1-\log(2\pi)]\frac{z^2}{2}+\frac{1}{4}\log\Gamma\left(\frac{1}{2}+z\right)\Gamma\left(\frac{1}{2}-z\right)\\&\qquad+\left(\frac{1}{2}+z\right)\log G\left(\frac{1}{2}+z\right)+\left(\frac{1}{2}-z\right)\log G\left(\frac{1}{2}-z\right)\\&\qquad-\int_0^z\log G\left(t+\frac{1}{2}\right)dt-\int_0^{-z}\log G\left(t+\frac{1}{2}\right)dt\quad(|z|<\frac{1}{2});\end{aligned}$$

(532)

$$\begin{aligned}&\sum_{k=1}^{\infty}\left(2^{2k+1}-1\right)\zeta(2k+1)\frac{z^{2k+3}}{2k+3}\\&\quad=\frac{z}{4}\log(2\pi)-(1+\gamma+2\log 2)\frac{z^3}{3}-\frac{1}{8}\log\frac{\Gamma\left(\frac{1}{2}+z\right)}{\Gamma\left(\frac{1}{2}-z\right)}\\&\qquad-\frac{1}{2}\left(\frac{1}{2}+z\right)\log G\left(\frac{1}{2}+z\right)+\frac{1}{2}\left(\frac{1}{2}-z\right)\log G\left(\frac{1}{2}-z\right)\\&\qquad+\frac{1}{2}\left\{\int_0^z\log G\left(t+\frac{1}{2}\right)dt-\int_0^{-z}\log G\left(t+\frac{1}{2}\right)dt\right\}\quad(|z|<\frac{1}{2}).\end{aligned}$$

By taking the limit in (530) as $z\to\frac{1}{2}$, and using 1.3(6), 1.3(55), 1.3(56), 1.3(73), and 1.3(77), we obtain

(533)

$$\sum_{k=2}^{\infty}(-1)^k\frac{2^k-1}{(k+2)\cdot 2^k}\zeta(k)=\frac{\gamma}{6}+\log\left(2^{\frac{23}{6}}\cdot\pi^{-\frac{1}{2}}\cdot A^4\cdot B^{-7}\right).$$

Integrating both sides of (1) through (4) from $t=0$ to $t=z$, in view of (444), we find that

(534)

$$\begin{aligned}&\sum_{k=2}^{\infty}\zeta(k,a)\frac{z^{k+1}}{k(k+1)}\\&\quad=-z\log\Gamma(a)+\psi(a)\frac{z^2}{2}-\int_0^{-z}\log\Gamma(a+t)\,dt\\&\quad=-z\log\Gamma(a)+\frac{\psi(a)}{2}z^2+\int_0^z\log\Gamma(a-t)\,dt\\&\quad=(a-1)\log\Gamma(a)-\log G(a)+\left(\frac{1}{2}+\frac{1}{2}\log(2\pi)-a-\log\Gamma(a)\right)z\\&\qquad+[1+\psi(a)]\frac{z^2}{2}+(z-a+1)\log\Gamma(a-z)+\log G(a-z)\quad(|z|<|a|);\end{aligned}$$

(535)
$$\sum_{k=2}^{\infty}(-1)^k\zeta(k,a)\frac{z^{k+1}}{k(k+1)} = -z\log\Gamma(a) - \psi(a)\frac{z^2}{2} + \int_0^z \log\Gamma(a+t)\,dt$$
$$= (1-a)\log\Gamma(a) + \log G(a) + \left(\frac{1}{2} + \frac{1}{2}\log(2\pi) - a - \log\Gamma(a)\right)z$$
$$- [1+\psi(a)]\frac{z^2}{2} + (z+a-1)\log\Gamma(z+a) - \log G(z+a) \quad (|z| < |a|);$$

(536)
$$\sum_{k=1}^{\infty}\zeta(2k,a)\frac{z^{2k+1}}{k(2k+1)}$$
$$= -2z\log\Gamma(a) + \int_{-z}^{z}\log\Gamma(a+t)\,dt$$
$$= -2z\log\Gamma(a) + \int_0^z \{\log\Gamma(a+t) + \log\Gamma(a-t)\}\,dt$$
$$= [1 - 2a + \log(2\pi) - 2\log\Gamma(a)]\,z + (z+a-1)\log\Gamma(a+z)$$
$$+ (z-a+1)\log\Gamma(a-z) + \log\frac{G(a-z)}{G(a+z)} \quad (|z| < |a|);$$

(537)
$$\sum_{k=1}^{\infty}\zeta(2k+1,a)\frac{z^{2k+2}}{(k+1)(2k+1)}$$
$$= \psi(a)z^2 - \int_0^z \log\Gamma(a+t)\,dt - \int_0^{-z}\log\Gamma(a+t)\,dt$$
$$= \psi(a)z^2 + \int_0^z \{\log\Gamma(a-t) - \log\Gamma(a+t)\}\,dt$$
$$= 2(a-1)\log\Gamma(a) - 2\log G(a) + [1+\psi(a)]z^2 + (1-a+z)\log\Gamma(a-z)$$
$$+ (1-a-z)\log\Gamma(a+z) + \log G(a-z)G(a+z) \quad (|z| < |a|).$$

If we set $a = 1$ in (534) through (537), by virtue of 1.3(6), 2.3(2), and 1.2(4), we get

(538)
$$\sum_{k=2}^{\infty}\zeta(k)\frac{z^{k+1}}{k(k+1)}$$
$$= [\log(2\pi) - 1]\,\frac{z}{2} + (1-\gamma)\frac{z^2}{2} + z\log\Gamma(1-z)$$
$$+ \log G(1-z) \quad (|z| < 1);$$

(539)
$$\sum_{k=2}^{\infty}(-1)^k\zeta(k)\frac{z^{k+1}}{k(k+1)}$$
$$= [\log(2\pi) - 1]\,\frac{z}{2} + (\gamma-1)\frac{z^2}{2} + z\log\Gamma(z+1)$$
$$- \log G(z+1) \quad (|z| < 1);$$

(540)
$$\sum_{k=1}^{\infty} \zeta(2k) \frac{z^{2k+1}}{k(2k+1)} = [\log(2\pi) - 1]\, z + z \log \Gamma(1-z)\Gamma(1+z) + \log \frac{G(1-z)}{G(1+z)} \quad (|z| < 1);$$

(541)
$$\sum_{k=1}^{\infty} \zeta(2k+1) \frac{z^{2k+2}}{(k+1)(2k+1)} = (1-\gamma)z^2 + z \log \frac{\Gamma(1-z)}{\Gamma(1+z)} + \log G(1-z)G(1+z) \quad (|z| < 1).$$

Taking the limit in (539) as $z \to 1$, and using 1.3(6), we obtain

(542)
$$\sum_{k=2}^{\infty} (-1)^k \frac{\zeta(k)}{k(k+1)} = -1 + \frac{\gamma}{2} + \frac{1}{2}\log(2\pi).$$

Setting $z = \frac{1}{2}$ in (538) through (541), and using 1.3(6) and 1.3(8), we obtain

(543)
$$\sum_{k=2}^{\infty} \frac{\zeta(k)}{k(k+1) \cdot 2^k} = -\frac{\gamma}{4} + \log\left(2^{\frac{7}{12}} \cdot \pi^{\frac{1}{2}} \cdot A^{-3}\right);$$

(544)
$$\sum_{k=2}^{\infty} (-1)^k \frac{\zeta(k)}{k(k+1) \cdot 2^k} = -1 + \frac{\gamma}{4} + \log\left(2^{-\frac{7}{12}} \cdot \pi^{\frac{1}{2}} \cdot A^{3}\right);$$

(545)
$$\sum_{k=1}^{\infty} \frac{\zeta(2k)}{k(2k+1) \cdot 2^{2k}} = -1 + \log \pi;$$

(546)
$$\sum_{k=1}^{\infty} \frac{\zeta(2k+1)}{(k+1)(2k+1) \cdot 2^{2k}} = 2 - \gamma + \log\left(2^{\frac{7}{3}} \cdot A^{-12}\right).$$

By putting $a = 2$ in (534) through (537), and using 1.3(6), 2.3(2), and 1.2(50), we obtain

(547)
$$\sum_{k=2}^{\infty} \{\zeta(k) - 1\} \frac{z^{k+1}}{k(k+1)} = [\log(2\pi) - 3]\, \frac{z}{2} + (2-\gamma)\frac{z^2}{2} + (z-1)\log\Gamma(2-z) + \log G(2-z) \quad (|z| < 2);$$

(548)
$$\sum_{k=2}^{\infty}(-1)^k\{\zeta(k)-1\}\frac{z^{k+1}}{k(k+1)}$$
$$= [\log(2\pi)-3]\,\frac{z}{2} + (\gamma-2)\frac{z^2}{2} + (z+1)\log\Gamma(z+2) - \log G(z+2) \quad (|z|<2);$$

(549)
$$\sum_{k=1}^{\infty}\{\zeta(2k)-1\}\frac{z^{2k+1}}{k(2k+1)}$$
$$= [\log(2\pi)-3]\,z + (z+1)\log\Gamma(2+z) + (z-1)\log\Gamma(2-z) + \log\frac{G(2-z)}{G(2+z)} \quad (|z|<2);$$

(550)
$$\sum_{k=1}^{\infty}\{\zeta(2k+1)-1\}\frac{z^{2k+2}}{(k+1)(2k+1)}$$
$$= (2-\gamma)z^2 + (z-1)\log\Gamma(2-z) - (z+1)\log\Gamma(2+z) + \log G(2-z)G(2+z) \quad (|z|<2).$$

Taking the limit in (548) as $z \to 2$, and using 1.3(6), we have

(551)
$$\sum_{k=2}^{\infty}(-1)^k\frac{\{\zeta(k)-1\}2^k}{k(k+1)} = -\frac{7}{2} + \gamma + \log\left(2^{\frac{3}{2}}\cdot 3^{\frac{3}{2}}\cdot \pi^{\frac{1}{2}}\right).$$

If we set $z=1$, $z=\frac{1}{2}$, and $z=\frac{3}{2}$ in (547) through (550), in view of 1.3(6) and 1.3(8), we obtain

(552)
$$\sum_{k=2}^{\infty}\frac{\zeta(k)-1}{k(k+1)} = -\frac{1}{2} - \frac{\gamma}{2} + \log\left(2^{\frac{1}{2}}\cdot\pi^{\frac{1}{2}}\right);$$

(553)
$$\sum_{k=2}^{\infty}(-1)^k\frac{\zeta(k)-1}{k(k+1)} = -\frac{5}{2} + \frac{\gamma}{2} + \log\left(2^{\frac{5}{2}}\cdot\pi^{\frac{1}{2}}\right);$$

(554)
$$\sum_{k=1}^{\infty}\frac{\zeta(2k)-1}{k(2k+1)} = -3 + \log(8\pi);$$

(555)
$$\sum_{k=1}^{\infty}\frac{\zeta(2k+1)-1}{(k+1)(2k+1)} = 2 - \gamma - 2\log 2;$$

(556) $$\sum_{k=2}^{\infty} \frac{\zeta(k)-1}{k(k+1)\cdot 2^k} = -\frac{3}{4} - \frac{\gamma}{4} + \log\left(2^{\frac{19}{12}} \cdot \pi^{\frac{1}{2}} \cdot A^{-3}\right);$$

(557) $$\sum_{k=2}^{\infty} (-1)^k \frac{\zeta(k)-1}{k(k+1)\cdot 2^k} = -\frac{9}{4} + \frac{\gamma}{4} + \log\left(2^{-\frac{43}{12}} \cdot 3^3 \cdot \pi^{\frac{1}{2}} \cdot A^3\right);$$

(558) $$\sum_{k=1}^{\infty} \frac{\zeta(2k)-1}{k(2k+1)\cdot 2^{2k}} = -3 + \log\left(2^{-2} \cdot 3^3 \cdot \pi\right);$$

(559) $$\sum_{k=1}^{\infty} \frac{\zeta(2k+1)-1}{(k+1)(2k+1)\cdot 2^{2k}} = 3 - \gamma + \log\left(2^{\frac{31}{3}} \cdot 3^{-6} \cdot A^{-12}\right);$$

(560) $$\sum_{k=2}^{\infty} \frac{\zeta(k)-1}{k(k+1)} \left(\frac{3}{2}\right)^k = \frac{1}{12} - \frac{3\gamma}{4} + \log\left(2^{\frac{19}{36}} \cdot \pi^{\frac{1}{2}} \cdot A^{-1}\right);$$

(561) $$\sum_{k=2}^{\infty} (-1)^k \frac{\zeta(k)-1}{k(k+1)} \left(\frac{3}{2}\right)^k = -\frac{37}{12} + \frac{3\gamma}{4} + \log\left(2^{-\frac{91}{36}} \cdot 3 \cdot 5^{\frac{5}{3}} \cdot \pi^{\frac{1}{2}} \cdot A\right);$$

(562) $$\sum_{k=1}^{\infty} \frac{\zeta(2k)-1}{k(2k+1)} \left(\frac{3}{2}\right)^{2k} = -3 + \log\left(2^{-2} \cdot 3 \cdot 5^{\frac{5}{3}} \cdot \pi\right);$$

(563) $$\sum_{k=1}^{\infty} \frac{\zeta(2k+1)-1}{(k+1)(2k+1)} \left(\frac{3}{2}\right)^{2k} = \frac{19}{9} - \gamma + \log\left(2^{\frac{55}{27}} \cdot 3^{-\frac{2}{3}} \cdot 5^{-\frac{10}{9}} \cdot A^{-\frac{4}{3}}\right).$$

If we take $a = \frac{1}{2}$ in (534) through (537), by virtue 1.3(8), 2.3(2) and 1.2(51), we obtain

(564) $$\begin{aligned} &\sum_{k=2}^{\infty} \left(2^k - 1\right) \zeta(k) \frac{z^{k+1}}{k(k+1)} \\ &= \log\left(2^{-\frac{1}{24}} \cdot e^{-\frac{1}{8}} \cdot A^{\frac{3}{2}}\right) + \frac{z}{2}\log 2 + (1-\gamma-2\log 2)\frac{z^2}{2} \\ &+ \left(\frac{1}{2}+z\right)\log\Gamma\left(\frac{1}{2}-z\right) + \log G\left(\frac{1}{2}-z\right) \quad \left(|z| < \frac{1}{2}\right); \end{aligned}$$

(565)
$$\sum_{k=2}^{\infty}(-1)^k\left(2^k-1\right)\zeta(k)\frac{z^{k+1}}{k(k+1)}$$
$$=\log\left(2^{\frac{1}{24}}\cdot e^{\frac{1}{8}}\cdot A^{-\frac{3}{2}}\right)+\frac{z}{2}\log 2+(\gamma+2\log 2-1)\frac{z^2}{2}$$
$$+\left(z-\frac{1}{2}\right)\log\Gamma\left(z+\frac{1}{2}\right)-\log G\left(z+\frac{1}{2}\right)\quad(|z|<\frac{1}{2});$$

(566)
$$\sum_{k=1}^{\infty}\left(2^{2k}-1\right)\zeta(2k)\frac{z^{2k+1}}{k(2k+1)}$$
$$=z\log 2+\left(z-\frac{1}{2}\right)\log\Gamma\left(\frac{1}{2}+z\right)+\left(\frac{1}{2}+z\right)\log\Gamma\left(\frac{1}{2}-z\right)$$
$$+\log\frac{G\left(\frac{1}{2}-z\right)}{G\left(\frac{1}{2}+z\right)}\quad(|z|<\frac{1}{2});$$

(567)
$$\sum_{k=1}^{\infty}\left(2^{2k+1}-1\right)\zeta(2k+1)\frac{z^{2k+2}}{(k+1)(2k+1)}$$
$$=\log\left(2^{-\frac{1}{12}}\cdot e^{-\frac{1}{4}}\cdot A^{3}\right)+(1-\gamma-2\log 2)z^2$$
$$+\left(\frac{1}{2}+z\right)\log\Gamma\left(\frac{1}{2}-z\right)+\left(\frac{1}{2}-z\right)\log\Gamma\left(\frac{1}{2}+z\right)$$
$$+\log G\left(\frac{1}{2}-z\right)G\left(\frac{1}{2}+z\right)\quad(|z|<\frac{1}{2}).$$

Taking the limit in (565) as $z\to\frac{1}{2}$, and using 1.3(6), we obtain

(568)
$$\sum_{k=2}^{\infty}(-1)^k\frac{2^k-1}{k(k+1)\cdot 2^k}\zeta(k)=\frac{\gamma}{4}+\log\left(2^{\frac{13}{12}}\cdot A^{-3}\right).$$

By setting $z=t$ in (534) through (537), integrating both sides of the resulting equations from $t=0$ to $t=z$, and using (444) and (452), we obtain

$$\sum_{k=2}^{\infty}\frac{\zeta(k,a)}{k(k+1)(k+2)}z^{k+2}$$
$$=\frac{1}{2}(a-1)\log G(a)-\frac{1}{2}(a-1)^2\log\Gamma(a)$$

(569)
$$+\left(\frac{1}{4}(a-1)[2a-1-\log(2\pi)]+(a-1)\log\Gamma(a)-\log G(a)\right)z$$
$$+[5-8a+3\log(2\pi)-4\log\Gamma(a)]\frac{z^2}{8}+[2+\psi(a)]\frac{z^3}{6}$$
$$+\frac{1}{2}(z-a+1)^2\log\Gamma(a-z)+\frac{1}{2}(z-a+1)\log G(a-z)$$
$$-\frac{1}{2}\int_0^{-z}\log G(t+a)\,dt\quad(|z|<|a|);$$

(570)
$$\sum_{k=2}^{\infty}(-1)^k\frac{\zeta(k,a)}{k(k+1)(k+2)}z^{k+2}$$
$$=\frac{1}{2}(a-1)\log G(a)-\frac{1}{2}(a-1)^2\log\Gamma(a)$$
$$+\left(\frac{1}{4}(a-1)[1-2a+\log(2\pi)]+(1-a)\log\Gamma(a)+\log G(a)\right)z$$
$$+[5-8a+3\log(2\pi)-4\log\Gamma(a)]\frac{z^2}{8}-[2+\psi(a)]\frac{z^3}{6}$$
$$+\frac{1}{2}(z+a-1)^2\log\Gamma(z+a)-\frac{1}{2}(z+a-1)\log G(z+a)$$
$$-\frac{1}{2}\int_0^{z}\log G(t+a)\,dt\quad(|z|<|a|);$$

(571)
$$\sum_{k=1}^{\infty}\frac{\zeta(2k,a)}{k(k+1)(2k+1)}z^{2k+2}$$
$$=2(a-1)\log G(a)-2(a-1)^2\log\Gamma(a)+[5-8a+3\log(2\pi)-4\log\Gamma(a)]\frac{z^2}{2}$$
$$+(z+a-1)^2\log\Gamma(a+z)+(z-a+1)^2\log\Gamma(a-z)$$
$$+(1-a-z)\log G(a+z)+(1-a+z)\log G(a-z)$$
$$-\int_0^{z}\log G(t+a)\,dt-\int_0^{-z}\log G(t+a)\,dt\quad(|z|<|a|);$$

$$\sum_{k=1}^{\infty}\frac{\zeta(2k+1,a)}{(k+1)(2k+1)(2k+3)}z^{2k+3}$$
$$=\left(2(a-1)\log\Gamma(a)-2\log G(a)+\frac{1}{2}(a-1)[2a-1-\log(2\pi)]\right)z$$

(572)
$$+[2+\psi(a)]\frac{z^3}{3}+\frac{1}{2}(z-a+1)^2\log\Gamma(a-z)-\frac{1}{2}(z+a-1)^2\log\Gamma(a+z)$$

$$+\frac{1}{2}(z-a+1)\log G(a-z)+\frac{1}{2}(z+a-1)\log G(a+z)$$
$$+\frac{1}{2}\int_0^z \log G(t+a)\,dt-\frac{1}{2}\int_0^{-z}\log G(t+a)\,dt \quad (|z|<|a|).$$

Taking $a = 1$ in (569) through (572), and using 1.3(6), 2.3(2) and 1.2(4), we obtain

(573)
$$\sum_{k=2}^{\infty}\frac{\zeta(k)}{k(k+1)(k+2)}z^{k+2}$$
$$=[\log(2\pi)-1]\frac{3z^2}{8}+(2-\gamma)\frac{z^3}{6}+\frac{z^2}{2}\log\Gamma(1-z)$$
$$+\frac{z}{2}\log G(1-z)-\frac{1}{2}\int_0^{-z}\log G(t+1)\,dt \quad (|z|<1);$$

(574)
$$\sum_{k=2}^{\infty}(-1)^k\frac{\zeta(k)}{k(k+1)(k+2)}z^{k+2}$$
$$=[\log(2\pi)-1]\frac{3z^2}{8}+(\gamma-2)\frac{z^3}{6}+\frac{z^2}{2}\log\Gamma(z+1)$$
$$-\frac{z}{2}\log G(z+1)-\frac{1}{2}\int_0^{z}\log G(t+1)\,dt \quad (|z|<1);$$

(575)
$$\sum_{k=1}^{\infty}\frac{\zeta(2k)}{k(k+1)(2k+1)}z^{2k+2}$$
$$=[\log(2\pi)-1]\frac{3z^2}{2}+z^2\log\Gamma(1+z)\Gamma(1-z)+z\log\frac{G(1-z)}{G(1+z)}$$
$$-\int_0^{z}\log G(t+1)\,dt-\int_0^{-z}\log G(t+1)\,dt \quad (|z|<1);$$

(576)
$$\sum_{k=1}^{\infty}\frac{\zeta(2k+1)}{(k+1)(2k+1)(2k+3)}z^{2k+3}$$
$$=(2-\gamma)\frac{z^3}{3}+\frac{z^2}{2}\log\frac{\Gamma(1-z)}{\Gamma(1+z)}+\frac{z}{2}\log G(1-z)G(1+z)$$
$$+\frac{1}{2}\int_0^{z}\log G(t+1)\,dt-\frac{1}{2}\int_0^{-z}\log G(t+1)\,dt \quad (|z|<1).$$

Taking the limit in (574) as $z \to 1$, and using 1.3(6) and 1.3(73), we obtain

(577)
$$\sum_{k=2}^{\infty}(-1)^k\frac{\zeta(k)}{k(k+1)(k+2)}=-\frac{3}{4}+\frac{\gamma}{6}+\log\left(2^{\frac{1}{4}}\cdot\pi^{\frac{1}{4}}\cdot A\right).$$

If we set $z = \frac{1}{2}$ in (573) through (576), in view of 1.3(6), 1.3(8), 1.3(77), and (498), we get

$$\sum_{k=2}^{\infty} \frac{\zeta(k)}{k(k+1)(k+2)\cdot 2^k} = -\frac{\gamma}{12} + \log\left(2^{\frac{1}{4}}\cdot\pi^{\frac{1}{4}}\cdot A^{-2}\cdot B^{\frac{7}{2}}\right); \tag{578}$$

$$\sum_{k=2}^{\infty} (-1)^k \frac{\zeta(k)}{k(k+1)(k+2)\cdot 2^k} = -\frac{3}{4} + \frac{\gamma}{12} + \log\left(2^{-\frac{1}{4}}\cdot\pi^{\frac{1}{4}}\cdot A^{2}\cdot B^{\frac{7}{2}}\right); \tag{579}$$

$$\sum_{k=1}^{\infty} \frac{\zeta(2k)}{k(k+1)(2k+1)\cdot 2^{2k}} = -\frac{3}{2} + \log\left(\pi\cdot B^{14}\right); \tag{580}$$

$$\sum_{k=1}^{\infty} \frac{\zeta(2k+1)}{(k+1)(2k+1)(2k+3)\cdot 2^{2k}} = \frac{3}{2} - \frac{\gamma}{3} + \log\left(2\cdot A^{-8}\right). \tag{581}$$

Taking $a = 2$ in (569) through (572), and using 1.3(6), 2.3(2), and 1.2(50), we have

$$\begin{aligned}\sum_{k=2}^{\infty} &\frac{\zeta(k)-1}{k(k+1)(k+2)} z^{k+2} \\ &= [3-\log(2\pi)]\frac{z}{4} + [3\log(2\pi)-11]\frac{z^2}{8} + (3-\gamma)\frac{z^3}{6} \\ &+ \frac{1}{2}(z-1)^2\log\Gamma(2-z) + \frac{1}{2}(z-1)\log G(2-z) \\ &- \frac{1}{2}\int_0^{-z}\log G(t+2)\,dt \quad (|z|<2);\end{aligned} \tag{582}$$

$$\begin{aligned}\sum_{k=2}^{\infty} &(-1)^k \frac{\zeta(k)-1}{k(k+1)(k+2)} z^{k+2} \\ &= [\log(2\pi)-3]\frac{z}{4} + [3\log(2\pi)-11]\frac{z^2}{8} + (\gamma-3)\frac{z^3}{6} \\ &+ \frac{1}{2}(z+1)^2\log\Gamma(z+2) - \frac{1}{2}(z+1)\log G(z+2) \\ &- \frac{1}{2}\int_0^{z}\log G(t+2)\,dt \quad (|z|<2);\end{aligned} \tag{583}$$

$$\begin{aligned}\sum_{k=1}^{\infty} &\frac{\zeta(2k)-1}{k(k+1)(2k+1)} z^{2k+2} \\ &= [3\log(2\pi)-11]\frac{z^2}{2} + (z+1)^2\log\Gamma(2+z) \\ &+ (z-1)^2\log\Gamma(2-z) - (z+1)\log G(z+2) + (z-1)\log G(2-z) \\ &- \int_0^{z}\log G(t+2)\,dt - \int_0^{-z}\log G(t+2)\,dt \quad (|z|<2);\end{aligned} \tag{584}$$

(585)

$$\sum_{k=1}^{\infty}\frac{\zeta(2k+1)-1}{(k+1)(2k+1)(2k+3)}z^{2k+3}$$
$$= [3-\log(2\pi)]\frac{z}{2}+(3-\gamma)\frac{z^3}{3}+\frac{1}{2}(z-1)^2\log\Gamma(2-z)$$
$$-\frac{1}{2}(z+1)^2\log\Gamma(2+z)+\frac{1}{2}(z-1)\log G(2-z)+\frac{1}{2}(z+1)\log G(2+z)$$
$$+\frac{1}{2}\int_0^z \log G(t+2)\,dt-\frac{1}{2}\int_0^{-z}\log G(t+2)\,dt \quad (|z|<2).$$

Taking the limit in (583) as $z\to 2$, and using 1.3(6) and (509), we obtain

(586) $$\sum_{k=2}^{\infty}(-1)^k\frac{\zeta(k)-1}{k(k+1)(k+2)}2^k = -\frac{109}{48}+\frac{\gamma}{3}+\log\left(2^{\frac{3}{4}}\cdot 3^{\frac{9}{8}}\cdot \pi^{\frac{1}{4}}\cdot A^{\frac{1}{2}}\right).$$

If we set $z=1$, $z=\frac{1}{2}$, and $z=\frac{3}{2}$ in (582) through (585), by virtue of 1.3(6), 1.3(8), 1.3(73), (497), (498), (509), (515), and (516), we get

(587) $$\sum_{k=2}^{\infty}\frac{\zeta(k)-1}{k(k+1)(k+2)} = -\frac{1}{12}-\frac{\gamma}{6}+\log\left(2^{\frac{1}{4}}\cdot \pi^{\frac{1}{4}}\cdot A^{-1}\right);$$

(588) $$\sum_{k=2}^{\infty}(-1)^k\frac{\zeta(k)-1}{k(k+1)(k+2)} = -\frac{13}{6}+\frac{\gamma}{6}+\log\left(2^{\frac{9}{4}}\cdot \pi^{\frac{1}{4}}\cdot A\right);$$

(589) $$\sum_{k=1}^{\infty}\frac{\zeta(2k)-1}{k(k+1)(2k+1)} = -\frac{9}{2}+5\log 2+\log\pi;$$

(590) $$\sum_{k=1}^{\infty}\frac{\zeta(2k+1)-1}{(k+1)(2k+1)(2k+3)} = \frac{25}{12}-\frac{\gamma}{3}+\log\left(2^{-2}\cdot A^{-2}\right);$$

(591) $$\sum_{k=2}^{\infty}\frac{\zeta(k)-1}{k(k+1)(k+2)\cdot 2^k} = \frac{1}{3}-\frac{\gamma}{12}+\log\left(2^{-\frac{1}{4}}\cdot \pi^{\frac{1}{4}}\cdot A^{-2}\cdot B^{\frac{7}{2}}\right);$$

(592) $$\sum_{k=2}^{\infty}(-1)^k\frac{\zeta(k)-1}{k(k+1)(k+2)\cdot 2^k} = -\frac{31}{12}+\frac{\gamma}{12}+\log\left(2^{-\frac{19}{4}}\cdot 3^{\frac{9}{2}}\cdot \pi^{\frac{1}{4}}\cdot A^2\cdot B^{\frac{7}{2}}\right);$$

(593) $$\sum_{k=1}^{\infty}\frac{\zeta(2k)-1}{k(k+1)(2k+1)\cdot 2^{2k}} = -\frac{9}{2}+\log\left(2^{-10}\cdot 3^9\cdot \pi\cdot B^{14}\right);$$

(594) $$\sum_{k=1}^{\infty} \frac{\zeta(2k+1)-1}{(k+1)(2k+1)(2k+3)\cdot 2^{2k}} = \frac{35}{6} - \frac{\gamma}{3} + \log\left(2^{\frac{59}{6}} \cdot 3^{-9} \cdot A^{-8}\right);$$

(595) $$\sum_{k=2}^{\infty} \frac{\zeta(k)-1}{k(k+1)(k+2)} \left(\frac{3}{2}\right)^{k} = -\frac{1}{12} - \frac{\gamma}{4} + \log\left(2^{\frac{1}{4}} \cdot \pi^{\frac{1}{4}} \cdot A^{-\frac{2}{3}} \cdot B^{\frac{7}{18}}\right);$$

(596) $$\begin{aligned}&\sum_{k=2}^{\infty} (-1)^k \frac{\zeta(k)-1}{k(k+1)(k+2)} \left(\frac{3}{2}\right)^{k} \\ &\quad = -\frac{13}{6} + \frac{\gamma}{4} + \log\left(2^{-\frac{61}{36}} \cdot 3^{\frac{1}{2}} \cdot 5^{\frac{25}{18}} \cdot \pi^{\frac{1}{4}} \cdot A^{\frac{2}{3}} \cdot B^{\frac{7}{18}}\right);\end{aligned}$$

(597) $$\sum_{k=1}^{\infty} \frac{\zeta(2k)-1}{k(k+1)(2k+1)} \left(\frac{3}{2}\right)^{2k} = -\frac{9}{2} + \log\left(2^{-\frac{26}{9}} \cdot 3 \cdot 5^{\frac{25}{9}} \cdot \pi \cdot B^{\frac{14}{9}}\right);$$

(598)
$$\sum_{k=1}^{\infty} \frac{\zeta(2k+1)-1}{(k+1)(2k+1)(2k+3)} \left(\frac{3}{2}\right)^{2k} = \frac{25}{18} - \frac{\gamma}{3} + \log\left(2^{\frac{35}{27}} \cdot 3^{-\frac{1}{3}} \cdot 5^{-\frac{25}{27}} \cdot A^{-\frac{8}{9}}\right).$$

Taking $a = \frac{1}{2}$ in (569) through (572), and using 1.3(6), 2.3(2), and 1.2(51), we get

(599) $$\begin{aligned}&\sum_{k=2}^{\infty} \frac{(2^k-1)\zeta(k)}{k(k+1)(k+2)} z^{k+2} = \log\left(2^{-\frac{1}{96}} \cdot e^{-\frac{1}{32}} \cdot A^{\frac{3}{8}}\right) \\ &\quad + \left(\frac{1}{12}\log 2 + \frac{1}{8}\log\pi + \frac{3}{2}\log A - \frac{1}{8}\right) z + [1 + 3\log 2 + \log\pi]\frac{z^2}{8} \\ &\quad + [2 - \gamma - 2\log 2]\frac{z^3}{6} + \frac{1}{2}\left(z + \frac{1}{2}\right)^2 \log\Gamma\left(\frac{1}{2} - z\right) \\ &\quad + \frac{1}{2}\left(\frac{1}{2} + z\right)\log G\left(\frac{1}{2} - z\right) - \frac{1}{2}\int_0^{-z} \log G\left(t + \frac{1}{2}\right) dt \quad (|z| < \frac{1}{2});\end{aligned}$$

(600) $$\begin{aligned}&\sum_{k=2}^{\infty} (-1)^k \frac{(2^k-1)\zeta(k)}{k(k+1)(k+2)} z^{k+2} = \log\left(2^{-\frac{1}{96}} \cdot e^{-\frac{1}{32}} \cdot A^{\frac{3}{8}}\right) \\ &\quad + \left(\frac{1}{8} - \frac{1}{12}\log 2 - \frac{1}{8}\log\pi - \frac{3}{2}\log A\right) z + [1 + 3\log 2 + \log\pi]\frac{z^2}{8} \\ &\quad + [\gamma - 2 + 2\log 2]\frac{z^3}{6} + \frac{1}{2}\left(z - \frac{1}{2}\right)^2 \log\Gamma\left(z + \frac{1}{2}\right)\end{aligned}$$

$$+\frac{1}{2}\left(\frac{1}{2}-z\right)\log G\left(\frac{1}{2}+z\right)-\frac{1}{2}\int_0^z \log G\left(t+\frac{1}{2}\right)\,dt \quad (|z|<\frac{1}{2});$$

(601)
$$\begin{aligned}&\sum_{k=1}^{\infty}\frac{(2^{2k}-1)\zeta(2k)}{k(k+1)(2k+1)}z^{2k+2}\\ &\quad=\log\left(2^{-\frac{1}{24}}\cdot e^{-\frac{1}{8}}\cdot A^{\frac{3}{2}}\right)+[1+3\log 2+\log\pi]\frac{z^2}{2}\\ &\quad+\left(\frac{1}{2}-z\right)^2\log\Gamma\left(\frac{1}{2}+z\right)+\left(\frac{1}{2}+z\right)^2\log\Gamma\left(\frac{1}{2}-z\right)\\ &\quad+\left(\frac{1}{2}-z\right)\log G\left(\frac{1}{2}+z\right)+\left(\frac{1}{2}+z\right)\log G\left(\frac{1}{2}-z\right)\\ &\quad-\int_0^z \log G\left(t+\frac{1}{2}\right)\,dt-\int_0^{-z}\log G\left(t+\frac{1}{2}\right)\,dt \quad (|z|<\frac{1}{2});\end{aligned}$$

(602)
$$\begin{aligned}&\sum_{k=1}^{\infty}\frac{(2^{2k+1}-1)\zeta(2k+1)}{(k+1)(2k+1)(2k+3)}z^{2k+3}\\ &\quad=\left(\frac{1}{6}\log 2+\frac{1}{4}\log\pi+3\log A-\frac{1}{4}\right)z+[2-\gamma-2\log 2]\frac{z^3}{3}\\ &\quad+\frac{1}{2}\left(\frac{1}{2}+z\right)^2\log\Gamma\left(\frac{1}{2}-z\right)-\frac{1}{2}\left(\frac{1}{2}-z\right)^2\log\Gamma\left(\frac{1}{2}+z\right)\\ &\quad+\frac{1}{2}\left(\frac{1}{2}+z\right)\log G\left(\frac{1}{2}-z\right)+\frac{1}{2}\left(z-\frac{1}{2}\right)\log G\left(z+\frac{1}{2}\right)\\ &\quad+\frac{1}{2}\int_0^z \log G\left(t+\frac{1}{2}\right)\,dt-\frac{1}{2}\int_0^{-z}\log G\left(t+\frac{1}{2}\right)\,dt \quad (|z|<\frac{1}{2}).\end{aligned}$$

By using 1.3(6), 1.3(55), 1.3(56), 1.3(73), and 1.3(77), we can show that

(603)
$$\begin{aligned}&\int_0^{\frac{1}{2}}\log G\left(t+\frac{1}{2}\right)\,dt=-\frac{1}{2}+\frac{1}{2}\log 2\\ &\quad+\int_0^1 \log G(t+1)\,dt-\int_0^1\log\Gamma(t+1)\,dt\\ &\quad-\int_0^{\frac{1}{2}}\log G(t+1)\,dt+\int_0^{\frac{1}{2}}\log\Gamma(t+1)\,dt\\ &\quad=\frac{1}{24}-\frac{13}{12}\log 2-\frac{1}{16}\log\pi-\frac{1}{4}\log A+\frac{7}{4}\log B.\end{aligned}$$

Taking the limit in (600) as $z \to \frac{1}{2}$, and applying 1.3(6) and (603), we obtain

$$\sum_{k=2}^{\infty}(-1)^k \frac{(2^k-1)\zeta(k)}{k(k+1)(k+2)\cdot 2^k} = \frac{\gamma}{12} + \log\left(2^{\frac{5}{2}} \cdot A^{-1} \cdot B^{-\frac{7}{2}}\right). \tag{604}$$

By setting $z = t$ in (459) through (462), integrating both sides of the resulting equations from $t = 0$ to $t = z$, and using (444), we obtain

$$\begin{aligned} &\sum_{k=2}^{\infty} \frac{\zeta(k,a)}{(k+1)(k+2)} z^{k+2} \\ &\quad = (1-a)^2 \log\Gamma(a) + (1-a)\log G(a) \\ &\quad + \left(\frac{1}{2}(a-1)[1+\log(2\pi)] + a(1-a) + (1-a)\log\Gamma(a) + \log G(a)\right) z \\ &\quad + [4a-3-\log(2\pi)]\,\frac{z^2}{4} + [\psi(a)-1]\frac{z^3}{6} \\ &\quad + (a-1)(z-a+1)\log\Gamma(a-z) + (a-1)\log G(a-z) \\ &\quad + \int_0^{-z} \log G(t+a)\,dt \qquad (|z|<|a|); \end{aligned} \tag{605}$$

$$\begin{aligned} &\sum_{k=2}^{\infty} (-1)^k \frac{\zeta(k,a)}{(k+1)(k+2)} z^{k+2} \\ &\quad = (1-a)^2 \log\Gamma(a) + (1-a)\log G(a) \\ &\quad + \left(\frac{1}{2}(1-a)[1+\log(2\pi)] + a(a-1) + (a-1)\log\Gamma(a) - \log G(a)\right) z \\ &\quad + [4a-3-\log(2\pi)]\,\frac{z^2}{4} + [1-\psi(a)]\frac{z^3}{6} \\ &\quad + (1-a)(z+a-1)\log\Gamma(z+a) + (a-1)\log G(z+a) \\ &\quad + \int_0^{z} \log G(t+a)\,dt \qquad (|z|<|a|); \end{aligned} \tag{606}$$

$$\begin{aligned} &\sum_{k=1}^{\infty} \frac{\zeta(2k,a)}{(k+1)(2k+1)} z^{2k+2} \\ &\quad = 2(1-a)^2 \log\Gamma(a) + 2(1-a)\log G(a) \\ &\quad + [4a-3-\log(2\pi)]\,\frac{z^2}{2} + (1-a)(z+a-1)\log\Gamma(a+z) \\ &\quad + (a-1)(z-a+1)\log\Gamma(a-z) + (a-1)\log G(a+z)G(a-z) \\ &\quad + \int_0^{z} \log G(t+a)\,dt + \int_0^{-z} \log G(t+a)\,dt \ \ (|z|<|a|); \end{aligned} \tag{607}$$

$$\sum_{k=1}^{\infty} \frac{\zeta(2k+1,a)}{(k+1)(2k+3)} z^{2k+3}$$
$$= \{(a-1)[1+\log(2\pi)] + 2a(1-a) + 2(1-a)\log\Gamma(a) + 2\log G(a)\} z$$
(608) $$+ [\psi(a)-1]\frac{z^3}{3} + (a-1)(z+a-1)\log\Gamma(a+z)$$
$$+ (a-1)(z-a+1)\log\Gamma(a-z) + (a-1)\log\frac{G(a-z)}{G(a+z)}$$
$$- \int_0^z \log G(t+a)\,dt + \int_0^{-z} \log G(t+a)\,dt \quad (|z|<|a|).$$

Setting $a=1$ in (605) through (608), and using 1.3(6), 2.3(2), and 1.2(4), we obtain

(609) $$\sum_{k=2}^{\infty} \frac{\zeta(k)}{(k+1)(k+2)} z^{k+2} = [1-\log(2\pi)]\frac{z^2}{4} - (1+\gamma)\frac{z^3}{6} + \int_0^{-z} \log G(t+1)\,dt \quad (|z|<1);$$

(610) $$\sum_{k=2}^{\infty} (-1)^k \frac{\zeta(k)}{(k+1)(k+2)} z^{k+2} = [1-\log(2\pi)]\frac{z^2}{4} + (1+\gamma)\frac{z^3}{6} + \int_0^{z} \log G(t+1)\,dt \quad (|z|<1);$$

(611) $$\sum_{k=1}^{\infty} \frac{\zeta(2k)}{(k+1)(2k+1)} z^{2k+2} = [1-\log(2\pi)]\frac{z^2}{2} + \int_0^{z} \log G(t+1)\,dt + \int_0^{-z} \log G(t+1)\,dt \quad (|z|<1);$$

(612) $$\sum_{k=1}^{\infty} \frac{\zeta(2k+1)}{(k+1)(2k+3)} z^{2k+3} = -(1+\gamma)\frac{z^3}{3} - \int_0^{z} \log G(t+1)\,dt + \int_0^{-z} \log G(t+1)\,dt \quad (|z|<1);$$

Taking the limit in (610) as $z \to 1$, and using 1.3(6) and 1.3(73), we obtain

(613) $$\sum_{k=2}^{\infty} (-1)^k \frac{\zeta(k)}{(k+1)(k+2)} = \frac{1}{2} + \frac{\gamma}{6} - 2\log A.$$

Setting $z = \frac{1}{2}$ in (609) through (612), and using 1.3(6), 1.3(8), 1.3(77), and (498), we obtain

(614) $$\sum_{k=2}^{\infty} \frac{\zeta(k)}{(k+1)(k+2)\cdot 2^k} = -\frac{\gamma}{12} + \log\left(2^{\frac{1}{12}} \cdot A \cdot B^{-7}\right);$$

(615) $$\sum_{k=2}^{\infty} (-1)^k \frac{\zeta(k)}{(k+1)(k+2)\cdot 2^k} = \frac{1}{2} + \frac{\gamma}{12} + \log\left(2^{-\frac{1}{12}} \cdot A^{-1} \cdot B^{-7}\right);$$

(616) $$\sum_{k=1}^{\infty} \frac{\zeta(2k)}{(k+1)(2k+1)\cdot 2^{2k}} = \frac{1}{2} - 14\log B;$$

(617) $$\sum_{k=1}^{\infty} \frac{\zeta(2k+1)}{(k+1)(2k+3)\cdot 2^{2k}} = -1 - \frac{\gamma}{3} + \log\left(2^{\frac{1}{3}} \cdot A^4\right).$$

If we take $a = 2$ in (605) through (608), in view of 1.3(6), 2.3(2), and 1.2(50), we get

(618) $$\begin{aligned} &\sum_{k=2}^{\infty} \frac{\zeta(k)-1}{(k+1)(k+2)} z^{k+2} \\ &= [\log(2\pi) - 3]\frac{z}{2} + [5 - \log(2\pi)]\frac{z^2}{4} - \frac{\gamma}{6}z^3 + (z-1)\log\Gamma(2-z) \\ &\quad + \log G(2-z) + \int_0^{-z} \log G(t+2)\,dt \quad (|z| < 2); \end{aligned}$$

(619) $$\begin{aligned} &\sum_{k=2}^{\infty} (-1)^k \frac{\zeta(k)-1}{(k+1)(k+2)} z^{k+2} \\ &= [3 - \log(2\pi)]\frac{z}{2} + [5 - \log(2\pi)]\frac{z^2}{4} + \frac{\gamma}{6}z^3 - (z+1)\log\Gamma(z+2) \\ &\quad + \log G(z+2) + \int_0^{z} \log G(t+2)\,dt \quad (|z| < 2); \end{aligned}$$

(620) $$\begin{aligned} &\sum_{k=1}^{\infty} \frac{\zeta(2k)-1}{(k+1)(2k+1)} z^{2k+2} \\ &= [5 - \log(2\pi)]\frac{z^2}{2} - (z+1)\log\Gamma(2+z) \\ &\quad + (z-1)\log\Gamma(2-z) + \log G(2+z)G(2-z) \\ &\quad + \int_0^{z} \log G(t+2)\,dt + \int_0^{-z} \log G(t+2)\,dt \quad (|z| < 2); \end{aligned}$$

$$\sum_{k=1}^{\infty} \frac{\zeta(2k+1)-1}{(k+1)(2k+3)} z^{2k+3}$$

(621) $$= [\log(2\pi) - 3]z - \frac{\gamma}{3}z^3 + (z-1)\log\Gamma(2-z) + (z+1)\log\Gamma(2+z)$$
$$+ \log\frac{G(2-z)}{G(2+z)} - \int_0^z \log G(t+2)\,dt + \int_0^{-z} \log G(t+2)\,dt \quad (|z|<2).$$

Taking the limit in (619) as $z \to 2$, and using 1.3(6), 1.3(73), and (509), we obtain

(622) $$\sum_{k=2}^{\infty} (-1)^k \frac{(\zeta(k)-1)2^k}{(k+1)(k+2)} = \frac{25}{24} + \frac{\gamma}{3} + \log\left(3^{-\frac{3}{4}} \cdot A^{-1}\right).$$

Setting $z = \frac{1}{2}$, $z = 1$, and $z = \frac{3}{2}$ in (618) through (621), and using 1.3(6), 1.3(8), 1.3(73), 1.3(77), (498), (509), and (515), we get

(623) $$\sum_{k=2}^{\infty} \frac{\zeta(k)-1}{(k+1)(k+2)\cdot 2^k} = -\frac{17}{12} - \frac{\gamma}{12} + \log\left(2^{\frac{25}{12}} \cdot A \cdot B^{-7}\right);$$

(624) $$\sum_{k=2}^{\infty} (-1)^k \frac{\zeta(k)-1}{(k+1)(k+2)\cdot 2^k} = \frac{35}{12} + \frac{\gamma}{12} + \log\left(2^{\frac{71}{12}} \cdot 3^{-6} \cdot A^{-1} \cdot B^{-7}\right);$$

(625) $$\sum_{k=1}^{\infty} \frac{\zeta(2k)-1}{(k+1)(2k+1)\cdot 2^{2k}} = \frac{3}{2} + \log\left(2^8 \cdot 3^{-6} \cdot B^{-14}\right);$$

(626) $$\sum_{k=1}^{\infty} \frac{\zeta(2k+1)-1}{(k+1)(2k+3)\cdot 2^{2k}} = -\frac{26}{3} - \frac{\gamma}{3} + \log\left(2^{-\frac{28}{3}} \cdot 3^{12} \cdot A^4\right);$$

(627) $$\sum_{k=2}^{\infty} \frac{\zeta(k)-1}{(k+1)(k+2)} = -\frac{1}{3} - \frac{\gamma}{6} + 2\log A;$$

(628) $$\sum_{k=2}^{\infty} (-1)^k \frac{\zeta(k)-1}{(k+1)(k+2)} = \frac{11}{6} + \frac{\gamma}{6} - 2\log(2A);$$

(629) $$\sum_{k=1}^{\infty} \frac{\zeta(2k)-1}{(k+1)(2k+1)} = \frac{3}{2} - 2\log 2;$$

$$\sum_{k=1}^{\infty} \frac{\zeta(2k+1)-1}{(k+1)(2k+3)} = -\frac{13}{6} - \frac{\gamma}{3} + \log\left(2^2 \cdot A^4\right). \tag{630}$$

$$\sum_{k=2}^{\infty} \frac{\zeta(k)-1}{(k+1)(k+2)} \left(\frac{3}{2}\right)^k = \frac{1}{4} - \frac{\gamma}{4} + \log\left(2^{\frac{1}{36}} \cdot A^{\frac{1}{3}} \cdot B^{-\frac{7}{9}}\right); \tag{631}$$

$$\sum_{k=2}^{\infty} (-1)^k \frac{\zeta(k)-1}{(k+1)(k+2)} \left(\frac{3}{2}\right)^k = \frac{5}{4} + \frac{\gamma}{4} + \log\left(2^{\frac{31}{36}} \cdot 5^{-\frac{10}{9}} \cdot A^{-\frac{1}{3}} \cdot B^{-\frac{7}{9}}\right); \tag{632}$$

$$\sum_{k=1}^{\infty} \frac{\zeta(2k)-1}{(k+1)(2k+1)} \left(\frac{3}{2}\right)^{2k} = \frac{3}{2} + \log\left(2^{\frac{8}{9}} \cdot 5^{-\frac{10}{9}} \cdot B^{-\frac{14}{9}}\right); \tag{633}$$

$$\sum_{k=1}^{\infty} \frac{\zeta(2k+1)-1}{(k+1)(2k+3)} \left(\frac{3}{2}\right)^{2k} = -\frac{2}{3} - \frac{\gamma}{3} + \log\left(2^{-\frac{5}{9}} \cdot 5^{\frac{20}{27}} \cdot A^{\frac{4}{9}}\right). \tag{634}$$

Taking $a = \frac{1}{2}$ in (605) through (608), and using 1.3(8), 2.3(2), and 1.2(51), we obtain

$$\begin{aligned} &\sum_{k=2}^{\infty} \frac{(2^k-1)\zeta(k)}{(k+1)(k+2)} z^{k+2} \\ &= \log\left(2^{\frac{1}{48}} \cdot e^{\frac{1}{16}} \cdot A^{-\frac{3}{4}}\right) - \left(\frac{5}{12}\log 2 + \frac{1}{2}\log\pi - \frac{1}{4} + 3\log A\right)\frac{z}{2} \\ &\quad - [1+\log(2\pi)]\frac{z^2}{4} - (1+\gamma+2\log 2)\frac{z^3}{6} - \frac{1}{2}\left(\frac{1}{2}+z\right)\log\Gamma\left(\frac{1}{2}-z\right) \\ &\quad - \frac{1}{2}\log G\left(\frac{1}{2}-z\right) + \int_0^{-z} \log G\left(t+\frac{1}{2}\right) dt \qquad (|z| < \frac{1}{2}); \end{aligned} \tag{635}$$

$$\begin{aligned} &\sum_{k=2}^{\infty} (-1)^k \frac{(2^k-1)\zeta(k)}{(k+1)(k+2)} z^{k+2} \\ &= \log\left(2^{\frac{1}{48}} \cdot e^{\frac{1}{16}} \cdot A^{-\frac{3}{4}}\right) + \left(\frac{5}{12}\log 2 + \frac{1}{2}\log\pi - \frac{1}{4} + 3\log A\right)\frac{z}{2} \\ &\quad - [1+\log(2\pi)]\frac{z^2}{4} + (1+\gamma+2\log 2)\frac{z^3}{6} + \frac{1}{2}\left(z-\frac{1}{2}\right)\log\Gamma\left(z+\frac{1}{2}\right) \\ &\quad - \frac{1}{2}\log G\left(z+\frac{1}{2}\right) + \int_0^{z} \log G\left(t+\frac{1}{2}\right) dt \qquad (|z| < \frac{1}{2}); \end{aligned} \tag{636}$$

$$\sum_{k=1}^{\infty} \frac{(2^{2k}-1)\zeta(2k)}{(k+1)(2k+1)} z^{2k+2}$$

(637)
$$\begin{aligned} &= \log\left(2^{\frac{1}{24}} \cdot e^{\frac{1}{8}} \cdot A^{-\frac{3}{2}}\right) - [1+\log(2\pi)]\frac{z^2}{2} + \frac{1}{2}\left(z-\frac{1}{2}\right)\log\Gamma\left(z+\frac{1}{2}\right) \\ &\quad - \frac{1}{2}\left(\frac{1}{2}+z\right)\log\Gamma\left(\frac{1}{2}-z\right) - \frac{1}{2}\log G\left(\frac{1}{2}+z\right)G\left(\frac{1}{2}-z\right) \\ &\quad + \int_0^z \log G\left(t+\frac{1}{2}\right)dt + \int_0^{-z} \log G\left(t+\frac{1}{2}\right)dt \quad (|z|<\frac{1}{2}); \end{aligned}$$

(638)
$$\begin{aligned} &\sum_{k=1}^{\infty} \frac{(2^{2k+1}-1)\zeta(2k+1)}{(k+1)(2k+3)} z^{2k+3} \\ &= -\left(\frac{5}{12}\log 2 + \frac{1}{2}\log\pi - \frac{1}{4} + 3\log A\right)z - (1+\gamma+2\log 2)\frac{z^3}{3} \\ &\quad + \frac{1}{2}\left(\frac{1}{2}-z\right)\log\Gamma\left(\frac{1}{2}+z\right) - \frac{1}{2}\left(\frac{1}{2}+z\right)\log\Gamma\left(\frac{1}{2}-z\right) + \frac{1}{2}\log\frac{G\left(\frac{1}{2}+z\right)}{G\left(\frac{1}{2}-z\right)} \\ &\quad - \int_0^z \log G\left(t+\frac{1}{2}\right)dt + \int_0^{-z} \log G\left(t+\frac{1}{2}\right)dt \quad (|z|<\frac{1}{2}). \end{aligned}$$

Taking the limit in (636) as $z \to \frac{1}{2}$, and using 1.3(6) and (603), we obtain

(639)
$$\sum_{k=2}^{\infty} (-1)^k \frac{(2^k-1)\zeta(k)}{(k+1)(k+2)\cdot 2^k} = \frac{\gamma}{12} + \log\left(2^{-\frac{47}{12}} \cdot A^{-1} \cdot B^7\right).$$

Evaluation in terms of Catalan's constant G

We now evaluate series associated with the Zeta function in terms of Catalan's constant G and $\Gamma\left(\frac{1}{4}\right)$ by using some of the formulas which we have already developed in this section.

Setting $z=\frac{1}{4}$ and $z=\frac{3}{4}$ in (456), and using 1.3(6), (1.12), 1.3(23), and 1.3(24), we obtain

(640)
$$\sum_{k=2}^{\infty} (-1)^k \frac{\zeta(k)}{(k+1)\cdot 2^{2k}} = 1 + \frac{\gamma}{8} - \frac{G}{\pi} + \log\left[2^{-\frac{1}{2}} \cdot \pi^{-\frac{1}{2}} \cdot A^{-\frac{9}{2}} \cdot \Gamma\left(\frac{1}{4}\right)\right];$$

(641)
$$\sum_{k=2}^{\infty} (-1)^k \frac{\zeta(k)}{k+1}\left(\frac{3}{4}\right)^k = 1 + \frac{3\gamma}{8} + \frac{G}{3\pi} + \log\left[\pi^{\frac{1}{2}} \cdot A^{-\frac{3}{2}} \cdot \left\{\Gamma\left(\frac{1}{4}\right)\right\}^{-1}\right].$$

Setting $z=\frac{1}{4}$ and $z=\frac{3}{4}$ in (463), (464), and (465), and using 1.3(6), 1.1(12), 1.3(23), and 1.3(24), we obtain

(642) $$\sum_{k=2}^{\infty} \frac{\zeta(k)}{(k+1)\cdot 2^{2k}} = -\frac{\gamma}{8} - \frac{G}{\pi} + \log\left[\pi^{\frac{1}{2}} \cdot A^{\frac{9}{2}} \cdot \left\{\Gamma\left(\frac{1}{4}\right)\right\}^{-1}\right];$$

(643) $$\sum_{k=2}^{\infty} \frac{\zeta(k)}{k+1}\left(\frac{3}{4}\right)^{k} = -\frac{3\gamma}{8} + \frac{G}{3\pi} + \log\left[2^{-\frac{1}{2}} \cdot \pi^{-\frac{1}{2}} \cdot A^{\frac{3}{2}} \cdot \Gamma\left(\frac{1}{4}\right)\right];$$

(644) $$\sum_{k=1}^{\infty} \frac{\zeta(2k)}{(2k+1)\cdot 2^{4k}} = \frac{1}{2} - \frac{G}{\pi} - \frac{1}{4}\log 2;$$

(645) $$\sum_{k=1}^{\infty} \frac{\zeta(2k)}{2k+1}\left(\frac{3}{4}\right)^{2k} = \frac{1}{2} + \frac{G}{3\pi} - \frac{1}{4}\log 2;$$

(646) $$\sum_{k=1}^{\infty} \frac{\zeta(2k+1)}{(k+1)\cdot 2^{4k}} = -4 - \gamma + \log\left[2^{2} \cdot \pi^{4} \cdot A^{36} \cdot \left\{\Gamma\left(\frac{1}{4}\right)\right\}^{-8}\right];$$

(647) $$\sum_{k=1}^{\infty} \frac{\zeta(2k+1)}{k+1}\left(\frac{3}{4}\right)^{2k} = -\frac{4}{3} - \gamma + \log\left[2^{-\frac{2}{3}} \cdot \pi^{-\frac{4}{3}} \cdot A^{4} \cdot \left\{\Gamma\left(\frac{1}{4}\right)\right\}^{\frac{8}{3}}\right].$$

If we set $z=\frac{1}{4}$, $z=\frac{3}{4}$, $z=\frac{5}{4}$, and $z=\frac{7}{4}$ in (469) through (472), in view of 1.3(6), 1.1(12), 1.3(23), and 1.3(24), we get

(648) $$\sum_{k=2}^{\infty}(-1)^{k} \frac{\zeta(k)-1}{(k+1)\cdot 2^{2k}} = \frac{15}{8} + \frac{\gamma}{8} - \frac{G}{\pi} + \log\left[2^{\frac{15}{2}} \cdot 5^{-4} \cdot \pi^{-\frac{1}{2}} \cdot A^{-\frac{9}{2}} \cdot \Gamma\left(\frac{1}{4}\right)\right];$$

(649) $$\sum_{k=2}^{\infty}(-1)^{k} \frac{\zeta(k)-1}{k+1}\left(\frac{3}{4}\right)^{k} = \frac{13}{8} + \frac{3\gamma}{8} + \frac{G}{3\pi} + \log\left[2^{\frac{8}{3}} \cdot 7^{-\frac{4}{3}} \cdot \pi^{\frac{1}{2}} \cdot A^{-\frac{3}{2}} \cdot \left\{\Gamma\left(\frac{1}{4}\right)\right\}^{-1}\right];$$

(650) $$\sum_{k=2}^{\infty} \frac{\zeta(k)-1}{(k+1)\cdot 2^{2k}} = \frac{9}{8} - \frac{\gamma}{8} - \frac{G}{\pi} + \log\left[2^{-8} \cdot 3^{4} \cdot \pi^{\frac{1}{2}} \cdot A^{\frac{9}{2}} \cdot \left\{\Gamma\left(\frac{1}{4}\right)\right\}^{-1}\right];$$

(651) $$\sum_{k=2}^{\infty} \frac{\zeta(k)-1}{k+1}\left(\frac{3}{4}\right)^k = \frac{11}{8} - \frac{3\gamma}{8} + \frac{G}{3\pi} + \log\left[2^{-\frac{19}{6}} \cdot \pi^{-\frac{1}{2}} \cdot A^{\frac{3}{2}} \cdot \Gamma\left(\frac{1}{4}\right)\right];$$

(652) $$\sum_{k=1}^{\infty} \frac{\zeta(2k)-1}{(2k+1)\cdot 2^{4k}} = \frac{3}{2} - \frac{G}{\pi} + \log\left[2^{-\frac{1}{4}} \cdot 3^2 \cdot 5^{-2}\right];$$

(653) $$\sum_{k=1}^{\infty} \frac{\zeta(2k)-1}{2k+1}\left(\frac{3}{4}\right)^{2k} = \frac{3}{2} + \frac{G}{3\pi} + \log\left[2^{-\frac{1}{4}} \cdot 7^{-\frac{2}{3}}\right];$$

(654) $$\sum_{k=1}^{\infty} \frac{\zeta(2k+1)-1}{(k+1)\cdot 2^{4k}} = -3 - \gamma + \log\left[2^{-62} \cdot 3^{16} \cdot 5^{16} \cdot \pi^4 \cdot A^{36} \cdot \left\{\Gamma\left(\frac{1}{4}\right)\right\}^{-8}\right];$$

(655) $$\sum_{k=1}^{\infty} \frac{\zeta(2k+1)-1}{k+1}\left(\frac{3}{4}\right)^{2k} = -\frac{1}{3} - \gamma + \log\left[2^{-\frac{70}{9}} \cdot 7^{\frac{16}{9}} \cdot \pi^{-\frac{4}{3}} \cdot A^4 \cdot \left\{\Gamma\left(\frac{1}{4}\right)\right\}^{\frac{8}{3}}\right];$$

(656) $$\sum_{k=2}^{\infty} (-1)^k \frac{\zeta(k)-1}{k+1}\left(\frac{5}{4}\right)^k = \frac{63}{40} + \frac{5\gamma}{8} - \frac{G}{5\pi} + \log\left[2^{-\frac{1}{2}} \cdot \pi^{-\frac{1}{2}} \cdot 3^{-\frac{8}{5}} \cdot A^{-\frac{9}{10}} \cdot \Gamma\left(\frac{1}{4}\right)\right];$$

(657) $$\sum_{k=2}^{\infty} (-1)^k \frac{\zeta(k)-1}{k+1}\left(\frac{7}{4}\right)^k = \frac{87}{56} + \frac{7\gamma}{8} + \frac{G}{7\pi} + \log\left[3^{\frac{4}{7}} \cdot 11^{-\frac{4}{7}} \cdot \pi^{\frac{1}{2}} \cdot A^{-\frac{9}{14}} \cdot \left\{\Gamma\left(\frac{1}{4}\right)\right\}^{-1}\right];$$

(658) $$\sum_{k=2}^{\infty} \frac{\zeta(k)-1}{k+1}\left(\frac{5}{4}\right)^k = \frac{57}{40} - \frac{5\gamma}{8} - \frac{G}{5\pi} + \log\left[\pi^{\frac{1}{2}} \cdot A^{\frac{9}{10}} \cdot \left\{\Gamma\left(\frac{1}{4}\right)\right\}^{-1}\right];$$

(659) $$\sum_{k=2}^{\infty}\frac{\zeta(k)-1}{k+1}\left(\frac{7}{4}\right)^{k}=\frac{81}{56}-\frac{7\gamma}{8}+\frac{G}{7\pi}+\log\left[2^{-\frac{1}{2}}\cdot\pi^{-\frac{1}{2}}\cdot A^{\frac{9}{14}}\cdot\Gamma\left(\frac{1}{4}\right)\right];$$

(660) $$\sum_{k=1}^{\infty}\frac{\zeta(2k)-1}{2k+1}\left(\frac{5}{4}\right)^{2k}=\frac{3}{2}-\frac{G}{5\pi}+\log\left[2^{-\frac{1}{4}}\cdot 3^{-\frac{4}{5}}\right];$$

(661) $$\sum_{k=1}^{\infty}\frac{\zeta(2k)-1}{2k+1}\left(\frac{7}{4}\right)^{2k}=\frac{3}{2}+\frac{G}{7\pi}+\log\left[2^{-\frac{1}{4}}\cdot 3^{\frac{2}{7}}\cdot 11^{-\frac{2}{7}}\right];$$

(662) $$\sum_{k=1}^{\infty}\frac{\zeta(2k+1)-1}{k+1}\left(\frac{5}{4}\right)^{2k}=-\frac{3}{25}-\gamma+\log\left[2^{\frac{2}{5}}\cdot 3^{\frac{32}{25}}\cdot\pi^{\frac{4}{5}}\cdot A^{\frac{36}{25}}\cdot\left\{\Gamma\left(\frac{1}{4}\right)\right\}^{-\frac{8}{5}}\right];$$

(663) $$\sum_{k=1}^{\infty}\frac{\zeta(2k+1)-1}{k+1}\left(\frac{7}{4}\right)^{2k}=-\frac{3}{49}-\gamma+\log\left[2^{-\frac{2}{7}}\cdot 3^{-\frac{16}{49}}\cdot 11^{\frac{16}{49}}\cdot\pi^{-\frac{4}{7}}\cdot A^{\frac{36}{49}}\cdot\left\{\Gamma\left(\frac{1}{4}\right)\right\}^{\frac{8}{7}}\right].$$

Setting $z=\frac{1}{4}$ in (486) through (489), and using 1.3(6), 1.1(12), 1.3(23), and 1.3(24), we obtain

(664) $$\sum_{k=2}^{\infty}\frac{(2^k-1)\zeta(k)}{(k+1)\cdot 2^{2k}}=-\frac{\gamma}{8}+\frac{G}{\pi}+\log\left[2^{-\frac{7}{12}}\cdot\pi^{-\frac{1}{2}}\cdot A^{-\frac{3}{2}}\cdot\Gamma\left(\frac{1}{4}\right)\right];$$

(665) $$\sum_{k=2}^{\infty}(-1)^k\frac{(2^k-1)\zeta(k)}{(k+1)\cdot 2^{2k}}=\frac{\gamma}{8}+\frac{G}{\pi}+\log\left[2^{\frac{1}{12}}\cdot\pi^{\frac{1}{2}}\cdot A^{\frac{3}{2}}\cdot\left\{\Gamma\left(\frac{1}{4}\right)\right\}^{-1}\right];$$

(666) $$\sum_{k=1}^{\infty}\frac{(2^{2k}-1)\zeta(2k)}{(2k+1)\cdot 2^{4k}}=\frac{G}{\pi}-\frac{1}{4}\log 2;$$

(667) $$\sum_{k=1}^{\infty}\frac{(2^{2k+1}-1)\zeta(2k+1)}{(k+1)\cdot 2^{4k}}=-\gamma+\log\left[2^{-\frac{8}{3}}\cdot\pi^{-4}\cdot A^{-12}\cdot\left\{\Gamma\left(\frac{1}{4}\right)\right\}^{8}\right].$$

If we set $z=\frac{1}{4}$ and $z=\frac{3}{4}$ in (538) through (541), by virtue of 1.3(6), 1.1(12), 1.3(23) and 1.3(24), we obtain

$$\sum_{k=2}^{\infty}(-1)^k\frac{\zeta(k)}{k(k+1)\cdot 2^{2k}}=-1+\frac{\gamma}{8}+\frac{G}{\pi}+\log\left[2^{-\frac{3}{2}}\cdot\pi^{\frac{1}{2}}\cdot A^{\frac{9}{2}}\right]; \tag{668}$$

$$\sum_{k=2}^{\infty}(-1)^k\frac{\zeta(k)}{k(k+1)}\left(\frac{3}{4}\right)^k=-1+\frac{3\gamma}{8}-\frac{G}{3\pi}+\log\left[2^{-\frac{3}{2}}\cdot 3\cdot\pi^{\frac{1}{2}}\cdot A^{\frac{3}{2}}\right]; \tag{669}$$

$$\sum_{k=2}^{\infty}\frac{\zeta(k)}{k(k+1)\cdot 2^{2k}}=-\frac{\gamma}{8}+\frac{G}{\pi}+\log\left[2^{\frac{1}{2}}\cdot\pi^{\frac{1}{2}}\cdot A^{-\frac{9}{2}}\right]; \tag{670}$$

$$\sum_{k=2}^{\infty}\frac{\zeta(k)}{k(k+1)}\left(\frac{3}{4}\right)^k=-\frac{3\gamma}{8}-\frac{G}{3\pi}+\log\left[2^{\frac{1}{2}}\cdot\pi^{\frac{1}{2}}\cdot A^{-\frac{3}{2}}\right]; \tag{671}$$

$$\sum_{k=1}^{\infty}\frac{\zeta(2k)}{k(2k+1)\cdot 2^{4k}}=-1+\frac{2G}{\pi}+\log\left(\frac{\pi}{2}\right); \tag{672}$$

$$\sum_{k=1}^{\infty}\frac{\zeta(2k)}{k(2k+1)}\left(\frac{3}{4}\right)^{2k}=-1-\frac{2G}{3\pi}+\log\left(\frac{3\pi}{2}\right); \tag{673}$$

$$\sum_{k=1}^{\infty}\frac{\zeta(2k+1)}{(k+1)(2k+1)\cdot 2^{4k}}=4-\gamma+\log\left[2^{8}\cdot A^{-36}\right]; \tag{674}$$

$$\sum_{k=1}^{\infty}\frac{\zeta(2k+1)}{(k+1)(2k+1)}\left(\frac{3}{4}\right)^{2k}=\frac{4}{3}-\gamma+\log\left[2^{\frac{8}{3}}\cdot 3^{-\frac{4}{3}}\cdot A^{-4}\right]. \tag{675}$$

Setting $z=\frac{1}{4}$, $z=\frac{3}{4}$, $z=\frac{5}{4}$, and $z=\frac{7}{4}$ in (547) through (550), and using 1.3(6), 1.1(12), 1.3(23), and 1.3(24), we get

$$\begin{aligned}&\sum_{k=2}^{\infty}(-1)^k\frac{\zeta(k)-1}{k(k+1)\cdot 2^{2k}}\\&\quad=-\frac{17}{8}+\frac{\gamma}{8}+\frac{G}{\pi}+\log\left[2^{-\frac{23}{2}}\cdot 5^{5}\cdot\pi^{\frac{1}{2}}\cdot A^{\frac{9}{2}}\right];\end{aligned} \tag{676}$$

(677)
$$\sum_{k=2}^{\infty}(-1)^k \frac{\zeta(k)-1}{k(k+1)}\left(\frac{3}{4}\right)^k = -\frac{19}{8}+\frac{3\gamma}{8}-\frac{G}{3\pi}+\log\left[2^{-\frac{37}{6}}\cdot 3\cdot 7^{\frac{7}{3}}\cdot \pi^{\frac{1}{2}}\cdot A^{\frac{3}{2}}\right];$$

(678)
$$\sum_{k=2}^{\infty}(-1)^k \frac{\zeta(k)-1}{k(k+1)}\left(\frac{5}{4}\right)^k = -\frac{113}{40}+\frac{5\gamma}{8}+\frac{G}{5\pi}+\log\left[2^{-\frac{11}{2}}\cdot 3^{\frac{18}{5}}\cdot 5\cdot \pi^{\frac{1}{2}}\cdot A^{\frac{9}{10}}\right];$$

(679)
$$\sum_{k=2}^{\infty}(-1)^k \frac{\zeta(k)-1}{k(k+1)}\left(\frac{7}{4}\right)^k = -\frac{185}{56}+\frac{7\gamma}{8}-\frac{G}{7\pi}+\log\left[2^{-\frac{11}{2}}\cdot 3^{\frac{3}{7}}\cdot 7\cdot 11^{\frac{11}{7}}\cdot \pi^{\frac{1}{2}}\cdot A^{\frac{9}{14}}\right];$$

(680)
$$\sum_{k=2}^{\infty}\frac{\zeta(k)-1}{k(k+1)\cdot 2^{2k}} = -\frac{7}{8}-\frac{\gamma}{8}+\frac{G}{\pi}+\log\left[2^{\frac{13}{2}}\cdot 3^{-3}\cdot \pi^{\frac{1}{2}}\cdot A^{-\frac{9}{2}}\right];$$

(681)
$$\sum_{k=2}^{\infty}\frac{\zeta(k)-1}{k(k+1)}\left(\frac{3}{4}\right)^k = -\frac{5}{8}-\frac{3\gamma}{8}-\frac{G}{3\pi}+\log\left[2^{\frac{7}{6}}\cdot \pi^{\frac{1}{2}}\cdot A^{-\frac{3}{2}}\right];$$

(682)
$$\sum_{k=2}^{\infty}\frac{\zeta(k)-1}{k(k+1)}\left(\frac{5}{4}\right)^k = -\frac{7}{40}-\frac{5\gamma}{8}+\frac{G}{5\pi}+\log\left[2^{\frac{1}{2}}\cdot \pi^{\frac{1}{2}}\cdot A^{-\frac{9}{10}}\right];$$

(683)
$$\sum_{k=2}^{\infty}\frac{\zeta(k)-1}{k(k+1)}\left(\frac{7}{4}\right)^k = \frac{17}{56}-\frac{7\gamma}{8}-\frac{G}{7\pi}+\log\left[2^{\frac{1}{2}}\cdot \pi^{\frac{1}{2}}\cdot A^{-\frac{9}{14}}\right];$$

(684)
$$\sum_{k=1}^{\infty}\frac{\zeta(2k)-1}{k(2k+1)\cdot 2^{4k}} = -3+\frac{2G}{\pi}+\log\left[2^{-5}\cdot 3^{-3}\cdot 5^{5}\cdot \pi\right];$$

(685)
$$\sum_{k=1}^{\infty}\frac{\zeta(2k)-1}{k(2k+1)}\left(\frac{3}{4}\right)^{2k} = -3-\frac{2G}{3\pi}+\log\left[2^{-5}\cdot 3\cdot 7^{\frac{7}{3}}\cdot \pi\right];$$

(686)
$$\sum_{k=1}^{\infty} \frac{\zeta(2k)-1}{k(2k+1)} \left(\frac{5}{4}\right)^{2k} = -3 + \frac{2G}{5\pi} + \log\left[2^{-5} \cdot 3^{\frac{18}{5}} \cdot 5 \cdot \pi\right];$$

(687)
$$\sum_{k=1}^{\infty} \frac{\zeta(2k)-1}{k(2k+1)} \left(\frac{7}{4}\right)^{2k} = -3 - \frac{2G}{7\pi} + \log\left[2^{-5} \cdot 3^{\frac{3}{7}} \cdot 7 \cdot 11^{\frac{11}{7}} \cdot \pi\right];$$

(688)
$$\sum_{k=1}^{\infty} \frac{\zeta(2k+1)-1}{(k+1)(2k+1) \cdot 2^{4k}} = 5 - \gamma + \log\left[2^{72} \cdot 3^{-12} \cdot 5^{-20} \cdot A^{-36}\right];$$

(689)
$$\sum_{k=1}^{\infty} \frac{\zeta(2k+1)-1}{(k+1)(2k+1)} \left(\frac{3}{4}\right)^{2k} = \frac{7}{3} - \gamma + \log\left[2^{\frac{88}{9}} \cdot 3^{-\frac{4}{3}} \cdot 7^{-\frac{28}{9}} \cdot A^{-4}\right];$$

(690)
$$\sum_{k=1}^{\infty} \frac{\zeta(2k+1)-1}{(k+1)(2k+1)} \left(\frac{5}{4}\right)^{2k} = \frac{53}{25} - \gamma + \log\left[2^{\frac{24}{5}} \cdot 3^{-\frac{72}{25}} \cdot 5^{-\frac{4}{5}} \cdot A^{-\frac{36}{25}}\right];$$

(691)
$$\sum_{k=1}^{\infty} \frac{\zeta(2k+1)-1}{(k+1)(2k+1)} \left(\frac{7}{4}\right)^{2k} = \frac{101}{49} - \gamma + \log\left[2^{\frac{24}{7}} \cdot 3^{-\frac{12}{49}} \cdot 7^{-\frac{4}{7}} \cdot 11^{-\frac{44}{49}} \cdot A^{-\frac{36}{49}}\right].$$

Setting $z = \frac{1}{4}$ in (564) through (567), and using 1.3(6), 1.1(12), 1.3(23), and 1.3(24), we obtain

(692)
$$\sum_{k=2}^{\infty} \frac{(2^k-1)\zeta(k)}{k(k+1) \cdot 2^{2k}} = -\frac{\gamma}{8} - \frac{G}{\pi} + \log\left[2^{\frac{1}{12}} \cdot A^{\frac{3}{2}}\right];$$

(693)
$$\sum_{k=2}^{\infty} (-1)^k \frac{(2^k-1)\zeta(k)}{k(k+1) \cdot 2^{2k}} = \frac{\gamma}{8} - \frac{G}{\pi} + \log\left[2^{\frac{11}{12}} \cdot A^{-\frac{3}{2}}\right];$$

(694)
$$\sum_{k=1}^{\infty} \frac{(2^{2k}-1)\zeta(2k)}{k(2k+1) \cdot 2^{4k}} = -\frac{2G}{\pi} + \log 2;$$

$$\sum_{k=1}^{\infty} \frac{(2^{2k+1}-1)\zeta(2k+1)}{(k+1)(2k+1)\cdot 2^{4k}} = -\gamma + \log\left[2^{-\frac{10}{3}} \cdot A^{12}\right]. \tag{695}$$

Further evaluation by using the triple Gamma function

We first rewrite the left-hand side of (462) as follows:

$$\begin{aligned}\sum_{n=2}^{\infty} \zeta(2n-1,a)\frac{z^{2n}}{n} &= [\psi(a)-1]z^2 + (a-1)\log[\Gamma(a+z)\Gamma(a-z)] \\ &- \log[G(a+z)G(a-z)] + 2(1-a)\log\Gamma(a) + 2\log G(a) \quad (|z|<|a|).\end{aligned} \tag{696}$$

By differentiating both sides of (4) with respect to t and multiplying the resulting equation by t^3, if we integrate the resulting equation with respect to t from $t=0$ to $t=z$, we obtain

(697)

$$\sum_{n=3}^{\infty} \zeta(2n-3,a)\frac{z^{2n}}{n} = -\int_0^z t^3\psi(a+t)dt - \int_0^{-z} t^3\psi(a+t)dt + \frac{\psi(a)}{2}z^4 \quad (|z|<|a|),$$

which, by virtue of (455), yields

(698)

$$\begin{aligned}\sum_{n=3}^{\infty} \zeta(2n-3,a)\frac{z^{2n}}{n} &= \left[\frac{3}{2}a^2 - \frac{7}{2}a + \frac{9}{4} + \left(\frac{3}{2}-a\right)\log(2\pi) - 2\log A\right] z^2 \\ &+ \frac{1}{4}[2\psi(a)-3]z^4 + (a-1)^3\log[\Gamma(a+z)\Gamma(a-z)] \\ &- (3a^2-9a+7)\log[G(a+z)G(a-z)] \\ &- 2(z-2a+3)\log\Gamma_3(a+z) + 2(z+2a-3)\log\Gamma_3(a-z) \\ &+ 2(1-a)^3\log\Gamma(a) + 2(3a^2-9a+7)\log G(a) \\ &+ 4(3-2a)\log\Gamma_3(a) + 2\int_0^z \log\Gamma_3(t+a)dt + 2\int_0^{-z}\log\Gamma_3(t+a)dt \quad (|z|<|a|).\end{aligned}$$

Setting $a=2$ in (696), and applying 1.2(50) and 2.3(9), we obtain

$$\begin{aligned}\sum_{n=3}^{\infty} \{\zeta(2n-1)-1\}\frac{z^{2n}}{n} &= -\gamma z^2 + \frac{1}{2}[1-\zeta(3)]z^4 \\ &+ \log[\Gamma(2+z)\Gamma(2-z)] - \log[G(2+z)G(2-z)] \quad (|z|<2).\end{aligned} \tag{699}$$

Setting $z=\frac{3}{2}$ in (699), and making use of 1.3(8) and 2.3(9), we obtain

$$\sum_{n=3}^{\infty} \frac{1}{n}\left(\frac{3}{2}\right)^{2n} \{\zeta(2n-1)-1\} = \frac{73}{32} - \frac{9}{4}\gamma + 3\log A - \frac{81}{32}\zeta(3) + \log\left(2^{-\frac{1}{12}}\cdot 5\right). \tag{700}$$

Setting $a = 4$ in (696), and applying 1.2(50), 2.3(9), and 2.3(12), we obtain
(701)
$$\sum_{n=3}^{\infty}\left(\zeta(2n-1)-1-\frac{1}{2^{2n-1}}-\frac{1}{3^{2n-1}}\right)\frac{z^{2n}}{n}=\left(\frac{5}{6}-\gamma\right)z^2+\frac{1}{2}\left(\frac{251}{216}-\zeta(3)\right)z^4$$
$$+3\log[\Gamma(4+z)\Gamma(4-z)]-\log[G(4+z)G(4-z)]+\log\left(2^{-4}\cdot 3^{-6}\right)\quad(|z|<4),$$

which, for $z = 3$, yields
(702)
$$\sum_{n=3}^{\infty}\frac{3^{2n}}{n}\left(\zeta(2n-1)-1-\frac{1}{2^{2n-1}}-\frac{1}{3^{2n-1}}\right)=\frac{873}{16}-9\gamma-\frac{81}{2}\zeta(3)+\log\left(3^{-3}\cdot 5^2\right).$$

Setting $a = 4$ in (698), and applying 1.2(50) and 2.3(9), we obtain
(703)
$$\sum_{n=3}^{\infty}\left(\zeta(2n-3)-1-\frac{1}{2^{2n-3}}-\frac{1}{3^{2n-3}}\right)\frac{z^{2n}}{n}=\left(\frac{49}{4}-\frac{5}{2}\log(2\pi)-2\log A\right)z^2$$
$$+\frac{1}{2}\left(\frac{1}{3}-\gamma\right)z^4+27\log[\Gamma(4+z)\Gamma(4-z)]-19\log[G(4+z)G(4-z)]$$
$$+2(5-z)\log\Gamma_3(4+z)+2(5+z)\log\Gamma_3(4-z)+\log\left(2^{-16}3^{-54}\right)$$
$$+2\int_0^z\log\Gamma_3(t+4)dt+2\int_0^{-z}\log\Gamma_3(t+4)dt\quad(|z|<4).$$

Using 1.2(39), 1.3(73), and (446), we find that

(704)
$$\begin{aligned}&\int_0^3\log\Gamma_3(t+4)\,dt\\&=10\int_0^1\log(t+1)\,dt+4\int_0^1\log(t+2)\,dt+\int_0^1\log(t+3)\,dt\\&\quad+19\int_0^1\log\Gamma(t+1)\,dt+12\int_0^1\log G(t+1)\,dt+3\int_0^1\log\Gamma_3(t+1)\,dt\\&=-33+\frac{259}{8}\log 2+9\log 3+\frac{99}{8}\log\pi-24\log A+\frac{9}{2}\log B\end{aligned}$$

and

(705)
$$\begin{aligned}&\int_0^{-3}\log\Gamma_3(t+4)\,dt\\&=-\int_0^1\log\Gamma(t+1)\,dt-3\int_0^1\log G(t+1)\,dt-3\int_0^1\log\Gamma_3(t+1)\,dt\\&=\frac{3}{4}-\frac{9}{8}\log(2\pi)+6\log A-\frac{9}{2}\log B.\end{aligned}$$

If we set $z = 3$ in (703), and make use of (704) and (705), we obtain

$$\begin{aligned}(706)\qquad &\sum_{n=3}^{\infty} \frac{3^{2n}}{n} \left(\zeta(2n-3) - 1 - \frac{1}{2^{2n-3}} - \frac{1}{3^{2n-3}} \right) \\ &= \frac{237}{4} - \frac{81}{2}\gamma - 54 \log A + \log\left(2^{12} \cdot 3^{-27} \cdot 5^{8}\right).\end{aligned}$$

Setting $a = 2$ in (698), and applying 1.2(50) and 2.3(9), we obtain

$$\begin{aligned}(707)\qquad &\sum_{n=3}^{\infty} [\zeta(2n-3) - 1] \frac{z^{2n}}{n} = \left(\frac{5}{4} - \frac{1}{2}\log(2\pi) - 2\log A \right) z^2 \\ &- \frac{1}{4}(1+2\gamma) z^4 + \log\left[\Gamma(2+z)\Gamma(2-z)\right] - \log\left[G(2+z)G(2-z)\right] \\ &+ 2(1-z)\log\Gamma_3(2+z) + 2(1+z)\log\Gamma_3(2-z) \\ &+ 2\int_0^{z} \log\Gamma_3(t+2)dt + 2\int_0^{-z} \log\Gamma_3(t+2)dt \quad (|z| < 2).\end{aligned}$$

Making use of 1.2(39), 1.3(55), 1.3(73), 1.3(77), (446), and (447), we find that

$$\begin{aligned}(708)\qquad &\int_0^{\frac{3}{2}} \log\Gamma_3(t+2)\,dt = \int_0^1 \log G(t+1)\,dt + \int_0^1 \log\Gamma_3(t+1)\,dt \\ &+ \int_0^{\frac{1}{2}} \log\Gamma(t+1)\,dt + 2\int_0^{\frac{1}{2}} \log G(t+1)\,dt + \int_0^{\frac{1}{2}} \log\Gamma_3(t+1)\,dt \\ &= -\frac{259}{768} - \frac{29}{1920}\log 2 + \frac{9}{16}\log\pi - \frac{15}{16}\log A - \frac{5}{4}\log B + \frac{15}{16}\log C\end{aligned}$$

and

$$\begin{aligned}(709)\qquad &\int_0^{-\frac{3}{2}} \log\Gamma_3(t+2)\,dt = -2\int_0^1 \log\Gamma_3(t+1)\,dt + \int_0^{\frac{1}{2}} \log\Gamma_3(t+1)\,dt \\ &+ \int_0^1 \log G(t+1)\,dt - \int_0^{\frac{1}{2}} \log G(t+1)\,dt - \int_0^1 \log\Gamma(t+1)\,dt \\ &+ \int_0^{\frac{1}{2}} \log\Gamma(t+1)\,dt - \int_0^{-\frac{1}{2}} \log(t+1)\,dt \\ &= \frac{29}{768} - \frac{29}{1920}\log 2 - \frac{3}{16}\log A - \frac{1}{2}\log B + \frac{15}{16}\log C.\end{aligned}$$

Setting $z = \frac{3}{2}$ in (707), and applying 1.3(8), 1.3(96), (708), and (709), we obtain

$$\begin{aligned}(710)\qquad &\sum_{n=3}^{\infty} \frac{\zeta(2n-3) - 1}{n} \left(\frac{3}{2}\right)^{2n} \\ &= -\frac{17}{96} - \frac{81}{32}\gamma + \frac{27}{4}\log A + \frac{15}{4}\log C + \log\left(2^{-\frac{269}{480}} \cdot 5\right).\end{aligned}$$

Applications of Theorem 3.1

Upon setting $n = 1, 2, 3,$ and 4 in 3.2(35), if we make use of the appropriate identities given in previous sections, we shall obtain the following known or new formulas for closed-form evaluations of several families of series involving the Hurwitz (or generalized) Zeta function $\zeta(s, a)$:

(711)
$$\sum_{k=2}^{\infty} \frac{\zeta(k,\, a)}{k(k+1)}\, t^{k+1} = \zeta'(-1,\, a) - \zeta'(-1,\, a-t) + \left(\frac{1}{2} - a - \log\Gamma(a) + \frac{1}{2}\log(2\pi)\right) t + [1 + \psi(a)]\, \frac{t^2}{2} \quad (|t| < |a|);$$

(712)
$$\sum_{k=2}^{\infty} \frac{\zeta(k,\, a)}{k(k+1)(k+2)}\, t^{k+2} = \frac{1}{2}\left[\zeta'(-2,\, a-t) - \zeta'(-2,\, a)\right] + \left[\frac{1}{4}\left(a^2 - a + \frac{1}{6}\right) + \zeta'(-1,\, a)\right] t + \left(\frac{3}{2} - 3a + \log(2\pi) - 2\log\Gamma(a)\right)\frac{t^2}{2} + [3 + 2\psi(a)]\,\frac{t^3}{12} \quad (|t| < |a|);$$

(713)
$$\sum_{k=2}^{\infty} \frac{\zeta(k,\, a)}{k(k+1)(k+2)(k+3)}\, t^{k+3} = \frac{1}{6}\left[\zeta'(-3,\, a) - \zeta'(-3,\, a-t)\right] - \left[\frac{1}{3}\left(a^3 - \frac{3}{2}a^2 + \frac{1}{2}a\right) + 3\,\zeta'(-2,\, a)\right]\frac{t}{6} + \left[\frac{5}{2}\left(a^2 - a + \frac{1}{6}\right) + 6\,\zeta'(-1,\, a)\right]\frac{t^2}{12} + \left(\frac{11}{2} - 11\,a + 3\log(2\pi) - 6\log\Gamma(a)\right)\frac{t^3}{36} + [11 + 6\,\psi(a)]\,\frac{t^4}{144} \quad (|t| < |a|);$$

(714)
$$\sum_{k=2}^{\infty} \frac{\zeta(k,\, a)}{k(k+1)(k+2)(k+3)(k+4)}\, t^{k+4} = \frac{1}{24}\left[\zeta'(-4,\, a-t) - \zeta'(-4,\, a)\right] + \left[\frac{1}{4}\left(a^4 - 2a^3 + a^2 - \frac{1}{30}\right) + 4\,\zeta'(-3,\, a)\right]\frac{t}{24} - \left[\frac{7}{3}\left(a^3 - \frac{3}{2}a^2 + \frac{1}{2}a\right) + 12\,\zeta'(-2,\, a)\right]\frac{t^2}{48} + \left[13\left(a^2 - a + \frac{1}{6}\right) + 24\,\zeta'(-1,\, a)\right]\frac{t^3}{144} + [25 - 50\,a + 12\log(2\pi) - 24\log\Gamma(a)]\,\frac{t^4}{576} + [25 + 12\,\psi(a)]\,\frac{t^5}{1440} \quad (|t| < |a|).$$

Setting $a = 1$ in (714) and applying the formulas presented in this and previous sections, we readily have

(715)
$$\begin{aligned}\sum_{k=2}^{\infty} \frac{\zeta(k)}{k(k+1)(k+2)(k+3)(k+4)} t^{k+4} &= \frac{1}{24}\left(\zeta'(-4,\, 1-t) - \frac{3\,\zeta(5)}{4\pi^4}\right) \\ &- \left(\frac{5}{72} + 4\,\log C\right)\frac{t}{24} + \frac{\zeta(3)}{16\,\pi^2}t^2 + \left(\frac{25}{6} - 24\,\log A\right)\frac{t^3}{144} \\ &+ [12\,\log(2\pi) - 25]\,\frac{t^4}{576} + (25 - 12\,\gamma)\frac{t^5}{1440} \quad (|t| < 1),\end{aligned}$$

which, for $t = \frac{1}{2}$, yields

(716)
$$\begin{aligned}&\sum_{k=2}^{\infty} \frac{\zeta(k)}{k(k+1)(k+2)(k+3)(k+4)\cdot 2^k} \\ &= \frac{1}{48}\,\log(2\pi) - \frac{\gamma}{240} - \frac{1}{3}\log A - \frac{4}{3}\log C + \frac{\zeta(3)}{4\,\pi^2} - \frac{31\,\zeta(5)}{32\,\pi^4}.\end{aligned}$$

Similarly, we obtain

(717)
$$\begin{aligned}&\sum_{k=2}^{\infty} \frac{\zeta(k)}{k(k+1)(k+2)(k+3)\cdot 2^k} \\ &= \frac{59}{720}\log 2 + \frac{1}{12}\log\pi - \frac{\gamma}{48} - \log A - \frac{5}{2}\log C + \frac{\zeta(3)}{2\,\pi^2};\end{aligned}$$

(718)
$$\begin{aligned}&\sum_{k=2}^{\infty} \frac{\zeta(k)}{k(k+1)(k+2)\cdot 2^k} \\ &= -\frac{3}{8} + \frac{1}{2}\log(2\pi) - \frac{\gamma}{12} - 2\,\log A + \frac{7\,\zeta(3)}{8\,\pi^2}.\end{aligned}$$

We remark in passing that the various series identities presented here are potentially useful in deriving further identities for series involving the Zeta and related functions. For example, if we replace t in (711) by $-t$, we get

(719)
$$\begin{aligned}&\sum_{k=2}^{\infty}(-1)^k\,\frac{\zeta(k,\,a)}{k(k+1)}\,t^{k+1} \\ &= \zeta'(-1,\,a-t) - \zeta'(-1,\,a) + \left(\frac{1}{2} - a - \log\Gamma(a) + \frac{1}{2}\log(2\pi)\right)\,t \\ &- [1 + \psi(a)]\,\frac{t^2}{2} \quad (|t| < |a|).\end{aligned}$$

Furthermore, by making use of 2.2(17) and 2.2(33), the series identity (719) can be written as follows in an equivalent form involving the double Gamma function Γ_2 [*cf.* Equation (535)]:

$$\sum_{k=2}^{\infty}(-1)^k \frac{\zeta(k,a)}{k(k+1)} t^{k+1}$$

$$(720) \qquad = (1-a)\log\Gamma(a) - \log\Gamma_2(a) + (t+a-1)\log\Gamma(a+t) + \log\Gamma_2(a+t)$$
$$+ \left(\frac{1}{2} - a + \frac{1}{2}\log(2\pi) - \log\Gamma(a)\right) t - [\psi(a)+1]\frac{t^2}{2} \quad (|t| < |a|),$$

which, in the form (535), was proven (in a markedly different way) by Choi and Srivastava (1997, p. 108, Eq. (2.14)).

3.5 Use of Hypergeometric Identities

Al-Saqabi *et al.* (1991) presented a systematic account of several interesting infinite series expressed in terms of the Psi (or Digamma) function. Aular de Durán *et al.* (1995) examined rather systematically the sums of numerous interesting families of infinite series with or without the use of fractional calculus. Shen (1995) investigated the connections between the Stirling numbers $s(n,k)$ of the first kind and the Riemann Zeta function $\zeta(n)$ by means of the Gauss summation formula 1.4(7) for the hypergeometric series. We show how the sums of certain interesting classes of infinite series can be evaluated by analyzing three known identities involving hypergeometric series. Our presentation in this section is based essentially upon the works of Shen (1995), Choi and Seo (1997a), Choi and Srivastava (1999b), and Choi, Srivastava, Seo, and Rathie (1999).

In addition to Gauss's summation formula 1.4(7), we first recall here some well-known summation formulas for hypergeometric series.

Gauss's formula:

$$(1) \qquad {}_2F_1\left(a, b; \frac{1}{2}(a+b+1); \frac{1}{2}\right) = \frac{\Gamma\left(\frac{1}{2}\right)\Gamma\left(\frac{1}{2}+\frac{1}{2}a+\frac{1}{2}b\right)}{\Gamma\left(\frac{1}{2}+\frac{1}{2}a\right)\Gamma\left(\frac{1}{2}+\frac{1}{2}b\right)}$$
$$(a+b \neq -1, -3, -5, \ldots).$$

Kummer's formulas:

$$(2) \qquad {}_2F_1\left(a, 1-a; c; \frac{1}{2}\right) = \frac{2^{1-c}\Gamma(c)\Gamma\left(\frac{1}{2}\right)}{\Gamma\left(\frac{1}{2}c+\frac{1}{2}a\right)\Gamma\left(\frac{1}{2}c-\frac{1}{2}a+\frac{1}{2}\right)} \qquad \left(c \notin \mathbb{Z}_0^-\right),$$

$$(3) \qquad {}_2F_1(a, b; 1+a-b; -1) = \frac{\Gamma(1+a-b)\Gamma\left(1+\frac{1}{2}a\right)}{\Gamma\left(1+\frac{1}{2}a-b\right)\Gamma(1+a)}$$
$$(\Re(b) < 1;\ 1+a-b \notin \mathbb{Z}_0^-).$$

Dixon's theorem:

$$(4) \qquad {}_3F_2\left[\begin{matrix} a, b, c; \\ 1+a-b, 1+a-c; \end{matrix} \; 1\right]$$
$$= \frac{\Gamma\left(1+\frac{1}{2}a\right)\Gamma(1+a-b)\Gamma(1+a-c)\Gamma\left(1+\frac{1}{2}a-b-c\right)}{\Gamma(1+a)\Gamma\left(1+\frac{1}{2}a-b\right)\Gamma\left(1+\frac{1}{2}a-c\right)\Gamma(1+a-b-c)}$$
$$\left(\Re\left(\frac{1}{2}a-b-c\right) > -1\right).$$

Watson's theorem:

(5)
$$
{}_3F_2\left[\begin{matrix} a,\ b,\ c; \\ \frac{1}{2}+\frac{1}{2}a+\frac{1}{2}b,\ 2c; \end{matrix}\ 1\right]
= \frac{\Gamma\left(\frac{1}{2}\right)\Gamma\left(c+\frac{1}{2}\right)\Gamma\left(\frac{1}{2}+\frac{1}{2}a+\frac{1}{2}b\right)\Gamma\left(\frac{1}{2}-\frac{1}{2}a-\frac{1}{2}b+c\right)}{\Gamma\left(\frac{1}{2}+\frac{1}{2}a\right)\Gamma\left(\frac{1}{2}+\frac{1}{2}b\right)\Gamma\left(\frac{1}{2}-\frac{1}{2}a+c\right)\Gamma\left(\frac{1}{2}-\frac{1}{2}b+c\right)}.
$$

Whipple's theorem:

(6)
$$
{}_3F_2\left[a,\ b,\ c;\ d,\ e;\ 1\right] = \pi 2^{1-2c} \frac{\Gamma(d)\Gamma(e)}{\Gamma\left(\frac{1}{2}a+\frac{1}{2}e\right)\Gamma\left(\frac{1}{2}a+\frac{1}{2}d\right)\Gamma\left(\frac{1}{2}b+\frac{1}{2}e\right)\Gamma\left(\frac{1}{2}b+\frac{1}{2}d\right)}
$$
$$
(a+b=1;\ d+e=1+2c).
$$

Series derivable from Gauss's summation formula 1.4(7)

Consider a as a variable in 1.4(7) and write

(7)
$$
f(z) := 1+\sum_{n=1}^{\infty} \frac{(b)_n\,(z)_n}{(c)_n\,n!} = \frac{\Gamma(c)\,\Gamma(c-b-z)}{\Gamma(c-b)\,\Gamma(c-z)}.
$$

By making use of the definition and properties of the Stirling numbers $s(n,k)$ of the first kind, we find that

(8)
$$
\begin{aligned}
f(z) =& 1+\sum_{n=1}^{\infty} \frac{(b)_n}{(c)_n}\left[\frac{z}{n}+\left(\frac{1}{n}\sum_{k=1}^{n-1}\frac{1}{k}\right)z^2+\frac{1}{2n}\left\{\left(\sum_{k=1}^{n-1}\frac{1}{k}\right)^2-\sum_{k=1}^{n-1}\frac{1}{k^2}\right\}z^3\right. \\
&\left.+\frac{1}{6n}\left\{\left(\sum_{k=1}^{n-1}\frac{1}{k}\right)^3-3\left(\sum_{k=1}^{n-1}\frac{1}{k}\right)\left(\sum_{k=1}^{n-1}\frac{1}{k^2}\right)+2\sum_{k=1}^{n-1}\frac{1}{k^3}\right\}z^4+\cdots\right].
\end{aligned}
$$

On the other hand, we let

(9)
$$
f(z) = \frac{\Gamma(c)\,\Gamma(c-b-z)}{\Gamma(c-b)\,\Gamma(c-z)} := 1+\sum_{n=1}^{\infty} a_n\,z^n,
$$

where the coefficients a_n are to be determined. Differentiating (9) logarithmically, we obtain

$$
\frac{f'(z)}{f(z)} := \sum_{n=0}^{\infty} c_n\,z^n,
$$

which, in view of 1.2(1) and 1.2(53), yields

(10)
$$
c_n = \begin{cases} \psi(c)-\psi(c-b) & (n=0), \\ \zeta(n+1,c-b)-\zeta(n+1,c) & (n\in\mathbb{N}), \end{cases}
$$

where $\psi(z)$ and $\zeta(s,a)$ denote the Digamma (or Psi) and Hurwitz (or generalized) Zeta functions, respectively (see Sections 1.2 and 2.2). Observe that

$$a_0 = f(0) = 1 \quad \text{and} \quad c_0 = \frac{f'(0)}{f(0)} = a_1.$$

Since

$$\sum_{n=0}^{\infty} (n+1)a_{n+1}\, z^n = f'(z) = f(z)\cdot\frac{f'(z)}{f(z)} = \sum_{n=0}^{\infty}\left(\sum_{k=0}^{n} a_{n-k}\, c_k\right) z^n,$$

we have the following recursion formula for a_n:

$$(n+1)a_{n+1} = \sum_{k=0}^{n} a_{n-k}\, c_k \quad (n\in\mathbb{N}). \tag{11}$$

Now, by comparing the power series in (8) and (9), we get

$$\sum_{n=1}^{\infty} \frac{(b)_n}{n\,(c)_n} = \psi(c) - \psi(c-b) \quad \left(\Re(c-b)>0;\ c\notin\mathbb{Z}_0^-\right), \tag{12}$$

which is a well-known (rather classical) result;

$$\sum_{n=1}^{\infty} \frac{(b)_n}{(c)_n}\left(\frac{1}{n}\sum_{k=1}^{n-1}\frac{1}{k}\right) = \frac{1}{2}\left[\{\psi(c)-\psi(c-b)\}^2 + \zeta(2,c-b) - \zeta(2,c)\right], \tag{13}$$

which is the *corrected* version of a result in Shen (1995, p. 1397, Eq. (31));

$$\begin{aligned}\sum_{n=1}^{\infty} \frac{1}{n}\frac{(b)_n}{(c)_n}\left\{\left(\sum_{k=1}^{n-1}\frac{1}{k}\right)^2 - \sum_{k=1}^{n-1}\frac{1}{k^2}\right\} &= \frac{1}{3}\left[\psi(c)-\psi(c-b)\right]^3 \\ &+ [\psi(c)-\psi(c-b)][\zeta(2,c-b)-\zeta(2,c)] + \frac{2}{3}\left[\zeta(3,c-b)-\zeta(3,c)\right].\end{aligned} \tag{14}$$

If use is made of 1.2(3), the classical result (12) is rewritten in its equivalent form:

$$\sum_{n=1}^{\infty} \frac{(b)_n}{n\,(c)_n} = \sum_{n=0}^{\infty}\left(\frac{1}{n+c-b} - \frac{1}{n+c}\right) \quad \left(\Re(c-b)>0;\ c\notin\mathbb{Z}_0^-\right). \tag{15}$$

Choose $c=1$ and $b=z$ in (15) and expand both sides as power series about $z=0$. We thus obtain

$$\sum_{k=1}^{\infty}\left(\sum_{n=k}^{\infty}\frac{(-1)^{n+k}\, s(n,k)}{n\cdot n!}\right) z^k = \sum_{n=1}^{\infty}\frac{(z)_n}{n\cdot n!} = \sum_{n=1}^{\infty}\left(\frac{1}{n-z} - \frac{1}{n}\right) = \sum_{k=1}^{\infty}\zeta(k+1)\, z^k,$$

which, upon equating the coefficients of z^k, yields an interesting relation between the Riemann Zeta function and the Stirling numbers of the first kind:

$$\zeta(k+1) = \sum_{n=k}^{\infty}\frac{(-1)^{n+k}\, s(n,k)}{n\cdot n!} \quad (k\in\mathbb{N}), \tag{16}$$

which is derived also in Problem 47 (Chapter 1). Similarly, we can rewrite (13) in the following form:

$$\sum_{n=1}^{\infty} \frac{(b)_n}{(c)_n} \left(\frac{1}{n} \sum_{k=1}^{n-1} \frac{1}{k} \right) = \frac{1}{2} \left[\left\{ \sum_{n=0}^{\infty} \left(\frac{1}{n+c-b} - \frac{1}{n+c} \right) \right\}^2 + \sum_{n=0}^{\infty} \left(\frac{1}{(n+c-b)^2} - \frac{1}{(n+c)^2} \right) \right], \tag{17}$$

which, upon choosing $c = 1$ and $b = z$ and expanding both sides as power series about $z = 0$, yields

$$\left\{ \sum_{n=1}^{\infty} \frac{1}{n^2} \left(\sum_{k=1}^{n-1} \frac{1}{k} \right) \right\} z + \left\{ \sum_{n=1}^{\infty} \left(\frac{1}{n} \sum_{k=1}^{n-1} \frac{1}{k} \right)^2 \right\} z^2 + \cdots$$
$$= \zeta(3)\, z + \frac{1}{2} \left[\{\zeta(2)\}^2 + 3\,\zeta(4) \right] z^2 + \cdots .$$

Therefore, we have [see Shen (1995, p. 1398)]

$$\sum_{n=1}^{\infty} \frac{1}{n^2} \left(\sum_{k=1}^{n-1} \frac{1}{k} \right) = \zeta(3), \tag{18}$$

which is an obvious special case of the classical result 2.3(54) when $n = 2$;

$$\sum_{n=1}^{\infty} \left(\frac{1}{n} \sum_{k=1}^{n-1} \frac{1}{k} \right)^2 = \frac{1}{2} \left[\{\zeta(2)\}^2 + 3\,\zeta(4) \right] = \frac{11}{4}\, \zeta(4). \tag{19}$$

If we set ($b = 1$ and $c = 2$) and ($b = \frac{1}{2}$ and $c = 1$) in (13) and (14) and use 1.2(50), 1.2(51), 2.3(2), and 2.3(9), we also obtain the following interesting formulas:

$$\sum_{n=1}^{\infty} \frac{1}{n(n+1)} \left(\sum_{k=1}^{n-1} \frac{1}{k} \right) = 1; \tag{20}$$

$$\sum_{n=1}^{\infty} \frac{1}{n(n+1)} \left\{ \left(\sum_{k=1}^{n-1} \frac{1}{k} \right)^2 - \sum_{k=1}^{n-1} \frac{1}{k^2} \right\} = 2; \tag{21}$$

$$\sum_{n=1}^{\infty} \frac{\left(\frac{1}{2}\right)_n}{n \cdot n!} \left(\sum_{k=1}^{n-1} \frac{1}{k} \right) = \zeta(2) + 2\,(\log 2)^2; \tag{22}$$

$$\sum_{n=1}^{\infty} \frac{\left(\frac{1}{2}\right)_n}{n \cdot n!} \left\{ \left(\sum_{k=1}^{n-1} \frac{1}{k} \right)^2 - \sum_{k=1}^{n-1} \frac{1}{k^2} \right\} = 4\,\zeta(3) + 4\,\zeta(2)\, \log 2 + \frac{8}{3}(\log 2)^3. \tag{23}$$

Series derivable from Kummer's formula (3)

Consider a as a variable in (3) and apply the same technique as above. We readily obtain

$$\sum_{n=1}^{\infty} \frac{(-1)^n (b)_n}{n(1-b)_n} = \frac{1}{2}\left[\gamma + \psi(1-b)\right], \tag{24}$$

which is, in fact, an immediate consequence of a known result recorded by Aular de Durán *et al.* (1995, p. 746, Eq. (2.9)) with $z = 1-b$;

$$\begin{aligned} &\sum_{n=1}^{\infty} \frac{(-1)^n (b)_n}{n(1-b)_n} \left\{ \sum_{k=1}^{n-1} \frac{1}{k} + \sum_{k=1}^{n} \frac{1}{b-k} \right\} \\ &= \frac{1}{8}[\gamma + \psi(1-b)]^2 + \frac{3}{8}[\zeta(2, 1-b) - \zeta(2)]; \end{aligned} \tag{25}$$

(26)

$$\begin{aligned} &\sum_{n=1}^{\infty} \frac{(-1)^n (b)_n}{n(1-b)_n} \left[\left(\sum_{k=1}^{n} \frac{1}{b-k} + \sum_{k=1}^{n-1} \frac{1}{k} \right)^2 + \sum_{k=1}^{n} \frac{1}{(b-k)^2} - \sum_{k=1}^{n-1} \frac{1}{k^2} \right] \\ &= \frac{1}{24}[\gamma + \psi(1-b)]^3 + \frac{3}{8}[\gamma + \psi(1-b)][\zeta(2, 1-b) - \zeta(2)] + \frac{7}{12}[\zeta(3) - \zeta(3, 1-b)]. \end{aligned}$$

By setting $b = \frac{1}{2}$ in (25) and applying some elementary identities given in Sections 1.2 and 2.3, we obtain

$$\begin{aligned} \frac{1}{2}(\log 2)^2 + \frac{3}{4}\zeta(2) &= \sum_{n=1}^{\infty} \frac{(-1)^n}{n} \left\{ \sum_{k=1}^{n-1} \frac{1}{k} - \sum_{k=0}^{n-1} \frac{2}{2k+1} \right\} \\ &= \sum_{n=1}^{\infty} \frac{(-1)^n}{n} \left\{ -2 + \sum_{k=1}^{n-1} \frac{1}{k(2k+1)} \right\}, \end{aligned}$$

from which, by using the Maclaurin series of $\log(1+x)$, we obtain

$$\sum_{n=1}^{\infty} \frac{(-1)^n}{n} \sum_{k=1}^{n-1} \frac{1}{k(2k+1)} = \frac{1}{2}(\log 2)^2 - 2\log 2 + \frac{3}{4}\zeta(2). \tag{27}$$

Setting $b = \frac{2}{3}$ in (25) yields

$$\begin{aligned} &\sum_{n=1}^{\infty} \frac{(-1)^n}{n} \cdot \frac{\left(\frac{2}{3}\right)_n}{\left(\frac{1}{3}\right)_n} \left\{ -3 + \sum_{k=1}^{n-1} \frac{1}{k(3k+1)} \right\} \\ &= \frac{1}{32} \left(\frac{\sqrt{3}}{3}\pi + 3\log 3 \right)^2 + \frac{3}{8}[\zeta(2, \tfrac{1}{3}) - \zeta(2)]. \end{aligned} \tag{28}$$

Finally, by setting $b = \frac{1}{2}$ in (26), we obtain the following series evaluation:

$$\begin{aligned} &\sum_{n=1}^{\infty} \frac{(-1)^{n+1}}{n} \left[\left(\frac{2}{2n-1} + \sum_{k=1}^{n-1} \frac{1}{k(2k-1)} \right)^2 + \sum_{k=1}^{n} \frac{4}{(2k-1)^2} - \sum_{k=1}^{n-1} \frac{1}{k^2} \right] \\ &= \frac{1}{3}(\log 2)^3 + \frac{3}{2}\zeta(2)\log 2 + \frac{7}{2}\zeta(3). \end{aligned} \tag{29}$$

Next, by considering b as a variable in (3), we also show that the sums of a different class of infinite series can be evaluated. For $b = z$, we find from (3) that

$$f(z) := 1 + \sum_{n=1}^{\infty} \frac{(-1)^n (a)_n}{n!} \cdot \frac{(z)_n}{(1+a-z)_n} = \frac{\Gamma\left(1+\frac{1}{2}a\right)}{\Gamma(1+a)} \cdot \frac{\Gamma(1+a-z)}{\Gamma\left(1+\frac{1}{2}a-z\right)}. \tag{30}$$

Now, since

$$\sum_{k=0}^{n}\sum_{\ell=0}^{k} B_{k,\ell} = \sum_{\ell=0}^{n}\sum_{k=\ell}^{n} B_{k,\ell},$$

we can apply 1.5(5) to obtain

$$\begin{aligned}(1+a-z)_n &= \sum_{k=0}^{n} (-1)^{n+k} s(n,k)(1+a-z)^k \\ &= \sum_{k=0}^{n} (-1)^{n+k} s(n,k) \left\{ \sum_{\ell=0}^{k} \binom{k}{\ell} (1+a)^{k-\ell} \cdot (-1)^{\ell} \cdot z^{\ell} \right\} \\ &= \sum_{\ell=0}^{n} \left[\sum_{k=\ell}^{n} (-1)^{n+k} s(n,k) \binom{k}{\ell} (-1)^{\ell} \cdot (1+a)^{k-\ell} \right] z^{\ell}.\end{aligned} \tag{31}$$

Expanding the first member of (30) in terms of the Stirling numbers $s(n,k)$, and using 1.5(9), 1.5(11), and (31), we have

$$f(z) = 1 + \sum_{n=1}^{\infty} \frac{(-1)^n (a)_n}{n!} \left(\alpha_1 z + \alpha_2 z^2 + \alpha_3 z^3 + \cdots\right), \tag{32}$$

where

$$\begin{aligned}\alpha_1 &= \frac{(n-1)!}{(1+a)_n}; \\ \alpha_2 &= \frac{(n-1)!}{(1+a)_n} \left(\sum_{k=1}^{n-1} \frac{1}{k} + \psi(1+a+n) - \psi(1+a) \right); \\ \alpha_3 &= \frac{(n-1)!}{2(1+a)_n} \left[-\sum_{k=1}^{n-1} \frac{1}{k^2} + \left(\sum_{k=1}^{n-1} \frac{1}{k} + \psi(1+a+n) - \psi(1+a) \right)^2 \right. \\ &\qquad \left. + \zeta(2,1+a) - \zeta(2,1+a+n) \right],\end{aligned}$$

and so on.

Now, using the same technique as above, we obtain

$$\sum_{n=1}^{\infty} \frac{(-1)^n a}{n(n+a)} = \psi\left(1+\frac{1}{2}a\right) - \psi(1+a), \tag{33}$$

which was recorded by Al-Saqabi *et al.* (1991, p. 366, Eq. (2.7)) who commented that (33) is essentially the same as the known result [*cf.* Hansen (1975, p. 102, Eq. (6.1.14))]:

$$\sum_{n=1}^{\infty} \frac{(-1)^n}{n(\lambda+\mu n)} = \frac{1}{\lambda}\left\{\beta\left(1+\frac{\lambda}{\mu}\right) - \log 2\right\}, \tag{34}$$

where $\beta(z)$ is defined by

$$\beta(z) := \frac{1}{2}\left\{\psi\left(\frac{1}{2}+\frac{1}{2}z\right) - \psi\left(\frac{1}{2}z\right)\right\} = \sum_{N=0}^{\infty} \frac{(-1)^N}{z+N}; \tag{35}$$

$$\begin{aligned}\sum_{n=0}^{\infty} \frac{(-1)^n\, a}{n(n+a)} &\left\{\sum_{k=1}^{n-1}\frac{1}{k} + \sum_{k=1}^{n}\frac{1}{k+a}\right\} \\ &= \frac{1}{2}\left[\left\{\psi\left(1+\frac{1}{2}a\right) - \psi(1+a)\right\}^2 + \zeta(2,\,1+a) - \zeta\left(2,\,1+\frac{1}{2}a\right)\right];\end{aligned} \tag{36}$$

(37)

$$\begin{aligned}\sum_{n=1}^{\infty} \frac{(-1)^n\, a}{n(n+1)} &\left[\sum_{k=1}^{n}\frac{1}{(k+a)^2} - \sum_{k=1}^{n-1}\frac{1}{k^2} + \left\{\sum_{k=1}^{n}\frac{1}{k+a} + \sum_{k=1}^{n-1}\frac{1}{k}\right\}^2\right] \\ = \frac{1}{3}&\left\{\psi\left(1+\frac{1}{2}a\right) - \psi(1+a)\right\}\left[\left\{\psi\left(1+\frac{1}{2}a\right) - \psi(1+a)\right\}^2\right. \\ &\left. + 3\left\{\zeta(2,\,1+a) - \zeta\left(2,\,1+\frac{1}{2}a\right)\right\}\right] + \frac{2}{3}\left\{\zeta(3,\,1+a) - \zeta\left(3,\,1+\frac{1}{2}a\right)\right\} \\ &\qquad (a \notin \mathbb{Z}_0^- \setminus \{0\}).\end{aligned}$$

Setting $a = 1$ in (36) and (37), and simplifying the resulting equations with the aid of various known (or easily derivable) identities including the following ones [*cf.*, *e.g.*, Hansen (1975, p. 116, Entry (6.2.14) for $m-2=n=1$ and $m=n-2=1$)]:

$$\begin{aligned}\sum_{n=1}^{\infty} \frac{(-1)^{n+1}}{n(n+1)^3} &= 2\log 2 - 3 + \frac{1}{2}\zeta(2) + \frac{3}{4}\zeta(3); \\ \sum_{n=1}^{\infty} \frac{(-1)^{n+1}}{n^3(n+1)} &= 2\log 2 - 1 - \frac{1}{2}\zeta(2) + \frac{3}{4}\zeta(3); \\ \sum_{n=1}^{\infty} \frac{(-1)^{n+1}}{n^2(n+1)^2}\left(\sum_{k=1}^{n}\frac{1}{k}\right) &= 4\sum_{n=1}^{\infty} \frac{(-1)^n}{n}\left(\sum_{k=1}^{n}\frac{1}{k}\right) + \zeta(2) + \frac{3}{4}\zeta(3),\end{aligned} \tag{38}$$

we readily obtain

$$\sum_{n=1}^{\infty} \frac{(-1)^n}{n(n+1)}\left(\sum_{k=1}^{n}\frac{1}{k}\right) = (\log 2)^2 - \frac{1}{2}\zeta(2), \tag{39}$$

which reduces immediately to

(40) $$\sum_{n=1}^{\infty} \frac{(-1)^{n+1}}{n} \left(\sum_{k=1}^{n} \frac{1}{k} \right) = \frac{1}{2} \left[\zeta(2) - (\log 2)^2 \right].$$

We can also readily obtain the following identity:

(41) $$\sum_{n=1}^{\infty} \frac{(-1)^n}{n(n+1)} \left(\sum_{k=1}^{n} \frac{1}{k} \right)^2 = -\frac{7}{9} - \frac{2}{3} (\log 2)^3 + \zeta(2) \log 2 - \frac{1}{3} \zeta(3).$$

Series derivable from other hypergeometric summation formulas

Consider a as a variable in (2) and apply the same technique as above. We thus find that

(42) $$\sum_{n=1}^{\infty} \frac{(n-1)!}{(c)_n \cdot 2^n} = \frac{1}{2} \left\{ \psi \left(\frac{1}{2} c + \frac{1}{2} \right) - \psi \left(\frac{1}{2} c \right) \right\};$$

(43)
$$\sum_{n=1}^{\infty} \frac{(n-1)!}{n\,(c)_n \cdot 2^n} = \frac{1}{8} \left[\zeta \left(2, \frac{1}{2} c \right) + \zeta \left(2, \frac{1}{2} c + \frac{1}{2} \right) - \left\{ \psi \left(\frac{1}{2} c + \frac{1}{2} \right) - \psi \left(\frac{1}{2} c \right) \right\}^2 \right],$$

which, upon setting $c = 1$, 2, and 3, yields

(44) $$\sum_{n=1}^{\infty} \frac{1}{n^2 \cdot 2^n} = \frac{1}{2} \left[\zeta(2) - (\log 2)^2 \right],$$

which may be compared with (40);

(45) $$\sum_{n=1}^{\infty} \frac{1}{n^2 (n+1) \cdot 2^{n-1}} = \zeta(2) - 1 - (\log 2 - 1)^2;$$

(46) $$\sum_{n=1}^{\infty} \frac{1}{n^2 (n+1)(n+2) \cdot 2^{n-1}} = \frac{1}{2} \zeta(2) - \frac{5}{8} - \frac{1}{8} (2 \log 2 - 1)^2.$$

By setting $c = z$ in (2) and 1.4(7), multiplying the resulting equations by z, and making use of the same technique as above, we obtain the following summation formulas:

(47) $$\sum_{n=1}^{\infty} \frac{\Gamma(a+n)\,\Gamma(1-a+n)}{n!\,(n-1)!\,2^n} = \frac{1}{\sqrt{\pi}} \Gamma \left(\frac{1}{2} + \frac{1}{2} a \right) \Gamma \left(1 - \frac{1}{2} a \right),$$

which, after some elementary manipulation, becomes

(48) $${}_2F_1 \left(1+a,\, 2-a;\, 2;\, \frac{1}{2} \right) = \frac{2}{\sqrt{\pi}} \, \frac{\Gamma\left(\frac{1}{2} + \frac{1}{2} a\right) \Gamma\left(1 - \frac{1}{2} a\right)}{\Gamma(1+a)\,\Gamma(2-a)}.$$

By using Legendre's duplication formula 1.1(49), (48) is rewritten in its *equivalent* form:

(49) $$ {}_2F_1\left(1+a,\ 2-a;\ 2;\ \frac{1}{2}\right) = \frac{\sqrt{\pi}}{\Gamma\left(1+\frac{1}{2}a\right)\Gamma\left(\frac{3}{2}-\frac{1}{2}a\right)}, $$

which can be shown to be a special case of Kummer's summation formula (2) in view of Euler's formula 1.4(21);

(50)
$$ \sum_{n=1}^{\infty} \frac{\Gamma(a+n)\,\Gamma(1-a+n)}{n!\,(n-1)!\,2^n} \left(\sum_{k=1}^{n-1} \frac{1}{k}\right) = \Gamma(a)\,\Gamma(1-a) $$
$$ + \frac{1}{\sqrt{\pi}}\Gamma\left(\frac{1}{2}+\frac{1}{2}a\right)\Gamma\left(1-\frac{1}{2}a\right)\left\{\log 2+\gamma+\frac{1}{2}\psi\left(\frac{1}{2}a\right)+\frac{1}{2}\psi\left(\frac{1}{2}-\frac{1}{2}a\right)\right\}; $$

(51)
$$ \sum_{n=1}^{\infty} \frac{\Gamma(a+n)\,\Gamma(1-a+n)}{n!\,(n-1)!\,2^n} \left\{\left(\sum_{k=1}^{n-1} \frac{1}{k}\right)^2 + \sum_{k=1}^{n-1}\frac{1}{k^2}\right\} $$
$$ = \frac{1}{\sqrt{\pi}}\Gamma\left(\frac{1}{2}+\frac{1}{2}a\right)\Gamma\left(1-\frac{1}{2}a\right)\left[\left\{\log 2+\gamma+\frac{1}{2}\psi\left(\frac{1}{2}a\right)+\frac{1}{2}\psi\left(\frac{1}{2}-\frac{1}{2}a\right)\right\}^2\right. $$
$$ \left. +\zeta(2)-\frac{1}{4}\left\{\zeta\left(2,\ \frac{1}{2}a\right)+\zeta\left(2,\ \frac{1}{2}-\frac{1}{2}a\right)\right\}\right]; $$

(52) $$ \sum_{n=1}^{\infty} \frac{(a)_n(b)_n}{(n-1)!\,n!} = \frac{\Gamma(-a-b)}{\Gamma(-a)\,\Gamma(-b)} \qquad (\Re(a+b)<0), $$

which, after some elementary manipulation, is seen to be the same as the following special case of Gauss's summation theorem 1.4(7):

(53) $$ {}_2F_1(a+1,\ b+1;\ 2;\ 1) = \frac{\Gamma(-a-b)}{\Gamma(1-a)\Gamma(1-b)} \qquad (\Re(a+b)<0); $$

(54)
$$ \sum_{n=1}^{\infty} \frac{(a)_n(b)_n}{(n-1)!\,n!}\left(\sum_{k=1}^{n-1}\frac{1}{k}\right) = 1+\frac{\Gamma(-a-b)}{\Gamma(-a)\,\Gamma(-b)}\left[\gamma+\psi(-a)+\psi(-b)-\psi(-a-b)\right]; $$

(55)
$$ \sum_{n=1}^{\infty} \frac{(a)_n(b)_n}{(n-1)!\,n!}\left\{\left(\sum_{k=1}^{n-1}\frac{1}{k}\right)^2+\sum_{k=1}^{n-1}\frac{1}{k^2}\right\} = \frac{\Gamma(-a-b)}{\Gamma(-a)\Gamma(-b)} $$
$$ \cdot\left[\{\gamma+\psi(-a)+\psi(-b)-\psi(-a-b)\}^2+\zeta(2)+\zeta(2,\ -a-b)-\zeta(2,\ -a)-\zeta(2,\ -b)\right]. $$

Setting $a=\frac{1}{2}$ in (47), (50), and (51), and simplifying with the aid of various known identities involving Gamma and Psi functions, we obtain

(56) $$ \sum_{n=1}^{\infty} \frac{\{(2n)!\}^2}{(n!)^3(n-1)!\,2^{5n}} = \frac{1}{\pi\sqrt{\pi}}\left\{\Gamma\left(\tfrac{3}{4}\right)\right\}^2; $$

(57) $$\sum_{n=1}^{\infty} \frac{\{(2n)!\}^2}{(n!)^3(n-1)!\,2^{5n}} \left(\sum_{k=1}^{n-1} \frac{1}{k}\right) = 1 - \frac{1}{\pi\sqrt{\pi}} \left\{\Gamma\left(\tfrac{3}{4}\right)\right\}^2 \left(\frac{\pi}{2} + 2\log 2\right);$$

(58) $$\sum_{n=1}^{\infty} \frac{\{(2n)!\}^2}{(n!)^3\,(n-1)!\,2^{5n}} \left\{\left(\sum_{k=1}^{n-1} \frac{1}{k}\right)^2 + \sum_{k=1}^{n-1} \frac{1}{k^2}\right\} = \frac{1}{\pi\sqrt{\pi}} \left\{\Gamma\left(\tfrac{3}{4}\right)\right\}^2 \left[\left(\frac{\pi}{2} + 2\log 2\right)^2 + \zeta(2) - \frac{1}{2}\zeta\left(2, \tfrac{1}{4}\right)\right].$$

In view of Euler's formula 1.4(21), the summation formula (56) can be deduced directly from (2) by setting $a = \frac{1}{2}$ and $c = 2$. Moreover, (56) follows readily from the special case of the identity (1) when $a = b = \frac{3}{2}$.

Considering (1) just as above, we obtain

(59) $$\sum_{n=1}^{\infty} \frac{(b)_n}{n \cdot 2^{n-1} \left(\frac{1}{2} + \frac{1}{2}b\right)_n} = \psi\left(\frac{1}{2} + \frac{1}{2}b\right) - \psi\left(\frac{1}{2}\right),$$

which is precisely the same as the aforementioned known result recorded (with correction) by Al-Saqabi *et al.* (1991, p. 362, Eq. (1.7));

(60) $$\sum_{n=1}^{\infty} \frac{(b)_n}{n \cdot 2^{n-1} \left(\frac{1}{2} + \frac{1}{2}b\right)_n} \left\{2\sum_{k=1}^{n-1} \frac{1}{k} - \sum_{k=0}^{n-1} \frac{1}{k + \frac{1}{2} + \frac{1}{2}b}\right\} = \frac{1}{4}\left\{\psi\left(\frac{1}{2} + \frac{1}{2}b\right) - \psi\left(\frac{1}{2}\right)\right\}^2 + \frac{1}{4}\left\{\zeta\left(2, \frac{1}{2} + \frac{1}{2}b\right) - \zeta\left(2, \frac{1}{2}\right)\right\}.$$

If we consider a as a variable in Dixon's theorem (4), similarly as above, we obtain

(61) $$\sum_{n=1}^{\infty} \frac{(b)_n(c)_n}{n(1-b)_n(1-c)_n} = \frac{1}{2}\left\{\gamma + \psi(1-b) + \psi(1-c) - \psi(1-b-c)\right\},$$

which is an immediate consequence of the special case $a = 0$ of a known result recorded by Srivastava (1992, p. 82, Eq. (3.2));

(62) $$\sum_{n=1}^{\infty} \frac{(b)_n(c)_n}{n\,(1-b)_n(1-c)_n} \left[\sum_{k=1}^{n} \left(\frac{1}{k} - \frac{1}{k-b} - \frac{1}{k-c}\right) - \frac{1}{n}\right] = \frac{1}{8}\left[\gamma + \psi(1-b) + \psi(1-c) - \psi(1-b-c)\right]^2 + \frac{3}{8}\left[\zeta(2, 1-b) + \zeta(2, 1-c) - \zeta(2) - \zeta(2, 1-b-c)\right];$$

$$\sum_{n=1}^{\infty} \frac{(b)_n (c)_n}{n\,(1-b)_n (1-c)_n} \left[\left\{ \sum_{k=1}^{n} \left(\frac{1}{k-b} + \frac{1}{k-c} - \frac{1}{k} \right) + \frac{1}{n} \right\}^2 + \sum_{k=1}^{n} \left\{ \frac{1}{(k-b)^2} + \frac{1}{(k-c)^2} - \frac{1}{k^2} \right\} + \frac{1}{n^2} \right]$$

(63)

$$= \frac{1}{24} [\gamma + \psi(1-b) + \psi(1-c) - \psi(1-b-c)] \cdot [\{\gamma + \psi(1-b) + \psi(1-c) - \psi(1-b-c)\}^2 + 3\{\zeta(2,1-b) + \zeta(2,1-c) - \zeta(2) - \zeta(2,1-b-c)\}] + \frac{7}{12} [\zeta(3) - \zeta(3,1-b) - \zeta(3,1-c) + \zeta(3,1-b-c)].$$

3.6 Other Methods and Their Applications

The Weierstrass canonical product form for the Gamma function

Starting from the Weierstrass canonical product formula 1.1(2), Zhang and Williams (1993a) proved the following well-known identity [*cf.* Hansen (1975, p. 46, Entry (54.3.4))]:

(1) $$\sum_{k=1}^{\infty} \zeta(2k+1)\, t^{2k+1} = -\frac{1}{2} [1 + \pi t \cot(\pi t)] - t\,[\psi(t) + \gamma] \quad (|t| < 1),$$

which may be compared with 3.4(1.6).

Now, for the generalized Euler constant $\gamma(r,n)$ defined by

(2) $$\gamma(r,n) := \lim_{m\to\infty} \left\{ \sum_{\substack{\ell=1 \\ \ell \equiv r (\mathrm{mod}\, n)}}^{m} \left(\frac{1}{\ell} - \frac{1}{n} \log m \right) \right\} \quad (r \in \mathbb{Z};\; n \in \mathbb{N}),$$

so that

(3) $$\gamma(0,1) = \gamma,$$

Lehmer (1975, Theorem 7) showed that

(4) $$\gamma(r,n) = -\frac{1}{n} \left[\psi\left(\frac{r}{n}\right) + \log n \right] \quad (0 < r \leq n;\; r,\, n \in \mathbb{N}).$$

A combination of (1) and (4) yields

(5) $$\sum_{k=1}^{\infty} \zeta(2k+1) \left(\frac{r}{n}\right)^{2k} = \log n - \frac{\pi}{2} \cot\left(\frac{r\pi}{n}\right) + n\,\gamma(r,n) - \gamma - \frac{n}{2r} \quad (0 < r \leq n;\; r,\, n \in \mathbb{N}),$$

which was given by Zhang and Williams (1993a, Corollary 4) who also considered its many special cases, included in Section 3.4, by evaluating $\gamma(r,n)$.

Next we rewrite (1) as

(6) $$\sum_{k=1}^{\infty} \zeta(2k+1)\, t^{2k} = \frac{1}{2t} - \frac{\pi}{2} \cot(\pi t) - \psi(t+1) - \gamma \quad (|t| < 1),$$

which, upon integrating and differentiating, yields the well-known result [see Hansen (1975, p. 356, Entry (54.5.8)); Zhang and Williams (1993a, Theorems 5 and 6)]:

$$(7)\quad \sum_{k=1}^{\infty} \frac{\zeta(2k+1)}{2k+1}\, t^{2k+1} = \frac{1}{2}\log\left(\frac{\pi t}{\sin \pi t}\right) - \log \Gamma(1+t) - \gamma t \quad (0<|t|<1);$$

$$(8)\quad \sum_{k=1}^{\infty} 2k\,\zeta(2k+1)\,t^{2k-1} = \frac{1}{2}\left(\frac{\pi^2 t}{\sin^2 \pi t} + \frac{1}{t^2}\right) - \psi'(t) \quad (0<|t|<1);$$

(9)

$$\sum_{k=1}^{\infty} (2k+1)\,\zeta(2k+1)\,t^{2k} = \frac{1}{2}\left[\frac{\pi^2 t}{\sin^2 \pi t} - \pi \cot(\pi t)\right] - \psi(t) - \gamma - \psi'(t)\,t \quad (0<|t|<1).$$

The special cases of (8) and (9) when $t=\frac{1}{2}$ yield

$$(10)\quad \sum_{k=1}^{\infty} \frac{2k\,\zeta(2k+1)}{2^{2k}} = 1;$$

$$(11)\quad \sum_{k=1}^{\infty} \frac{(2k+1)\,\zeta(2k+1)}{2^{2k+1}} = \log 2.$$

By replacing t in 3.4(18) by it, we obtain

$$(12)\quad \sum_{k=1}^{\infty} (-1)^{k-1}\,\zeta(2k)\,t^{2k-1} = \frac{\pi}{2}\coth(\pi t) - \frac{1}{2t} \quad (0<|t|<1);$$

$$(13)\quad \sum_{k=1}^{\infty} (-1)^{k-1}\,\{\zeta(2k)-1\}\,t^{2k-1} = \frac{\pi}{2}\coth(\pi t) - \frac{3t^2+1}{2t\,(t^2+1)} \quad (0<|t|\leq 1).$$

The special cases of (12) and (13) when $t=\frac{1}{2}$ and $t=1$ give us the following sums:

$$(14)\quad \sum_{k=1}^{\infty} \frac{(-1)^{k-1}\,\zeta(2k)}{2^{2k}} = \frac{\pi}{4}\coth\left(\frac{\pi}{2}\right) - \frac{1}{2};$$

$$(15)\quad \sum_{k=1}^{\infty} (-1)^{k-1}\,\{\zeta(2k)-1\}\,t^{2k-1} = \frac{\pi}{2}\coth \pi - 1.$$

By integrating (12) (or, alternatively, replacing t in 3.4(17) by it), we find that

$$(16)\quad \sum_{k=1}^{\infty} \frac{(-1)^{k-1}\,\zeta(2k)}{k}\, t^{2k} = \log\left(\frac{\sinh \pi t}{\pi t}\right) \quad (0<|t|\leq 1),$$

which, for $t = 1$ and $t = \frac{1}{2}$, yields

$$\sum_{k=1}^{\infty} \frac{(-1)^{k-1}\,\zeta(2k)}{k} = \log\left(\frac{\sinh\pi}{\pi}\right); \tag{17}$$

$$\sum_{k=1}^{\infty} \frac{(-1)^{k-1}\,\zeta(2k)}{k\cdot 2^{2k}} = \log\left(\frac{2\sinh\frac{\pi}{2}}{\pi}\right). \tag{18}$$

Evaluation by using infinite products

Let us begin by rewriting Stirling's formula 1.1(52) in the following form:

$$\lim_{n\to\infty}\left\{2\sum_{k=1}^{n}\log k - (2n+1)\log n + 2n\right\} = \log(2\pi). \tag{19}$$

Using the infinite product of $\cos z$ [see Spiegel (1968, p. 188)], we write

$$e^{\frac{\pi}{2}} + e^{-\frac{\pi}{2}} = 2\cos\left(\frac{1}{2}\pi i\right) = 2\prod_{k=1}^{\infty}\left(1+\frac{1}{(2k-1)^2}\right),$$

or, equivalently,

$$\begin{aligned}\log\left(e^{\frac{\pi}{2}}+e^{-\frac{\pi}{2}}\right) &= \log 2 + \lim_{n\to\infty}\sum_{k=1}^{n}\log\left(1+\frac{1}{(2k-1)^2}\right)\\ &= \log 2 + \lim_{n\to\infty}\left\{n\log 8 + \sum_{k=1}^{n}\log(2k^2-2k+1) - 2\sum_{k=1}^{2n}\log k + 2\sum_{k=1}^{n}\log k\right\},\end{aligned}$$

which, upon applying Stirling's formula (19), yields

$$\log\left(e^{\frac{\pi}{2}}+e^{-\frac{\pi}{2}}\right) = \lim_{n\to\infty}\left[2n - n\log 2 - 2n\log n\sum_{k=1}^{n}\log\,(2k^2-2k+1)\right]. \tag{20}$$

Now we let

$$P := \prod_{k=1}^{\infty}\left\{e^{-\frac{1}{k}}\left(1+\frac{1}{k}+\frac{1}{2k^2}\right)\right\},$$

which, upon taking logarithms, and using 1.1(3) and (20), gives

$$\begin{aligned}\log P &= \lim_{n\to\infty}\sum_{k=1}^{n}\left\{-\frac{1}{k}+\log\left(1+\frac{1}{k}+\frac{1}{2k^2}\right)\right\}\\ &= -\gamma + \log\left(e^{\frac{\pi}{2}}+e^{-\frac{\pi}{2}}\right) + \lim_{n\to\infty}\left[(2n-1)\log n - 2n - 2\sum_{k=1}^{n}\log k\right.\\ &\qquad \left.+\sum_{k=1}^{n}\left\{\log\,(2k^2+2k+1)-\log(2k^2-2k+1)\right\}\right] \qquad (21)\\ &= -\gamma + \log\left(e^{\frac{\pi}{2}}+e^{-\frac{\pi}{2}}\right) + \lim_{n\to\infty} S_n,\end{aligned}$$

where, for convenience,

$$S_n := (2n-1)\log n - 2n + \log\left(2n^2+2n+1\right) - 2\sum_{k=1}^{n}\log k.$$

By applying (19) again, we find that

$$\begin{aligned}\lim_{n\to\infty} S_n &= -\log(2\pi) + \lim_{n\to\infty}\left[\log\left(2n^2+2n+1\right) - \log n^2\right] \\ &= -\log\pi,\end{aligned}$$

which, upon substituting into (21) and exponentiating the resulting equation, leads us to an interesting identity posed by Wilf (1997) as a problem:

$$\prod_{k=1}^{\infty}\left\{e^{-\frac{1}{k}}\left(1+\frac{1}{k}+\frac{1}{2k^2}\right)\right\} = \frac{e^{\frac{\pi}{2}}+e^{-\frac{\pi}{2}}}{\pi e^{\gamma}}. \tag{22}$$

Taking logarithms on both sides of (22) and using the Maclaurin expansion of $\log(1+x)$, we obtain

$$\frac{\zeta(2)}{2} + \sum_{n=2}^{\infty}\frac{(-1)^{n-1}}{n}\sum_{k=0}^{n}\binom{n}{k}\frac{\zeta(n+k)-1}{2^k} = \frac{3}{2} + \log\left(\frac{2\left(e^{\frac{\pi}{2}}+e^{-\frac{\pi}{2}}\right)}{5\pi}\right) - \gamma. \tag{23}$$

Next, by similarly using the infinite product of $\sin\left(-\frac{1}{2}\pi i\right)$, that is,

$$e^{\frac{\pi}{2}} - e^{-\frac{\pi}{2}} = \pi\prod_{k=1}^{\infty}\left(1+\frac{1}{4k^2}\right),$$

we also obtain the interesting identities:

$$\prod_{k=1}^{\infty}\left\{e^{-\frac{2}{k}}\left(1+\frac{2}{k}+\frac{5}{4k^2}\right)\right\} = \frac{4}{5}\cdot\frac{e^{\frac{\pi}{2}}-e^{-\frac{\pi}{2}}}{\pi e^{2\gamma}} \tag{24}$$

and

$$\begin{aligned}&\frac{5}{4}\zeta(2) + \sum_{n=2}^{\infty}(-1)^{n-1}\frac{2^n}{n}\sum_{k=0}^{n}\binom{n}{k}\left(\frac{5}{8}\right)^k\left(\zeta(n+k)-1-\frac{1}{2^{n+k}}\right) \\ &\quad = \frac{73}{16} + \log\left(\frac{256\left(e^{\frac{\pi}{2}}-e^{-\frac{\pi}{2}}\right)}{3145\,\pi}\right) - 2\gamma.\end{aligned} \tag{25}$$

If, instead, we consider the infinite products of $\sinh z$ and $\cosh z$, we similarly obtain the following identities:

$$\prod_{k=1}^{\infty}\left\{e^{-\frac{2}{k}}\left(1+\frac{2}{k}+\frac{2}{k^2}\right)\right\} = \frac{e^{\pi}-e^{-\pi}}{4\pi\,e^{2\gamma}}; \tag{26}$$

$$
\begin{aligned}
&2\zeta(2)+\sum_{n=2}^{\infty}(-1)^{n-1}\frac{2^n}{n}\sum_{k=0}^{n}\binom{n}{k}\left(\zeta(n+k)-1-\frac{1}{2^{n+k}}\right)\\
&\qquad=\frac{11}{2}+\log\left(\frac{e^{\pi}-e^{-\pi}}{50\,\pi}\right)-2\gamma
\end{aligned}
\tag{27}
$$

and

$$
\prod_{k=1}^{\infty}\left\{e^{-\frac{1}{k}}\left(1+\frac{1}{k}+\frac{5}{4\,k^2}\right)\right\}=\frac{2}{5}\cdot\frac{e^{\pi}+e^{-\pi}}{\pi\,e^{\gamma}};
\tag{28}
$$

$$
\begin{aligned}
&\frac{5}{4}\zeta(2)+\sum_{n=2}^{\infty}\frac{(-1)^{n-1}}{n}\sum_{k=0}^{n}\binom{n}{k}\left(\frac{5}{4}\right)^k[\zeta(n+k)-1]\\
&\qquad=\frac{9}{4}+\log\left(\frac{8\,(e^{\pi}+e^{-\pi})}{65\,\pi}\right)-\gamma.
\end{aligned}
\tag{29}
$$

The summation formulas (23), (25), (27), and (28), which we have presented here, are obviously different from the various classes of series involving Zeta functions, which we considered elsewhere in this chapter.

Higher-Order derivatives of the Gamma Function

Here we present explicit formulas for the evaluation of higher-order derivatives of the familiar Gamma function and consider several applications of these explicit formulas including evaluation of some families of definite integrals [see Choi and Srivastava (2000c)].

It follows from 2.3(19) that

$$
\begin{aligned}
\{\zeta(2)\}^2&=\frac{5}{2}\zeta(4), &\zeta(2)\zeta(4)&=\frac{7}{4}\zeta(6),\\
\zeta(2)\zeta(6)&=\frac{5}{3}\zeta(8), &\{\zeta(4)\}^2&=\frac{7}{6}\zeta(8),\\
\zeta(2)\zeta(8)&=\frac{33}{20}\zeta(10), &\zeta(4)\zeta(6)&=\frac{11}{10}\zeta(10).
\end{aligned}
\tag{30}
$$

In addition to the Maclaurin series expansion 3.4(10) for $\log\Gamma(1+z)$, it is also known that [*cf.*, *e.g.*, Luke (1969, Vol. I, p. 27)]

$$
\Gamma(1+z)=\sum_{n=0}^{\infty}a_n\,z^n\quad(|z|<1),
\tag{31}
$$

where

$$
a_0=1\quad\text{and}\quad n\,a_n=-\gamma\,a_{n-1}+\sum_{k=2}^{n}(-1)^k\,a_{n-k}\,\zeta(k).
\tag{32}
$$

We first evaluate some higher-order derivatives of the Gamma function $\Gamma(z)$ at $z=1$ and $z=\frac{1}{2}$. Upon replacing z by $z-1$ in (31), we readily obtain

$$
a_n=\frac{\Gamma^{(n)}(1)}{n!}\quad(n\in\mathbb{N}_0).
\tag{33}
$$

Formula (32) and (33), together, immediately yield the recursion formula:

$$\Gamma^{(n+1)}(1) = -\gamma\,\Gamma^{(n)}(1) + n!\sum_{k=1}^{n}\frac{(-1)^{k+1}}{(n-k)!}\zeta(k+1)\,\Gamma^{(n-k)}(1) \quad (n \in \mathbb{N}_0), \tag{34}$$

which can alternatively be derived directly from the definition 1.2(52) with $n = 1$ (and z replaced by $z + 1$) by applying such results as the relationship 1.2(53) with $z = 1$.

Similarly, we obtain another recursion formula:

$$\Gamma^{(n+1)}\left(\frac{1}{2}\right) = -\delta\,\Gamma^{(n)}\left(\frac{1}{2}\right) + n!\sum_{k=1}^{n}\frac{(-1)^{k+1}}{(n-k)!}\left(2^{k+1}-1\right)\zeta(k+1)\,\Gamma^{(n-k)}\left(\frac{1}{2}\right) \quad (n \in \mathbb{N}_0;\ \delta := \gamma + 2\log 2). \tag{35}$$

Now, using the formulas in earlier sections, we compute $\Gamma^{(n)}(1)$ explicitly for the first ten derivatives:

$$\Gamma'(1) = -\gamma = \psi(1);$$

$$\Gamma^{(2)}(1) = \gamma^2 + \zeta(2);$$

$$\Gamma^{(3)}(1) = -\gamma^3 - 3\gamma\zeta(2) - 2\zeta(3),$$

which is the *corrected* version of a formula given by Campbell (1966, p. 25);

$$\Gamma^{(4)}(1) = \gamma^4 + 6\gamma^2\zeta(2) + 8\gamma\zeta(3) + \frac{27}{2}\zeta(4);$$

$$\Gamma^{(5)}(1) = -\gamma^5 - 10\gamma^3\zeta(2) - 20\gamma^2\zeta(3) - \frac{135}{2}\gamma\zeta(4) - 20\zeta(2)\zeta(3) - 24\zeta(5);$$

$$\Gamma^{(6)}(1) = \gamma^6 + 15\gamma^4\zeta(2) + 40\gamma^3\zeta(3) + \frac{405}{2}\gamma^2\zeta(4) + 120\gamma\zeta(2)\zeta(3) + 144\gamma\zeta(5) + 40\{\zeta(3)\}^2 + \frac{2745}{8}\zeta(6);$$

$$\Gamma^{(7)}(1) = -\gamma^7 - 21\gamma^5\zeta(2) - 70\gamma^4\zeta(3) - \frac{945}{2}\gamma^3\zeta(4) - 420\gamma^2\zeta(2)\zeta(3) - 504\gamma^2\zeta(5) - 280\gamma\{\zeta(3)\}^2 - \frac{19215}{8}\gamma\zeta(6) - 945\zeta(3)\zeta(4) - 504\zeta(2)\zeta(5) - 720\zeta(7);$$

$$\Gamma^{(8)}(1) = \gamma^8 + 28\gamma^6\zeta(2) + 112\gamma^5\zeta(3) + 945\gamma^4\zeta(4) + 1120\gamma^3\zeta(2)\zeta(3) + 1344\gamma^3\zeta(5) + 1120\gamma^2\{\zeta(3)\}^2 + \frac{19215}{2}\gamma^2\zeta(6) + 7560\gamma\zeta(3)\zeta(4) + 4032\gamma\zeta(2)\zeta(5) + 5760\gamma\zeta(7) + 1120\zeta(2)\{\zeta(3)\}^2 + 2688\zeta(3)\zeta(5) + \frac{132405}{8}\zeta(8);$$

$$\Gamma^{(9)}(1) = -\gamma^9 - 36\gamma^7\zeta(2) - 168\gamma^6\zeta(3) - 1701\gamma^5\zeta(4)$$
$$- 2520\gamma^4\zeta(2)\zeta(3) - 3024\gamma^4\zeta(5) - 3360\gamma^3\{\zeta(3)\}^2 - \frac{57645}{2}\gamma^3\zeta(6)$$
$$- 34020\gamma^2\zeta(3)\zeta(4) - 18144\gamma^2\zeta(2)\zeta(5) - 25920\gamma^2\zeta(7) - 10080\gamma\zeta(2)\{\zeta(3)\}^2$$
$$- 24192\gamma\zeta(3)\zeta(5) - \frac{1191645}{8}\gamma\zeta(8) - 57645\zeta(3)\zeta(6) - 40824\zeta(4)\zeta(5)$$
$$- 25920\zeta(2)\zeta(7) - 2240\{\zeta(3)\}^3 - 40320\zeta(9);$$

$$\Gamma^{(10)}(1)$$
$$= \gamma^{10} + 45\gamma^8\zeta(2) + 240\gamma^7\zeta(3) + 2835\gamma^6\zeta(4) + 5040\gamma^5\zeta(2)\zeta(3)$$
$$+ 6048\gamma^5\zeta(5) + 8400\gamma^4\{\zeta(3)\}^2 + \frac{288225}{4}\gamma^4\zeta(6) + 113400\gamma^3\zeta(3)\zeta(4)$$
$$+ 60480\gamma^3\zeta(2)\zeta(5) + 86400\gamma^3\zeta(7) + 50400\gamma^2\zeta(2)\{\zeta(3)\}^2 + 120960\gamma^2\zeta(3)\zeta(5)$$
$$+ \frac{5958225}{8}\gamma^2\zeta(8) + 576450\gamma\zeta(3)\zeta(6) + 408240\gamma\zeta(4)\zeta(5) + 259200\gamma\zeta(2)\zeta(7)$$
$$+ 22400\gamma\{\zeta(3)\}^3 + 403200\gamma\zeta(9) + 113400\{\zeta(3)\}^2\zeta(4) + 120960\zeta(2)\zeta(3)\zeta(5)$$
$$+ 72576\{\zeta(5)\}^2 + 172800\zeta(3)\zeta(7) + \frac{42329385}{32}\zeta(10).$$

The corresponding problem for $\Gamma^{(n)}\left(\frac{1}{2}\right)$ $(n = 1, \ldots, 10)$ yields

$$\Gamma'\left(\frac{1}{2}\right) = -\delta\sqrt{\pi},$$
$$\frac{1}{\sqrt{\pi}}\Gamma^{(2)}\left(\frac{1}{2}\right) = \delta^2 + 3\zeta(2),$$
$$\frac{1}{\sqrt{\pi}}\Gamma^{(3)}\left(\frac{1}{2}\right) = -\delta^3 - 9\delta\zeta(2) - 14\zeta(3),$$

all three of which are also recorded by Campbell (1966, p. 25);

$$\frac{1}{\sqrt{\pi}}\Gamma^{(4)}\left(\frac{1}{2}\right) = \delta^4 + 18\delta^2\zeta(2) + 56\delta\zeta(3) + \frac{315}{2}\zeta(4);$$
$$\frac{1}{\sqrt{\pi}}\Gamma^{(5)}\left(\frac{1}{2}\right) = -\delta^5 - 30\delta^3\zeta(2) - 140\delta^2\zeta(3)$$
$$- \frac{1575}{2}\delta\zeta(4) - 420\zeta(2)\zeta(3) - 744\zeta(5);$$
$$\frac{1}{\sqrt{\pi}}\Gamma^{(6)}\left(\frac{1}{2}\right) = \delta^6 + 45\delta^4\zeta(2) + 280\delta^3\zeta(3) + \frac{4725}{2}\delta^2\zeta(4)$$
$$+ 2520\delta\zeta(2)\zeta(3) + 4464\delta\zeta(5) + 1960\{\zeta(3)\}^2 + \frac{131355}{8}\zeta(6);$$

$$\frac{1}{\sqrt{\pi}}\Gamma^{(7)}\left(\frac{1}{2}\right) = -\delta^7 - 63\delta^5\zeta(2) - 490\delta^4\zeta(3) - \frac{11025}{2}\delta^3\zeta(4) - 8820\delta^2\zeta(2)\zeta(3)$$
$$- 15624\delta^2\zeta(5) - 13720\delta\{\zeta(3)\}^2 - \frac{919485}{8}\delta\zeta(6)$$
$$- 77175\zeta(3)\zeta(4) - 46872\zeta(2)\zeta(5) - 91440\zeta(7);$$

$$\frac{1}{\sqrt{\pi}}\Gamma^{(8)}\left(\frac{1}{2}\right) = \delta^8 + 84\delta^6\zeta(2) + 784\delta^5\zeta(3) + 11025\delta^4\zeta(4) + 23520\delta^3\zeta(2)\zeta(3) + 41664\delta^3\zeta(5) + 54880\delta^2\{\zeta(3)\}^2 + 377055\delta^2\zeta(6) + 617400\delta\zeta(3)\zeta(4) + 374976\delta\zeta(2)\zeta(5) + 731520\delta\zeta(7) + 164640\zeta(2)\{\zeta(3)\}^2 + 583296\zeta(3)\zeta(5) + \frac{25859925}{8}\zeta(8);$$

$$\frac{1}{\sqrt{\pi}}\Gamma^{(9)}\left(\frac{1}{2}\right) = -\delta^9 - 108\,\delta^7\,\zeta(2) - 1176\,\delta^6\,\zeta(3) - 19845\,\delta^5\,\zeta(4) - 52920\,\delta^4\,\zeta(2)\zeta(3) - 93744\,\delta^4\,\zeta(5) - 164640\,\delta^3\,\{\zeta(3)\}^2 - \frac{2758455}{2}\,\delta^3\,\zeta(6) - 2778300\,\delta^2\,\zeta(3)\zeta(4) - 1687392\,\delta^2\,\zeta(2)\zeta(5) - 3291840\,\delta^2\,\zeta(7) - 1481760\,\delta\,\zeta(2)\,\{\zeta(3)\}^2 - 5249664\,\delta\,\zeta(3)\zeta(5) - \frac{232739325}{8}\,\delta\,\zeta(8) - 19309185\,\zeta(3)\zeta(6) - 14764680\,\zeta(4)\zeta(5) - 768320\,\{\zeta(3)\}^3 - 9875520\,\zeta(2)\zeta(7) - 20603520\,\zeta(9);$$

$$\frac{1}{\sqrt{\pi}}\Gamma^{(10)}\left(\frac{1}{2}\right) = \delta^{10} + 135\,\delta^8\,\zeta(2) + 1680\,\delta^7\,\zeta(3) + 33075\,\delta^6\,\zeta(4) + 105840\,\delta^5\,\zeta(2)\,\zeta(3) + 187488\,\delta^5\,\zeta(5) + 411600\,\delta^4\,\{\zeta(3)\}^2 + \frac{13792275}{4}\,\delta^4\,\zeta(6) + 9261000\,\delta^3\,\zeta(3)\,\zeta(4) + 5624640\,\delta^3\,\zeta(2)\zeta(5) + 10972800\,\delta^3\,\zeta(7) + 7408800\,\delta^2\,\zeta(2)\,\{\zeta(3)\}^2 + 26248320\,\delta^2\,\zeta(3)\zeta(5) + \frac{1163696625}{8}\,\delta^2\,\zeta(8) + 193091850\,\delta\,\zeta(3)\zeta(6) + 147646800\,\delta\,\zeta(4)\zeta(5) + 7683200\,\delta\,\{\zeta(3)\}^3 + 98755200\,\delta\,\zeta(2)\zeta(7) + 206035200\,\delta\,\zeta(9) + 64827000\,\{\zeta(3)\}^2\,\zeta(4) + 78744960\,\zeta(2)\zeta(3)\zeta(5) + 153619200\,\zeta(3)\zeta(7) + 69745536\,\{\zeta(5)\}^2 + \frac{33466588155}{32}\,\zeta(10),$$

δ being defined already with (35).

For applications of our evaluations for $\Gamma^{(n)}(1)$ and $\Gamma^{(n)}\left(\frac{1}{2}\right)$, by replacing z by $z+1$ in 1.1(1), differentiating both sides of the resulting equation n times, and then setting $z=0$, we obtain

$$\Gamma^{(n)}(1) = \int_0^\infty e^{-t}\,(\log t)^n\,dt \quad (n \in \mathbb{N}_0). \tag{36}$$

Similarly, we have

$$\Gamma^{(n)}\left(\frac{1}{2}\right) = \int_0^\infty \frac{e^{-t}}{\sqrt{t}}\,(\log t)^n\,dt \quad (n \in \mathbb{N}_0). \tag{37}$$

Both (36) and (37) are recorded by Campbell (1966, p. 25).

Making use of 1.1(8) instead of 1.1(1) [or, alternatively, by setting $t = \log(1/\tau)$ in (36) and (37)], we obtain

$$\Gamma^{(n)}(1) = \int_0^1 \left[\log\left(\log\frac{1}{t}\right)\right]^n dt \quad (n \in \mathbb{N}_0) \tag{38}$$

and

$$\Gamma^{(n)}\left(\frac{1}{2}\right) = \int_0^1 \left(\log\frac{1}{t}\right)^{-\frac{1}{2}} \left[\log\left(\log\frac{1}{t}\right)\right]^n dt \quad (n \in \mathbb{N}_0). \tag{39}$$

By using the computations presented here, each of the integral formulas (36), (37), (38), and (39) can be expressed explicitly in terms of γ and $\zeta(n)$, at least for $n = 1, \ldots, 10$.

Differentiating both sides of 1.1(66) n times by employing the Leibniz rule for differentiating the Γ-product and setting $\lambda = \frac{1}{2}$ in the resulting equation, we obtain

$$\begin{aligned}\int_0^\infty \frac{(\log t)^n}{(1+t)\sqrt{t}} dt &:= I(n) \\ &= \sum_{k=0}^{n} (-1)^k \binom{n}{k} \Gamma^{(n-k)}\left(\frac{1}{2}\right) \Gamma^{(k)}\left(\frac{1}{2}\right) \quad (n \in \mathbb{N}_0).\end{aligned} \tag{40}$$

It is interesting to observe that

$$\int_0^\infty \frac{(\log t)^{2n+1}}{(1+t)\sqrt{t}} dt = 0 \quad (n \in \mathbb{N}_0), \tag{41}$$

which can also be shown as follows by separating the sum on the right-hand side of (40) into two parts:

$$\begin{aligned}I(2n+1) &= \left(\sum_{k=0}^{n} + \sum_{k=n+1}^{2n+1}\right) (-1)^k \binom{2n+1}{k} \Gamma^{(2n+1-k)}\left(\frac{1}{2}\right) \Gamma^{(k)}\left(\frac{1}{2}\right) \\ &= \sum_{k=0}^{n} (-1)^k \binom{2n+1}{k} \Gamma^{(2n+1-k)}\left(\frac{1}{2}\right) \Gamma^{(k)}\left(\frac{1}{2}\right) \\ &\quad + \sum_{k=0}^{n} (-1)^{n+1-k} \binom{2n+1}{n-k} \Gamma^{(n-k)}\left(\frac{1}{2}\right) \Gamma^{(n+1+k)}\left(\frac{1}{2}\right),\end{aligned}$$

which, upon reversing the order of the latter sum, yields

$$\begin{aligned}I(2n+1) &= \{1 + (-1)^{2n+1}\} \sum_{k=0}^{n} (-1)^k \binom{2n+1}{k} \Gamma^{(2n+1-k)}\left(\frac{1}{2}\right) \Gamma^{(k)}\left(\frac{1}{2}\right) \\ &= 0.\end{aligned} \tag{42}$$

On the other hand, in the case of even integers, we have

$$\begin{aligned}I(2n) &= \int_0^\infty \frac{(\log t)^{2n}}{(1+t)\sqrt{t}} dt \\ &= 2\sum_{k=0}^{n-1} (-1)^k \binom{2n}{k} \Gamma^{(2n-k)}\left(\frac{1}{2}\right) \Gamma^{(k)}\left(\frac{1}{2}\right) \\ &\quad + (-1)^n \binom{2n}{n} \left[\Gamma^{(n)}\left(\frac{1}{2}\right)\right]^2 \quad (n \in \mathbb{N}_0),\end{aligned} \tag{43}$$

which, in view of the results already presented here, readily yields the following special cases:

(44) $$\int_0^\infty \frac{(\log t)^2}{(1+t)\sqrt{t}}dt = \pi^3,$$

(45) $$\int_0^\infty \frac{(\log t)^4}{(1+t)\sqrt{t}}dt = 5\,\pi^5,$$

(46) $$\int_0^\infty \frac{(\log t)^6}{(1+t)\sqrt{t}}dt = 46\,\pi^7,$$

(47) $$\int_0^\infty \frac{(\log t)^8}{(1+t)\sqrt{t}}dt = 1385\,\pi^9,$$

and

(48) $$\int_0^\infty \frac{(\log t)^{10}}{(1+t)\sqrt{t}}dt = 50521\,\pi^{11}.$$

We conclude by remarking that the evaluations given here may find further applications involving (for example) computation and evaluation of several families of definite integrals.

Problems

1. Evaluate the following special values of the G-function given in Section 1.3:

$$G\left(\tfrac{1}{3}\right) = \exp\left[\frac{1}{9} + \frac{\pi\sqrt{3}}{54} - \frac{\sqrt{3}\,\zeta\left(2,\frac{1}{3}\right)}{36\pi}\right] \cdot 3^{\frac{1}{72}} \cdot A^{-\frac{4}{3}} \cdot \left\{\Gamma\left(\tfrac{1}{3}\right)\right\}^{-\frac{2}{3}};$$

$$G\left(\tfrac{2}{3}\right) = \exp\left[\frac{1}{9} - \frac{\pi\sqrt{3}}{54} + \frac{\sqrt{3}\,\zeta\left(2,\frac{1}{3}\right)}{36\pi}\right] \cdot 2^{-\frac{1}{3}} \cdot 3^{\frac{13}{72}} \cdot \pi^{-\frac{1}{3}} \cdot A^{-\frac{4}{3}} \cdot \left\{\Gamma\left(\tfrac{1}{3}\right)\right\}^{\frac{1}{3}};$$

$$G\left(\tfrac{1}{6}\right) = \exp\left[\frac{5}{72} + \frac{\pi\sqrt{3}}{36} - \frac{\sqrt{3}\,\zeta\left(2,\frac{1}{3}\right)}{24\pi}\right] \cdot 2^{\frac{19}{72}} \cdot 3^{-\frac{61}{144}} \cdot \pi^{\frac{5}{12}} \cdot A^{-\frac{5}{6}} \cdot \left\{\Gamma\left(\tfrac{1}{3}\right)\right\}^{-\frac{5}{3}};$$

$$G\left(\tfrac{5}{6}\right) = \exp\left[\frac{5}{72} - \frac{\pi\sqrt{3}}{36} + \frac{\sqrt{3}\,\zeta\left(2,\frac{1}{3}\right)}{24\pi}\right] \cdot 2^{-\frac{17}{72}} \cdot 3^{\frac{11}{144}} \cdot \pi^{-\frac{1}{4}} \cdot A^{-\frac{5}{6}} \cdot \left\{\Gamma\left(\tfrac{1}{3}\right)\right\}^{\frac{1}{3}}.$$

[Gosper (1997); Choi and Srivastava (1999a and 2000d)]

2. Applying the results of Problem 1, derive the following closed-form evaluations of series involving the Riemann Zeta function:

$$\sum_{k=2}^{\infty}(-1)^k\frac{\zeta(k)}{(k+1)\cdot 3^k}=1+\frac{\gamma}{6}+\frac{\pi\sqrt{3}}{18}-\frac{\sqrt{3}}{12\pi}\zeta\left(2,\frac{1}{3}\right)+\log\left(2^{-\frac{1}{2}}\cdot 3^{\frac{1}{24}}\cdot\pi^{-\frac{1}{2}}\cdot A^{-4}\cdot\Gamma\left(\tfrac{1}{3}\right)\right);$$

$$\sum_{k=2}^{\infty}(-1)^k\frac{\zeta(k)-1}{(k+1)\cdot 3^k}=\frac{11}{6}+\frac{\gamma}{6}+\frac{\pi\sqrt{3}}{18}-\frac{\sqrt{3}}{12\pi}\zeta\left(2,\frac{1}{3}\right)+\log\left(2^{-\frac{13}{2}}\cdot 3^{\frac{73}{24}}\cdot\pi^{-\frac{1}{2}}\cdot A^{-4}\cdot\Gamma\left(\tfrac{1}{3}\right)\right);$$

$$\sum_{k=2}^{\infty}\frac{\zeta(k)}{(k+1)\cdot 3^k}=-\frac{\gamma}{6}+\frac{\pi\sqrt{3}}{18}-\frac{\sqrt{3}}{12\pi}\zeta\left(2,\frac{1}{3}\right)+\log\left(2^{\frac{1}{2}}\cdot 3^{-\frac{13}{24}}\cdot\pi^{\frac{1}{2}}\cdot A^{4}\cdot\left\{\Gamma\left(\tfrac{1}{3}\right)\right\}^{-1}\right);$$

$$\sum_{k=2}^{\infty}\frac{\zeta(k)-1}{(k+1)\cdot 3^k}=\frac{7}{6}-\frac{\gamma}{6}+\frac{\pi\sqrt{3}}{18}-\frac{\sqrt{3}}{12\pi}\zeta\left(2,\frac{1}{3}\right)+\log\left(2^{\frac{7}{2}}\cdot 3^{-\frac{85}{24}}\cdot\pi^{\frac{1}{2}}\cdot A^{4}\cdot\left\{\Gamma\left(\tfrac{1}{3}\right)\right\}^{-1}\right);$$

$$\sum_{k=1}^{\infty}\frac{\zeta(2k)}{(2k+1)\cdot 3^{2k}}=\frac{1}{2}+\frac{\pi\sqrt{3}}{18}-\frac{\sqrt{3}}{12\pi}\zeta\left(2,\frac{1}{3}\right)-\frac{1}{4}\log 3;$$

$$\sum_{k=1}^{\infty}\frac{\zeta(2k)-1}{(2k+1)\cdot 3^{2k}}=\frac{11}{6}+\frac{\pi\sqrt{3}}{18}-\frac{\sqrt{3}}{12\pi}\zeta\left(2,\frac{1}{3}\right)+\log\left(2^{-\frac{3}{2}}\cdot 3^{-\frac{1}{4}}\right);$$

$$\sum_{k=1}^{\infty}\frac{\zeta(2k+1)}{(k+1)\cdot 3^{2k}}=-3-\gamma+\log\left(2^{3}\cdot 3^{-\frac{7}{4}}\cdot\pi^{3}\cdot A^{24}\cdot\left\{\Gamma\left(\tfrac{1}{3}\right)\right\}^{-6}\right);$$

$$\sum_{k=1}^{\infty}\frac{\zeta(2k+1)-1}{(k+1)\cdot 3^{2k}}=-2-\gamma+\log\left(2^{30}\cdot 3^{-\frac{79}{4}}\cdot\pi^{3}\cdot A^{24}\cdot\left\{\Gamma\left(\tfrac{1}{3}\right)\right\}^{-6}\right);$$

$$\sum_{k=2}^{\infty}(-1)^k\frac{\zeta(k)}{k+1}\left(\frac{2}{3}\right)^k=1+\frac{\gamma}{3}-\frac{\pi\sqrt{3}}{36}+\frac{\sqrt{3}}{24\pi}\zeta\left(2,\frac{1}{3}\right)+\log\left(2^{\frac{1}{2}}\cdot 3^{-\frac{23}{48}}\cdot\pi^{\frac{1}{2}}\cdot A^{-2}\cdot\left\{\Gamma\left(\tfrac{1}{3}\right)\right\}^{-1}\right);$$

$$\sum_{k=2}^{\infty}(-1)^k\frac{\zeta(k)-1}{k+1}\left(\frac{2}{3}\right)^k=\frac{5}{3}+\frac{\gamma}{3}-\frac{\pi\sqrt{3}}{36}+\frac{\sqrt{3}}{24\pi}\zeta\left(2,\frac{1}{3}\right)+\log\left(2^{\frac{1}{2}}\cdot 3^{\frac{49}{48}}\cdot 5^{-\frac{3}{2}}\cdot\pi^{\frac{1}{2}}\cdot A^{-2}\cdot\left\{\Gamma\left(\tfrac{1}{3}\right)\right\}^{-1}\right);$$

$$\sum_{k=2}^{\infty}\frac{\zeta(k)}{k+1}\left(\frac{2}{3}\right)^k=-\frac{\gamma}{3}-\frac{\pi\sqrt{3}}{36}+\frac{\sqrt{3}}{24\pi}\zeta\left(2,\frac{1}{3}\right)+\log\left(2^{-\frac{1}{2}}\cdot 3^{-\frac{1}{48}}\cdot\pi^{-\frac{1}{2}}\cdot A^{2}\cdot\Gamma\left(\tfrac{1}{3}\right)\right);$$

$$\sum_{k=2}^{\infty}\frac{\zeta(k)-1}{k+1}\left(\frac{2}{3}\right)^{k}=\frac{4}{3}-\frac{\gamma}{3}-\frac{\pi\sqrt{3}}{36}+\frac{\sqrt{3}}{24\pi}\zeta\left(2,\frac{1}{3}\right)+\log\left(2^{-\frac{1}{2}}\cdot 3^{-\frac{73}{48}}\cdot\pi^{-\frac{1}{2}}\cdot A^{2}\cdot\Gamma\left(\tfrac{1}{3}\right)\right);$$

$$\sum_{k=1}^{\infty}\frac{\zeta(2k)}{2k+1}\left(\frac{2}{3}\right)^{2k}=\frac{1}{2}-\frac{\pi\sqrt{3}}{36}+\frac{\sqrt{3}}{24\pi}\zeta\left(2,\frac{1}{3}\right)-\frac{1}{4}\log 3;$$

$$\sum_{k=1}^{\infty}\frac{\zeta(2k)-1}{2k+1}\left(\frac{2}{3}\right)^{2k}=\frac{3}{2}-\frac{\pi\sqrt{3}}{36}+\frac{\sqrt{3}}{24\pi}\zeta\left(2,\frac{1}{3}\right)+\log\left(3^{-\frac{1}{4}}\cdot 5^{-\frac{3}{4}}\right);$$

$$\sum_{k=1}^{\infty}\frac{\zeta(2k+1)}{k+1}\left(\frac{2}{3}\right)^{2k}=-\frac{3}{2}-\gamma+\log\left(2^{-\frac{3}{2}}\cdot 3^{\frac{11}{16}}\cdot\pi^{-\frac{3}{2}}\cdot A^{6}\cdot\left\{\Gamma\left(\tfrac{1}{3}\right)\right\}^{3}\right);$$

$$\sum_{k=1}^{\infty}\frac{\zeta(2k+1)-1}{k+1}\left(\frac{2}{3}\right)^{2k}=-\frac{1}{2}-\gamma+\log\left(2^{-\frac{3}{2}}\cdot 3^{-\frac{61}{16}}\cdot 5^{\frac{9}{4}}\cdot\pi^{-\frac{3}{2}}\cdot A^{6}\cdot\left\{\Gamma\left(\tfrac{1}{3}\right)\right\}^{3}\right);$$

$$\sum_{k=2}^{\infty}(-1)^{k}\frac{\zeta(k)}{(k+1)\cdot 6^{k}}=1+\frac{\gamma}{12}+\frac{\pi\sqrt{3}}{6}-\frac{\sqrt{3}}{4\pi}\zeta\left(2,\frac{1}{3}\right)+\log\left(2^{-\frac{11}{12}}\cdot 3^{\frac{11}{24}}\cdot\pi^{-1}\cdot A^{-5}\cdot\left\{\Gamma\left(\tfrac{1}{3}\right)\right\}^{2}\right);$$

$$\sum_{k=2}^{\infty}(-1)^{k}\frac{\zeta(k)-1}{(k+1)\cdot 6^{k}}=\frac{23}{12}+\frac{\gamma}{12}+\frac{\pi\sqrt{3}}{6}-\frac{\sqrt{3}}{4\pi}\zeta\left(2,\frac{1}{3}\right)+\log\left(2^{\frac{61}{12}}\cdot 3^{\frac{155}{24}}\cdot 7^{-6}\cdot\pi^{-1}\cdot A^{-5}\cdot\left\{\Gamma\left(\tfrac{1}{3}\right)\right\}^{2}\right);$$

$$\sum_{k=2}^{\infty}\frac{\zeta(k)}{(k+1)\cdot 6^{k}}=-\frac{\gamma}{12}+\frac{\pi\sqrt{3}}{6}-\frac{\sqrt{3}}{4\pi}\zeta\left(2,\frac{1}{3}\right)+\log\left(2^{\frac{11}{12}}\cdot 3^{-\frac{11}{24}}\cdot\pi\cdot A^{5}\cdot\left\{\Gamma\left(\tfrac{1}{3}\right)\right\}^{-2}\right);$$

$$\sum_{k=2}^{\infty}\frac{\zeta(k)-1}{(k+1)\cdot 6^{k}}=\frac{13}{12}-\frac{\gamma}{12}+\frac{\pi\sqrt{3}}{6}-\frac{\sqrt{3}}{4\pi}\zeta\left(2,\frac{1}{3}\right)+\log\left(2^{-\frac{61}{12}}\cdot 3^{-\frac{155}{24}}\cdot 5^{6}\cdot\pi\cdot A^{5}\cdot\left\{\Gamma\left(\tfrac{1}{3}\right)\right\}^{-2}\right);$$

$$\sum_{k=1}^{\infty}\frac{\zeta(2k)}{(2k+1)\cdot 6^{2k}}=\frac{1}{2}+\frac{\pi\sqrt{3}}{6}-\frac{\sqrt{3}}{4\pi}\zeta\left(2,\frac{1}{3}\right);$$

$$\sum_{k=1}^{\infty}\frac{\zeta(2k)-1}{(2k+1)\cdot 6^{2k}}=\frac{3}{2}+\frac{\pi\sqrt{3}}{6}-\frac{\sqrt{3}}{4\pi}\zeta\left(2,\frac{1}{3}\right)+\log\left(5^{3}\cdot 7^{-3}\right);$$

$$\sum_{k=1}^{\infty}\frac{\zeta(2k+1)}{(k+1)\cdot 6^{2k}}=-6-\gamma+\log\left(2^{11}\cdot 3^{-\frac{11}{2}}\cdot\pi^{12}\cdot A^{60}\cdot\left\{\Gamma\left(\tfrac{1}{3}\right)\right\}^{-24}\right);$$

$$\sum_{k=1}^{\infty} \frac{\zeta(2k+1)-1}{(k+1)\cdot 6^{2k}} = -5-\gamma + \log\left(2^{-61}\cdot 3^{-\frac{155}{2}}\cdot 5^{36}\cdot 7^{36}\cdot \pi^{12}\cdot A^{60}\cdot \left\{\Gamma\left(\tfrac{1}{3}\right)\right\}^{-24}\right);$$

$$\sum_{k=2}^{\infty} (-1)^k \frac{\zeta(k)}{k+1}\left(\frac{5}{6}\right)^k = 1+\frac{5}{12}\gamma - \frac{\pi\sqrt{3}}{30} + \frac{\sqrt{3}}{20\pi}\zeta\left(2,\frac{1}{3}\right) + \log\left(2^{\frac{49}{60}}\cdot 3^{-\frac{61}{120}}\cdot \pi\cdot A^{-1}\cdot \left\{\Gamma\left(\tfrac{1}{3}\right)\right\}^{-2}\right);$$

$$\sum_{k=2}^{\infty} (-1)^k \frac{\zeta(k)-1}{k+1}\left(\frac{5}{6}\right)^k = \frac{19}{12}+\frac{5}{12}\gamma - \frac{\pi\sqrt{3}}{30} + \frac{\sqrt{3}}{20\pi}\zeta\left(2,\frac{1}{3}\right) + \log\left(2^{\frac{121}{60}}\cdot 3^{\frac{83}{120}}\cdot 11^{-\frac{6}{5}}\cdot \pi\cdot A^{-1}\cdot \left\{\Gamma\left(\tfrac{1}{3}\right)\right\}^{-2}\right);$$

$$\sum_{k=2}^{\infty} \frac{\zeta(k)}{k+1}\left(\frac{5}{6}\right)^k = -\frac{5}{12}\gamma - \frac{\pi\sqrt{3}}{30} + \frac{\sqrt{3}}{20\pi}\zeta\left(2,\frac{1}{3}\right) + \log\left(2^{-\frac{49}{60}}\cdot 3^{\frac{61}{120}}\cdot \pi^{-1}\cdot A\cdot \left\{\Gamma\left(\tfrac{1}{3}\right)\right\}^{2}\right);$$

$$\sum_{k=2}^{\infty} \frac{\zeta(k)-1}{k+1}\left(\frac{5}{6}\right)^k = \frac{17}{12}-\frac{5}{12}\gamma - \frac{\pi\sqrt{3}}{30} + \frac{\sqrt{3}}{20\pi}\zeta\left(2,\frac{1}{3}\right) + \log\left(2^{-\frac{121}{60}}\cdot 3^{-\frac{83}{120}}\cdot \pi^{-1}\cdot A\cdot \left\{\Gamma\left(\tfrac{1}{3}\right)\right\}^{2}\right);$$

$$\sum_{k=1}^{\infty} \frac{\zeta(2k)}{2k+1}\left(\frac{5}{6}\right)^{2k} = \frac{1}{2} - \frac{\pi\sqrt{3}}{30} + \frac{\sqrt{3}}{20\pi}\zeta\left(2,\frac{1}{3}\right);$$

$$\sum_{k=1}^{\infty} \frac{\zeta(2k)-1}{2k+1}\left(\frac{5}{6}\right)^{2k} = \frac{3}{2} - \frac{\pi\sqrt{3}}{30} + \frac{\sqrt{3}}{20\pi}\zeta\left(2,\frac{1}{3}\right) - \frac{3}{5}\log 11;$$

$$\sum_{k=1}^{\infty} \frac{\zeta(2k+1)}{k+1}\left(\frac{5}{6}\right)^{2k} = -\frac{6}{5}-\gamma+\log\left(2^{-\frac{49}{25}}\cdot 3^{\frac{61}{50}}\cdot \pi^{-\frac{12}{5}}\cdot A^{\frac{12}{5}}\cdot \left\{\Gamma\left(\tfrac{1}{3}\right)\right\}^{\frac{24}{5}}\right);$$

$$\sum_{k=1}^{\infty} \frac{\zeta(2k+1)-1}{k+1}\left(\frac{5}{6}\right)^{2k} = -\frac{1}{5}-\gamma + \log\left(2^{-\frac{121}{25}}\cdot 3^{-\frac{83}{50}}\cdot 11^{\frac{36}{25}}\cdot \pi^{-\frac{12}{5}}\cdot A^{\frac{12}{5}}\cdot \left\{\Gamma\left(\tfrac{1}{3}\right)\right\}^{\frac{24}{5}}\right);$$

$$\sum_{k=2}^{\infty} (-1)^k \frac{\zeta(k)-1}{k+1}\left(\frac{4}{3}\right)^k = \frac{19}{12}+\frac{2}{3}\gamma + \frac{\pi\sqrt{3}}{72} - \frac{\sqrt{3}}{48\pi}\zeta\left(2,\frac{1}{3}\right) + \log\left(2^{-\frac{1}{2}}\cdot 3^{\frac{1}{96}}\cdot 7^{-\frac{3}{4}}\cdot \pi^{-\frac{1}{2}}\cdot A^{-1}\cdot \Gamma\left(\tfrac{1}{3}\right)\right);$$

$$\sum_{k=2}^{\infty} \frac{\zeta(k)-1}{k+1}\left(\frac{4}{3}\right)^k = \frac{17}{12}-\frac{2}{3}\gamma + \frac{\pi\sqrt{3}}{72} - \frac{\sqrt{3}}{48\pi}\zeta\left(2,\frac{1}{3}\right) + \log\left(2^{\frac{1}{2}}\cdot 3^{-\frac{49}{96}}\cdot \pi^{\frac{1}{2}}\cdot A\cdot \left\{\Gamma\left(\tfrac{1}{3}\right)\right\}^{-1}\right);$$

$$\sum_{k=1}^{\infty} \frac{\zeta(2k)-1}{2k+1}\left(\frac{4}{3}\right)^{2k} = \frac{3}{2} + \frac{\pi\sqrt{3}}{72} - \frac{\sqrt{3}}{48\pi}\zeta\left(2,\frac{1}{3}\right) + \log\left(3^{-\frac{1}{4}}\cdot 7^{-\frac{3}{8}}\right);$$

$$\sum_{k=1}^{\infty} \frac{\zeta(2k+1)-1}{k+1}\left(\frac{4}{3}\right)^{2k} = -\frac{1}{8} - \gamma + \log\left(2^{\frac{3}{4}}\cdot 3^{-\frac{25}{64}}\cdot 7^{\frac{9}{16}}\cdot \pi^{\frac{3}{4}}\cdot A^{\frac{3}{2}}\cdot \left\{\Gamma\left(\tfrac{1}{3}\right)\right\}^{-\frac{3}{2}}\right);$$

$$\sum_{k=2}^{\infty} (-1)^k \frac{\zeta(k)-1}{k+1}\left(\frac{5}{3}\right)^{k} = \frac{47}{30} + \frac{5}{6}\gamma - \frac{\pi\sqrt{3}}{90} + \frac{\sqrt{3}}{60\pi}\zeta\left(2,\frac{1}{3}\right) + \log\left(2^{-\frac{7}{10}}\cdot 3^{-\frac{59}{120}}\cdot \pi^{\frac{1}{2}}\cdot A^{-\frac{4}{5}}\cdot \left\{\Gamma\left(\tfrac{1}{3}\right)\right\}^{-1}\right);$$

$$\sum_{k=2}^{\infty} \frac{\zeta(k)-1}{k+1}\left(\frac{5}{3}\right)^{k} = \frac{43}{30} - \frac{5}{6}\gamma - \frac{\pi\sqrt{3}}{90} + \frac{\sqrt{3}}{60\pi}\zeta\left(2,\frac{1}{3}\right) + \log\left(2^{-\frac{1}{2}}\cdot 3^{-\frac{1}{120}}\cdot \pi^{-\frac{1}{2}}\cdot A^{\frac{4}{5}}\cdot \Gamma\left(\tfrac{1}{3}\right)\right);$$

$$\sum_{k=1}^{\infty} \frac{\zeta(2k)-1}{2k+1}\left(\frac{5}{3}\right)^{2k} = \frac{3}{2} - \frac{\pi\sqrt{3}}{90} + \frac{\sqrt{3}}{60\pi}\zeta\left(2,\frac{1}{3}\right) + \log\left(2^{-\frac{3}{5}}\cdot 3^{-\frac{1}{4}}\right);$$

$$\sum_{k=1}^{\infty} \frac{\zeta(2k+1)-1}{k+1}\left(\frac{5}{3}\right)^{2k} = -\frac{2}{25} - \gamma + \log\left(2^{\frac{3}{25}}\cdot 3^{\frac{29}{100}}\cdot \pi^{-\frac{3}{5}}\cdot A^{\frac{24}{25}}\cdot \left\{\Gamma\left(\tfrac{1}{3}\right)\right\}^{\frac{6}{5}}\right);$$

$$\sum_{k=2}^{\infty} (-1)^k \frac{\zeta(k)-1}{k+1}\left(\frac{7}{6}\right)^{k} = \frac{131}{84} + \frac{7}{12}\gamma + \frac{\pi\sqrt{3}}{42} - \frac{\sqrt{3}}{28\pi}\zeta\left(2,\frac{1}{3}\right) + \log\left(2^{-\frac{71}{84}}\cdot 3^{\frac{83}{168}}\cdot 13^{-\frac{6}{7}}\cdot \pi^{-1}\cdot A^{-\frac{5}{7}}\cdot \left\{\Gamma\left(\tfrac{1}{3}\right)\right\}^{2}\right);$$

$$\sum_{k=2}^{\infty} \frac{\zeta(k)-1}{k+1}\left(\frac{7}{6}\right)^{k} = \frac{121}{84} - \frac{7}{12}\gamma + \frac{\pi\sqrt{3}}{42} - \frac{\sqrt{3}}{28\pi}\zeta\left(2,\frac{1}{3}\right) + \log\left(2^{\frac{71}{84}}\cdot 3^{-\frac{83}{168}}\cdot \pi\cdot A^{\frac{5}{7}}\cdot \left\{\Gamma\left(\tfrac{1}{3}\right)\right\}^{-2}\right);$$

$$\sum_{k=1}^{\infty} \frac{\zeta(2k)-1}{2k+1}\left(\frac{7}{6}\right)^{2k} = \frac{3}{2} + \frac{\pi\sqrt{3}}{42} - \frac{\sqrt{3}}{28\pi}\zeta\left(2,\frac{1}{3}\right) - \frac{3}{7}\log 13;$$

$$\sum_{k=1}^{\infty} \frac{\zeta(2k+1)-1}{k+1}\left(\frac{7}{6}\right)^{2k} = -\frac{5}{49} - \gamma + \log\left(2^{\frac{71}{49}}\cdot 3^{-\frac{83}{98}}\cdot \pi^{\frac{12}{7}}\cdot A^{\frac{60}{49}}\cdot \left\{\Gamma\left(\tfrac{1}{3}\right)\right\}^{-\frac{24}{7}}\right);$$

$$\sum_{k=2}^{\infty} (-1)^k \frac{\zeta(k)-1}{k+1}\left(\frac{11}{6}\right)^{k} = \frac{203}{132} + \frac{11}{12}\gamma - \frac{\pi\sqrt{3}}{66} + \frac{\sqrt{3}}{44\pi}\zeta\left(2,\frac{1}{3}\right) + \log\left(2^{\frac{109}{132}}\cdot 3^{-\frac{133}{264}}\cdot 5^{\frac{6}{11}}\cdot 17^{-\frac{6}{11}}\cdot \pi\cdot A^{-\frac{5}{11}}\cdot \left\{\Gamma\left(\tfrac{1}{3}\right)\right\}^{-2}\right);$$

$$\sum_{k=2}^{\infty} \frac{\zeta(k)-1}{k+1}\left(\frac{11}{6}\right)^{k} = \frac{193}{132} - \frac{11}{12}\gamma - \frac{\pi\sqrt{3}}{66} + \frac{\sqrt{3}}{44\pi}\zeta\left(2,\frac{1}{3}\right)$$
$$+ \log\left(2^{-\frac{109}{132}} \cdot 3^{\frac{133}{264}} \cdot \pi^{-1} \cdot A^{\frac{5}{11}} \cdot \left\{\Gamma\left(\tfrac{1}{3}\right)\right\}^{2}\right);$$

$$\sum_{k=1}^{\infty} \frac{\zeta(2k)-1}{2k+1}\left(\frac{11}{6}\right)^{2k} = \frac{3}{2} - \frac{\pi\sqrt{3}}{66} + \frac{\sqrt{3}}{44\pi}\zeta\left(2,\frac{1}{3}\right) + \log\left(5^{\frac{3}{11}} \cdot 17^{-\frac{3}{11}}\right);$$

$$\sum_{k=1}^{\infty} \frac{\zeta(2k+1)-1}{k+1}\left(\frac{11}{6}\right)^{2k} = -\frac{5}{121} - \gamma$$
$$+ \log\left(2^{-\frac{109}{121}} \cdot 3^{\frac{133}{242}} \cdot 5^{-\frac{36}{121}} \cdot 17^{\frac{36}{121}} \cdot \pi^{-\frac{12}{11}} \cdot A^{\frac{60}{121}} \cdot \left\{\Gamma\left(\tfrac{1}{3}\right)\right\}^{\frac{24}{11}}\right).$$

[Choi and Srivastava (1999a and 2000d)]

3. Prove that

$$\sum_{m=2}^{\infty} \frac{\zeta(m,a)}{m+\lambda} z^{m+\lambda} = \sum_{k=0}^{\lambda} \binom{\lambda}{k} \zeta'(-k, a-z)\, z^{\lambda-k} - \zeta'(-k,a)$$
$$- \sum_{\ell=0}^{\lambda-1} \frac{\zeta(-\ell,a)}{\lambda-\ell} z^{\lambda-\ell} - \frac{z^{\lambda+1}}{\lambda+1}\left[\psi(\lambda+1) - \psi(a) + \gamma\right] \quad (\lambda \in \mathbb{N}_0;\ |z| < |a|).$$

[Kanemitsu *et al.* (2000)]

4. Prove the following identity:

$$\sum_{k=2}^{\infty} \frac{(-1)^k}{(k+1)(k+2)} \zeta(k) = \frac{\gamma}{6} + \frac{1}{2} + \frac{5}{18}\log 2 + \frac{1}{3}\log\pi$$
$$- \frac{4}{3}\int_0^{\frac{1}{2}} \log\Gamma(t)\, dt,$$

where γ denotes the Euler-Mascheroni constant defined by 1.1(2).

[Janos and Srivastava (1999)]

5. Prove the following companion of the summation formulas deduced in Problems 46 and 47 (Chapter 1):

$$\zeta(k-1) = \sum_{n=k}^{\infty} \frac{(-1)^{n+k}}{(n-1)(n-2)\cdot(n-1)!} s(n,k).$$

[*Cf.* Jordan (1965, p. 339); see also Equation 3.5(16) *and* Hansen (1975, p. 348)]

6. For $\Phi(z,s,a)$ defined by 2.5(1), derive the following generalization of the expansion formula 3.2(7):

$$\sum_{n=0}^{\infty} \frac{(\lambda)_n}{n!} \Phi(z, \lambda+n, a) t^n = \Phi(z, \lambda, a-t) \quad (|t| < |a|;\ \lambda \neq 1).$$

[*Cf.* Equation 2.5(33)]

CHAPTER 4

EVALUATIONS AND SERIES REPRESENTATIONS

Based essentially upon some of the most recent works on the subject by Srivastava (1999a and 2000a), Srivastava and Tsumura (2000a), and others, this chapter aims at investigating rather systematically several interesting evaluations and representations of the Riemann Zeta function $\zeta(s)$ when $s \in \mathbb{N} \setminus \{1\}$. We begin by presenting some of the methods of evaluation of $\zeta(2n)$ $(n \in \mathbb{N})$. We then proceed to develop various families of rapidly convergent series representations for $\zeta(2n+1)$ $(n \in \mathbb{N})$. Finally, in one of many computationally useful special cases considered here, we observe that $\zeta(3)$ can be represented by means of a series which converges much more rapidly than that in Euler's celebrated formula 3.1(8) as well as the series 4.2(2) used by Apéry (1979) in his proof of the irrationality of $\zeta(3)$. Symbolic and numerical computations using *Mathematica* (Version 4.0) for Linux show, among other things, that only 50 terms of this series are capable of producing an accuracy of seven decimal places.

4.1 Evaluation of $\zeta(2n)$

The solution of the so-called *Basler problem* [*cf.*, *e.g.*, Spiess (1955, p. 66)]:

$$\zeta(2) = \sum_{k=1}^{\infty} \frac{1}{k^2} = \frac{\pi^2}{6} \tag{1}$$

was first found in 1736 by Leonhard Euler (1707-1783), although Jakob Bernoulli (1654-1705) and Johann Bernoulli (1667-1748) did their utmost to sum the series in (1). In fact, the former of these Bernoulli brothers did not live to see the solution of the problem, and the solution became known to the latter soon after Euler found it [see, for details, Knopp (1951, p. 238)].

The Basler problem (1) has been solved in the mathematical literature in many different ways. In addition to numerous papers containing elementary proofs of (1), many of which are referenced by Stark (1974), there are a fairly large number of books on complex analysis and advanced calculus in which (1) is proven by using Cauchy's residue calculus, Weierstrass' product theorem, Parseval's theorem, Fourier series expansions, and so on. Some of these books were referred to by Choe (1987) who seems to have rediscovered one of Euler's remarkably elementary proofs of (1) [*cf.*, *e.g.*, Ayoub (1974, p. 1079)], which is based upon the integration of the following Taylor series expansion of $\arcsin x$ near $x = 0$:

$$\arcsin x = x + \sum_{k=1}^{\infty} \frac{1 \cdot 3 \cdot 5 \cdots (2k-1)}{2 \cdot 4 \cdot 6 \cdots (2k)} \frac{x^{2k+1}}{2k+1} \qquad (-1 \leq x \leq 1) \tag{2}$$

with respect to t (where $x = \sin t$), by means of Wallis' integral formula in the form:

$$\int_0^{\pi/2} \sin^{2k+1} t \, dt = \frac{2 \cdot 4 \cdot 6 \cdots (2k)}{1 \cdot 3 \cdot 5 \cdots (2k+1)} \qquad (k \in \mathbb{N}). \tag{3}$$

Motivated largely by the aforementioned (Euler's) proof of (1) detailed by Choe (1987), Choi and Rathie (1997) gave an essentially analogous derivation of (1) by appealing to the Gauss summation theorem 1.4(7). Subsequently, Choi, Rathie, and Srivastava (1999) showed how the derivation of (1) by Choi and Rathie (1997) can be accomplished *without* using the Gauss summation theorem 1.4(7). They also presented several other evaluations of $\zeta(2)$ and related sums including (for example) a fairly straightforward evaluation based upon the theory of hypergeometric series.

Elementary and hypergeometric evaluations of $\zeta(2)$

Starting from the elementary integral:

$$\frac{1}{2}(\arcsin x)^2 = \int_0^x \frac{\arcsin t}{\sqrt{1-t^2}}\,dt \tag{4}$$

and replacing $\arcsin t$ by its Taylor series expansion given by (2), it is not difficult to obtain the representation:

$$\frac{1}{2}(\arcsin x)^2 = \sum_{k=0}^{\infty} \binom{2k}{k} \frac{2^{-2k}}{2k+1} \int_0^x \frac{t^{2k+1}}{\sqrt{1-t^2}}\,dt \qquad (-1 \leq x \leq 1) \tag{5}$$

or, equivalently,

$$\frac{1}{2}(\arcsin x)^2 = \sum_{k=0}^{\infty} \frac{\left(\frac{1}{2}\right)_k}{k!} \frac{1}{2k+1} \int_0^x \frac{t^{2k+1}}{\sqrt{1-t^2}}\,dt \qquad (-1 \leq x \leq 1). \tag{6}$$

Choi and Rathie (1997, p. 396) express the second member of (6) as a hypergeometric ${}_2F_1$ series by using the binomial expansion of $\left(1-t^2\right)^{-1/2}$ in order to evaluate the integral, set $x = 1$, and then apply the Gauss summation theorem 1.4(7).

The use of the Gauss summation theorem 1.4(7) in the above derivation of (1) by Choi and Rathie (1997) can easily be avoided. Indeed, in the special case of (6) when $x = 1$, if we set $t = \sqrt{u}$ and apply the known integral formula 1.1(59) and 1.1(62), we readily find from (6) that

$$\sum_{k=0}^{\infty} \frac{1}{(2k+1)^2} = \frac{\pi^2}{8}, \tag{7}$$

which, in view of 2.3(1), is equivalent to (1).

By virtue of the relationships 2.5(26) and

$$\Phi\left(1, s, \frac{1}{2}\right) = \zeta\left(s, \frac{1}{2}\right) = (2^s - 1)\,\zeta(s), \tag{8}$$

2.5(17) immediately yields the hypergeometric representations:

$$\zeta(n, a) = a^{-n}\ {}_{n+1}F_n\,(1, a, \cdots, a;\ a+1, \cdots, a+1;\ 1) \quad (n \in \mathbb{N} \setminus \{1\}) \tag{9}$$

and

$$(10)\qquad \zeta(n) = \left(\frac{2^n}{2^n-1}\right) {}_{n+1}F_n\left(1, \frac{1}{2}, \cdots, \frac{1}{2};\ \tfrac{3}{2}, \cdots, \tfrac{3}{2};\ 1\right) \qquad (n \in \mathbb{N} \setminus \{1\}).$$

In particular, by setting $n = 2$ in (10), we have

$$(11)\qquad \zeta(2) = \frac{4}{3}\ {}_3F_2\left(1, \frac{1}{2}, \frac{1}{2};\ \tfrac{3}{2}, \tfrac{3}{2};\ 1\right),$$

which can be evaluated by applying Dixon's summation theorem 3.5(4) with

$$a = 1 \qquad \text{and} \qquad b = c = \frac{1}{2}$$

or by applying Whipple's summation theorem 3.5(6) with

$$a = b = \frac{1}{2}, \qquad c = 1, \qquad \text{and} \qquad d = e = \frac{3}{2}.$$

We thus find from (11) and either one of the above applications that

$$(12)\qquad \zeta(2) = \frac{4}{3}\ \left\{\Gamma\left(\tfrac{3}{2}\right)\right\}^3 \Gamma\left(\frac{1}{2}\right) = \frac{\pi^2}{6}.$$

Next, by appealing directly to the definition 2.3(1) or (9) with $a = 1$, we obtain the hypergeometric representation:

$$(13)\qquad \zeta(n) = {}_{n+1}F_n(1, \cdots, 1;\ 2, \cdots, 2;\ 1) \qquad (n \in \mathbb{N} \setminus \{1\}).$$

The representations (10) and (13), together, yield the hypergeometric transformation:

$$(14)\qquad \begin{aligned} &{}_{n+1}F_n\left(1, \frac{1}{2}, \cdots, \frac{1}{2};\ \tfrac{3}{2}, \cdots, \tfrac{3}{2};\ 1\right) \\ &\quad = \left(1 - 2^{-n}\right) {}_{n+1}F_n(1, \cdots, 1;\ 2, \cdots, 2;\ 1) \qquad (n \in \mathbb{N} \setminus \{1\}), \end{aligned}$$

which, for $n = 2$, assumes the form:

$$(15)\qquad {}_3F_2\left(1, \frac{1}{2}, \frac{1}{2};\ \tfrac{3}{2}, \tfrac{3}{2};\ 1\right) = \frac{3}{4}\ {}_3F_2(1, 1, 1;\ 2, 2;\ 1).$$

We remark in passing that, by making use of contiguous analogues of Dixon's summation theorem 3.5(4) and Whipple's summation theorem 3.5(6), which were given recently by Lavoie *et al.* [*cf.* (1994, p. 268, Eq. (2)) and (1996, p. 294, Eq. (4))], we can easily express the sums of many series [analogous to those occurring in (1) and (7)] in terms of $\zeta(2)$. The details are fairly straightforward.

Since

$$(16)\qquad \zeta(n) = \int_0^1 \cdots \int_0^1 \frac{dt_1 \cdots dt_n}{1 - t_1 \cdots t_n} \qquad (n \in \mathbb{N} \setminus \{1\}),$$

it is not difficult to provide yet another solution of the Basler problem (1) by using the following change of variables:

$$t_1 = \frac{v+u}{\sqrt{2}} \quad \text{and} \quad t_2 = \frac{v-u}{\sqrt{2}}$$

in the double integral resulting from (16) when $n = 2$ [*cf.*, *e.g.*, Ojha and Singh (1999)].

The general case of $\zeta(2n)$

Following Srivastava (2000a), let us begin by recalling the following summation formula for the Bernoulli polynomials [*cf.* Hansen (1975, p. 342, Entry (50.11.10))]:

$$\begin{aligned}(17)\qquad &\sum_{k=0}^{[n/2]} \frac{(-n)_{2k}}{(2k)!} B_{2k}(x)B_{n-2k}(y) \\ &= \frac{1}{2}\{(x+y-1)nB_{n-1}(x+y) + (-1)^n(x-y)nB_{n-1}(x-y) \\ &-(n-1)\left[B_n(x+y) + (-1)^n B_n(x-y)\right]\},\end{aligned}$$

which, upon setting $x = 1$ and $y = 0$ (and replacing n by $2n$), yields

$$(18)\qquad \sum_{k=0}^{n} \binom{2n}{2k} B_{2k}B_{2n-2k} = -nB_{2n-1} - (2n-1)B_{2n} \quad (n \in \mathbb{N}_0).$$

Finally, by transposing to the right-hand side the terms for $k = 0$ and $k = n$ in (18), and making use of 1.6(7), we obtain

$$(19)\qquad \sum_{k=1}^{n-1} \binom{2n}{2k} B_{2k}B_{2n-2k} = -(2n+1)B_{2n} \quad (n \in \mathbb{N} \setminus \{1\}),$$

which is, in view of the relationship 2.3(18), equivalent to 2.3(20) which (as noted there) can also be used to evaluate $\zeta(2n)$ $(n \in \mathbb{N} \setminus \{1\})$ with (1).

In order to give a *direct* proof of the summation formula 2.3(20), let us put [*cf.* Prudnikov (1996)]

$$(20)\qquad \Omega(x) := \sum_{k=1}^{\infty} \exp\left(-\pi^2 k^2 x\right) \quad (x > 0),$$

so that

$$(21)\qquad \int_0^x \Omega(x-t)\Omega(t)dt = x\Omega(x) + \frac{3}{2}\sum_{k=1}^{\infty} \frac{\exp\left(-\pi^2 k^2 x\right)}{\pi^2 k^2},$$

which, by virtue of (1), readily yields

$$(22)\qquad \lim_{x\to 0} \int_0^x \Omega(x-t)\Omega(t)dt = \frac{1}{4}$$

and

$$\int_0^x \Omega(x-t)\Omega(t)dt = O\left(xe^{-x}\right) \quad (x \to \infty). \tag{23}$$

Making use of (21) and (23), it is easily seen that

$$\begin{aligned}\mathcal{I}_s :&= \int_0^\infty \int_0^\infty (x+y)^{s-2}\,\Omega(x)\Omega(y)dx\,dy \\ &= \int_0^\infty x^{s-2}dx \int_0^x \Omega(x-t)\Omega(t)dt \\ &= \left(s+\tfrac{1}{2}\right)\frac{\Gamma(s-1)}{\pi^{2s}}\zeta(2s) \quad (\Re(s)>1).\end{aligned} \tag{24}$$

On the other hand, since

$$\int_0^\infty t^{s-1}\Omega(t)dt = \frac{\Gamma(s)}{\pi^{2s}}\zeta(2s) \quad \left(\Re(s)>0;\ s\neq\tfrac{1}{2}\right), \tag{25}$$

which is an immediate consequence of the definition (20) and 2.3(1), an appeal to the binomial theorem for $(x+y)^{n-2}$ in the form:

$$(x+y)^{n-2} = \sum_{k=1}^{n-1}\binom{n-2}{k-1}x^{k-1}y^{n-k-1} \quad (n\in\mathbb{N}\setminus\{1\}) \tag{26}$$

would lead us also to

$$\mathcal{I}_n = \frac{(n-2)!}{\pi^{2n}}\sum_{k=1}^{n-1}\zeta(2k)\zeta(2n-2k) \quad (n\in\mathbb{N}\setminus\{1\}). \tag{27}$$

Now the summation formula 2.3(20) follows upon equating the two values of $\mathcal{I}_n$ given by (24) (*with* $s=n$) and (27).

Remark 1. The recursion formula (19) is a well-known (rather classical) result for Bernoulli numbers. It appears (for example) in Nielsen's book (1923) and was also derived independently by Underwood (1928). The equivalent result 2.3(20) was proven, by elementary methods, by Williams (1953, p. 20, Theorem I) [see also Apostol (1973, p. 427)].

Remark 2. In terms of the $\mathcal{L}$-function defined by [*cf.* Eq. 2.5(1)]

$$\mathcal{L}(s) = \sum_{k=0}^\infty \frac{(-1)^k}{(2k+1)^s} = 2^{-s}\,\Phi\left(-1,s,\tfrac{1}{2}\right) \quad (\Re(s)>0), \tag{28}$$

Williams (1953, p. 22, Theorem II) gave an interesting companion of the result 2.3(20) in the form:

$$\sum_{k=1}^n \mathcal{L}(2k-1)\mathcal{L}(2n-2k+1) = \left(n-\tfrac{1}{2}\right)\left(1-2^{-2n}\right)\zeta(2n), \tag{29}$$

which appears erroneously in Hansen (1975, p. 357, Entry (54.7.1)). Since $\mathcal{L}(1)$ is the well-known Gregory series for $\frac{\pi}{4}$ (with $\mathcal{L}(2)$ being the familiar Catalan constant G), by setting $n=1$ in (29), we immediately obtain

$$\zeta(2) = \frac{8}{3}\left\{\mathcal{L}(1)\right\}^2 = \frac{\pi^2}{6}. \tag{30}$$

Remark 3. In view of the constraint $\Re(s) > 0$ $\left(s \neq \frac{1}{2}\right)$ associated with the Mellin transform in (25), an earlier attempt by Kalla and Villalobos (1991) to extend the summation formula 2.3(20) by expressing $\mathcal{I}_s$ as a non-terminating (infinite) sum analogous to (27) cannot be justified. Thus the *main* result of Kalla and Villalobos (1991, p. 17, Eq. (17)) holds true only in the finite form given already by the well-known result 2.3(20).

4.2 Rapidly Convergent Series for $\zeta(2n+1)$

On the subject of series representations for $\zeta(3)$, in addition to Euler's result 3.1(8), Chen and Srivastava (1998) gave many series expressions for $\zeta(3)$, which are more rapidly convergent than that in 3.1(8), including

$$\zeta(3) = -\frac{8\pi^2}{5}\sum_{k=0}^{\infty}\frac{\zeta(2k)}{(2k+1)(2k+2)(2k+3)\,2^{2k}}. \tag{1}$$

And, as pointed out by (for example) Chen and Srivastava (1998, pp. 180-181), another interesting series representation:

$$\zeta(3) = \frac{5}{2}\sum_{k=1}^{\infty}\frac{(-1)^{k-1}}{k^3\binom{2k}{k}}, \tag{2}$$

which played a key rôle in Apéry's proof of the irrationality of $\zeta(3)$ [see Apéry (1979)], was proven independently by (among others) Hjortnaes (1954), Gosper (1976), and Apéry (1979) [see also Problem 26(b) (Chapter 2)]. Furthermore, in their recent work on the Ray-Singer torsion and topological field theories, Nash and O'Connor (1993 and 1995) obtained a number of remarkable integral expressions for $\zeta(3)$, including (for example) the following result (1995, p. 1489 *et seq.*):

$$\zeta(3) = \frac{2\pi^2}{7}\log 2 - \frac{8}{7}\int_0^{\pi/2} z^2 \cot z\, dz. \tag{3}$$

Since [Erdélyi *et al.* (1953, p. 51, Eq. 1.20(3))]

$$z\cot z = -2\sum_{k=0}^{\infty}\zeta(2k)\left(\frac{z}{\pi}\right)^{2k} \qquad (|z| < \pi), \tag{4}$$

the result (3) is obviously equivalent to the series representation [*cf.* Dąbrowski (1996, p. 202); see also Chen and Srivastava (1998, p. 191, Eq. (3.19))]:

$$\zeta(3) = \frac{2\pi^2}{7}\left(\log 2 + \sum_{k=0}^{\infty}\frac{\zeta(2k)}{(k+1)2^{2k}}\right). \tag{5}$$

Moreover, by integrating by parts, it is easily seen that

$$\int_0^{\pi/2} z^2\cot z\, dz = -2\int_0^{\pi/2} z\log\sin z\, dz, \tag{6}$$

so that the result (3) is equivalent also to the integral representation:

$$\zeta(3) = \frac{2\pi^2}{7}\log 2 + \frac{16}{7}\int_0^{\pi/2} z \log \sin z \, dz, \tag{7}$$

which was proven in the aforementioned 1772 paper by Euler [*cf.*, *e.g.*, Ayoub (1974, p. 1084); see also Srivastava (2000a)].

It follows from 3.2(7) that [*cf.*, *e.g.*, Hansen (1975, p. 359); see also Srivastava (1999) where other references are also cited]:

$$\begin{aligned}&\sum_{k=0}^{\infty} \frac{(s)_{2k}}{(2k)!}\,\zeta(s+2k,a)\,t^{2k}\\ &\qquad = \frac{1}{2}\left[\zeta(s,a-t)+\zeta(s,a+t)\right] \quad (|t|<|a|).\end{aligned} \tag{8}$$

In view of 2.3(3), the special case of the identity (8) when $a = 1$ and $t = \frac{1}{m}$ can be rewritten in the form:

$$\begin{aligned}&\sum_{k=0}^{\infty} \frac{(s)_{2k}}{(2k)!}\,\frac{\zeta(s+2k)}{m^{2k}}\\ &= \begin{cases} (2^s-1)\,\zeta(s) - 2^{s-1} & (m=2)\\ \dfrac{1}{2}\left[(m^s-1)\zeta(s) - m^s - \displaystyle\sum_{j=2}^{m-2} \zeta\left(s, \frac{j}{m}\right)\right] & (m \in \mathbb{N}\setminus\{1,2\}),\end{cases}\end{aligned} \tag{9}$$

where (as usual) an empty sum is to be interpreted as nil. In addition to the case $m = 2$, the formula (9) simplifies also in the cases when $m = 3$, 4, and 6, and we thus obtain the identities:

$$\sum_{k=0}^{\infty} \frac{(s)_{2k}}{(2k)!}\,\frac{\zeta(s+2k)}{3^{2k}} = \frac{1}{2}\left[(3^s-1)\,\zeta(s) - 3^s\right], \tag{10}$$

$$\sum_{k=0}^{\infty} \frac{(s)_{2k}}{(2k)!}\,\frac{\zeta(s+2k)}{4^{2k}} = \frac{1}{2}\left[(4^s-2^s)\,\zeta(s) - 4^s\right], \tag{11}$$

and

$$\begin{aligned}&\sum_{k=0}^{\infty} \frac{(s)_{2k}}{(2k)!}\,\frac{\zeta(s+2k)}{6^{2k}}\\ &\qquad = \frac{1}{2}\left[(6^s-3^s-2^s+1)\,\zeta(s) - 6^s\right],\end{aligned} \tag{12}$$

respectively.

Identities of this kind seem to have first appeared in the work of Ramaswami (1934), who actually proved the cases $m = 2$, 3, and 6 of the general result in

(9). Each of these three identities of Ramaswami (1934) can also be found in the work of Hansen (1975, p. 357), who referred to Apostol (1954) as his source for the identities (10) and (12) only. As a matter of fact, Apostol (1954) reproduced the identities (10) and (12) from Ramaswami's work (1934) and then proved an interesting *arithmetical* generalization of these identities [see also Klusch (1992, p. 520)].

In its slightly variant form:

$$\sum_{k=1}^{\infty} \frac{(s+1)_{2k}}{(2k)!} \frac{\zeta(s+2k)}{2^{2k}} = (2^s - 2)\,\zeta(s), \tag{13}$$

the case $m = 2$ of the general result (9) was applied by Zhang and Williams (1993b) [and, subsequently, by Cvijović and Klinowski (1997)] with a view to finding two seemingly different series representations for $\zeta(2n+1)$ $(n \in \mathbb{N})$. Srivastava (1999a) obtained much more rapidly converging series representations for $\zeta(2n+1)$ $(n \in \mathbb{N})$ chiefly by appealing appropriately to each of the aforementioned cases ($m = 2, 3, 4,$ and 6) of the general result (9). Following his work [Srivastava (1999a)], we begin here with the case $m = 2$ of the general result (9). Upon separating the first $n+1$ terms of the series occurring on the left-hand side, if we transpose the terms for $k = 0$ and $k = n$ to the right-hand side, we obtain

$$\begin{aligned} &\sum_{k=1}^{n-1} \frac{(s)_{2k}}{(2k)!} \frac{\zeta(s+2k)}{2^{2k}} + \sum_{k=n+1}^{\infty} \frac{(s)_{2k}}{(2k)!} \frac{\zeta(s+2k)}{2^{2k}} \\ &\qquad = (2^s - 2)\zeta(s) - 2^{s-1} - \frac{(s)_{2n}}{(2n)!} \frac{\zeta(s+2n)}{2^{2n}}, \end{aligned} \tag{14}$$

which readily yields the identity:

$$\begin{aligned} &\sum_{k=1}^{n-1} \frac{(s)_{2k}}{(2k)!} 2^{2(n-k)} \zeta(s+2k) + \sum_{k=1}^{\infty} \frac{(s)_{2n+2k}}{(2n+2k)!} \frac{\zeta(s+2n+2k)}{2^{2k}} \\ &\qquad = 2^{2n}(2^s - 2)\,\zeta(s) - 2^{s+2n-1} - \frac{(s)_{2n}}{(2n)!}\,\zeta(s+2n) \quad (n \in \mathbb{N}). \end{aligned} \tag{15}$$

Now we apply the functional equation 2.3(12) in the first term on the right-hand side of (15) and divide both sides by $s + 2n$. We thus find that

$$\begin{aligned} &\sum_{k=1}^{n-1} \frac{(s)_{2k}}{(2k)!} 2^{2(n-k)} \left\{ \frac{\zeta(s+2k)}{s+2n} \right\} + \sum_{k=1}^{\infty} \frac{(s)_{2n}(s+2n+1)_{2k-1}}{(2n+2k)!} \frac{\zeta(s+2n+2k)}{2^{2k}} \\ &\qquad = 2^{s+2n}(2^s - 2)\,\pi^{s-1}\,\Gamma(1-s)\,\zeta(1-s) \left\{ \frac{\sin\left(\frac{1}{2}\pi s\right)}{s+2n} \right\} \\ &\qquad\quad - \left\{ \frac{2^{s+2n-1} + \dfrac{(s)_{2n}}{(2n)!}\,\zeta(s+2n)}{s+2n} \right\} \quad (s \neq -2n;\ n \in \mathbb{N}). \end{aligned} \tag{16}$$

It is easy to see from 1.5(10) that

(17)
$$\frac{d}{ds}\{(s)_{2n}\}\bigg|_{s=-2n} = -(2n)!\; H_{2n} \quad (n \in \mathbb{N}),$$

where H_n denotes the familiar harmonic numbers defined by 3.2(36). We observe also that the following limit formula:

(18)
$$\lim_{s\to -2n}\left\{\frac{\zeta(s+2k)}{s+2n}\right\} = \frac{(-1)^{n-k}}{2(2\pi)^{2(n-k)}}\,(2n-2k)!\,\zeta(2n-2k+1)$$
$$(k=1,\cdots,n-1; \quad n \in \mathbb{N}\setminus\{1\})$$

is needed in the first sum on the left-hand side of (16) only when this sum is nonzero (that is, only when $n \in \mathbb{N}\setminus\{1\}$). Furthermore, by l'Hôpital's rule, we have

(19)
$$\lim_{s\to -2n}\left\{\frac{2^{s+2n-1} + \dfrac{(s)_{2n}}{(2n)!}\,\zeta(s+2n)}{s+2n}\right\}$$
$$= \left[2^{s+2n-1}\log 2 + \frac{d}{ds}\{(s)_{2n}\}\cdot\frac{\zeta(s+2n)}{(2n)!} + \frac{(s)_{2n}}{(2n)!}\,\zeta'(s+2n)\right]\bigg|_{s=-2n}$$
$$= \frac{1}{2}\,(H_{2n} - \log\pi) \quad (n \in \mathbb{N}).$$

Finally, letting $s \to -2n$ in (16), and making use of the limit relationships (18) and (19), we obtain Srivastava's first series representation for $\zeta(2n+1)$:

(20)
$$\zeta(2n+1) = (-1)^{n-1}\,\frac{(2\pi)^{2n}}{2^{2n+1}-1}\left[\frac{H_{2n}-\log\pi}{(2n)!} + \sum_{k=1}^{n-1}\frac{(-1)^k}{(2n-2k)!}\,\frac{\zeta(2k+1)}{\pi^{2k}} + 2\sum_{k=1}^{\infty}\frac{(2k-1)!}{(2n+2k)!}\,\frac{\zeta(2k)}{2^{2k}}\right] \quad (n \in \mathbb{N}).$$

In precisely the same manner, we can apply the identities (10), (11), and (12) in order to prove the following *additional* series representations for $\zeta(2n+1)$ [Srivastava (1999a, p. 389)]:

(21)
$$\zeta(2n+1) = (-1)^{n-1}\,\frac{2(2\pi)^{2n}}{3^{2n+1}-1}\left[\frac{H_{2n}-\log\left(\frac{2}{3}\pi\right)}{(2n)!} + \sum_{k=1}^{n-1}\frac{(-1)^k}{(2n-2k)!}\,\frac{\zeta(2k+1)}{\left(\frac{2}{3}\pi\right)^{2k}} + 2\sum_{k=1}^{\infty}\frac{(2k-1)!}{(2n+2k)!}\,\frac{\zeta(2k)}{3^{2k}}\right] \quad (n \in \mathbb{N});$$

(22)
$$\zeta(2n+1) = (-1)^{n-1}\,\frac{2(2\pi)^{2n}}{2^{4n+1}+2^{2n}-1}\left[\frac{H_{2n}-\log\left(\frac{1}{2}\pi\right)}{(2n)!} + \sum_{k=1}^{n-1}\frac{(-1)^k}{(2n-2k)!}\,\frac{\zeta(2k+1)}{\left(\frac{1}{2}\pi\right)^{2k}} + 2\sum_{k=1}^{\infty}\frac{(2k-1)!}{(2n+2k)!}\,\frac{\zeta(2k)}{4^{2k}}\right] \quad (n \in \mathbb{N});$$

(23)

$$\zeta(2n+1) = (-1)^{n-1} \frac{2(2\pi)^{2n}}{3^{2n}(2^{2n}+1)+2^{2n}-1} \left[\frac{H_{2n} - \log\left(\frac{1}{3}\pi\right)}{(2n)!} + \sum_{k=1}^{n-1} \frac{(-1)^k}{(2n-2k)!} \frac{\zeta(2k+1)}{\left(\frac{1}{3}\pi\right)^{2k}} + 2\sum_{k=1}^{\infty} \frac{(2k-1)!}{(2n+2k)!} \frac{\zeta(2k)}{6^{2k}} \right] \qquad (n \in \mathbb{N}).$$

Remarks and observations

The series representation (20) is markedly different from each of the series representations for $\zeta(2n+1)$, which were given earlier by Zhang and Williams (1993b, p. 1590, Eq. (3.13)) and (more recently) by Cvijović and Klinowski (1997, p. 1265, Theorem A). Since $\zeta(2k) \to 1$ as $k \to \infty$, the general term in the series representation (20) has the order estimate:

$$O\left(2^{-2k} \cdot k^{-2n-1}\right) \qquad (k \to \infty; \quad n \in \mathbb{N}),$$

whereas the general term in each of these earlier series representations has the order estimate:

$$O\left(2^{-2k} \cdot k^{-2n}\right) \qquad (k \to \infty; \quad n \in \mathbb{N}).$$

By suitably combining (20) and (22), it is fairly straightforward to obtain the series representation:

(24)

$$\begin{aligned}\zeta(2n+1) = &(-1)^{n-1} \frac{2(2\pi)^{2n}}{(2^{2n}-1)(2^{2n+1}-1)} \left[\frac{\log 2}{(2n)!} \right. \\ &+ \sum_{k=1}^{n-1} \frac{(-1)^k(2^{2k}-1)}{(2n-2k)!} \frac{\zeta(2k+1)}{\pi^{2k}} \\ &\left. -2\sum_{k=1}^{\infty} \frac{(2k-1)!\,(2^{2k}-1)}{(2n+2k)!} \frac{\zeta(2k)}{2^{4k}} \right] \qquad (n \in \mathbb{N}).\end{aligned}$$

Making use of the relationship given in Problem 11 of Chapter 2, the series representation (24) can immediately be put in the form:

(25)

$$\begin{aligned}\zeta(2n+1) = &(-1)^{n-1} \frac{2(2\pi)^{2n}}{(2^{2n}-1)(2^{2n+1}-1)} \left[\frac{\log 2}{(2n)!} \right. \\ &+ \sum_{k=1}^{n-1} \frac{(-1)^k(2^{2k}-1)}{(2n-2k)!} \frac{\zeta(2k+1)}{\pi^{2k}} \\ &\left. +\frac{1}{2}\sum_{k=1}^{\infty} \frac{(-1)^{k-1}}{(2n+2k)!} \left(\frac{\pi}{2}\right)^{2k} E_{2k-1}(0) \right] \qquad (n \in \mathbb{N}),\end{aligned}$$

which is a slightly modified (and corrected) version of a result proven in a significantly different way by Tsumura (1994, p. 383, Theorem B).

Another interesting combination of the series representations (20) and (22) leads to the following variant of Tsumura's result (24) or (25):

$$\begin{aligned}\zeta(2n+1) =&(-1)^{n-1}\frac{\pi^{2n}}{2^{2n+1}-1}\left[\frac{H_{2n}-\log\left(\frac{1}{4}\pi\right)}{(2n)!}\right.\\ &+\sum_{k=1}^{n-1}\frac{(-1)^k(2^{2k+1}-1)}{(2n-2k)!}\frac{\zeta(2k+1)}{\pi^{2k}}\\ &\left.-4\sum_{k=1}^{\infty}\frac{(2k-1)!\,(2^{2k-1}-1)}{(2n+2k)!}\frac{\zeta(2k)}{2^{4k}}\right] \qquad (n\in\mathbb{N}),\end{aligned} \tag{26}$$

which is essentially the same as the *determinantal* expression for $\zeta(2n+1)$ derived recently by Ewell (1995, p. 1010, Corollary 3) by employing an entirely different technique from Srivastava's (1999a).

Other similar combinations of the series representations (20) to (23) would yield the following companions of Ewell's result (26):

$$\begin{aligned}\zeta(2n+1) =&(-1)^{n-1}\frac{2(2\pi)^{2n}}{(2^{2n+1}-1)(3^{2n}+1)}\left[\frac{H_{2n}-\log\left(\frac{1}{6}\pi\right)}{(2n)!}\right.\\ &+\sum_{k=1}^{n-1}\frac{(-1)^k(2^{2k+1}-1)}{(2n-2k)!}\frac{\zeta(2k+1)}{\left(\frac{2}{3}\pi\right)^{2k}}\\ &\left.-4\sum_{k=1}^{\infty}\frac{(2k-1)!(2^{2k-1}-1)}{(2n+2k)!}\frac{\zeta(2k)}{6^{2k}}\right] \qquad (n\in\mathbb{N});\end{aligned} \tag{27}$$

$$\begin{aligned}\zeta(2n+1) =&(-1)^{n-1}\frac{2(2\pi)^{2n}}{(2^{2n}+1)(3^{2n+1}-1)}\left[\frac{2H_{2n}-\log\left(\frac{\pi^2}{27}\right)}{(2n)!}\right.\\ &+\sum_{k=1}^{n-1}\frac{(-1)^k(3^{2k+1}-1)}{(2n-2k)!}\frac{\zeta(2k+1)}{\pi^{2k}}\\ &\left.-6\sum_{k=1}^{\infty}\frac{(2k-1)!\,(3^{2k-1}-1)}{(2n+2k)!}\frac{\zeta(2k)}{6^{2k}}\right] \qquad (n\in\mathbb{N});\end{aligned} \tag{28}$$

$$\begin{aligned}\zeta(2n+1) =&(-1)^{n-1}\frac{2(2\pi)^{2n}}{3^{2n+2}-2^{2n+3}+1}\left[\frac{H_{2n}-\log\left(\frac{8\pi}{27}\right)}{(2n)!}\right.\\ &+\sum_{k=1}^{n-1}\frac{(-1)^k(3^{2k+1}-2^{2k+1})}{(2n-2k)!}\frac{\zeta(2k+1)}{(2\pi)^{2k}}\\ &\left.-12\sum_{k=1}^{\infty}\frac{(2k-1)!\,(3^{2k-1}-2^{2k-1})}{(2n+2k)!}\frac{\zeta(2k)}{6^{2k}}\right] \qquad (n\in\mathbb{N});\end{aligned} \tag{29}$$

$$
\begin{aligned}
\zeta(2n+1) =&(-1)^{n-1}\frac{2(2\pi)^{2n}}{2^{4n+3}+2^{2n+2}-3^{2n+2}-1}\left[\frac{H_{2n}-\log\left(\frac{27\pi}{128}\right)}{(2n)!}\right.\\
&+\sum_{k=1}^{n-1}\frac{(-1)^k(4^{2k+1}-3^{2k+1})}{(2n-2k)!}\frac{\zeta(2k+1)}{(2\pi)^{2k}}\\
&\left.-24\sum_{k=1}^{\infty}\frac{(2k-1)!\,(4^{2k-1}-3^{2k-1})}{(2n+2k)!}\frac{\zeta(2k)}{12^{2k}}\right]\qquad(n\in\mathbb{N}),
\end{aligned}\tag{30}
$$

and

$$
\begin{aligned}
\zeta(2n+1) =&(-1)^{n-1}\frac{2(2\pi)^{2n}}{3^{2n+1}(2^{2n}+1)-2^{4n+2}+2^{2n}-1}\left[\frac{H_{2n}-\log\left(\frac{4\pi}{27}\right)}{(2n)!}\right.\\
&+\sum_{k=1}^{n-1}\frac{(-1)^k(3^{2k+1}-2^{2k+1})}{(2n-2k)!}\frac{\zeta(2k+1)}{\pi^{2k}}\\
&\left.-12\sum_{k=1}^{\infty}\frac{(2k-1)!\,(3^{2k-1}-2^{2k-1})}{(2n+2k)!}\frac{\zeta(2k)}{12^{2k}}\right]\qquad(n\in\mathbb{N}).
\end{aligned}\tag{31}
$$

Next we recall another identity similar to (8) [see Srivastava (1999a, p. 386, Eq. (1.7)]:

$$
\begin{aligned}
&\sum_{k=0}^{\infty}\frac{(s)_{2k+1}}{(2k+1)!}\zeta(s+2k+1,a)\,t^{2k+1}\\
&\qquad=\frac{1}{2}\left[\zeta(s,a-t)-\zeta(s,a+t)\right]\qquad(|t|<|a|).
\end{aligned}\tag{32}
$$

By setting $t=1/m$ and differentiating both sides with respect to s, we find from (32) that

$$
\begin{aligned}
&\sum_{k=0}^{\infty}\frac{(s)_{2k+1}}{(2k+1)!\,m^{2k}}\left[\zeta'(s+2k+1,a)+\zeta(s+2k+1,a)\sum_{j=0}^{2k}\frac{1}{s+j}\right]\\
&\qquad=\frac{m}{2}\frac{d}{ds}\left\{\zeta\left(s,\,a-\frac{1}{m}\right)-\zeta\left(s,\,a+\frac{1}{m}\right)\right\}\qquad(m\in\mathbb{N}\setminus\{1\}).
\end{aligned}\tag{33}
$$

In particular, when $m=2$, (33) immediately yields

$$
\begin{aligned}
&\sum_{k=0}^{\infty}\frac{(s)_{2k+1}}{(2k+1)!\,2^{2k}}\left[\zeta'(s+2k+1,\,a)+\zeta(s+2k+1,\,a)\sum_{j=0}^{2k}\frac{1}{s+j}\right]\\
&\qquad=-\left(a-\frac{1}{2}\right)^{-s}\log\left(a-\frac{1}{2}\right).
\end{aligned}\tag{34}
$$

By letting $s \to -2n-1$ ($n \in \mathbb{N}$) in the *further* special of this last identity (34) when $a = 1$, Wilton (1922/1923, p. 92) obtained the following series representation for $\zeta(2n+1)$ [see also Hansen (1975, p. 357, Entry (54.6.9))]:

$$\begin{aligned}\zeta(2n+1) = (-1)^{n-1}\,\pi^{2n}\Biggl[&\frac{H_{2n+1}-\log\pi}{(2n+1)!}+\sum_{k=1}^{n-1}\frac{(-1)^k}{(2n-2k+1)!}\,\frac{\zeta(2k+1)}{\pi^{2k}}\\ &+2\sum_{k=1}^{\infty}\frac{(2k-1)!}{(2n+2k+1)!}\,\frac{\zeta(2k)}{2^{2k}}\Biggr] \qquad (n\in\mathbb{N}),\end{aligned} \tag{35}$$

which may be compared with the series representation (20). As a matter of fact, since

$$\frac{(2k)!}{(2n+2k)!} = \frac{(2k-1)!}{(2n+2k-1)!} - 2n\,\frac{(2k-1)!}{(2n+2k)!} \qquad (n,k\in\mathbb{N}), \tag{36}$$

it is not difficult to deduce from (20) and (35) (with n replaced by $n-1$) that

$$\begin{aligned}\zeta(2n+1) = (-1)^n\,\frac{(2\pi)^{2n}}{n(2^{2n+1}-1)}\Biggl[&\sum_{k=1}^{n-1}\frac{(-1)^{k-1}\,k}{(2n-2k)!}\,\frac{\zeta(2k+1)}{\pi^{2k}}\\ &+\sum_{k=0}^{\infty}\frac{(2k)!}{(2n+2k)!}\,\frac{\zeta(2k)}{2^{2k}}\Biggr] \qquad (n\in\mathbb{N}),\end{aligned} \tag{37}$$

which is precisely the aforementioned *main* result of Cvijović and Klinowski (1997, p. 1265, Theorem A). As a matter of fact, in view of the derivative formula 2.3(22), the series representation (37) is essentially the same as a result given earlier by Zhang and Williams (1993b, p. 1590, Eq. (3.13)) [see also Zhang and Williams (1993b, p. 1591, Eq. (3.16)), where an obviously more complicated (*asymptotic*) version of (37) was proven by applying the same identity (13) above].

Observing also that

$$\frac{(2k)!}{(2n+2k+1)!} = \frac{(2k-1)!}{(2n+2k)!} - (2n+1)\,\frac{(2k-1)!}{(2n+2k+1)!} \qquad (n,k\in\mathbb{N}), \tag{38}$$

we obtain yet another series representation for $\zeta(2n+1)$ by applying (20) and (35):

$$\begin{aligned}\zeta(2n+1) = (-1)^n\,\frac{2(2\pi)^{2n}}{(2n-1)\,2^{2n}+1}\Biggl[&\sum_{k=1}^{n-1}\frac{(-1)^{k-1}\,k}{(2n-2k+1)!}\,\frac{\zeta(2k+1)}{\pi^{2k}}\\ &+\sum_{k=0}^{\infty}\frac{(2k)!}{(2n+2k+1)!}\,\frac{\zeta(2k)}{2^{2k}}\Biggr] \qquad (n\in\mathbb{N}),\end{aligned} \tag{39}$$

which provides a significantly simpler (and much more rapidly convergent) version of the other *main* result of Cvijović and Klinowski (1997, p. 1265, Theorem B):

$$\zeta(2n+1) = (-1)^n\,\frac{2(2\pi)^{2n}}{(2n)!}\sum_{k=0}^{\infty}\Omega_{n,k}\,\frac{\zeta(2k)}{2^{2k}} \qquad (n\in\mathbb{N}), \tag{40}$$

where the coefficients $\Omega_{n,k}$ are given explicitly by

$$\Omega_{n,k} := \sum_{j=0}^{2n} \binom{2n}{j} \frac{B_{2n-j}}{(j+2k+1)(j+1)\,2^j} \qquad (n \in \mathbb{N};\ k \in \mathbb{N}_0)\,, \tag{41}$$

in terms of the Bernoulli numbers (see Section 1.6). The definition (41) can be rewritten at once in the form:

$$\Omega_{n,k} = \sum_{j=0}^{n} \binom{2n}{2j} \frac{B_{2n-2j}}{(2j+2k+1)(2j+1)\,2^{2j}} - \frac{1}{(2n+2k)\,2^{2n}} \qquad (n \in \mathbb{N};\ k \in \mathbb{N}_0)\,, \tag{42}$$

or, equivalently,

$$\begin{aligned}\Omega_{n,k} = (-1)^{n-1} \frac{2(2n)!}{(2\pi)^{2n}} \sum_{j=0}^{n} \frac{(-\pi^2)^j\, \zeta(2n-2j)}{(2j+1)!(2j+2k+1)} \\ - \frac{1}{(2n+2k)\,2^{2n}} \quad (n \in \mathbb{N};\ k \in \mathbb{N}_0),\end{aligned} \tag{43}$$

by virtue of the relationship 2.3(18). Combining the partial fractions occurring in (42) or (43), it is easily seen that

$$\begin{aligned}\Omega_{n,k} = \frac{\prod_{\ell=0}^{n} \{(2k+2\ell+1)^{-1}\}}{(2n+2k)\,2^{2n}} \Bigg[\sum_{j=0}^{n} \binom{2n}{2j} \frac{2n+2k}{2j+1}\, 2^{2n-2j}\, B_{2n-2j} \\ \cdot \prod_{\substack{\ell=0 \\ (\ell \neq j)}}^{n} (2k+2\ell+1) - \prod_{\ell=0}^{n} (2k+2\ell+1)\Bigg] \quad (n \in \mathbb{N};\ k \in \mathbb{N}_0).\end{aligned} \tag{44}$$

In view of the identity:

$$\sum_{j=0}^{n} \binom{2n}{2j} \frac{2^{2n-2j}\, B_{2n-2j}}{2j+1} = 1 = \sum_{j=0}^{n} \binom{2n}{2j} \frac{2^{2j}\, B_{2j}}{2n-2j+1}, \tag{45}$$

which is due essentially to Euler [*cf.*, *e.g.*, Riordan (1968, p. 123, Problem 12)], the expression inside brackets in (44) is a polynomial in k of degree n (not $n+1$), and therefore

$$\Omega_{n,k} = O(k^{-2}) \qquad (k \to \infty;\ n \in \mathbb{N}). \tag{46}$$

It follows from (46) that the general term in (40) has the order estimate:

$$O(2^{-2k} \cdot k^{-2}) \qquad (k \to \infty), \tag{47}$$

whereas the general term in our series representation (39) has precisely the same order estimate:

$$O(2^{-2k} \cdot k^{-2n-1}) \qquad (k \to \infty;\ n \in \mathbb{N}) \tag{48}$$

as that in (20). Thus, even in the special case when $n = 1$, the series representing $\zeta(3)$ converges faster in (39) than in (40).

Various known series representations for $\zeta(2n+1)$ $(n \in \mathbb{N})$ of other types include those given (for example) by Ramanujan (1957) [see also Berndt (1977)], Glaisher (1912/1913) [see also Hansen (1975, p. 359)], Koshliakov (1928/1929), Leshchiner (1981), Grosswald (1970 and 1972), Terras (1976), Cohen (1981), Butzer *et al.* (1991 and 1992), Dąbrowski (1996), and others [see, *e.g.*, Berndt (1989, pp. 275 and 276)].

We conclude this section by remarking that a particular case of the series representation (23) when $n = 1$ was proven, by an entirely different method, by Zhang and Williams (1993b, p. 707, Theorem 9). Furthermore, the following particular case of (37) when $n = 1$ [see 3.1(8)]:

$$\zeta(3) = -\frac{4\pi^2}{7}\sum_{k=0}^{\infty}\frac{\zeta(2k)}{(2k+1)(2k+2)\,2^{2k}}, \tag{49}$$

which is contained in a 1772 paper entitled *Exercitationes Analyticae* by Euler [see, *e.g.*, Ayoub (1974, pp. 1084–1085)], was rediscovered by Ramaswami (1934) and (more recently) by Ewell (1990). In fact, Euler's formula (49) was reproduced by Srivastava (1988a, p. 7, Eq. (2.23)) from the work of Ramaswami (1934). In the *current* mathematical literature, however, Euler's formula (49) is being attributed to Ewell (1990).

4.3 Further Series Representations

There are many further known series representations for $\zeta(2n+1)$ $(n \in \mathbb{N})$, which converge much more rapidly than those given by the defining series in 2.3(1). For example, we have the series representation [see also Equation 4.2(35)]:

$$\begin{aligned}\zeta(2n+1) =&(-1)^{n-1}\,\pi^{2n}\left[\frac{H_{2n+1}-\log\pi}{(2n+1)!}+\sum_{k=1}^{n-1}\frac{(-1)^k}{(2n-2k+1)!}\,\frac{\zeta(2k+1)}{\pi^{2k}}\right.\\ &\left.+2\sum_{k=1}^{\infty}\frac{(2k-1)!}{(2n+2k+1)!}\,\frac{\zeta(2k)}{2^{2k}}\right]\qquad (n\in\mathbb{N}),\end{aligned} \tag{1}$$

which was given, over seven decades ago, by Wilton (1922/1923, p. 92) [see also Hansen (1975, p. 357, Entry (54.6.9))], and the following result given recently by Srivastava (1998, p. 10, Eq. (42)) [see also Srivastava (1997, p. 5, Eq. (3.4))]:

$$\begin{aligned}\zeta(2n+1) = (-1)^{n-1}\left(\frac{\pi}{2}\right)^{2n}&\left[\frac{H_{2n+1}-\log\left(\frac{1}{2}\pi\right)}{(2n+1)!}+\frac{2(4^n-1)}{(2n+2)!}B_{2n+2}\log 2\right.\\ &-\frac{2^{2n+1}-1}{(2n+1)!}\zeta'(-2n-1)-\frac{2^{4n+3}}{(2n+1)!}\zeta'\left(-2n-1,\,\frac{1}{4}\right)\\ &\left.+\sum_{k=1}^{n-1}\frac{(-1)^k}{(2n-2k+1)!}\,\frac{\zeta(2k+1)}{\left(\frac{1}{2}\pi\right)^{2k}}+2\sum_{k=1}^{\infty}\frac{(2k-1)!}{(2n+2k+1)!}\,\frac{\zeta(2k)}{4^{2k}}\right]\qquad (n\in\mathbb{N}),\end{aligned} \tag{2}$$

where (*and in what follows*) a prime denotes the derivative of $\zeta(s)$ or $\zeta(s,a)$ with respect to s, an empty sum is to be interpreted as nil, and H_n denotes the familiar harmonic numbers defined by 3.2(36).

Of the two seemingly analogous representations (1) and (2), the infinite series in (2) would obviously converge more rapidly, with their general terms having the order estimates:

$$O\left(k^{-2n-2}\cdot m^{-2k}\right) \qquad (k\to\infty;\ n\in\mathbb{N};\ m=2 \text{ and } 4)\,. \tag{3}$$

Srivastava and Tsumura (2000a) derived three (presumably new) members of the class of the series representations (1) and (2). The general terms of the infinite series occurring in these three members [given in (29), (30), and (31) below] have the order estimates:

$$O\left(k^{-2n-2}\cdot m^{-2k}\right) \qquad (k\to\infty;\ n\in\mathbb{N};\ m=3,4,6)\,, \tag{4}$$

which exhibit the fact that *each* of the three series representations derived here for $\zeta(2n+1)$ converges more rapidly than Wilton's result (1) and two of them [*cf.* Equations (30) and (31) below] at least as rapidly as Srivastava's result (2).

Srivastava and Tsumura (2000a) begin by defining the sequence $\{\gamma_n(x)\}_{n=0}^{\infty}$ by means of the generating function [*cf.* Eq. 1.6(2)]:

$$F(x,t) := \frac{t-\log x}{e^t - x} = \sum_{n=0}^{\infty} \gamma_n(x)\,\frac{t^n}{n!} \tag{5}$$

$$(1 \leqq x \leqq 1+c;\ c>0)\,,$$

so that, clearly,

$$\gamma_n(1) = B_n \quad (n\in N_0)\,. \tag{6}$$

Since the zeros of $e^t - x$ are given by

$$t = 2n\pi i + \log x \quad (n\in\mathbb{Z})\,, \tag{7}$$

the radius of convergence of the series in (5) is at least 2π. Hence, by the Cauchy-Hadamard theorem for absolute convergence [*cf.*, *e.g.*, Whittaker and Watson (1963, p. 30)], we have

Lemma 1 [Srivastava and Tsumura (2000a)]. *Let the sequence* $\{\gamma_n(x)\}_{n=0}^{\infty}$ *be defined by* (5). *Then there exists some nonnegative real number* κ *such that*

$$\liminf_{n\to\infty}\left(\frac{|\gamma_n(x)|}{n!}\right)^{-1/n} = 2\pi+\kappa \qquad (\kappa \geqq 0). \tag{8}$$

Following Srivastava and Tsumura (2000a), we now consider the following Dirichlet series [*cf.* Eq. 2.3(1)]:

$$\omega(s,x) := \sum_{n=1}^{\infty}\frac{x^{-n-1}}{n^s} + \frac{\log x}{s-1}\sum_{n=1}^{\infty}\frac{x^{-n-1}}{n^{s-1}} \tag{9}$$

$$(1 \leq x \leq 1 + c;\ c > 0)\,,$$

so that, obviously,

$$\omega(s, 1) = \zeta(s). \tag{10}$$

In case $1 < x \leq 1 + c$ ($c > 0$), we can see that the function $\omega(s, x)$ is meromorphic, that is, holomorphic on the whole complex s-plane except for a simple pole at $s = 1$ with residue

$$\frac{\log x}{x - 1}.$$

Lemma 2 below provides an interesting generalization of the familiar relatioship 2.3(10).

Lemma 2 [Srivastava and Tsumura (2000a)]. *Let $\gamma_n(x)$ and $\omega(s, x)$ be defined by* (5) *and* (9), *respectively. Then*

$$\omega(1 - n, x) = -\frac{\gamma_n(x)}{n} \qquad (n \in \mathbb{N} \setminus \{1\})\,. \tag{11}$$

Making use of the generating function (5), we also have

Lemma 3 [Srivastava and Tsumura (2000a)]. *For $n \in \mathbb{N}$ and $|\theta| < 2\pi$ ($\theta \in \mathbb{R}$),*

$$\begin{aligned}(2n+1)\sum_{k=1}^{\infty}\frac{x^{-k-1}}{k^{2n+2}}\sin(k\theta) &+ \sum_{k=1}^{\infty}\frac{x^{-k-1}}{k^{2n+1}}\left[\sin(k\theta)\log x - \theta\cos(k\theta)\right]\\ = \frac{(-1)^n\theta^{2n+1}}{(2n+1)!}\,\frac{\log x}{x(x-1)} &+ \sum_{k=0}^{n-1}\frac{(-1)^k\,\theta^{2k+1}}{(2k+1)!}(2n-2k)\,\omega(2n-2k+1, x)\\ &+ \sum_{k=n+1}^{\infty}\frac{(-1)^k\,\theta^{2k+1}}{(2k+1)!}\,\gamma_{2k-2n}(x)\\ &(1 < x \leq 1 + c,\ c > 0).\end{aligned} \tag{12}$$

By applying Lemma 3, Srivastava and Tsumura (2000a) first proved that

$$\begin{aligned}(2n+1)&\sum_{k=1}^{\infty}\frac{\sin(k\theta)}{k^{2n+2}} - \theta\sum_{k=1}^{\infty}\frac{\cos(k\theta)}{k^{2n+1}} - 2n\theta\,\zeta(2n+1)\\ &= 2(-1)^n\,\theta^{2n+1}\left(\sum_{k=1}^{n-1}\frac{(-1)^k\cdot k}{(2n-2k+1)!}\,\frac{\zeta(2k+1)}{\theta^{2k}} - \sum_{k=0}^{\infty}\frac{(2k)!}{(2n+2k+1)!}\,\frac{\zeta(2k)}{(2\pi/\theta)^{2k}}\right)\\ &\qquad(|\theta| < 2\pi;\ \theta \in \mathbb{R};\ n \in \mathbb{N}).\end{aligned} \tag{13}$$

Since each series in (13) is uniformly convergent with respect to θ on the open interval $(-2\pi, 2\pi)$, by executing termwise differentiation in (13) with respect to θ, they showed that [Srivastava and Tsumura (2000a, p. 328, Theorem 2)]

$$
\begin{aligned}
(14)\qquad & 2n\sum_{k=1}^{\infty}\frac{\cos(k\theta)}{k^{2n+1}}+\theta\sum_{k=1}^{\infty}\frac{\sin(k\theta)}{k^{2n}}-2n\,\zeta(2n+1)\\
&=2(-1)^n\,\theta^{2n}\left(\sum_{k=1}^{n-1}\frac{(-1)^k\cdot k}{(2n-2k)!}\,\frac{\zeta(2k+1)}{\theta^{2k}}-\sum_{k=0}^{\infty}\frac{(2k)!}{(2n+2k)!}\,\frac{\zeta(2k)}{(2\pi/\theta)^{2k}}\right)\\
&\qquad(|\theta|<2\pi;\ \theta\in\mathbb{R};\ n\in\mathbb{N}).
\end{aligned}
$$

Formula (14) was proved *independently* by Katsurada (1999, p. 81, Theorem 1) by using the Mellin transform techniques.

In their special cases when $\theta=\pi$, Equations (13) and (14) readily yield a result given by Srivastava (1999a, p. 393, Eq. (3.20)):

$$
\begin{aligned}
(15)\qquad \zeta(2n+1)=(-1)^n\,\frac{2(2\pi)^{2n}}{(2n-1)\,2^{2n}+1}&\left[\sum_{k=1}^{n-1}\frac{(-1)^{k-1}\cdot k}{(2n-2k+1)!}\,\frac{\zeta(2k+1)}{\pi^{2k}}\right.\\
&\left.+\sum_{k=0}^{\infty}\frac{(2k)!}{(2n+2k+1)!}\,\frac{\zeta(2k)}{2^{2k}}\right]\qquad(n\in\mathbb{N})
\end{aligned}
$$

and one of the two *main* results of Cvijović and Klinowski (1997, p. 1265, Theorem A):

$$
\begin{aligned}
(16)\qquad \zeta(2n+1)=(-1)^n\,\frac{(2\pi)^{2n}}{n(2^{2n+1}-1)}&\left[\sum_{k=1}^{n-1}\frac{(-1)^{k-1}\cdot k}{(2n-2k)!}\,\frac{\zeta(2k+1)}{\pi^{2k}}\right.\\
&\left.+\sum_{k=0}^{\infty}\frac{(2k)!}{(2n+2k)!}\,\frac{\zeta(2k)}{2^{2k}}\right]\qquad(n\in\mathbb{N}).
\end{aligned}
$$

Remark 1. The series representation (15) was given by Srivastava (1997, p. 4, Eq. (2.5)) [see also Srivastava (1999a, p. 393, Eq. (3.20))]. Indeed, as already observed by Srivastava [*op. cit.*], (15) provides a significantly simpler (and much more rapidly convergent) version of one of the two *main* results of Cvijović and Klinowski (1997, p. 1265, Theorem B).

Remark 2. The series representation (16) is the other *main* result of Cvijović and Klinowski (1997, p. 1265, Theorem A), whose companion was referred to in Remark 1 above.

Remark 3. Since

$$
(17)\qquad \frac{(2k-1)!}{(2n+2k+1)!}=\frac{1}{2n+1}\left[\frac{(2k-1)!}{(2n+2k)!}-\frac{(2k)!}{(2n+2k+1)!}\right]\qquad(n,k\in\mathbb{N}),
$$

it is not difficult to obtain Wilton's result (1) by combining the series representation (15) with another result 4.2(20) of Srivastava (1997, p. 1, Eq. (1.3)) [see also Srivastava (1999a, p. 389, Eq. (2.9))].

With a view to applying (13) and (14) in their other special cases when

$$\theta = \frac{2}{3}\pi, \quad \frac{1}{2}\pi, \quad \text{and} \quad \frac{1}{3}\pi, \tag{18}$$

we now recall several trigonometric sums given by

Lemma 4 [Srivastava and Tsumura (2000a)]. *For* $\Re(s) > 1$,

$$\sum_{n=1}^{\infty} \frac{\cos\left(\frac{2}{3}n\pi\right)}{n^s} = \frac{3^{1-s}-1}{2}\,\zeta(s), \tag{19}$$

$$\sum_{n=1}^{\infty} \frac{\sin\left(\frac{2}{3}n\pi\right)}{n^s} = \sqrt{3}\left\{\frac{3^{-s}-1}{2}\,\zeta(s) + 3^{-s}\,\zeta\left(s,\frac{1}{3}\right)\right\}, \tag{20}$$

$$\sum_{n=1}^{\infty} \frac{\cos\left(\frac{1}{2}n\pi\right)}{n^s} = 2^{-s}(2^{1-s}-1)\,\zeta(s), \tag{21}$$

$$\sum_{n=1}^{\infty} \frac{\sin\left(\frac{1}{2}n\pi\right)}{n^s} = (2^{-s}-1)\,\zeta(s) + 2^{1-2s}\,\zeta\left(s,\frac{1}{4}\right), \tag{22}$$

$$\sum_{n=1}^{\infty} \frac{\cos\left(\frac{1}{3}n\pi\right)}{n^s} = \frac{1}{2}\left(6^{1-s} - 3^{1-s} - 2^{1-s} + 1\right)\,\zeta(s), \tag{23}$$

and

$$\sum_{n=1}^{\infty} \frac{\sin\left(\frac{1}{3}n\pi\right)}{n^s} = \sqrt{3}\left[\frac{3^{-s}-1}{2}\,\zeta(s) + 6^{-s}\left\{\zeta\left(s,\frac{1}{6}\right) + \zeta\left(s,\frac{1}{3}\right)\right\}\right]. \tag{24}$$

Srivastava and Tsumura (2000a) first proved the following results:

$$\begin{aligned}\zeta(2n+1) = (-1)^n \frac{2(2\pi)^{2n}}{n(3^{2n+1}-1)} &\left[\frac{(3^{2n}-1)\pi}{4\sqrt{3}\,(2n)!}\,B_{2n} + \frac{(-1)^n\,\pi}{\sqrt{3}\,(2\pi)^{2n}}\,\zeta\left(2n,\frac{1}{3}\right)\right.\\ &\left. + \sum_{k=1}^{n-1} \frac{(-1)^{k-1}\cdot k}{(2n-2k)!}\,\frac{\zeta(2k+1)}{\left(\frac{2}{3}\pi\right)^{2k}} + \sum_{k=0}^{\infty} \frac{(2k)!}{(2n+2k)!}\,\frac{\zeta(2k)}{3^{2k}}\right] \qquad (n \in \mathbb{N})\end{aligned} \tag{25}$$

(26)

$$\begin{aligned}\zeta(2n+1) =(-1)^n \frac{2(2\pi)^{2n}}{n(2^{4n+1}+2^{2n}-1)} &\left[\frac{2^{2n-3}(2^{2n}-1)\pi}{(2n)!}\,B_{2n} + \frac{(-1)^n\,\pi}{2(2\pi)^{2n}}\,\zeta\left(2n,\frac{1}{4}\right)\right.\\ &\left. + \sum_{k=1}^{n-1} \frac{(-1)^{k-1}\cdot k}{(2n-2k)!}\,\frac{\zeta(2k+1)}{\left(\frac{1}{2}\pi\right)^{2k}} + \sum_{k=0}^{\infty} \frac{(2k)!}{(2n+2k)!}\,\frac{\zeta(2k)}{4^{2k}}\right] \qquad (n \in \mathbb{N}),\end{aligned}$$

and

$$\begin{aligned}\zeta(2n+1) &= (-1)^n \frac{2(2\pi)^{2n}}{n(6^{2n}+3^{2n}+2^{2n}-1)} \left[\frac{2^{2n-3}(3^{2n}-1)\pi}{\sqrt{3}\,(2n)!} B_{2n} \right. \\ &\quad + \frac{(-1)^n\,\pi}{2\sqrt{3}\,(2\pi)^{2n}} \left\{ \zeta\left(2n, \frac{1}{3}\right) + \zeta\left(2n, \frac{1}{6}\right) \right\} \\ &\quad \left. + \sum_{k=1}^{n-1} \frac{(-1)^{k-1}\cdot k}{(2n-2k)!} \frac{\zeta(2k+1)}{\left(\frac{1}{3}\pi\right)^{2k}} + \sum_{k=0}^{\infty} \frac{(2k)!}{(2n+2k)!} \frac{\zeta(2k)}{6^{2k}} \right] \quad (n \in \mathbb{N}), \end{aligned} \tag{27}$$

in terms of the Bernoulli numbers B_n defined by 1.6(2).

By comparing the series representation (26) with a known result due to Srivastava (1998, p. 9, Eq. (41)) [see also Srivastava (1997, p. 5, Eq. (3.3))], Srivastava and Tsumura (2000a) derived an interesting identity involving the Zeta functions $\zeta(s)$ and $\zeta(s,a)$ (and their derivatives with respect to s), which is given below:

$$\begin{aligned}\frac{(-1)^n\pi}{2(2\pi)^{2n}}\,\zeta\left(2n, \frac{1}{4}\right) &= \left\{(2^{2n-2}-1)\log 2 - 2^{2n-3}(2^{2n}-1)\pi\right\} \frac{B_{2n}}{(2n)!} \\ &\quad - \frac{2^{2n-1}-1}{2(2n-1)!}\,\zeta'(1-2n) - \frac{4^{2n-1}}{(2n-1)!}\,\zeta'\left(1-2n, \frac{1}{4}\right) \quad (n \in \mathbb{N}).\end{aligned} \tag{28}$$

By applying (25), (26), and (27), and Srivastava's series representations 4.2(21), 4.2(22), and 4.2(23), Srivastava and Tsumura (2000a) proved their *main* results given by

$$\begin{aligned}\zeta(2n+1) &= (-1)^{n-1}\left(\frac{2\pi}{3}\right)^{2n} \left[\frac{H_{2n+1} - \log\left(\frac{2}{3}\pi\right)}{(2n+1)!} + \frac{(3^{2n+2}-1)\pi}{2\sqrt{3}\,(2n+2)!} B_{2n+2} \right. \\ &\quad + \frac{(-1)^{n-1}}{\sqrt{3}\,(2\pi)^{2n+1}}\,\zeta\left(2n+2, \frac{1}{3}\right) \\ &\quad \left. + \sum_{k=1}^{n-1} \frac{(-1)^k}{(2n-2k+1)!} \frac{\zeta(2k+1)}{\left(\frac{2}{3}\pi\right)^{2k}} + 2\sum_{k=1}^{\infty} \frac{(2k-1)!}{(2n+2k+1)!} \frac{\zeta(2k)}{3^{2k}} \right] \quad (n \in \mathbb{N}),\end{aligned} \tag{29}$$

$$\begin{aligned}\zeta(2n+1) &= (-1)^{n-1}\left(\frac{\pi}{2}\right)^{2n} \left[\frac{H_{2n+1} - \log\left(\frac{1}{2}\pi\right)}{(2n+1)!} + \frac{2^{2n}(2^{2n+2}-1)\pi}{(2n+2)!} B_{2n+2} \right. \\ &\quad + \frac{(-1)^{n-1}}{2(2\pi)^{2n+1}}\,\zeta\left(2n+2, \frac{1}{4}\right) \\ &\quad \left. + \sum_{k=1}^{n-1} \frac{(-1)^k}{(2n-2k+1)!} \frac{\zeta(2k+1)}{\left(\frac{1}{2}\pi\right)^{2k}} + 2\sum_{k=1}^{\infty} \frac{(2k-1)!}{(2n+2k+1)!} \frac{\zeta(2k)}{4^{2k}} \right] \quad (n \in \mathbb{N}),\end{aligned} \tag{30}$$

and

$$\begin{aligned}\zeta(2n+1) &= (-1)^{n-1}\left(\frac{\pi}{3}\right)^{2n} \left[\frac{H_{2n+1} - \log\left(\frac{1}{3}\pi\right)}{(2n+1)!} + \frac{2^{2n}(3^{2n+2}-1)\pi}{\sqrt{3}\,(2n+2)!} B_{2n+2} \right. \\ &\quad + \frac{(-1)^{n-1}}{2\sqrt{3}\,(2\pi)^{2n+1}} \left\{ \zeta\left(2n+2, \frac{1}{3}\right) + \zeta\left(2n+2, \frac{1}{6}\right) \right\} \\ &\quad \left. + \sum_{k=1}^{n-1} \frac{(-1)^k}{(2n-2k+1)!} \frac{\zeta(2k+1)}{\left(\frac{1}{3}\pi\right)^{2k}} + 2\sum_{k=1}^{\infty} \frac{(2k-1)!}{(2n+2k+1)!} \frac{\zeta(2k)}{6^{2k}} \right] \quad (n \in \mathbb{N}).\end{aligned} \tag{31}$$

Each of the main series representations (29), (30), and (31) belongs to the class containing Wilton's formula (1) and Srivastava's formula (2). If, for convenience, we denote the summands of the *infinite* series in the representations (1), (2), (29), (30), and (31) by $\mathcal{S}_k^{(j)}$ $(j = 1,2,3,4,5)$, respectively, and apply Stirling's formula 1.1(54) and the fact that $\zeta(2k) \to 1$ as $k \to \infty$, we easily obtain the following order estimates [*cf.* Equations (3) and (4)]:

$$\mathcal{S}_k^{(j)} = O\left(k^{-2n-2} \cdot m^{-2k}\right) \quad (k \to \infty;\ n \in \mathbb{N}) \tag{32}$$

($m = 2$ when $j = 1$; $m = 3$ when $j = 3$; $m = 4$ when $j = 2$ and $j = 4$; $m = 6$ when $j = 5$).

Clearly, therefore, of these five series representations for $\zeta(2n+1)$, the result (31) involves the most rapidly convergent series. On the other hand, the rate of convergence of the series involved in each of the Srivastava-Tsumura results (29), (30), and (31) is obviously much better than that of the series involved in Wilton's result (1). And, in particular, the rate of convergence of the series involved in the Srivastava-Tsumura result (30) is as good as that of the series involved in Srivastava's result (2).

4.4 Computational Results

Srivastava (2000a) investigated rather systematically several interesting evaluations and representations of $\zeta(s)$ when $s \in \mathbb{N}\setminus\{1\}$. In one of many computationally useful special cases considered by him, it is observed that $\zeta(3)$ can be represented by means of a series which converges much more rapidly than that in Euler's celebrated formula 3.1(8) as well as the series 4.2(2) used recently by Apéry (1979) in his proof of the irrationality of $\zeta(3)$. Symbolic and numerical computations using *Mathematica* (Version 4.0) for Linux show, among other things, that only 50 terms of the series (38) below are capable of producing an accuracy of seven decimal places.

Since

$$i \cot iz = \coth z = \frac{2}{e^{2z}-1} + 1 \qquad \left(i := \sqrt{-1}\right), \tag{1}$$

by replacing z in the known expansion 4.2(4) by $\frac{1}{2}i\pi z$, it is easily seen that [*cf.*, *e.g.*, Koblitz (1977, p. 25); see also Erdélyi *et al.* (1953, p. 51, Eq. 1.20(1))]

$$\frac{\pi z}{e^{\pi z}-1} + \frac{\pi z}{2} = \sum_{k=0}^{\infty} \frac{(-1)^{k+1}\zeta(2k)}{2^{2k-1}} z^{2k} \qquad (|z| < 2). \tag{2}$$

By setting $z = it$ in (2), multiplying both sides by t^{m-1} $(m \in \mathbb{N})$, and then integrating the resulting equation from $t = 0$ to $t = \tau$ $(0 < \tau < 2)$, if we apply the (readily derivable) integral formula:

$$\int_0^\tau t^m e^{\lambda t} dt = \frac{(-1)^m m! e^{\lambda\tau}}{\lambda^{m+1}} \left(\sum_{j=0}^{m} \frac{(-\lambda\tau)^j}{j!} - e^{-\lambda\tau}\right) \qquad (\lambda \neq 0), \tag{3}$$

we shall obtain

$$
\begin{aligned}
&m!\left(\frac{i}{\pi\tau}\right)^m\left\{\sum_{k=1}^{\infty}\frac{\cos(k\pi\tau)+i\sin(k\pi\tau)}{k^{m+1}}-\zeta(m+1)\right\}\\
&\quad+\sum_{k=1}^{\infty}\frac{\cos(k\pi\tau)}{k}+i\left(\frac{\pi\tau}{2(m+1)}+\sum_{k=1}^{\infty}\frac{\sin(k\pi\tau)}{k}\right)\\
&\quad+\sum_{j=1}^{m-1}\frac{m!}{(m-j)!}\left(\frac{i}{\pi\tau}\right)^j\sum_{k=1}^{\infty}\frac{\cos(k\pi\tau)+i\sin(k\pi\tau)}{k^{j+1}}
\end{aligned}
$$

$$
\tag{4} =\sum_{k=0}^{\infty}\frac{\zeta(2k)}{k+\frac{1}{2}m}\left(\frac{\tau}{2}\right)^{2k}\qquad (m\in\mathbb{N};\ 0<\tau<2),
$$

which, in the special case when $\tau = 1$, would immediately simplify to the form:

$$
\tag{5}
\begin{aligned}
&\frac{i\pi}{2(m+1)}-\left(2^{m+1}-1\right)\left(\frac{i}{2\pi}\right)^m m!\,\zeta(m+1)\\
&\quad=\log 2+\sum_{j=1}^{m-1}\frac{m!}{(m-j)!}\left(2^j-1\right)\left(\frac{i}{2\pi}\right)^j\zeta(j+1)\\
&\quad+\sum_{k=0}^{\infty}\frac{\zeta(2k)}{\left(k+\frac{1}{2}m\right)2^{2k}}\qquad (m\in\mathbb{N}),
\end{aligned}
$$

where we have also applied the last part of the definition 2.3(1). Furthermore, in (4), (5), and elsewhere in this section, an empty sum is interpreted (as usual) to be nil.

Now, for a given sequence $\{\Omega_n\}_{n=1}^{\infty}$, it is easily verified that

$$
\tag{6} \sum_{j=1}^{m-1}\Omega_j=\sum_{j=1}^{[(m-1)/2]}\Omega_{2j}+\sum_{j=1}^{[m/2]}\Omega_{2j-1},
$$

where, just as in 4.1(17), $[\kappa]$ denotes the greatest integer in κ. Thus we find from (5) that

$$
\tag{7}
\begin{aligned}
&\frac{i\pi}{2(m+1)}-\left(2^{m+1}-1\right)\left(\frac{i}{2\pi}\right)^m m!\,\zeta(m+1)\\
&\quad=\log 2+\sum_{j=1}^{[(m-1)/2]}(-1)^j\binom{m}{2j}\frac{(2j)!\,\left(2^{2j}-1\right)}{(2\pi)^{2j}}\zeta(2j+1)\\
&\quad+i\sum_{j=1}^{[m/2]}(-1)^{j-1}\binom{m}{2j-1}\frac{(2j-1)!\,\left(2^{2j-1}-1\right)}{(2\pi)^{2j-1}}\zeta(2j)\\
&\quad+\sum_{k=0}^{\infty}\frac{\zeta(2k)}{\left(k+\frac{1}{2}m\right)2^{2k}}\qquad (m\in\mathbb{N}).
\end{aligned}
$$

Setting $m = 2n \quad (n \in \mathbb{N})$ in (7), and then equating the real and imaginary parts in the resulting equation, we obtain [Srivastava (2000a)]

(8)
$$\zeta(2n+1) = (-1)^{n-1} \frac{(2\pi)^{2n}}{(2n)!\,(2^{2n+1}-1)} \cdot \left[\log 2 + \sum_{j=1}^{n-1} (-1)^j \binom{2n}{2j} \frac{(2j)!\,(2^{2j}-1)}{(2\pi)^{2j}} \zeta(2j+1) + \sum_{k=0}^{\infty} \frac{\zeta(2k)}{(k+n)2^{2k}} \right] \quad (n \in \mathbb{N})$$

and

(9)
$$\zeta(2n) = (-1)^{n-1} \frac{(2\pi)^{2n-1}}{(2n)!(2^{2n-1}-1)} \cdot \left[\frac{\pi}{2(2n+1)} + \sum_{j=1}^{n-1} (-1)^j \binom{2n}{2j-1} \frac{(2j-1)!\,(2^{2j-1}-1)}{(2\pi)^{2j-1}} \zeta(2j) \right] \quad (n \in \mathbb{N}).$$

Similarly, if we set $m = 2n+1$ $(n \in \mathbb{N}_0)$ in (7), we shall obtain the following results which can provide series representations for $\zeta(2n+1)$ and $\zeta(2n)$ when $n \in \mathbb{N}$:

(10)
$$\sum_{j=1}^{n} (-1)^{j-1} \binom{2n+1}{2j} \frac{(2j)!\,(2^{2j}-1)}{(2\pi)^{2j}} \zeta(2j+1) = \log 2 + \sum_{k=0}^{\infty} \frac{\zeta(2k)}{\left(k+n+\frac{1}{2}\right) 2^{2k}} \quad (n \in \mathbb{N}_0)$$

and

(11)
$$\zeta(2n) = (-1)^{n-1} \frac{(2\pi)^{2n-1}}{(2n-1)!\,(2^{2n}-1)} \cdot \left[\frac{\pi}{4n} + \sum_{j=1}^{n-1} (-1)^j \binom{2n-1}{2j-1} \frac{(2j-1)!\,(2^{2j-1}-1)}{(2\pi)^{2j-1}} \zeta(2j) \right] \quad (n \in \mathbb{N}).$$

Indeed, in its particular cases when $n = 0$ and $n \in \mathbb{N}$, the summation formula (10) immediately yields the known sum 3.4(467) and [Srivastava (2000a)]

(12)
$$\zeta(2n+1) = (-1)^{n-1} \frac{(2\pi)^{2n}}{(2n+1)!\,(2^{2n}-1)} \cdot \left[\log 2 + \sum_{j=1}^{n-1} (-1)^j \binom{2n+1}{2j} \frac{(2j)!\,(2^{2j}-1)}{(2\pi)^{2j}} \zeta(2j+1) + \sum_{k=0}^{\infty} \frac{\zeta(2k)}{\left(k+n+\frac{1}{2}\right) 2^{2k}} \right] \quad (n \in \mathbb{N}),$$

respectively.

Each of the recursion formulas (9) and (11) can be used to evaluate $\zeta(2n)$ $(n \in \mathbb{N})$. Formula (9) was proven, in a markedly different way, by (for example) Stark (1974,

p. 199, Eq. (5)). Formulas (9) and (11), together, yield the series identity [Srivastava (2000a)]:

(13)
$$\sum_{j=1}^{n-1}(-1)^{j-1}\binom{2n-1}{2j-1}\frac{(2j-1)!\,(2^{2j-1}-1)}{(2\pi)^{2j-1}}\zeta(2j)=\frac{\pi\left\{(2n-1)2^{2n-1}-2n\right\}}{4n(2n+1)\,(2^{2n-1}-1)}$$
$$+\frac{2^{2n}-1}{2n\,(2^{2n-1}-1)}\sum_{j=1}^{n-1}(-1)^{j-1}\binom{2n}{2j-1}\frac{(2j-1)!\,(2^{2j-1}-1)}{(2\pi)^{2j-1}}\zeta(2j)\quad(n\in\mathbb{N})$$

or, equivalently,

(14)
$$\sum_{j=1}^{n-1}(-1)^{j-1}\binom{2n-1}{2j-2}\left\{2(n-j)\,(2^{2n-1}-1)-2^{2n-1}\right\}\frac{(2j-2)!\,(2^{2j-1}-1)}{(2\pi)^{2j-1}}\zeta(2j)$$
$$=\frac{\pi\left\{(2n-1)2^{2n-1}-2n\right\}}{4n(2n+1)}\qquad(n\in\mathbb{N}),$$

which can also be used as a recurrence relation for evaluating $\zeta(2n)$ $(n\in\mathbb{N})$.

An integral representation for $\zeta(2n+1)$, which is equivalent to the series representation (8), was given earlier by Dąbrowski (1996, p. 203, Eq. (16)), who (1996, p. 206) also mentioned the existence of (but did not fully state) the series representation (12). The series representation (8) is derived also in an interesting paper by Borwein *et al.* (2000, p. 269, Eq. (57)).

For $n=1$, (8) immediately yields the series representation 4.2(5) which, in conjunction with the known sum 3.4(467), would lead us readily to Euler's formula 3.1(8). On the other hand, by setting $n=1$ in the series representation (12), we obtain

$$\zeta(3)=\frac{4\pi^2}{9}\left(\frac{1}{2}\log 2+\sum_{k=0}^{\infty}\frac{\zeta(2k)}{(2k+3)2^{2k}}\right),\tag{15}$$

which was derived *independently* by (for example) Glasser (1968, p. 446, Eq. (12)), Zhang and Williams (1993b, p. 1585, Eq. (2.13)), and Dąbrowski (1996, p. 206) [see also Chen and Srivastava (1998, p. 183, Eq. (2.15))].

By suitably combining the series occurring in 4.2(5), 3.4(467), and (15), it is not difficult to derive several other series representations for $\zeta(3)$, which are analogous to Euler's formula 3.1(8). More generally, since

$$\begin{aligned}&\frac{\lambda k^2+\mu k+\nu}{(2k+2n-1)(2k+2n)(2k+2n+1)}\\&\qquad=\frac{\mathcal{A}}{2k+2n-1}+\frac{\mathcal{B}}{2k+2n}+\frac{\mathcal{C}}{2k+2n+1},\end{aligned}\tag{16}$$

where, for convenience,

$$\mathcal{A}=\mathcal{A}_n(\lambda,\mu,\nu):=\frac{1}{2}\left[\lambda n^2-(\lambda+\mu)n+\frac{1}{4}(\lambda+2\mu+4\nu)\right],\tag{17}$$

$$\mathcal{B}=\mathcal{B}_n(\lambda,\mu,\nu):=-\left(\lambda n^2-\mu n+\nu\right),\tag{18}$$

and

$$\mathcal{C} = \mathcal{C}_n(\lambda, \mu, \nu) := \frac{1}{2}\left[\lambda n^2 + (\lambda - \mu)n + \frac{1}{4}(\lambda - 2\mu + 4\nu)\right], \tag{19}$$

by appealing to (8), (10) *with* n replaced by $n-1$, and (12), we can derive the following unification of a large number of known (or new) series representations including (for example) Euler's result 3.1(8)[Srivastava (2000a)]:

$$\begin{aligned}\zeta(2n+1) &= \frac{(-1)^{n-1}(2\pi)^{2n}}{(2n)!\left\{(2^{2n+1}-1)\,\mathcal{B} + (2n+1)\,(2^{2n}-1)\,\mathcal{C}\right\}} \\ &\cdot \left[\frac{1}{4}\lambda \log 2 + \sum_{j=1}^{n-1}(-1)^j\binom{2n-1}{2j-2}\left\{2j(2j-1)\mathcal{A} + [\lambda(4n-1) - 2\mu]\,nj\right.\right. \\ &\qquad \left. + \lambda n\left(n+\frac{1}{2}\right)\right\}\frac{(2j-2)!\,(2^{2j}-1)}{(2\pi)^{2j}}\zeta(2j+1) \\ &\quad \left. + \sum_{k=0}^{\infty}\frac{(\lambda k^2 + \mu k + \nu)\,\zeta(2k)}{(2k+2n-1)(k+n)(2k+2n+1)2^{2k}}\right] \qquad (n \in \mathbb{N};\ \lambda, \mu, \nu \in \mathbb{C}),\end{aligned} \tag{20}$$

where $\mathcal{A}$, $\mathcal{B}$, and $\mathcal{C}$ are given by (17), (18), and (19), respectively.

For $\lambda = 0$, the series representation (20) simplifies to the form:

(21)

$$\begin{aligned}\zeta(2n+1) &= \frac{(-1)^{n-1}(2\pi)^{2n}}{(2n)!\left\{(2^{2n+1}-1)\,(\mu n - \nu) - (2^{2n}-1)\,\left(n+\frac{1}{2}\right)\left[\mu(n+\frac{1}{2}) - \nu\right]\right\}} \\ &\cdot \left[\sum_{j=1}^{n-1}(-1)^j\binom{2n-1}{2j-2}\left\{j(2j-1)\left[\nu - \mu\left(n-\frac{1}{2}\right)\right] - 2\mu nj\right\}\right. \\ &\cdot \left.\frac{(2j-2)!\,(2^{2j}-1)}{(2\pi)^{2j}}\zeta(2j+1) + \sum_{k=0}^{\infty}\frac{(\mu k+\nu)\zeta(2k)}{(2k+2n-1)(k+n)(2k+2n+1)2^{2k}}\right] \\ &\qquad (n \in \mathbb{N};\ \mu, \nu \in \mathbb{C}).\end{aligned}$$

Furthermore, by setting

$$\lambda = \mu = 0 \qquad \text{and} \qquad \nu = 1$$

in (20) or (alternatively) by setting

$$\mu = 0 \qquad \text{and} \qquad \nu = 1$$

in (21), we immediately obtain the series representation [Srivastava (2000a)]:

(22)

$$\begin{aligned}&\zeta(2n+1) \\ &= \frac{(-1)^{n-1}(2\pi)^{2n}}{(2n)!\left\{2^{2n}(2n-3) - 2n + 1\right\}}\left[\sum_{j=1}^{n-1}(-1)^j\binom{2n-1}{2j-2}\frac{(2j)!\,(2^{2j}-1)}{(2\pi)^{2j}}\zeta(2j+1)\right. \\ &\qquad \left. + 2\sum_{k=0}^{\infty}\frac{\zeta(2k)}{(2k+2n-1)(k+n)(2k+2n+1)2^{2k}}\right] \qquad (n \in \mathbb{N}),\end{aligned}$$

which, in the special case when $n = 1$, was given by Chen and Srivastava (1998, p. 189, Eq. (2.45)].

Of the three representations (20), (21), and (22) for $\zeta(2n+1)$ $(n \in \mathbb{N})$, the infinite series in (22) converges most rapidly. Many other families of rapidly convergent series representations for $\zeta(2n+1)$ $(n \in \mathbb{N})$ can be found in the recent works of Srivastava (1998 and 1999a) (and indeed also in the numerous references already cited in each of these earlier works), including the results presented in previous sections (especially Section 4.2).

For various suitable special values of the parameters λ, μ, and ν, we can easily deduce from (20) and (21) several known (or new) series representations for $\zeta(2n+1)$ $(n \in \mathbb{N})$. For example, if we set

$$\mu = 2 \quad \text{and} \quad \nu = 2n+1$$

in our series representation (21), we shall obtain [Srivastava (2000a)]

$$\begin{aligned}\zeta(2n+1) = (-1)^{n-1} \frac{(2\pi)^{2n}}{(2n)!\,(2^{2n+1}-1)} &\left[\sum_{j=1}^{n-1} (-1)^j \binom{2n-1}{2j-1} \right. \\ \cdot \frac{(2j)!\,(2^{2j}-1)}{(2\pi)^{2j}} \zeta(2j+1) - &\left. \sum_{k=0}^{\infty} \frac{\zeta(2k)}{(2k+2n-1)(k+n)2^{2k}} \right] \quad (n \in \mathbb{N}),\end{aligned} \tag{23}$$

which, in the special case when $n = 1$, immediately yields Euler's formula 3.1(8).

The following additional series representations for $\zeta(2n+1)$ $(n \in \mathbb{N})$, which are analogous to (23), can also be deduced similarly from (21) [Srivastava (2000a)]:

$$\begin{aligned}\zeta(2n+1) = (-1)^{n-1} \frac{(2\pi)^{2n}}{(2n)!\,\{(2n-1)2^{2n}-2n\}} &\left[2n \sum_{j=1}^{n-1} (-1)^j \binom{2n-1}{2j-2} \right. \\ \cdot \frac{(2j)!\,(2^{2j}-1)}{(2\pi)^{2j}} \zeta(2j+1) - &\left. \sum_{k=0}^{\infty} \frac{\zeta(2k)}{(k+n)(2k+2n+1)2^{2k}} \right] \quad (n \in \mathbb{N})\end{aligned} \tag{24}$$

and

(25)

$$\begin{aligned}\zeta(2n+1) = (-1)^{n-1} \frac{(2\pi)^{2n}}{(2n+1)!\,(2^{2n}-1)} &\left[\sum_{j=1}^{n-1} (-1)^j \left(\frac{4nj-2j+1}{2j-1} \right) \binom{2n-1}{2j-2} \right. \\ \cdot \frac{(2j)!\,(2^{2j}-1)}{(2\pi)^{2j}} \zeta(2j+1) - 4 &\left. \sum_{k=0}^{\infty} \frac{\zeta(2k)}{(2k+2n-1)(2k+2n+1)2^{2k}} \right] \quad (n \in \mathbb{N}).\end{aligned}$$

The special case of each of the last two series representations (24) and (25) when $n = 1$ was given by Zhang and Williams (1993b, p. 1586). For $n = 2$, Srivastava's formulas (22) to (25) would readily yield the following series representations for $\zeta(5)$:

$$\zeta(5) = \frac{\pi^2}{13} \zeta(3) - \frac{8\pi^4}{13} \sum_{k=0}^{\infty} \frac{\zeta(2k)}{(2k+3)(2k+4)(2k+5)2^{2k}}, \tag{26}$$

(27) $$\zeta(5)=\frac{3\pi^2}{31}\zeta(3)+\frac{4\pi^4}{93}\sum_{k=0}^{\infty}\frac{\zeta(2k)}{(2k+3)(2k+4)2^{2k}},$$

(28) $$\zeta(5)=\frac{\pi^2}{11}\zeta(3)+\frac{\pi^4}{33}\sum_{k=0}^{\infty}\frac{\zeta(2k)}{(2k+4)(2k+5)2^{2k}},$$

and

(29) $$\zeta(5)=\frac{7\pi^2}{75}\zeta(3)+\frac{8\pi^4}{225}\sum_{k=0}^{\infty}\frac{\zeta(2k)}{(2k+3)(2k+5)2^{2k}}.$$

Next, with a view to *further* improving the rate of convergence in the most rapidly convergent series representations (22) and (26) considered in this work, we observe that

(30) $$\begin{aligned}&\frac{1}{(2k+2n-1)(2k+2n)(2k+2n+1)(2k+2n+2)}\\&\quad=\frac{1}{6}\left(\frac{1}{2k+2n-1}-\frac{1}{2k+2n+2}\right)-\frac{1}{2}\frac{1}{(2k+2n)(2k+2n+1)}.\end{aligned}$$

Thus, by applying the series representations (10) *with* n replaced by $n-1$, (8) *with* n replaced by $n+1$, and (24), we obtain [Srivastava (2000a)]

(31)
$$\begin{aligned}\zeta(2n+3)&=\frac{2\pi^2\left\{2^{2n+2}+n(2n-3)\left(2^{2n}-1\right)-1\right\}}{(n+1)(2n+1)\left(2^{2n+3}-1\right)}\zeta(2n+1)\\&+(-1)^{n-1}\frac{(2\pi)^{2n+2}}{(2n+2)!\left(2^{2n+3}-1\right)}\left[\sum_{j=1}^{n-1}(-1)^j\left\{\binom{2n-1}{2j}-\binom{2n+2}{2j}\right.\right.\\&\qquad\left.+6n\binom{2n-1}{2j-2}\right\}\frac{(2j)!\left(2^{2j}-1\right)}{(2\pi)^{2j}}\zeta(2j+1)\\&+12\sum_{k=0}^{\infty}\left.\frac{\zeta(2k)}{(2k+2n-1)(2k+2n)(2k+2n+1)(2k+2n+2)2^{2k}}\right]\qquad(n\in\mathbb{N}),\end{aligned}$$

where the series converges faster than that in (22).

In its special case when $n=1$, (31) readily yields the following improved version of the series representation (26) above [*cf.* Zhang and Williams (1993b , p. 1590, Eq. (3.14))]:

(32) $$\zeta(5)=\frac{4\pi^2}{31}\zeta(3)+\frac{8\pi^4}{31}\sum_{k=0}^{\infty}\frac{\zeta(2k)}{(2k+1)(2k+2)(2k+3)(2k+4)2^{2k}},$$

in which $\zeta(3)$ can be replaced by its known value $-4\pi^2\,\zeta'(-2)$ given by 2.3(22) for $n=1$.

Yet another rapidly convergent series representation for $\zeta(2n+3)$ $(n \in \mathbb{N})$, analogous to (31), can be derived by means of the identity:

$$\frac{1}{(2k+2n)(2k+2n+1)(2k+2n+2)(2k+2n+3)} = \frac{1}{6}\left(\frac{1}{2k+2n} - \frac{1}{2k+2n+3}\right) - \frac{1}{2}\frac{1}{(2k+2n+1)(2k+2n+2)}, \tag{33}$$

together with the series representations (8), (12) *with* n replaced by $n+1$, and (23) *with* n replaced by $n+1$. We thus obtain the series representation [Srivastava (2000a)]:

(34)

$$\begin{aligned}\zeta(2n+3) &= \frac{2\pi^2\left\{\frac{1}{3}(2n+1)\left(2n^2-4n+3\right)\left(2^{2n}-1\right)-2^{2n+1}+1\right\}}{(n+1)(2n+1)\left\{(2n-3)2^{2n+2}-2n\right\}}\zeta(2n+1) \\ &+ (-1)^{n-1}\frac{(2\pi)^{2n+2}}{(2n+2)!\left\{(2n-3)2^{2n+2}-2n\right\}}\left[\sum_{j=1}^{n-1}(-1)^j\left\{\binom{2n}{2j}-\binom{2n+3}{2j}\right.\right. \\ &\qquad \left.+3\binom{2n+1}{2j-1}\right\}\frac{(2j)!\left(2^{2j}-1\right)}{(2\pi)^{2j}}\zeta(2j+1) \\ &\left.+12\sum_{k=0}^{\infty}\frac{\zeta(2k)}{(2k+2n)(2k+2n+1)(2k+2n+2)(2k+2n+3)2^{2k}}\right] \qquad (n \in \mathbb{N}),\end{aligned}$$

which, in the special case when $n=1$, yields

$$\zeta(5) = \frac{2\pi^2}{27}\zeta(3) - \frac{4\pi^4}{9}\sum_{k=0}^{\infty}\frac{\zeta(2k)}{(2k+2)(2k+3)(2k+4)(2k+5)2^{2k}}, \tag{35}$$

where the series obviously converges faster than that in (26).

Lastly, by applying the identity:

$$\begin{aligned}\frac{1}{2k(2k+2n-1)(2k+2n)(2k+2n+1)} &= \frac{1}{2n(2n-1)(2n+1)}\frac{1}{2k} \\ -\frac{1}{2(2n-1)}\frac{1}{2k+2n-1} &+ \frac{1}{2n}\frac{1}{2k+2n} - \frac{1}{2(2n+1)}\frac{1}{2k+2n+1}\end{aligned} \tag{36}$$

in conjunction with the series representations (10) *with* n replaced by $n-1$, (8), (12), and the known result 3.4(26), we arrive at the following series representation for $\zeta(2n+1)$ $(n \in \mathbb{N})$ [Srivastava (2000a)]:

$$\begin{aligned}\zeta(2n+1) &= (-1)^{n-1}\frac{(2\pi)^{2n}}{2\cdot(2n-1)!\left\{n-1-(n-2)2^{2n}\right\}}\left[\frac{12n^2-1}{2n^2\left(4n^2-1\right)^2}\right. \\ &- \frac{\log\pi}{n\left(4n^2-1\right)} - \sum_{j=2}^{n-1}(-1)^j\binom{2n-2}{2j-3}\frac{(2j-1)!\left(2^{2j}-1\right)}{(2\pi)^{2j}}\zeta(2j+1) \\ &\left.+\sum_{k=1}^{\infty}\frac{\zeta(2k)}{k(k+n)(2k+2n-1)(2k+2n+1)2^{2k}}\right] \qquad (n \in \mathbb{N}),\end{aligned} \tag{37}$$

where we have also applied the fact that $\zeta(0) = -\frac{1}{2}$.

For $n = 1$, (37) reduces immediately to Wilton's formula [*cf.* Hansen (1975, p. 357, Entry (54.5.9)); see also Chen and Srivastava (1998, p. 181, Eq. (2.1))]:

$$\zeta(3) = \frac{\pi^2}{2}\left(\frac{11}{18} - \frac{1}{3}\log\pi + \sum_{k=1}^{\infty}\frac{\zeta(2k)}{k(k+1)(2k+1)(2k+3)2^{2k}}\right). \tag{38}$$

Furthermore, in its special case when $n = 2$, (37) would yield the following companion of the series representations (32) and (35):

$$\zeta(5) = \frac{2\pi^4}{45}\left(\log\pi - \frac{47}{60} - 30\sum_{k=1}^{\infty}\frac{\zeta(2k)}{k(k+2)(2k+3)(2k+5)2^{2k}}\right), \tag{39}$$

which does *not* contain a term involving $\zeta(3)$ on the right-hand side.

By eliminating $\zeta(2n+3)$ between the results (31) and (34), we can obtain a series representation for $\zeta(2n+1)$ $(n \in \mathbb{N})$, which would converge as rapidly as the series in (37). For $n \in \mathbb{N}$, we thus find that [Srivastava (2000a)]

$$\begin{aligned}\zeta(2n+1) = (-1)^{n-1}\frac{(2\pi)^{2n}}{(2n)!\Delta_n}\Bigg[&\sum_{j=1}^{n-1}(-1)^j\Bigg(\left\{(2n-3)2^{2n+2}-2n\right\}\\ &\left\{\binom{2n-1}{2j} - \binom{2n+2}{2j} + 6n\binom{2n-1}{2j-2}\right\} - \left(2^{2n+3}-1\right)\\ &\cdot\left\{\binom{2n}{2j} - \binom{2n+3}{2j} + 3\binom{2n+1}{2j-1}\right\}\Bigg)\frac{(2j)!\,(2^{2j}-1)}{(2\pi)^{2j}}\zeta(2j+1)\\ &+ 12\sum_{k=0}^{\infty}\frac{(\xi_n k + \eta_n)\,\zeta(2k)}{(2k+2n-1)(2k+2n)(2k+2n+1)(2k+2n+2)(2k+2n+3)2^{2k}}\Bigg],\end{aligned} \tag{40}$$

where, for convenience,

$$\begin{aligned}\Delta_n := \left(2^{2n+3}-1\right)&\left\{\frac{1}{3}(2n+1)\left(2n^2-4n+3\right)\left(2^{2n}-1\right) - 2^{2n+1} + 1\right\}\\ &- \left\{(2n-3)2^{2n+2}-2n\right\}\left\{2^{2n+2} + n(2n-3)\left(2^{2n}-1\right) - 1\right\},\end{aligned} \tag{41}$$

$$\xi_n := 2\left\{(2n-5)2^{2n+2} - 2n + 1\right\}, \tag{42}$$

and

$$\eta_n := \left(4n^2 - 4n - 7\right)2^{2n+2} - (2n+1)^2. \tag{43}$$

In its special case when $n = 1$, (40) yields the following series representation [Srivastava (2000a)]:

$$\zeta(3) = -\frac{6\pi^2}{23}\sum_{k=0}^{\infty}\frac{(98k+121)\zeta(2k)}{(2k+1)(2k+2)(2k+3)(2k+4)(2k+5)2^{2k}}, \tag{44}$$

where the series obviously converges much more rapidly than that in *each* of the celebrated results 3.1(8) and 4.2(2).

We conclude this section by summarizing below the results of some symbolic and numerical computations with the series in (44) using *Mathematica* (Version 4.0) for Linux:

In[1] :=(98k + 121)Zeta[2k]/ ((2k + 1)(2k + 2)(2k + 3)(2k + 4)(2k + 5)2^(2k))

$$\text{Out[1]} = \frac{(121 + 98k)\text{Zeta}[2k]}{2^{2k}(1 + 2k)(2 + 2k)(3 + 2k)(4 + 2k)(5 + 2k)}$$

In[2] :=Sum [%, {k, 1, Infinity}] //Simplify

$$\text{Out[2]} = \frac{121}{240} - \frac{23\,\text{Zeta}[3]}{6\,\text{Pi}^2}$$

In[3] :=N[%]

Out[3] =0.0372903

In[4] :=Sum [N[%1]//Evaluate, {k, 1, 50}]

Out[4] =0.0372903

In[5] :=N Sum [%1//Evaluate, {k, 1, Infinity}]

Out [5] =0.0372903

Since $\zeta(0) = -\frac{1}{2}$, Out [2] evidently validates the series representation (44) symbolically. Furthermore, our numerical computations in Out [3], Out [4], and Out [5], together, exhibit the fact that only 50 terms ($k = 1$ to $k = 50$) of the series in (44) can produce an accuracy of seven decimal places [see, for details, Srivastava (2000a)].

Problems

1. Prove that

$$\sum_{n=1}^{\infty} \frac{H_n}{n^2} = 2\sum_{n=1}^{\infty} \frac{1}{n^3} = 2\sum_{n=1}^{\infty} \frac{H_n}{(n+1)^2},$$

where H_n denotes the harmonic numbers defined by 3.2(36).

[*Cf.* Equations 2.3(54) *and* 3.5(18); see also Klamkin (1951, p. 195)]

2. Prove that

$$\zeta(4) = \frac{36}{17}\sum_{k=1}^{\infty} \frac{1}{k^4\binom{2k}{k}}.$$

[Comtet (1974, p. 89); Cohen (1981, p. 280)]

3. Prove that

$$\zeta(3) = \frac{\pi^2}{6}\left\{\frac{3}{4} - \frac{1}{2}\log\frac{\pi}{3} + \sum_{k=1}^{\infty} \frac{\zeta(2k)}{k(2k+1)(2k+2)\,6^{2k}}\right\}.$$

[Zhang and Williams (1993b, Section 2); Chen and Srivastava (1998, p. 184)]

4. Prove that

$$\zeta(3) = \frac{\pi^2}{14}\left\{\frac{3}{2} - \log\frac{\pi}{4} - 4\sum_{k=1}^{\infty}\frac{(2^{2k-1}-1)\zeta(2k)}{k(2k+1)(2k+2)\,2^{4k}}\right\}.$$

[Ewell (1995, p. 1011); Chen and Srivastava (1998, p. 187)]

5. Prove the following series representations for $\zeta(3)$:

$$\zeta(3) = \frac{2\pi^2}{35}\left\{\frac{3}{2} - \log\frac{\pi}{6} - 4\sum_{k=1}^{\infty}\frac{(2^{2k-1}-1)\zeta(2k)}{k(2k+1)(2k+2)\,6^{2k}}\right\};$$

$$\zeta(3) = \frac{2\pi^2}{65}\left\{3 - \log\left(\frac{\pi^2}{27}\right) - 6\sum_{k=1}^{\infty}\frac{(3^{2k-1}-1)\zeta(2k)}{k(2k+1)(2k+2)\,6^{2k}}\right\};$$

$$\zeta(3) = \frac{4\pi^2}{25}\left\{\frac{3}{4} - \frac{1}{2}\log\left(\frac{8\pi}{27}\right) - 6\sum_{k=1}^{\infty}\frac{(3^{2k-1}-2^{2k-1})\zeta(2k)}{k(2k+1)(2k+2)\,6^{2k}}\right\};$$

$$\zeta(3) = \frac{4\pi^2}{31}\left\{\frac{3}{4} - \frac{1}{2}\log\left(\frac{27\pi}{128}\right) - 12\sum_{k=1}^{\infty}\frac{(4^{2k-1}-3^{2k-1})\zeta(2k)}{k(2k+1)(2k+2)\,12^{2k}}\right\},$$

and

$$\zeta(3) = \frac{4\pi^2}{37}\left\{\frac{3}{4} - \frac{1}{2}\log\left(\frac{4\pi}{27}\right) - 6\sum_{k=1}^{\infty}\frac{(3^{2k-1}-2^{2k-1})\zeta(2k)}{k(2k+1)(2k+2)\,12^{2k}}\right\}.$$

[Chen and Srivastava (1998, p. 187)]

6. For $n \in \mathbb{N}\setminus\{1\}$, show that

$$\begin{aligned}\zeta(2n+1) = &\frac{(-1)^n\,(2\pi)^{2n}}{(2^{2n+1}-1)\,n}\left\{\sum_{k=1}^{n-1}\frac{2^{2k}}{(2k-1)!\,(2n-2k)!\,(2k+1)^2}\right.\\ &+\sum_{k=1}^{n-1}\frac{2^{2k}}{(2k-1)!\,(2n-2k)!}\sum_{j=1}^{\infty}\frac{(-1)^j\,(2j-1)!}{2^{2j-1}\,\pi^{2j}}\\ &\cdot\left.\sum_{i=0}^{\min(2j-2,2k)}\binom{2k}{i}\frac{(-1)^i\,\zeta(2j)}{2j-i-1}+\sum_{k=0}^{\infty}\frac{(2k)!\,\zeta(2k)}{(2k+4)!\,2^{2k}}\right\}.\end{aligned}$$

[Zhang and Williams (1993b, p. 1591)]

7. Let

$$C_{p,q} := \sum_{j=2}^{\infty}\sum_{k=1}^{j-1}\frac{1}{j^p\,k^q} \qquad (p \in \mathbb{N}\setminus\{1\};\ q \in \mathbb{N})$$

and

$$C_{p,q,r} := \sum_{j=3}^{\infty} \sum_{k=2}^{j-1} \sum_{i=1}^{k-1} \frac{1}{j^p\, k^q\, i^r} \quad (p \in \mathbb{N} \setminus \{1\};\ q,\, r \in \mathbb{N}).$$

Show that

$$\begin{aligned} C_{p,q,r} = (-1)^q \sum_{n=0}^{p-2} \Bigg\{ &\sum_{m=0}^{q+n-1} \frac{(q)_n\,(r)_m}{n!m!} C_{p-n,r+m,q+n-m} \\ &+ \sum_{m=0}^{r-1} \frac{(q)_{n+m}}{n!m!} C_{p-n,q+n+m,r-m} \Bigg\} \\ + \sum_{n=0}^{q-2} \frac{(p)_n}{n!} (-1)^n\, \zeta(p+n)\, C_{q-n,r} &+ (-1)^{q-1} \frac{(p+q-2)!}{(p-1)!\,(q-1)!} \\ \cdot \Bigg\{ \sum_{m=0}^{r-2} (-1)^m\, \zeta(r-m)\, &[C_{p+q-1,m+1} + \zeta(p+q+m)] \\ + (-1)^{r-1}\, [C_{p+q-1,r,1} + C_{p+q+r-1,1} &+ C_{p+q-1,r+1} + \zeta(p+q+r)] \Bigg\}, \end{aligned}$$

where empty sums are interpreted (as usual) to be nil.

[*Cf.* Berndt (1985, p. 253); Markett (1994, p. 122)]

8. Prove that

$$\zeta(2n+1) = (-1)^n \frac{4(2\pi)^{2n}}{(2n+1)!} \sum_{k=0}^{\infty} \mathcal{R}_{2n+1,k}\, \zeta(2k) \quad (n \in \mathbb{N}),$$

where the coefficients $\mathcal{R}_{2n+1,k}$ are rational and given by

$$\mathcal{R}_{2n+1,k} := \sum_{m=0}^{2n} \binom{2n}{m} \frac{(2n+1)\, B_{2n-m}}{2^{2k+m+1}\,(2k+m+1)\,(m+1)} \quad (k \in \mathbb{N}_0),$$

and by the family of generating functions

$$\begin{aligned} &\sum_{n=0}^{\infty} \mathcal{R}_{2n+1,k} \frac{t^{2n+1}}{(2n+1)!} \\ &= \frac{(2k-1)!}{t^{2k-1}\,(e^t-1)} \left(1 - e^{\frac{1}{2}t} \sum_{m=0}^{2k-1} \frac{(-1)^m}{m!} \left(\frac{t}{2}\right)^m \right) \quad (k \in \mathbb{N};\ |t| < 2\pi). \end{aligned}$$

[Cvijović and Klinowski (1997, p. 1265); see also 4.2(39) *et seq.*]

9. Prove that

$$\zeta(n) = \sum_{j=1}^{n-2} \sum_{p,\,q=1}^{\infty} \frac{1}{p^j\,(p+q)^{n-j}} \quad (n \in \mathbb{N} \setminus \{1,\, 2\}).$$

[*Cf.* Briggs (1989); see also Markett (1994, p. 115)]

10. Let a, b, c be real numbers greater than 1 and

$$P(a,b,c)=\sum_{r=1}^{\infty} r^{-a}\sum_{k=1}^{r} k^{-b}\sum_{l=1}^{k} l^{-c}.$$

Then show that

$$\begin{aligned}P(a,b,c)+P(a,c,b)+P(b,c,a)+P(b,a,c)&\\ +P(c,a,b)+P(c,b,a)&\\ =\zeta(a)\,\zeta(b)\,\zeta(c)+\zeta(a)\,\zeta(b+c)+\zeta(b)\,\zeta(c+a)&\\ +\zeta(c)\,\zeta(a+b)+2\,\zeta(a+b+c),&\end{aligned}$$

where ζ is the Riemann Zeta function defined by 2.3(1).

[Sitaramachandrarao and Subbarao (1984, p. 471)]

11. For each pair $(m,n)\in\mathbb{N}_0\times\mathbb{N}$, define $A_{2m}(n)$ as follows:
(i) $A_{2m}(1):=B_{2m}$, and
(ii) for $n\in\mathbb{N}\setminus\{1\}$,

$$A_{2m}(n)$$
$$:=\sum\frac{\binom{2m}{2j_1,\,\ldots,\,2j_n}}{\{2j_1+1\}\{2\,(j_1+j_2)+1\}\cdots\{2\,(j_1+\cdots+j_{n-1})+1\}}\,B_{2j_1}\cdots B_{2j_n},$$

where the sum is taken over all $(j_1,\ldots,j_n)\in{\mathbb{N}_0}^n$ such that $j_1+\cdots+j_n=m$, $\binom{2m}{2j_1,\,\ldots,\,2j_n}$ is a multinomial coefficient, and the B_j are Bernoulli numbers defined by 1.6(2). Show that

$$\zeta(n)=\frac{2^{n-2}}{2^n-1}\,\pi^2\sum_{k=0}^{\infty}(-1)^k\,A_{2k}(n-2)\,\frac{\pi^{2k}}{(2k+2)!}\qquad(n\in\mathbb{N}\setminus\{1,2\}).$$

[Ewell (1991, p. 61)]

12. Show that

$$\zeta(3)=\frac{2}{45}\pi^3-4\sum_{n=1}^{\infty}e^{-2\pi n}\,\sigma_{-3}(n)\left(2\,\pi^2\,n^2+\pi n+\frac{1}{2}\right)$$

and

$$\zeta(3)=\frac{7}{180}\pi^3-2\sum_{n=1}^{\infty}e^{-2\pi n}\,\sigma_{-3}(n),$$

where $\sigma_\nu(n)$ is given in Problem 20 of Chapter 2.

[*Cf.* Terras (1976, p. 181) and Grosswald (1970); see also Problem 21 of Chapter 2]

13. Consider the cosecant integral:

$$\mathcal{I}_s(\omega):=\int_0^{\pi/\omega}t^s\csc^2 t\,dt\qquad(\Re(s)>1;\quad\omega>1).$$

Prove the following identities:

$$\mathcal{I}_s(\omega) = -\left(\frac{\pi}{\omega}\right)^s \cot\left(\frac{\pi}{\omega}\right) + s\int_0^{\pi/\omega} t^{s-1}\cot t\,dt \qquad (\Re(s) > 1;\quad \omega > 1);$$

$$\mathcal{I}_s(\omega) = -\left(\frac{\pi}{\omega}\right)^s \cot\left(\frac{\pi}{\omega}\right) - 2s\left(\frac{\pi}{\omega}\right)^{s-1}\sum_{k=0}^{\infty}\frac{\zeta(2k)}{(s+2k-1)\omega^{2k}}$$
$$(\Re(s) > 1;\quad \omega > 1);$$

$$\int_0^{\pi/2} t^s \csc^2 t\,dt = -2s\left(\frac{\pi}{2}\right)^{s-1}\sum_{k=0}^{\infty}\frac{\zeta(2k)}{(s+2k-1)2^{2k}} \qquad (\Re(s) > 1);$$

$$\int_0^{\pi/4} t^s \csc^2 t\,dt = -\left(\frac{\pi}{4}\right)^s - 2s\left(\frac{\pi}{4}\right)^{s-1}\sum_{k=0}^{\infty}\frac{\zeta(2k)}{(s+2k-1)4^{2k}} \qquad (\Re(s) > 1);$$

$$\int_0^{\pi/3} t^s \csc^2 t\,dt = -\frac{1}{\sqrt{3}}\left(\frac{\pi}{3}\right)^s - 2s\left(\frac{\pi}{3}\right)^{s-1}\sum_{k=0}^{\infty}\frac{\zeta(2k)}{(s+2k-1)3^{2k}} \qquad (\Re(s) > 1);$$

$$\int_0^{\pi/6} t^s \csc^2 t\,dt = -\sqrt{3}\left(\frac{\pi}{6}\right)^{s-1} - 2s\left(\frac{\pi}{6}\right)^{s-1}\sum_{k=0}^{\infty}\frac{\zeta(2k)}{(s+2k-1)6^{2k}} \qquad (\Re(s) > 1);$$

$$\mathcal{I}_s(\omega) = -\left(\frac{\pi}{\omega}\right)^s \left[\cot\left(\frac{\pi}{\omega}\right) + i\right] - 2is\sum_{k=1}^{\infty}\frac{\gamma(s, -2k\pi i/\omega)}{(-2ki)^s} \; (\Re(s) > 1;\ \omega > 1),$$

where $\gamma(z,\alpha)$ denotes the incomplete Gamma function defined by 1.1(76);

$$\gamma(m+1,\alpha) = m!\left(1 - e^{-\alpha}\sum_{j=0}^{m}\frac{\alpha^j}{j!}\right) \qquad (m \in \mathbb{N}_0);$$

$$\mathcal{I}_{2n}(\omega) = -\left(\frac{\pi}{\omega}\right)^{2n}\cot\left(\frac{\pi}{\omega}\right) - (2n)!\left(\frac{\pi}{\omega}\right)^{2n-1}\left[\frac{1}{(2n-1)!}\sum_{k=1}^{\infty}\frac{\cos(2k\pi/\omega)}{k}\right.$$
$$+\sum_{j=1}^{n-1}\frac{(-1)^j}{(2n-2j-1)!}\left(\frac{\omega}{2\pi}\right)^{2j}\sum_{k=1}^{\infty}\frac{\cos(2k\pi/\omega)}{k^{2j+1}}$$
$$\left.+\sum_{j=1}^{n}\frac{(-1)^j}{(2n-2j)!}\left(\frac{\omega}{2\pi}\right)^{2j-1}\sum_{k=1}^{\infty}\frac{\sin(2k\pi/\omega)}{k^{2j}}\right] \qquad (n \in \mathbb{N};\ \omega > 1);$$

$$\mathcal{I}_{2n+1}(\omega) = -\left(\frac{\pi}{\omega}\right)^{2n+1}\cot\left(\frac{\pi}{\omega}\right) + (-1)^n\frac{(2n+1)!}{2^{2n}}\zeta(2n+1)$$
$$-(2n+1)!\left(\frac{\pi}{\omega}\right)^{2n}\left[\frac{1}{(2n)!}\sum_{k=1}^{\infty}\frac{\cos(2k\pi/\omega))}{k} + \sum_{j=1}^{n}\frac{(-1)^j}{(2n-2j)!}\left(\frac{\omega}{2\pi}\right)^{2j}\right.$$
$$\left.\cdot\sum_{k=1}^{\infty}\frac{\cos(2k\pi/\omega)}{k^{2j+1}} + \sum_{j=1}^{n}\frac{(-1)^j}{(2n-2j+1)!}\left(\frac{\omega}{2\pi}\right)^{2j-1}\sum_{k=1}^{\infty}\frac{\sin(2k\pi/\omega)}{k^{2j}}\right] \qquad (n \in \mathbb{N};\ \omega > 1);$$

$$\sum_{k=0}^{\infty} \frac{\zeta(2k)}{(2k+2n-1)\omega^{2k}} = -\frac{1}{2}\log\left[2\sin\left(\frac{\pi}{\omega}\right)\right]$$
$$+\frac{(2n-1)!}{2}\left[\sum_{j=1}^{n-1} \frac{(-1)^j}{(2n-2j-1)!}\left(\frac{\omega}{2\pi}\right)^{2j} \mathrm{C}\ell_{2j+1}\left(\frac{2\pi}{\omega}\right)\right.$$
$$\left.+\sum_{j=1}^{n} \frac{(-1)^j}{(2n-2j)!}\left(\frac{\omega}{2\pi}\right)^{2j-1} \mathrm{C}\ell_{2j}\left(\frac{2\pi}{\omega}\right)\right] \qquad (n \in \mathbb{N};\ \omega > 1),$$

and

$$\sum_{k=0}^{\infty} \frac{\zeta(2k)}{(2k+2n)\omega^{2k}} = (-1)^{n-1}\frac{(2n)!}{2^{2n+1}}\left(\frac{\omega}{\pi}\right)^{2n} \zeta(2n+1) - \frac{1}{2}\log\left[2\sin\left(\frac{\pi}{\omega}\right)\right]$$
$$+\frac{(2n)!}{2}\sum_{j=1}^{n} \frac{(-1)^j}{(2n-2j+1)!}\left(\frac{\omega}{2\pi}\right)^{2j}\left[(2n-2j+1)\mathrm{C}\ell_{2j+1}\left(\frac{2\pi}{\omega}\right) + \frac{2\pi}{\omega}\mathrm{C}\ell_{2j}\left(\frac{2\pi}{\omega}\right)\right]$$
$$(n \in \mathbb{N};\ \omega > 1)$$

in terms of the generalized Clausen functions $\mathrm{C}\ell_{2n}$ and $\mathrm{C}\ell_{2n+1}$ defined by 2.4(80) and 2.4(81).

[Srivastava, Glasser, and Adamchik (2000, p. 832 *et seq.*)]

14. Continuing Problem 13, show that

$$\zeta(2n+1) = (-1)^{n-1}\frac{(2\pi)^{2n}}{(2n+1)!\,(2^{2n}-1)}\left[\log 2 + 2\sum_{k=0}^{\infty} \frac{\zeta(2k)}{(2k+2n+1)2^{2k}}\right.$$
$$\left.+(2n+1)!\sum_{j=1}^{n-1} \frac{(-1)^j}{(2n-2j+1)!}\left(\frac{2^{2j}-1}{(2\pi)^{2j}}\right)\zeta(2j+1)\right] \qquad (n \in \mathbb{N});$$

$$\zeta(2n+1) = (-1)^{n-1}\frac{(2\pi)^{2n}}{(2n)!\,(2^{2n+1}-1)}\left[\log 2 + \sum_{k=0}^{\infty} \frac{\zeta 2k)}{(k+n)2^{2k}}\right.$$
$$\left.+(2n)!\sum_{j=1}^{n-1} \frac{(-1)^j}{(2n-2j)!}\left(\frac{2^{2j}-1}{(2\pi)^{2j}}\right)\zeta(2j+1)\right] \qquad (n \in \mathbb{N});$$

$$\zeta(2n+1) = (-1)^{n-1}\frac{(2\pi)^{2n}}{(2n+1)!\,(3^{2n}-1)}\left[\log 3 + 4\sum_{k=0}^{\infty} \frac{\zeta(2k)}{(2k+2n+1)3^{2k}}\right.$$
$$+ (2n+1)!\sum_{j=1}^{n-1} \frac{(-1)^j}{(2n-2j+1)!}\left(\frac{3^{2j}-1}{(2\pi)^{2j}}\right)\zeta(2j+1)$$
$$\left.-\frac{(2n+1)!}{\sqrt{3}}\sum_{j=1}^{n+1} \frac{(-1)^j}{(2n-2j+2)!}\,\frac{2\zeta\left(2j,\frac{1}{3}\right) - \left(3^{2j}-1\right)\zeta(2j)}{(2\pi)^{2j-1}}\right] \qquad (n \in \mathbb{N});$$

$$\zeta(2n+1)=(-1)^{n-1}\frac{(2\pi)^{2n}}{(2n)!\,(3^{2n+1}-1)}\left[\log 3+2\sum_{k=0}^{\infty}\frac{\zeta(2k)}{(k+n)3^{2k}}\right.$$
$$+(2n)!\sum_{j=1}^{n-1}\frac{(-1)^j}{(2n-2j)!}\left(\frac{3^{2j}-1}{(2\pi)^{2j}}\right)\zeta(2j+1)$$
$$\left.-\frac{(2n)!}{\sqrt{3}}\sum_{j=1}^{n}\frac{(-1)^j}{(2n-2j+1)!}\frac{2\zeta\left(2j,\frac{1}{3}\right)-\left(3^{2j}-1\right)\zeta(2j)}{(2\pi)^{2j-1}}\right]\qquad(n\in\mathbb{N});$$

$$\zeta(3)=\frac{\pi^2}{12}\left[\log 3+4\sum_{k=0}^{\infty}\frac{\zeta(2k)}{(2k+3)3^{2k}}\right.$$
$$\left.+2\sqrt{3}\sum_{j=1}^{2}\frac{(-1)^{j-1}}{(4-2j)!}\frac{2\zeta\left(2j,\frac{1}{3}\right)-\left(3^{2j}-1\right)\zeta(2j)}{(2\pi)^{2j-1}}\right];$$

$$\zeta(3)=\frac{\pi^2}{13}\left[\log 3+2\sum_{k=0}^{\infty}\frac{\zeta(2k)}{(k+1)3^{2k}}+\frac{2}{\pi\sqrt{3}}\left\{\zeta\left(2,\frac{1}{3}\right)-4\zeta(2)\right\}\right];$$

$$\zeta(2n+1)=(-1)^{n-1}\frac{(2\pi)^{2n}}{(2n+1)!\,(2^{2n}-1)}\left[\log 2+4\sum_{k=0}^{\infty}\frac{\zeta(2k)}{(2k+2n+1)4^{2k}}\right.$$
$$+(2n+1)!\sum_{j=1}^{n-1}\frac{(-1)^j}{(2n-2j+1)!}\left(\frac{2^{2j}-1}{(2\pi)^{2j}}\right)\zeta(2j+1)$$
$$\left.-(2n+1)!\sum_{j=1}^{n+1}\frac{(-1)^j}{(2n-2j+2)!}\frac{\zeta\left(2j,\frac{1}{4}\right)-2^{2j-1}\left(2^{2j}-1\right)\zeta(2j)}{(2\pi)^{2j-1}}\right]\qquad(n\in\mathbb{N});$$

$$\zeta(2n+1)=(-1)^{n-1}\frac{(2\pi)^{2n}}{(2n)!\,(2^{4n+1}+2^{2n}-1)}\left[\log 2+2\sum_{k=0}^{\infty}\frac{\zeta(2k)}{(k+n)4^{2k}}\right.$$
$$+(2n)!\sum_{j=1}^{n-1}\frac{(-1)^j}{(2n-2j)!}\left(\frac{2^{2j}-1}{(2\pi)^{2j}}\right)\zeta(2j+1)$$
$$\left.-(2n)!\sum_{j=1}^{n}\frac{(-1)^j}{(2n-2j+1)!}\frac{\zeta\left(2j,\frac{1}{4}\right)-2^{2j-1}\left(2^{2j}-1\right)\zeta(2j)}{(2\pi)^{2j-1}}\right]\qquad(n\in\mathbb{N});$$

$$\zeta(3)=\frac{2\pi^2}{9}\left[\log 2+4\sum_{k=0}^{\infty}\frac{\zeta 2k)}{(2k+3)4^{2k}}\right.$$
$$\left.+6\sum_{j=1}^{2}\frac{(-1)^{j-1}}{(4-2j)!}\frac{\zeta\left(2j,\frac{1}{4}\right)-2^{2j-1}\left(2^{2j}-1\right)\zeta(2j)}{(2\pi)^{2j-1}}\right];$$

$$\zeta(3)=\frac{2\pi^2}{35}\left[\log 2+2\sum_{k=0}^{\infty}\frac{\zeta(2k)}{(k+1)4^{2k}}+\frac{1}{\pi}\left\{\zeta\left(2,\frac{1}{4}\right)-6\zeta(2)\right\}\right];$$

$$\zeta(2n+1) = (-1)^{n-1}\frac{(2\pi)^{2n}}{(2^{2n}-1)(3^{2n}-1)}\left[-\frac{4}{(2n+1)!}\sum_{k=0}^{\infty}\frac{\zeta(2k)}{(2k+2n+1)6^{2k}}\right.$$
$$+\sum_{j=1}^{n-1}\frac{(-1)^j}{(2n-2j+1)!}\left(\frac{(2^{2j}-1)(3^{2j}-1)}{(2\pi)^{2j}}\right)\zeta(2j+1)$$
$$\left.+\frac{1}{\sqrt{3}}\sum_{j=1}^{n+1}\frac{(-1)^j}{(2n-2j+2)!}\frac{\zeta\left(2j,\frac{1}{3}\right)+\zeta\left(2j,\frac{1}{6}\right)-2^{2j-1}\left(3^{2j}-1\right)\zeta(2j)}{(2\pi)^{2j-1}}\right]$$
$$(n\in\mathbb{N});$$

$$\zeta(2n+1) = (-1)^{n-1}\frac{(2\pi)^{2n}}{2^{2n}+3^{2n}+6^{2n}-1}\left[\frac{2}{(2n)!}\sum_{k=0}^{\infty}\frac{\zeta(2k)}{(k+n)6^{2k}}\right.$$
$$-\sum_{j=1}^{n-1}\frac{(-1)^j}{(2n-2j)!}\left(\frac{(2^{2j}-1)(3^{2j}-1)}{(2\pi)^{2j}}\right)\zeta(2j+1)$$
$$\left.-\frac{1}{\sqrt{3}}\sum_{j=1}^{n}\frac{(-1)^j}{(2n-2j+1)!}\frac{\zeta\left(2j,\frac{1}{3}\right)+\zeta\left(2j,\frac{1}{6}\right)-2^{2j-1}\left(3^{2j}-1\right)\zeta(2j)}{(2\pi)^{2j-1}}\right]$$
$$(n\in\mathbb{N});$$

$$\zeta(3) = -\frac{\pi^2}{18}\left[2\sum_{k=0}^{\infty}\frac{\zeta(2k)}{(2k+3)6^{2k}}\right.$$
$$\left.+\sqrt{3}\sum_{j=1}^{2}\frac{(-1)^{j-1}}{(4-2j)!}\frac{\zeta\left(2j,\frac{1}{3}\right)+\left(2j,\frac{1}{6}\right)-2^{2j-1}\left(3^{2j}-1\right)\zeta(2j)}{(2\pi)^{2j-1}}\right],$$

and

$$\zeta(3) = \frac{\pi^2}{12}\left[\sum_{k=0}^{\infty}\frac{\zeta(2k)}{(k+1)6^{2k}}+\frac{1}{2\pi\sqrt{3}}\left\{\zeta\left(2,\frac{1}{3}\right)+\zeta\left(2,\frac{1}{6}\right)-16\zeta(2)\right\}\right].$$

[Srivastava, Glasser, and Adamchik (2000, p. 836 *et seq.*)]

15. Continuing Problems 13 and 14, show that

$$\sum_{k=0}^{\infty}\frac{\zeta(2k)}{(2k+2n+1)2^{2k}} = \frac{1}{2}\sum_{j=0}^{n}\frac{(-1)^{j-1}}{2j+1}\binom{2n+1}{2j}\left(\frac{2}{\pi}\right)^{2j}\int_0^\infty t^{2j+1}\,\mathrm{sech}^2 t\,dt$$
$$(n\in\mathbb{N}_0);$$

$$\sum_{k=0}^{\infty}\frac{\zeta(2k)}{(2k+2n+1)2^{2k}}$$
$$=\frac{1}{2}\sum_{j=1}^{n}(-1)^{j-1}\binom{2n+1}{2j}\frac{(2j)!\,(2^{2j}-1)}{(2\pi)^{2j}}\zeta(2j+1)-\frac{1}{2}\log 2$$
$$(n\in\mathbb{N}_0),$$

and

$$\zeta(2n+1) = (-1)^{n-1}\frac{(2\pi)^{2n}}{(2n+1)!\,(2^{2n}-1)}\left[\log 2 + 2\sum_{k=0}^{\infty}\frac{\zeta(2k)}{(2k+2n+1)2^{2k}}\right.$$
$$\left. + \sum_{j=1}^{n-1}(-1)^{j-1}\binom{2n+1}{2j}\frac{(2j)!\,(2^{2j}-1)}{(2\pi)^{2j}}\zeta\,(2j+1)\right] \quad (n \in \mathbb{N}_0).$$

[Srivastava, Glasser, and Adamchik (2000, p. 842 *et seq.*)]

16. Continuing Problems 13, 14, and 15, define

$$\mathcal{S}_p := \sum_{k=0}^{\infty}\frac{\zeta(2k)}{(2k+p)\omega^{2k}} = -\frac{\pi}{2\omega}\int_0^1 t^p \cot\left(\frac{\pi t}{\omega}\right) dt \quad (p \in \mathbb{N};\ |\omega| > 1).$$

Show that

$$\mathcal{S}_p = \frac{\pi i}{2(p+1)\omega} + \frac{1}{2}\left(-\frac{i\omega}{2\pi}\right)^p \int_1^{\Omega}\frac{(\log z)^p}{1-z}dz \quad (p \in \mathbb{N};\ |\omega| > 1);$$

$$\mathcal{S}_p = \frac{\pi i}{2(p+1)\omega} - \frac{1}{2}\left(-\frac{i\omega}{2\pi}\right)^p \int_0^1 \{\log(1-(1-\Omega)t)\}^p\frac{dt}{t}$$
$$(p \in \mathbb{N};\ |\omega| > 1),$$

where, for convenience,

$$\Omega := \exp\left(\frac{2\pi i}{\omega}\right) \quad (|\omega| > 1);$$

Or, equivalently,

$$\sum_{k=0}^{\infty}\frac{\zeta(2k)}{(2k+p)\omega^{2k}} = \frac{\pi i}{2(p+1)\omega} - \frac{p!}{2}\left(\frac{i\omega}{2\pi}\right)^p S_{1,p}\left(1-e^{2\pi i/\omega}\right)$$
$$(p \in \mathbb{N};\ |\omega| > 1)$$

in terms of Nielsen's generalized Polylogarithmic function $S_{n,p}(z)$ defined by [*cf.*, *e.g.*, Kölbig (1986, p. 1233, Eq. (1.3)]

$$S_{n,p}(z) := \frac{(-1)^{n+p-1}}{(n-1)!p!}\int_0^1 (\log t)^{n-1}\{\log(1-zt)\}^p\frac{dt}{t} \quad (n,p \in \mathbb{N};\ z \in \mathbb{C}),$$

so that [*cf.* Kölbig (1986, p. 1240, Eq. (5.6))]

$$S_{1,p}(z) = \zeta(p+1) + \sum_{k=0}^{p}\frac{(-1)^{k-1}}{k!}\{\log(1-z)\}^k\,\mathrm{Li}_{p-k+1}(1-z);$$

$$\sum_{k=0}^{\infty}\frac{\zeta(2k)}{(2k+p)\omega^{2k}} = \frac{\pi i}{2(p+1)\omega} - \frac{1}{2}\log\left(1-e^{2\pi i/\omega}\right)$$
$$-\frac{p!}{2}\left(\frac{i\omega}{2\pi}\right)^p\zeta(p+1) + \frac{1}{2}\sum_{k=1}^{p}\binom{p}{k}k!\left(\frac{i\omega}{2\pi}\right)^k \mathrm{Li}_k\left(e^{2\pi i/\omega}\right)$$
$$(p \in \mathbb{N};\ |\omega| > 1);$$

$$\int_0^1 t^p \cot(\nu t)dt = -\frac{i}{p+1} + \frac{1}{\nu}\log\left(1-e^{2\nu i}\right)$$
$$+\frac{p!}{\nu}\left(\frac{i}{2\nu}\right)^p \zeta(p+1) - \frac{1}{\nu}\sum_{k=1}^{p}\binom{p}{k}k!\left(\frac{i}{2\nu}\right)^k \mathrm{Li}_{k+1}\left(e^{2\nu i}\right)$$
$$(p \in \mathbb{N};\ \nu \in \mathbb{C}\setminus\{0\}).$$

[Srivastava, Glasser, and Adamchik (2000, p. 843 *et seq.*)]

17. Prove the following series representations for the values of $L(s,\chi)$:
(a) If $\chi(-1) = 1$ and $\chi \neq 1$, then

$$nL(2n+1,\chi) - n\sum_{l=1}^{\infty}\frac{\chi(l)}{l^{2n+1}}\cos\left(\frac{2l\pi x}{p}\right) - \frac{\pi x}{p}\sum_{l=1}^{\infty}\frac{\chi(l)}{l^{2n}}\sin\left(\frac{2l\pi x}{p}\right)$$
$$= (-1)^n\left(\frac{2\pi x}{p}\right)^{2n}\left[\sum_{k=1}^{n-1}\frac{(-1)^{k-1}\cdot k}{(2n-2k)!}\frac{L(2k+1,\chi)}{(2\pi x/p)^{2k}}\right.$$
$$\left.+\frac{\tau(\chi)}{p}\sum_{k=1}^{\infty}\frac{(2k)!}{(2n+2k)!}L(2k,\overline{\chi})\,x^{2k}\right] \quad (n \in \mathbb{N});$$

(b) If $\chi(-1) = -1$, then

$$L(2n,\chi) \quad \sum_{l=1}^{\infty}\frac{\chi(l)}{l^{2n}}\cos\left(\frac{2l\pi x}{p}\right)$$
$$= (-1)^n\left(\frac{2\pi x}{p}\right)^{2n-1}\left[\sum_{k=1}^{n-1}\frac{(-1)^{k-1}}{(2n-2k)!}\frac{L(2k,\chi)}{(2\pi x/p)^{2k-1}}\right.$$
$$\left.+\frac{2\tau(\chi)\,i}{p}\sum_{k=0}^{\infty}\frac{(2k)!}{(2n+2k)!}L(2k+1,\overline{\chi})\,x^{2k+1}\right] \quad (n \in \mathbb{N}),$$

where $L(s,\chi)$ is the Dirichlet L-function associated with a non-trivial Dirichlet character χ of modulus p, and $\tau(\chi)$ denotes the Gauss sum defined by

$$\tau(\chi) := \sum_{k=1}^{p}\chi(k)\exp\left(\frac{2k\pi i}{p}\right) \quad \left(i := \sqrt{-1}\right).$$

[Katsurada (1999, p. 82, Theorem 3)]

18. Prove the following series representation:

$$\zeta(2n+1) + \frac{1}{2\pi x}\sum_{l=1}^{\infty}\frac{\sin(2\pi l x)}{l^{2n+2}}$$
$$= (-1)^{n-1}(2\pi x)^{2n}\left[\frac{1}{(2n+1)!}\left(\sum_{m=1}^{2n+1}\frac{1}{m} - \log(2\pi x)\right)\right.$$
$$\left.+\sum_{k=1}^{n-1}(-1)^k\frac{\zeta(2k+1)}{(2n-2k+1)!\,(2\pi x)^{2k}} + 2\sum_{k=1}^{\infty}\frac{(2k-1)!\,\zeta(2k)}{(2n+2k+1)!}x^{2k}\right]$$
$$(n \in \mathbb{N};\ x \in \mathbb{R};\ |x| \leq 1).$$

[Katsurada (1999, p. 81, Theorem 2)]

19. Prove the following series representations for the values of $L(s,\chi)$:
(a) If $\chi(-1)=1$ and $\chi\neq 1$, then

$$L(2n+1,\chi)-\frac{p}{2\pi x}\sum_{l=1}^{\infty}\frac{\chi(l)}{l^{2n+2}}\sin\left(\frac{2l\pi x}{p}\right)$$
$$=(-1)^{n+1}\left(\frac{2\pi x}{p}\right)^{2n}\left[\sum_{k=0}^{n-1}\frac{(-1)^k}{(2n-2k+1)!}\frac{L(2k+1,\chi)}{(2\pi x/p)^{2k}}\right.$$
$$\left.+\frac{2\tau(\chi)}{p}\sum_{k=1}^{\infty}\frac{(2k-1)!}{(2n+2k+1)!}L(2k,\overline{\chi})\,x^{2k}\right]\qquad(n\in\mathbb{N});$$

(b) If $\chi(-1)=-1$, then

$$L(2n,\chi)-\frac{p}{2\pi x}\sum_{l=1}^{\infty}\frac{\chi(l)}{l^{2n+1}}\sin\left(\frac{2l\pi x}{p}\right)$$
$$=(-1)^{n+1}\left(\frac{2\pi x}{p}\right)^{2n-1}\left[\sum_{k=1}^{n-1}\frac{(-1)^k}{(2n+2k+1)!}\frac{L(2k,\chi)}{(2\pi x/p)^{2k-1}}\right.$$
$$\left.-\frac{2\tau(\chi)\,i}{p}\sum_{k=0}^{\infty}\frac{(2k)!}{(2n+2k+1)!}L(2k+1,\overline{\chi})\,x^{2k+1}\right]\qquad(n\in\mathbb{N}),$$

where $L(s,\chi)$ and $\tau(\chi)$ are already given in Problem 17.

[Srivastava and Tsumura (2000b, p. 23)]

20. Prove the following series representations for $\zeta(3)$:

$$\zeta(3)=-\frac{120}{1573}\pi^2$$
$$\cdot\sum_{k=0}^{\infty}\frac{8576k^2+24286k+17283}{(2k+1)(2k+2)(2k+3)(2k+4)(2k+5)(2k+6)(2k+7)}\frac{\zeta(2k)}{2^{2k}};$$

$$\zeta(3)=-\frac{6}{17}\pi^2\left[\frac{1}{120}(80H_4-31H_5-49\log\pi)\right.$$
$$\left.+2\sum_{k=1}^{\infty}\frac{32k+49}{(2k)(2k+1)(2k+2)(2k+3)(2k+4)(2k+5)}\frac{\zeta(2k)}{2^{2k}}\right]\quad(n\in\mathbb{N}),$$

where H_k denotes the harmonic numbers defined by 3.2(36).

[Srivastava and Tsumura (2000b, pp. 20 and 22)]

CHAPTER 5

DETERMINANTS OF THE LAPLACIANS

During the last two decades, the problem of evaluating the determinants of the Laplacians on Riemann manifolds has received considerable attention by many authors including D'Hoker and Phong (1986a and 1989), Sarnak (1987), and Voros (1987), who computed the determinants of the Laplacians on compact Riemann surfaces of constant curvature in terms of special values of the Selberg Zeta function. Although the first interest in the determinants of the Laplacians arose mainly for Riemann surfaces, it is also interesting to compute these determinants for classical Riemannian manifolds of higher dimensions, such as spheres. In this chapter we are particularly concerned with the evaluation of the functional determinant for the n-dimensional sphere $\mathbf{S}^n$ with the standard metric.

In computations of the determinants of the Laplacians on manifolds of constant curvature, an important rôle is played by the closed-form evaluations of the series involving the Zeta function given in Section 3.4 [*cf.*, *e.g.*, Choi and Srivastava (1999c and 2000a)] as well as the theory of multiple Gamma functions given in Section 1.3 [*cf.*, *e.g.*, Voros (1987), Vardi (1988b), Choi (1994a), and Quine and Choi (1996)].

5.1 The n-Dimensional Problem

Let $\{\lambda_n\}$ be a sequence such that

$$0 = \lambda_0 < \lambda_1 \leq \lambda_2 \leq \cdots \leq \lambda_n \leq \cdots ; \ \lambda_n \uparrow \infty \ (n \to \infty); \tag{1}$$

henceforth we consider only such nonnegative increasing sequences. Then we can show that

$$Z(s) := \sum_{n=1}^{\infty} \frac{1}{\lambda_n^s}, \tag{2}$$

which is known to converge absolutely in a half-plane $\Re(s) > \sigma$ for some $\sigma \in \mathbb{R}$.

Definition 5.1 [*cf.* Osgood *et al.* (1988)]. The determinant of the Laplacian Δ on the compact manifold M is defined to be

$$\det{}'\Delta := \prod_{\lambda_k \neq 0} \lambda_k, \tag{3}$$

where $\{\lambda_k\}$ is the sequence of eigenvalues of the Laplacian Δ on M. The sequence $\{\lambda_k\}$ is known to satisfy the condition as in (1), but the product in (3) is always divergent; so, in order for the expression (3) to make sense, some sort of regularization procedure must be used. It is easily seen that, formally, $e^{-Z'(0)}$ is the product of nonzero eigenvalues of Δ. This product does not converge, but $Z(s)$ can be continued analytically to a neighborhood of $s = 0$. Therefore, we can give a *meaningful* definition:

$$\det{}'\Delta := e^{-Z'(0)}, \tag{4}$$

which is called the *Functional Determinant of the Laplacian* Δ on M.

Definition 5.2. The *order* μ of the sequence $\{\lambda_k\}$ is defined by

$$\mu := \inf\left\{\alpha > 0 \,\middle|\, \sum_{k=1}^{\infty} \frac{1}{\lambda_k^\alpha} < \infty\right\}. \tag{5}$$

The analogous and shifted analogous Weierstrass canonical products $E(\lambda)$ and $E(\lambda, a)$ of the sequence $\{\lambda_k\}$ are defined, respectively, by

$$E(\lambda) := \prod_{k=1}^{\infty}\left\{\left(1-\frac{\lambda}{\lambda_k}\right)\exp\left(\frac{\lambda}{\lambda_k}+\frac{\lambda^2}{2\lambda_k^2}+\cdots+\frac{\lambda^{[\mu]}}{[\mu]\lambda_k^{[\mu]}}\right)\right\} \tag{6}$$

and

$$E(\lambda, a) := \prod_{k=1}^{\infty}\left\{\left(1-\frac{\lambda}{\lambda_k+a}\right)\exp\left(\frac{\lambda}{\lambda_k+a}+\cdots+\frac{\lambda^{[\mu]}}{[\mu]\,(\lambda_k+a)^{[\mu]}}\right)\right\}, \tag{7}$$

where $[\mu]$ denotes the greatest integer part in the order μ of the sequence $\{\lambda_k\}$.

There exists the following relationship between $E(\lambda)$ and $E(\lambda, a)$ [see Voros (1987)]:

$$E(\lambda, a) = \exp\left(\sum_{m=1}^{[\mu]} \mathcal{R}_{m-1}(-a)\frac{\lambda^m}{m!}\right)\frac{E(\lambda-a)}{E(-a)}, \tag{8}$$

where, for convenience,

$$\mathcal{R}_{[\mu]}(\lambda - a) := \frac{d^{[\mu]+1}}{d\lambda^{[\mu]+1}}\left\{-\log E(\lambda, a)\right\}. \tag{9}$$

The shifted series $Z(s, a)$ of $Z(s)$ in (2) by a is given by

$$Z(s, a) := \sum_{k=1}^{\infty}\frac{1}{(\lambda_k + a)^s}. \tag{10}$$

Formally, indeed, we have

$$Z'(0, -\lambda) = -\sum_{k=1}^{\infty}\log(\lambda_k - \lambda),$$

which, if we define

$$D(\lambda) := \exp\left[-Z'(0, -\lambda)\right], \tag{11}$$

immediately implies that

$$D(\lambda) = \prod_{k=1}^{\infty} (\lambda_k - \lambda).$$

In fact, Voros (1987) gave the relationship between $D(\lambda)$ and $E(\lambda)$ as follows:

$$\begin{aligned} D(\lambda) = & \exp[-Z'(0)] \exp\left[- \sum_{m=1}^{[\mu]} \mathrm{FP} Z(m) \frac{\lambda^m}{m} \right] \\ & \cdot \exp\left[- \sum_{m=2}^{[\mu]} C_{-m} \left(\sum_{k=1}^{m-1} \frac{1}{k} \right) \frac{\lambda^m}{m!} \right] E(\lambda), \end{aligned} \tag{12}$$

where an empty sum is interpreted to be nil and the *finite part* prescription is applied (as usual) as follows [*cf.* Voros (1987, p. 446)]:

$$\mathrm{FP} f(s) := \begin{cases} f(s), & \text{if } s \text{ is not a pole,} \\ \lim\limits_{\epsilon \to 0} \left(f(s+\epsilon) - \dfrac{\text{Residue}}{\epsilon} \right), & \text{if } s \text{ is a simple pole,} \end{cases} \tag{13}$$

and

$$Z(-m) = (-1)^m \, m! \, C_{-m}. \tag{14}$$

Now consider the sequence of eigenvalues on the standard Laplacian Δ_n on $\mathbf{S}^n$. It is known from the work of Vardi (1988b) [see also Terras (1985)] that the standard Laplacian Δ_n $(n \in \mathbb{N})$ has eigenvalues

$$\mu_k := k(k+n-1) \tag{15}$$

with multiplicity

$$\binom{k+n}{n} - \binom{k+n-2}{n} = \frac{(2k+n-1)\,(k+n-2)!}{k!\,(n-1)!} \qquad (k \in \mathbb{N}_0). \tag{16}$$

From now on we consider the shifted sequence $\{\lambda_k\}$ of $\{\mu_k\}$ in (15) by $\left(\frac{n-1}{2}\right)^2$ as a fundamental sequence. Then the sequence $\{\lambda_k\}$ is written in the following simple and tractable form:

$$\lambda_k = \mu_k + \left(\frac{n-1}{2}\right)^2 = \left(k + \frac{n-1}{2}\right)^2 \tag{17}$$

with the same multiplicity as in (16).

We will exclude the zero mode, that is, start the sequence at $k = 1$ for later use. Furthermore, with a view to emphasizing n on $\mathbf{S}^n$, we choose the notations $Z_n(s)$, $Z_n(s,a)$, $E_n(\lambda)$, $E_n(\lambda, a)$, and $D_n(\lambda)$ instead of $Z(s)$, $Z(s,a)$, $E(\lambda)$, $E(\lambda, a)$, and $D(\lambda)$, respectively.

We readily observe from (11) that

$$D_n\left(\left(\frac{n-1}{2}\right)^2\right)=\det{}'\Delta_n, \tag{18}$$

where $\det{}'\,\Delta_n$ denote the *determinants of the Laplacians* on $\mathbf{S}^n$ $(n\in\mathbb{N})$.

Choi (1994a) computed the determinants of the Laplacians on the n-dimensional unit sphere $\mathbf{S}^n$ $(n=1,\,2,\,3)$ by factorizing the analogous Weierstrass canonical product of a shifted sequence $\{\lambda_k\}$ in (17) of eigenvalues of the Laplacians on $\mathbf{S}^n$ into multiple Gamma functions, whereas Choi and Srivastava (1999c and 2000a) made use of some closed-form evaluations of the series involving Zeta function given in Section 3.4 for the computation of the determinants of the Laplacians on $\mathbf{S}^n$ $(n=2,\,3,\,4)$. In following two sections we compute the determinants of the Laplacians on $\mathbf{S}^n$ $(n=1,\,2,\,3,\,4)$ by using the abovementioned methods.

5.2 Computations Using the Simple and Multiple Gamma Functions

Factorizations into simple and multiple Gamma functions

We begin by expressing $E_n(\lambda)$ $(n=1,\,2,\,3)$ as the simple and multiple Gamma functions. Our results are summarized in the following proposition [see Choi (1994a); see also Voros (1987)]

Proposition 5.1. *The analogous Weierstrass canonical products* $E_n(\lambda)$ $(n=1,\,2,\,3)$ *of the shifted sequence* $\{\lambda_k\}$ *in* 5.1(17) *can be expressed in terms of the simple and multiple Gamma functions as follows:*

$$E_1(\lambda)=\frac{1}{\left\{\Gamma\left(1-\sqrt{\lambda}\right)\Gamma\left(1+\sqrt{\lambda}\right)\right\}^2}, \tag{1}$$

$$E_2(\lambda)=\frac{\left\{\Gamma_2\left(\frac{1}{2}\right)\right\}^4 e^{c_1\lambda}\,\Gamma\left(\frac{1}{2}-\sqrt{\lambda}\right)\Gamma\left(\frac{1}{2}+\sqrt{\lambda}\right)}{\pi\left(1-2\sqrt{\lambda}\right)\left(1+2\sqrt{\lambda}\right)\left\{\Gamma_2\left(\frac{1}{2}-\sqrt{\lambda}\right)\Gamma_2\left(\frac{1}{2}+\sqrt{\lambda}\right)\right\}^2}, \tag{2}$$

and

$$E_3(\lambda)=\frac{e^{c_2\lambda}}{1-\lambda}\,\frac{\Gamma_2\left(1-\sqrt{\lambda}\right)\Gamma_2\left(1+\sqrt{\lambda}\right)}{\left\{\Gamma_3\left(1-\sqrt{\lambda}\right)\Gamma_3\left(1+\sqrt{\lambda}\right)\right\}^2}, \tag{3}$$

where, for convenience,

$$c_1:=2(\gamma-1+2\log 2)\quad \textit{and}\quad c_2:=\log(2\pi)-\tfrac{3}{2},$$

and γ *is the Euler-Mascheroni constant defined by* 1.1(3).

Proof. Here we will prove only the representation (2). First of all, it follows from 5.1(15) and 5.1(16) with $n=2$ that the eigenvalues of Δ_2 on $\mathbf{S}^2$ have the sequence

$$\mu_k=k(k+1)\quad\text{with multiplicity}\quad 2k+1\quad(k\in\mathbb{N}_0) \tag{4}$$

and the corresponding shifted sequence $\{\lambda_k\}$ of $\{\mu_k\}$ in (4) by $\frac{1}{4}$ [or 5.1(17) with $n=2$] is

$$\lambda_k=\left(k+\frac{1}{2}\right)^2\quad\text{with multiplicity}\quad 2k+1\quad(k\in\mathbb{N}_0). \tag{5}$$

It is easily seen from the definition 5.1(5) that both of the sequences (4) and (5) are of order $\mu = 1$.

From 5.1(6) we find that the analogous Weierstrass canonical product of the shifted sequence $\{\lambda_k\}$ in (5) is

$$E_2(\lambda) = \prod_{k=1}^{\infty} \left[\left(1 - \frac{\lambda}{(k+\frac{1}{2})^2}\right) \exp\left(\frac{\lambda}{(k+\frac{1}{2})^2}\right) \right]^{2k+1}. \tag{6}$$

In order to express $E_2(\lambda)$ in terms of simple and multiple Gamma functions, we consider the associated product E_2^+ defined by

$$E_2^+(z) := \prod_{k=1}^{\infty} \left[\left(1 - \frac{z}{k+\frac{1}{2}}\right) \exp\left(\frac{z}{k+\frac{1}{2}} + \frac{z^2}{2\,(k+\frac{1}{2})^2}\right) \right]^{2k+1}. \tag{7}$$

Letting $\lambda = -w^2$ in (6) and using the definition (7), we obtain

$$E_2(\lambda) = E_2^+(iw)\, E_2^+(-iw). \tag{8}$$

It is seen from 1.3(3) that the G-function itself is related to the Barnes product $E^B(\lambda)$ given by

$$\begin{aligned} E^B(\lambda) :=& \prod_{k=1}^{\infty} \left[\left(1 - \frac{\lambda}{k}\right) \exp\left(\frac{\lambda}{k} + \frac{\lambda^2}{2k^2}\right) \right]^k \\ =&(2\pi)^{\frac{\lambda}{2}}\, e^{\frac{1}{2}[(1+\gamma)\lambda^2 - \lambda]}\, G(1-\lambda). \end{aligned} \tag{9}$$

Thus $E_2^+(\lambda)$ can be further decomposed to involve the Barnes G-function, or rather the shifted Barnes product:

$$E^B\left(z, -\frac{1}{2}\right) = \prod_{k=1}^{\infty} \left[\left(1 - \frac{z}{k-\frac{1}{2}}\right) \exp\left(\frac{z}{k-\frac{1}{2}} + \frac{z^2}{2\,(k-\frac{1}{2})^2}\right) \right]^k. \tag{10}$$

It readily follows from (7) and (10) that

(11)

$$E_2^+(z) = \frac{\{E^B(z, -\frac{1}{2})\}^2}{(1-2z)^2 \exp(4z+4z^2)} \left\{ \prod_{k=1}^{\infty} \left[\left(1 - \frac{z}{k+\frac{1}{2}}\right) \exp\left(\frac{z}{k+\frac{1}{2}} + \frac{z^2}{2\,(k+\frac{1}{2})^2}\right) \right] \right\}^{-1}.$$

Obviously, the sequence $\{\lambda_k\}$ with $\lambda_k = k$ is of order $\mu = 1$. It follows from 1.2(2) and 5.1(6) that the analogous Weierstrass canonical product $E(\lambda)$ of this sequence is

$$E(\lambda) = \prod_{k=1}^{\infty} \left(1 - \frac{\lambda}{k}\right) e^{\frac{\lambda}{k}} = \frac{e^{\gamma\lambda}}{\Gamma(1-\lambda)}, \tag{12}$$

which, in view of 5.1(9), yields

$$\mathcal{R}_0(\lambda) = -\frac{d}{d\lambda}\{\log E(\lambda)\} = -\psi(1-\lambda) - \gamma, \tag{13}$$

where ψ is the Psi (or Digamma) function defined by 1.2(1).

Now, the shift equation (8) applied to (12) with $a = -\frac{1}{2}$ and $\mu = 1$ yields the classical result:

$$(14) \qquad \prod_{k=0}^{\infty}\left(1-\frac{z}{k+\frac{1}{2}}\right)\exp\left(\frac{z}{k+\frac{1}{2}}\right) = e^{(\gamma+2\log 2)z}\frac{\Gamma\left(\frac{1}{2}\right)}{\Gamma\left(\frac{1}{2}-z\right)},$$

which, by virtue of 1.1(14), immediately becomes

$$(15) \qquad \prod_{k=1}^{\infty}\left(1-\frac{z}{k+\frac{1}{2}}\right)\exp\left(\frac{z}{k+\frac{1}{2}}\right) = \frac{\sqrt{\pi}\, e^{(\gamma+2\log 2)z}}{(1-2z)\, e^{2z}\, \Gamma\left(\frac{1}{2}-z\right)}.$$

If we apply 1.1(14), 1.2(53), 2.2(4), 2.3(2), and (15) to (11), we easily find that

$$(16) \qquad E_2^{+}(z) = \frac{\Gamma\left(\frac{1}{2}-z\right)\left\{E^B\left(z,-\frac{1}{2}\right)\right\}^2}{\sqrt{\pi}\,(1-2z)\,2^{2z}\,\exp\left[(2+\gamma)z+\left(2+\frac{\pi^2}{4}\right)z^2\right]}.$$

Similarly, the shift equation 5.1(8) applied to (9) with $a = -\frac{1}{2}$ and $\mu = 2$ yields

$$(17) \qquad \begin{aligned} E^B\left(z,-\tfrac{1}{2}\right) =& (2\pi)^{\frac{1}{2}z}\exp\left[(\log 2)z+\left(\log 2+\frac{\pi^2}{8}\right)z^2\right] \\ & \cdot \exp\left[\tfrac{1}{2}\left\{(1+\gamma)z^2+\gamma z\right\}\right]\frac{G\left(\frac{1}{2}-z\right)}{G\left(\frac{1}{2}\right)}. \end{aligned}$$

A combination of (16) and (17) gives

$$(18) \quad E_2^{+}(z) = \frac{(2\pi)^z\,\Gamma\left(\frac{1}{2}-z\right)}{\sqrt{\pi}\,(1-2z)}\left\{\frac{G\left(\frac{1}{2}-z\right)}{G\left(\frac{1}{2}\right)}\right\}^2 \exp\left[-2z+(\gamma-1+2\log 2)\,z^2\right].$$

Finally, it follows from (8) and (18) that

$$(19) \qquad \begin{aligned} E_2(\lambda) =& \frac{\Gamma\left(\frac{1}{2}-iw\right)\Gamma\left(\frac{1}{2}+iw\right)\left\{G\left(\frac{1}{2}-iw\right)G\left(\frac{1}{2}+iw\right)\right\}^2}{\pi\left\{G\left(\frac{1}{2}\right)\right\}^2(1-2iw)(1+2iw)} \\ & \cdot \exp\left[-2(\gamma-1+2\log 2)\,w^2\right]. \end{aligned}$$

Since $\lambda = -w^2$, we have $w = \pm i\sqrt{\lambda}$. We see that the second member of (19) remains the same with w replaced by either $i\sqrt{\lambda}$ or $-i\sqrt{\lambda}$. In view of this observation and the relationship:

$$G(z) = \frac{1}{\Gamma_2(z)},$$

(19) is seen to be equivalent to (2). This completes the proof of the proposition (2).

Propositions (1) and (3) can be proven similarly.

Evaluations of $\det' \Delta_n$ $(n = 1, 2, 3)$

By making use of Proposition 5.1, we can compute the determinants of the Laplacians on $\mathbf{S}^n$ $(n = 1, 2, 3)$ explicitly. The results of these computations (as well as our computation of $\det' \Delta_4$) are summarized by

Proposition 5.2. *The following evaluations hold true:*

$$\det{}'\Delta_1 = 4\pi^2, \tag{20}$$

$$\det{}'\Delta_2 = \exp\left[\tfrac{1}{2} - 4\,\zeta'(-1)\right] = e^{\frac{1}{6}}\, A^4 = 3.195311496\ldots, \tag{21}$$

$$\det{}'\Delta_3 = \pi \exp\left[\frac{\zeta(3)}{2\pi^2}\right] = 3.338851215\ldots, \tag{22}$$

and

$$\begin{aligned} \det{}'\Delta_4 =& \frac{1}{3}\exp\left(\frac{35,639,301}{2^{17}\cdot 5\cdot 7} - \frac{13}{3}\zeta'(-1) - \frac{2}{3}\zeta'(-3)\right) \\ =& \frac{1}{3}\cdot A^{\frac{13}{3}}\cdot C^{\frac{2}{3}}\cdot \exp\left(\frac{183,758,875}{2^{17}\cdot 3^3\cdot 7}\right), \end{aligned} \tag{23}$$

where A is the Glaisher-Kinkelin constant defined by 1.3(2) *and C is the mathematical constant defined by* 1.3(70).

Proof. $\det{}'\Delta_1$: It follows from 5.1(15) and 5.1(16) with $n = 1$ that the sequence of eigenvalues of Δ_1 on $\mathbf{S}^1$ is

$$\mu_k = k^2 \quad \text{with multiplicity} \quad 2 \quad (k \in \mathbb{N}_0), \tag{24}$$

which corresponds to the shifted sequence $\{\lambda_k\}$ in 5.1(17). It is obvious from the definition 5.1(5) that the sequence $\{\mu_k\}$ in (24) (and also its shifted sequence $\{\lambda_k\}$) is of order $\mu = \frac{1}{2}$. Thus its corresponding Zeta series $Z_1(s)$ is given by

$$Z_1(s) = \sum_{k=1}^{\infty} \frac{2}{k^{2s}} = 2\,\zeta(2s), \tag{25}$$

where $\zeta(s)$ is the Riemann Zeta function defined by 2.3(1). It is seen from 5.1(4) and 2.2(20) that

$$\det{}'\Delta_1 = \exp\left[-4\,\zeta'(0)\right] = 4\pi^2,$$

which proves the evaluation (20).

Alternatively, even though it is not necessary to do so in the case of $\det{}'\Delta_1$, we can also prove (20) by making use of 5.1(12), 5.1(18), and (1). Indeed, since the shifted sequence considered above is of order $\mu = \frac{1}{2}$, it is found from 5.1(12), 5.1(18), and 1.1(13) that

$$\det{}'\Delta_1 = D_1(0) = \exp\left[-Z_1'(0)\right] \lim_{\lambda\to 0} E_1(\lambda) = \exp\left[-Z_1'(0)\right],$$

where $Z_1(s)$ is the same as in (25), and the evaluation (20) follows immediately.

det$'\Delta_2$: In this case we saw that the shifted sequence $\{\lambda_k\}$ in (5) is of order $\mu = 1$. Therefore, we find from 5.1(12) and 5.1(18) that

$$\det{}'\Delta_2 = D_2\left(\tfrac{1}{4}\right) = \exp\left[-Z_2{}'(0) - \tfrac{1}{4}\mathrm{FP}Z_2(1)\right] \lim_{\lambda\to\frac{1}{4}} E_2(\lambda). \tag{26}$$

From (5) and 5.1(2), we have

$$Z_2(s) = \sum_{k=1}^{\infty} \frac{2k+1}{\left(k+\frac{1}{2}\right)^{2s}} = \left(2^{2s}-2\right)\zeta(2s-1) - 4^s, \tag{27}$$

where $\zeta(s)$ is the Riemann Zeta function defined by 2.3(1). Since $Z_2(s)$ has a simple pole only at $s = 1$ with its residue 1 (see Section 2.3), it is seen from 5.1(13), 2.2(19) with $a = 1$, and (27) that

$$\begin{aligned}
\mathrm{FP}Z_2(1) &= \lim_{\epsilon\to0}\left\{Z_2(1+\epsilon) - \frac{1}{\epsilon}\right\} \\
&= -4 + \lim_{\epsilon\to0}\left\{\left(2^{2+2\epsilon}-2\right)\zeta(2\epsilon+1) - \frac{1}{\epsilon}\right\} \\
&= -4 + \lim_{\epsilon\to0}\left[\left(2^{2+2\epsilon}-2\right)\left\{\zeta(2\epsilon+1) - \frac{1}{2\epsilon}\right\} + \frac{2^{1+2\epsilon}-2}{\epsilon}\right] \\
&= -4 + 2\gamma + \lim_{\epsilon\to0}\left\{\frac{2^{1+2\epsilon}-2}{\epsilon}\right\},
\end{aligned}$$

which, upon employing l'Hôpital's rule, immediately yields

$$\mathrm{FP}Z_2(1) = 2\gamma + 4\log 2 - 4. \tag{28}$$

It is easy to see from 2.3(10), 1.6(7), and 2.1(31) that

$$Z_2{}'(0) = -\frac{13}{6}\log 2 - 2\zeta'(-1) = -\frac{1}{6} - \frac{13}{6}\log 2 + 2\log A. \tag{29}$$

It follows from (2) and 1.1(12) (with z replaced by $\frac{1}{2}+z$) that

$$\begin{aligned}
\lim_{\lambda\to\frac{1}{4}} E_2(\lambda) = &\frac{\left\{\Gamma_2\left(\frac{1}{2}\right)\right\}^4}{4}\exp\left[\frac{1}{2}(\gamma - 1 + 2\log 2)\right] \\
&\cdot \lim_{\lambda\to\frac{1}{4}} \frac{G\left(\frac{1}{2}-\sqrt{\lambda}\right) G\left(\frac{1}{2}-\sqrt{\lambda}\right)}{\left(\frac{1}{2}-\sqrt{\lambda}\right)\cosh\left(\pi i\sqrt{\lambda}\right)},
\end{aligned}$$

which, upon considering the following limit relationships:

$$\lim_{\lambda\to\frac{1}{4}} \frac{G\left(\frac{1}{2}-\sqrt{\lambda}\right)}{\cosh\left(\pi i\sqrt{\lambda}\right)} = \frac{1}{\pi} \quad\text{and}\quad \lim_{\lambda\to\frac{1}{4}} \frac{G\left(\frac{1}{2}-\sqrt{\lambda}\right)}{\frac{1}{2}-\sqrt{\lambda}} = 1, \tag{30}$$

yields

$$\lim_{\lambda \to \frac{1}{4}} E_2(\lambda) = \frac{\{\Gamma_2\left(\frac{1}{2}\right)\}^4}{4\pi} \exp\left[\frac{1}{2}(\gamma - 1 + 2\log 2)\right]. \tag{31}$$

Finally, if we substitute from (28), (29), and (31) into (26), and make use of 2.1(31), we are led at once to the evaluation (21).

det$'\Delta_3$: From 5.1(17) with $n = 3$, we find that the shifted sequence $\{\lambda_k\}$ (by 1) of eigenvalues of the Laplacian Δ_3 on $\mathbf{S}^3$ is

$$\lambda_k = (k+1)^2 \quad \text{with multiplicity} \quad (k+1)^2 \quad (k \in \mathbb{N}_0), \tag{32}$$

which is easily seen to be of order $\mu = \frac{3}{2}$. Therefore, the analogous Weierstrass canonical product and its accompanying Zeta series for this sequence are

$$E_3(\lambda) = \prod_{k=1}^{\infty} \left[\left(1 - \frac{\lambda}{(k+1)^2}\right) \exp\left(\frac{\lambda}{(k+1)^2}\right)\right]^{(k+1)^2} \tag{33}$$

and

$$Z_3(s) = \sum_{k=1}^{\infty} \frac{(k+1)^2}{(k+1)^{2s}} = \zeta(2s-2) - 1. \tag{34}$$

It follows from 5.1(12) and 5.1(18) that

$$\det{}'\Delta_3 = D_3\,(1) = \exp\left[-Z_3{}'(0) - \mathrm{FP}Z_3(1)\right] \lim_{\lambda \to 1} E_3(\lambda). \tag{35}$$

It is also seen that $Z_3(s)$ has a simple pole only at $s = \frac{3}{2}$ with its residue $\frac{1}{2}$. Thus we have from 5.1(13), 2.2(13) (with $a = 1$), and 2.3(22) (with $n = 1$) that

$$\mathrm{FP}Z_3(1) = Z_3(1) = \zeta(0) - 1 = -\tfrac{3}{2} \tag{36}$$

and

$$Z_3{}'(0) = 2\,\zeta'(-2) = -\frac{\zeta(3)}{2\,\pi^2}. \tag{37}$$

If we substitute from (36) and (37) into (35), we obtain

$$\det{}'\Delta_3 = \exp\left(\frac{3}{2} + \frac{\zeta(3)}{2\,\pi^2}\right) \lim_{\lambda \to 1} E_3(\lambda). \tag{38}$$

By taking the limit as $\lambda \to 1$ on both sides of (3), and applying 1.3(6) and Theorem 1.5(a), we obtain

$$\lim_{\lambda \to 1} E_3(\lambda) = 2\pi \exp\left(-\tfrac{3}{2}\right) \mathcal{A}, \tag{39}$$

where, for convenience,

$$\mathcal{A} = \lim_{\lambda\to 1}\left\{\frac{\Gamma_2\left(1-\sqrt{\lambda}\right)\left\{\Gamma_3\left(1-\sqrt{\lambda}\right)\right\}^{-2}}{1-\lambda}\right\}. \tag{40}$$

It is readily seen from 2.1(27), 2.1(28), 2.1(24), and 2.2(17) that

$$\mathcal{A} = \exp\left[\zeta'(0) + 2\,\zeta'(-1) + \zeta'(-2)\right]\cdot \mathcal{B}\cdot\mathcal{C}, \tag{41}$$

where, for convenience, $\mathcal{B}$ and $\mathcal{C}$ are defined by

$$\mathcal{B} := \lim_{\lambda\to 1}\left\{\frac{1}{1-\lambda}\left[\frac{(2\pi)^{\frac{1}{2}}}{\Gamma\left(1-\sqrt{\lambda}\right)}\right]^{\lambda}\right\}$$

and

$$\mathcal{C} := \lim_{\lambda\to 1}\left\{\exp\left[-\zeta'(-2, 1-\sqrt{\lambda}) - 2\sqrt{\lambda}\,\zeta'(-1, 1-\sqrt{\lambda})\right]\right\}.$$

First, in order to determine the limit value $\mathcal{B}$, consider

$$\mathcal{B} = \lim_{\lambda\to 1}\left\{\frac{\left[(2\pi)^{\frac{1}{2}}\right]^{\lambda}}{1+\sqrt{\lambda}}\cdot\frac{1}{\left(1-\sqrt{\lambda}\right)\left\{\Gamma\left(1-\sqrt{\lambda}\right)\right\}^{\lambda}}\right\},$$

which, upon setting $z = 1-\sqrt{\lambda}$, and using 1.1(9) and 1.1(13), immediately becomes

$$\mathcal{B} = \left(\frac{\pi}{2}\right)^{\frac{1}{2}}\lim_{z\to 0}\left\{\frac{1}{z\left\{\Gamma(z)\right\}^{1-2z+z^2}}\right\} = \left(\frac{\pi}{2}\right)^{\frac{1}{2}}\lim_{z\to 0}\left\{\left(\frac{1}{\Gamma(z)}\right)^{z^2-2z}\right\}.$$

We thus find from 1.1(2) that

$$\mathcal{B} = \left(\frac{\pi}{2}\right)^{\frac{1}{2}}\lim_{z\to 0} z^{z^2-2z} = \left(\frac{\pi}{2}\right)^{\frac{1}{2}}. \tag{42}$$

Next, in order to evaluate the limit value $\mathcal{C}$, by setting $n = 1$ in 2.2(4), we have

$$\zeta(s, a) = \zeta(s, 1+a) + a^{-s}, \tag{43}$$

which, upon differentiating with respect to s, yields

$$\zeta'(s, a) = \zeta'(s, 1+a) - a^{-s}\,\log a. \tag{44}$$

Therefore, we see from (44) that

$$\zeta'\left(-1, 1-\sqrt{\lambda}\right) = \zeta'\left(-1, 2-\sqrt{\lambda}\right) - \left(1-\sqrt{\lambda}\right)\log\left(1-\sqrt{\lambda}\right)$$

and

$$\zeta'\left(-2, 1-\sqrt{\lambda}\right) = \zeta'\left(-2, 2-\sqrt{\lambda}\right) - \left(1-\sqrt{\lambda}\right)^2 \log\left(1-\sqrt{\lambda}\right),$$

and so

$$\lim_{\lambda\to 1} \zeta'\left(-1, 1-\sqrt{\lambda}\right) = \zeta'(-1) \quad \text{and} \quad \lim_{\lambda\to 1} \zeta'\left(-2, 1-\sqrt{\lambda}\right) = \zeta'(-2),$$

which yield

$$\mathcal{C} = \exp\left[-2\,\zeta'(-1) - \zeta'(-2)\right]. \tag{45}$$

Finally, by substituting from (42) and (45) into (41) and using 2.2(20), we find from (39) that

$$\lim_{\lambda\to 1} E_3(\lambda) = \pi \exp\left(-\tfrac{3}{2}\right), \tag{46}$$

which, when substituted into (38), proves the evaluation (22).

A proof of the evaluation (23) will be presented in Section 5.3 in which each of the results (21), (22), and (23) is obtained by means of the summation formulas of Section 3.4 for series of Zeta functions (see also Problem 3 at the end of this chapter).

5.3 Computations Using Series of Zeta Functions

In this section, we compute $\det'\Delta_n$ (n = 2, 3, 4) in a markedly different way from that detailed in Section 5.2 for n = 1, 2, 3. Our evaluations here are based largely upon the summation formulas of Section 3.4 for series of Zeta functions.

det′$\boldsymbol{\Delta_2}$: In view of 5.2(26), 5.2(28), and 5.2(29), it suffices to evaluate $E_2\left(\frac{1}{4}\right)$. It follows from 5.2(6) that

$$E_2\left(\frac{1}{4}\right) = \prod_{k=1}^{\infty}\left[\left(1-\frac{1}{(2k+1)^2}\right)^{2k+1} \exp\left(\frac{1}{2k+1}\right)\right], \tag{1}$$

which, upon taking logarithms and applying the Maclaurin series expansion of $\log(1+x)$, yields

$$\begin{aligned}\log E_2\left(\frac{1}{4}\right) &= -\sum_{n=2}^{\infty}\frac{1}{n}\left(\sum_{n=2}^{\infty}\frac{1}{(2k-1)^{2n-1}}-1\right)\\ &= -\sum_{n=2}^{\infty}\frac{\zeta(2n-1)-1}{n}+\sum_{n=2}^{\infty}\frac{\zeta(2n-1)}{n\cdot 2^{2n-1}}.\end{aligned} \tag{2}$$

If we apply 3.4(468) and 3.4(477) on the right-hand side of (2), we obtain

$$\log E_2\left(\frac{1}{4}\right) = -1 + \frac{\gamma}{2} - \frac{7}{6}\log 2 + 6\log A, \tag{3}$$

which, upon combining 5.2(26) with 5.2(8), 5.2(29), and 2.1(31), proves the evaluation 5.2(21).

det$'\Delta_3$: By virtue of 5.2(38), it suffices to compute $E_3(1)$. Indeed, letting $\lambda = 1$ in 5.2(33) and taking logarithms of the resulting equation with the aid of the Maclaurin series expansion of $\log(1+x)$, we get

$$\log E_3(1) = -\sum_{k=2}^{\infty}\left(\sum_{n=1}^{\infty}\frac{1}{(n+1)k^{2n}}\right) = -\sum_{n=1}^{\infty}\frac{\zeta(2n)-1}{n+1}, \tag{4}$$

which, in view of 3.4(519), immediately yields

$$\log E_3(1) = \log\pi - \frac{3}{2}. \tag{5}$$

Now, if we make use of (5) in 5.2(38), we are led to the evaluation 5.2(22).

det$'\Delta_4$: Letting $n = 4$ in the shifted sequence 5.1(17) of eigenvalues of Δ_4 on $\mathbf{S}^4$, we obtain the following sequence:

$$\lambda_k = \left(k + \tfrac{3}{2}\right)^2 \quad \text{with multiplicity} \quad \tfrac{1}{6}(k+1)(k+2)(2k+3) \quad (k \in \mathbb{N}_0), \tag{6}$$

which obviously has the order $\mu = 2$. Now it follows from 5.1(12) and 5.1(18) that

$$\begin{aligned} \det{}'\Delta_4 &= D_4\left(\frac{9}{4}\right) \\ &= \exp\left(-Z_4'(0) - \frac{9}{4}\mathrm{FP}\,Z_4(1) - \frac{81}{32}\mathrm{FP}\,Z_4(2) - \frac{81}{32}C_{-2}\right)E_4\left(\frac{9}{4}\right). \end{aligned} \tag{7}$$

We can express $Z_4(s)$ for the sequence (6) in terms of the Riemann Zeta function as follows:

$$\begin{aligned} Z_4(s) &= \frac{1}{6}\sum_{k=1}^{\infty}\frac{(k+1)(k+2)(2k+3)}{\left(k+\frac{3}{2}\right)^{2s}} \\ &= \frac{2^{2s}}{6}\sum_{k=1}^{\infty}\frac{(k+1)(k+2)}{(2k+3)^{2s-1}} = \frac{2^{2s}}{6}\sum_{k=2}^{\infty}\frac{k(k+1)}{(2k+1)^{2s-1}} \\ &= \frac{2^{2s}}{24}\left(\sum_{k=2}^{\infty}\frac{1}{(2k+1)^{2s-3}} - \sum_{k=2}^{\infty}\frac{1}{(2k+1)^{2s-1}}\right) \\ &= \frac{2^{2s}}{24}\left(\sum_{k=1}^{\infty}\frac{1}{(2k-1)^{2s-3}} - \sum_{k=1}^{\infty}\frac{1}{(2k-1)^{2s-1}} - \frac{1}{3^{2s-3}} + \frac{1}{3^{2s-1}}\right), \end{aligned} \tag{8}$$

which, in view of 2.3(1), becomes

$$\begin{aligned} Z_4(s) = \frac{1}{3}\left(2^{2s-3}-1\right)\zeta(2s-3) - \frac{1}{3}\left(2^{2s-3}-\frac{1}{4}\right)\zeta(2s-1) \\ -\frac{1}{3}\left(\frac{2}{3}\right)^{2s-3} + \frac{1}{8}\left(\frac{2}{3}\right)^{2s}. \end{aligned} \tag{9}$$

We observe from Section 2.3 that $Z_4(s)$ has simple poles at $s=1$ and $s=2$ with their residues $-\frac{1}{24}$ and $\frac{1}{6}$, respectively.

Using 2.3(10) and 5.1(14), we obtain

$$C_{-2} = \frac{1}{2} Z_4(-2) = -\frac{9,801,047}{2^{12}\cdot 3^3 \cdot 5\cdot 7} \tag{10}$$

and

$$Z_4'(0) = \log\left(2^{-\frac{2869}{1440}}\cdot 3^2\right) + \frac{1}{12}\zeta'(-1) - \frac{7}{12}\zeta'(-3). \tag{11}$$

Now we evaluate $\mathrm{FP}Z_4(1)$ and $\mathrm{FP}Z_4(2)$. Since $Z_4(s)$ has simple poles at $s=1$ and $s=2$, we have to use the second case of the definition of $\mathrm{FP}f(s)$ in 5.1(13) to compute the finite parts of $Z_4(s)$ for $s=1$ and $s=2$.

Using the expression in (9) for $Z_4(s)$ and 2.3(6) [or 2.2(19) with $a=1$], we easily see that

$$\begin{aligned}\mathrm{FP}Z_4(1) &= \lim_{\epsilon\to 0}\left(Z_4(1+\epsilon)+\frac{1}{24\epsilon}\right)\\ &= -\frac{31}{72} - \frac{1}{3}\lim_{\epsilon\to 0}\left[\left(2^{2\epsilon-1}-\frac{1}{4}\right)\left\{\zeta(2\epsilon+1)-\frac{1}{2\epsilon}\right\}+\frac{2^{2\epsilon}-1}{4\epsilon}\right]\\ &= -\frac{31}{72}-\frac{\gamma}{12}-\frac{1}{6}\log 2.\end{aligned} \tag{12}$$

Similarly, we have

(13)

$$\begin{aligned}\mathrm{FP}Z_4(2) &= \lim_{\epsilon\to 0}\left(Z_4(2+\epsilon)-\frac{1}{6\epsilon}\right)\\ &= -\frac{16}{81}-\frac{7}{12}\zeta(3)+\frac{1}{3}\lim_{\epsilon\to 0}\left[\left(2^{2\epsilon+1}-1\right)\left\{\zeta(2\epsilon+1)-\frac{1}{2\epsilon}\right\}+\frac{2^{2\epsilon+1}-2}{2\epsilon}\right]\\ &= -\frac{16}{81}+\frac{\gamma}{3}+\frac{2}{3}\log 2-\frac{7}{12}\zeta(3).\end{aligned}$$

Since the sequence in (6) is of order $\mu=2$, its analogous Weierstrass canonical product $E_4(\lambda)$ is

$$\begin{aligned}E_4(\lambda) &= \prod_{k=1}^{\infty}\left(1-\frac{\lambda}{\left(k+\frac{3}{2}\right)^2}\right)^{\frac{1}{6}(k+1)(k+2)(2k+3)}\\ &\quad\cdot\exp\left[\frac{1}{6}(k+1)(k+2)(2k+3)\left(\frac{\lambda}{\left(k+\frac{3}{2}\right)^2}+\frac{\lambda^2}{2\left(k+\frac{3}{2}\right)^4}\right)\right].\end{aligned} \tag{14}$$

Upon setting $\lambda=\frac{9}{4}$ in (14) and taking the logarithms on both sides of the resulting equation, if we make use of 2.3(1) and the Maclaurin series expansion of

$\log(1+x)$, we obtain

(15)

$$\log E_4\left(\frac{9}{4}\right)=-\frac{1}{6}\sum_{k=1}^{\infty}(k+1)(k+2)\left\{\sum_{n=3}^{\infty}\frac{3^{2n}}{n(2k+3)^{2n-1}}\right\}$$

$$=-\frac{1}{24}\sum_{n=3}^{\infty}\frac{3^{2n}}{n}\left[\sum_{k=1}^{\infty}\frac{1}{(2k-1)^{2n-3}}-\sum_{k=1}^{\infty}\frac{1}{(2k-1)^{2n-1}}-\frac{1}{3^{2n-3}}+\frac{1}{3^{2n-1}}\right]$$

$$=-\frac{1}{24}\sum_{n=3}^{\infty}\frac{1}{n}\left[3^{2n}\zeta(2n-3)-8\left(\frac{3}{2}\right)^{2n}\zeta(2n-3)-3^{2n}\zeta(2n-1)\right.$$

$$\left.+2\left(\frac{3}{2}\right)^{2n}\zeta(2n-1)-24\right].$$

Now let α_1, α_2, α_3, and α_4 denote the sums of the Zeta series occurring in 3.4(700), 3.4(703), 3.4(706), and 3.4(710), respectively. We then find from (15) that

(16)

$$\log E_4\left(\frac{9}{4}\right)=-\frac{1}{24}(2\alpha_1-\alpha_2+\alpha_3-8\alpha_4)$$

$$=-\frac{4}{9}+\frac{21}{32}\gamma-\frac{189}{128}\zeta(3)$$

$$+\frac{17}{4}\log A+\frac{5}{4}\log C+\log\left(2^{-\frac{979}{1440}}\cdot 3\right).$$

Finally, in view of 2.1(31) and 2.3(27), we are easily led from (7) and (9) through (16) to the evaluation 5.2(23).

It should be *noted* that we need to compare the evaluation of $\det'\Delta_4$ in this section with those in other works (see Section 5.4).

5.4 Remarks and Observations

Even though each of the determinantal expressions 5.2(20), 5.2(21), and 5.2(22) has been considered by other workers on this subject, our novel approach in Section 5.3 to the familiar problem of computation of the determinants of the Laplacians on the n-sphere S^n ($n=1,2,3$) is relatively more natural than the earlier approaches to these and analogous problems.

The evaluations of $\det'\Delta_n$ ($n=1$, 2, 3) in Choi (1994a), Quine and Choi (1996), Choi and Srivastava (1999c), and Kumagai (1999) have the same results, whereas the value of $\det'\Delta_4$ in Quine and Choi (1996) is equal to that in Kumagai (1999), and is not equal to that in Choi and Srivastava (2000a) (see also Section 5.3). The only difference is that the number $35,639,301/(2^{17}\cdot5\cdot7)$ appearing in the evaluation of $\det'\Delta_4$ in Choi and Srivastava (2000a) (see also Proposition 5.2) was assumed to be 15/16 in Quine and Choi (1996, p. 726) and in Kumagai (1999, p. 201, Corollary (iv)). Thus we need to examine C_{-m} in 5.1(14) [see Voros (1987, p. 444, Eq. (3.3))] once again.

Vardi (1988b) also computed $\det'\Delta_n$ by employing a direct analysis of the involved Zeta series in 5.1(2) with the original (or unshifted) sequence $\{\mu_k\}$ in 5.1(15).

The results for $\det'\Delta_n$ $(n = 1, 2)$ in Proposition 5.2 agree with those of Vardi (1988b). Yet Vardi's ingenious computation for $\det'\Delta_n$ seems to have some gaps (see Problem 5 below). Kumagai (1999) evaluated $\det'\Delta_n$ by filling these gaps in Vardi's work (1988b) including a rather detailed discussion of the computations of $\det'\Delta_n$ presented by several earlier authors [*cf.* Choi (1994), Efrat (1991), Quine and Choi (1996), and Vardi (1988b)].

Quine and Choi (1996) computed $\det'\Delta_n$ and $\det(\Delta_n + n(n-2)/4)$ (which are the determinants of the Laplacians and the conformal Laplacians, respectively) on sphere by using techniques due to Weisberger (1987). These techniques offer a unified approach to such problems. Branson and Ørsted (1991) computed the conformal determinant $\det(\Delta_4 + 2)$ as follows:

$$\det(\Delta_4 + 2) = \exp\left(-\frac{1}{144} - \frac{1}{3}\zeta'(-1) - \frac{2}{3}\zeta'(-3)\right),$$

which agrees with the result in Quine and Choi (1996, p. 727).

Quine *et al.* (1993) introduced the Zeta regularized product for the general sequences of nonzero complex numbers instead of the increasing sequences of non-negative real numbers given in 5.1(1), in which case their Zeta regularized product corresponds to the determinant of the Laplacian defined by 5.1(4). They gave many examples for this extended concept of the determinant of the Laplacian.

Osgood *et al.* (1988) noted that the modern quantum geometry of strings is concerned (to a large extent) with the set of all surfaces, with varying metrics on those surfaces, and with determinants of the associated Laplacians. They investigated this determinant as a function of the metric on a given surface and, in particular, its extreme values when the metric is suitably restricted. They also observed an interesting fact related to the study of this chapter: *For all metrics σ on $\mathbf{S}^2$ of area 4π,*

$$\det'\Delta_\sigma \leq \exp\left[\tfrac{1}{2} - 4\,\zeta'(-1)\right] = \det'\Delta_2, \tag{4}$$

where the equality holds true if and only if σ is the standard round metric, Δ_σ being the Laplacian on $\mathbf{S}^2$ with the given metric σ.

Problems

1. Suppose that the series $\sum |f(n)\,n^{-s}|$ does not converge for all s or is divergent for all s. Show that there exists a real number σ_a, called the *abscissa of absolute convergence*, such that the series $\sum |f(n)\,n^{-s}|$ converges absolutely if $\Re(s) = \sigma > \sigma_a$, but does not converge absolutely if $\sigma < \sigma_a$.
[*Cf.* Apostol (1976a, p. 225)]

2. Show that the analogous Weierstrass canonical product $E_n(\lambda)$ of the shifted sequence $\{\lambda_k\}$ in 5.1(17) for the standard Laplacian Δ_n on the n-dimensional unit sphere $\mathbf{S}^n$ can be expressed, in terms of the multiple Gamma functions Γ_n given in Section 1.3, as follows:

$$E_n(\lambda) = \mathcal{R}_n(\lambda) \frac{\Gamma_{n-1}\left(\frac{n-1}{2} - \sqrt{\lambda}\right) \Gamma_{n-1}\left(\frac{n-1}{2} + \sqrt{\lambda}\right)}{\left\{\Gamma_n\left(\frac{n-1}{2} - \sqrt{\lambda}\right) \Gamma_n\left(\frac{n-1}{2} + \sqrt{\lambda}\right)\right\}^2},$$

where $\mathcal{R}_n(\lambda)$ is a meromorphic function with a simple pole only at $\lambda = \left(\frac{n-1}{2}\right)^2$ and $\Gamma_1(z) = \Gamma(z)$.

Show also that the above formula for $E_n(\lambda)$ holds true for all $n \in \mathbb{N}$ if we define $\Gamma_0(z) := \frac{1}{z}$.

[*Cf.* Choi (1992); see also Theorem 1.4]

3. Let $\mathcal{R}_n(\lambda)$ be as in Problem 2 above. By using Proposition 5.1, show that

$$\mathcal{R}_1(\lambda) = -\frac{1}{\lambda},$$

$$\mathcal{R}_2(\lambda) = \frac{\left\{\Gamma_2\left(\frac{1}{2}\right)\right\}^4}{\pi\,(1-4\,\lambda)} \exp\left[2(\gamma - 1 + 2\,\log 2)\lambda\right],$$

and

$$\mathcal{R}_3(\lambda) = \frac{1}{1-\lambda} \exp\left[\left(\log(2\pi) - \tfrac{3}{2}\right)\lambda\right].$$

Also determine $\mathcal{R}_4(\lambda)$ explicitly and evaluate $\det{}'\Delta_4$ with the explicit form of $E_4(\lambda)$ by using of the method presented in Section 5.2.

4. Compute $\det{}'\Delta_n$ $(n = 5,\, 6)$ by employing the technique illustrated in Section 5.3.

5. Show why Vardi's arguments leading to the evaluation:

$$\det{}'\Delta_3 = \exp\left[-2\,\zeta'(-2) + \zeta'(-1) - 3\,\zeta'(0)\right]$$

cannot be justified.

[*Cf.* Vardi (1988b)]

6. Prove the inequality 5.4(2).

[*Cf.* Osgood *et al.* (1988)]

7. If λ_k is a sequence of nonzero complex numbers, consider the Zeta function:

$$Z(s) = \sum_{j=1}^{\infty} \lambda_j^{-s}.$$

If $Z(s)$ converges for $\Re(s) > s_0$ and has meromorphic continuation to a function meromorphic in $\Re(s) > s_1$ $(s_1 < 0)$ with at most simple poles, then we say that the sequence is *Zeta regularizable*. We define $\lambda_j^{-s} = \exp(-s\,\log\lambda_j)$, and the definition of Z depends on the choice of $\arg \lambda_j$. For a Zeta regularizable sequence, we define the Zeta regularized product:

$$\prod_z \lambda_j = \exp(-Z'(0)).$$

We also define the product $\prod_z (\lambda_j - \lambda)$, where (for $\lambda \neq \lambda_j$) we adopt the convention that $\arg(\lambda_j - \lambda) \approx \arg \lambda_j$ for $|\lambda_j|$ large.

Let λ_j be a Zeta regularizable sequence, and let h be an integer such that $\sum_{j=1}^{\infty} |\lambda_j|^{-h-1} < \infty$. For $\Re(s)$ sufficiently large and for $\lambda \neq \lambda_j$, let

$$F(s) = \sum_{j=1}^{\infty} \left[\left(\lambda_j^2 - \lambda^2\right)^{-s} - (\lambda_j - \lambda)^{-s} - (\lambda_j + \lambda)^{-s} \right].$$

Show that

$$-F'(0) = \sum_{j=1}^{[h/2]} \operatorname*{Res}_{s=2j} Z(s) \left(1 + \frac{1}{3} + \cdots + \frac{1}{2j-1}\right) \frac{\lambda^{2j}}{j},$$

which, in terms of Zeta regularized products, can be used to state the result as

$$\prod_z \left(\lambda_j^2 - \lambda^2\right) = e^{F'(0)} \prod_z (\lambda_j - \lambda) \prod_z (\lambda_j + \lambda).$$

[*Cf.* Quine and Choi (1996)]

8. For $n \in \mathbb{N} \setminus \{1\}$ and a an integer such that $0 \leq a \leq n-1$, and $\Re(s) > n$, let

$$Z(s,a) = \sum_{\ell=1}^{\infty} \beta_\ell^n \left[(\ell + a)(\ell + n - 1 - a)\right]^{-s},$$

where

$$\beta_\ell^n = \binom{n+\ell}{n} - \binom{n+\ell-2}{n}.$$

For $k = 1, \ldots, n$, define the functions $t_k^n(a)$ by the expansion:

$$\beta_{\ell-a}^n = \sum_{k=1}^{n} t_k^n(a)\, \ell^{k-1}.$$

Show that the derivative at zero of $Z(s,a)$ is given by

$$-Z'(0,a) = \tau_n(a) - \sum_{k=1}^{n} \left(1 + (-1)^{n+k}\right) t_k^n(a)\, \zeta'(1-k) - \mathcal{A},$$

where

$$\tau_n(a) = (1 + (-1)^n) \sum_{j=1}^{[n/2]} \frac{1}{2j} t_{2j}^n \left(\frac{n-1}{2}\right) \cdot \left(1 + \frac{1}{3} + \cdots \frac{1}{2j-1}\right) \left(\frac{n-1}{2} - a\right)^{2j},$$

and

$$\mathcal{A} := \begin{cases} \log(n-1) & \text{if} \quad a(n-1-a) = 0, \\ \log(a(n-1-a)) & \text{otherwise.} \end{cases}$$

[*Cf.* Quine and Choi (1996)]

9. Let **M** be a compact Riemann surface of genus $g \geq 2$. **M** can be identified with H/Γ, the action of a Fuchsian group Γ on the upper complex half-plane $\mathrm{H} = \{z = x + iy \mid y > 0\}$ endowed with the Poincaré metric $\mathrm{d}s^2 = y^{-2}\left(\mathrm{d}x^2 + \mathrm{d}y^2\right)$. This is the classical model for hyperbolic geometry of constant negative curvature, $\mathcal{R} = -1$. Here Γ is a discrete subgroup of $\mathbf{PSL}(2,\mathbb{R}) = \mathbf{SL}(2,\mathbb{R})/\{\pm 1\}$, the group of Möbius transformations. From the Gauss-Bonnet theorem, one infers that

$$\mathcal{R}\mathcal{A} = 2\pi\chi = 4\pi(1-g),$$

where $\mathcal{A}$ denotes the area of **M** and χ its Euler characteristic, *i.e.*, $\mathcal{A} = 4\pi(g-1)$. In the Poincaré metric the Laplacian on **M** (Laplace-Beltrami operator) is given by $\Delta = y^2\left(\partial^2/\partial x^2 + \partial^2/\partial y^2\right)$, and one is interested in the eigenvalue problem $-\Delta u = \lambda u$. The spectrum of Δ on **M** is discrete and real, $0 = \lambda_0 < \lambda_1 \leq \lambda_2 \leq \cdots$ with $\#(\text{eigenvalues } \lambda_n \text{ with } \lambda_n \leq \lambda) \sim (\mathcal{A}/4\pi)\,\lambda$ asymptotically (Weyl's law).

Since the elements $\tau \in \left(\begin{smallmatrix} a & b \\ c & d \end{smallmatrix}\right) \in \Gamma$ are hyperbolic, *i.e.*, $|\mathrm{Tr}\,\tau| = |a + d| > 2$, they are conjugate with Γ to a Möbius transformation of the form $z \to N(\tau)\, z$ $(1 < N(\tau) < \infty)$, where $N(\tau)$ is called the norm of τ. (We say that $\tau_1, \tau_2 \in \Gamma$ are conjugate within Γ if there exists a $\tau_3 \in \Gamma$ such that $\tau_1 = \tau_3\,\tau_2\,\tau_3^{-1}$.) The class of all elements in Γ which are conjugate to a given τ is called the conjugacy class of τ in Γ and denoted by $\{\tau\}$. The number $N(\tau)$ is, of course, the same within a conjugacy class and measures the "magnification". The number $N(\tau)$ has, however, another striking geometric interpretation, since there exists a unique relationship between the conjugacy classes in Γ and the homotopy classes of closed paths on the surface **M**. In each class one defines a length $l(\tau)$ by the length of the shortest closed path measured by means of the Poincaré distance. One then obtains $N(\tau) = \exp[l(\tau)]$ $(l(\tau) > 0)$. Thus the conjugacy classes in Γ can be uniquely parametrized by their length spectrum $\{l(\tau)\}$. Given any $\tau \in \Gamma$, there is a unique τ_0 such that $\tau = \tau_0^n$ $(n \in \mathbb{N})$; here τ_0 is called the primitive element of Γ, since it cannot be expressed as a power of any other element of Γ. The corresponding closed orbit with length $l(\tau_0)$ is called a prime geodesic on **M**. Obviously, $l(\tau) = l\left(\tau_0^n\right) = n\,l(\tau_0)$, since (in this case) the prime geodesic is traversed n times. For the length spectrum of **M** one has Huber's law $\nu(x) \sim e^x/x$ $(x \to \infty)$, where $\nu(x)$ is the number of inconjugate primitive τ's with $l(\tau) \leq x$.

(a) Prove the *Selberg trace formula*:

$$\sum_{n=0}^{\infty} h\left(\tau_n\right) = \frac{\mathcal{A}}{2\pi}\int_{-\infty}^{\infty} t\,\tanh(\pi\,t)\,h(t)\,dt$$
$$+ \sum_{\{\tau\}_p}\sum_{n=1}^{\infty} \frac{l(\tau)}{\sinh\left(\frac{1}{2}nl(\tau)\right)}\,g(n\,l(\tau)),$$

where all series and the integral converge absolutely under the following conditions on the function $h(t)$: (i) $h(-t) = h(t)$, (ii) $h(t)$ is analytic in a strip $|\Im t| \leq \frac{1}{2}+\epsilon$ ($\epsilon > 0$), (iii) $|h(t)| \leq a\left(1+|t|^2\right)^{-1-\epsilon}$ ($a > 0$); the function $g(u)$ is the Fourier transform of $h(t)$:

$$g(u) = \frac{1}{2\pi}\int_{-\infty}^{\infty} \exp(-iut)\, h(t)\, dt;$$

the sum on the left-hand side runs over the eigenvalues of Δ parametrized in the form $\lambda_n = \frac{1}{4} + \tau_n^2$, *i.e.*, over the pairs $(\tau_n, -\tau_n)$ ($\tau_n \in \mathbb{C}$) ($\tau = 0$ has to be counted twice if $\frac{1}{4}$ happens to be an eigenvalue); the sum on the right-hand side is taken over all primitive conjugacy classes in Γ, denoted by $\{\tau\}_p$.

(b) By choosing

$$h(t) = \frac{1}{t^2 + \left(s-\frac{1}{2}\right)^2} - \frac{1}{t^2 + \left(z-\frac{1}{2}\right)^2} \qquad (\min\{\Re(s),\, \Re(z)\} > 1)$$

in the Selberg trace formula in Part (a), prove that

$$\begin{aligned}\sum_{n=0}^{\infty}\left(\frac{1}{\lambda_n + s(s-1)} - \frac{1}{\lambda_n + z(z-1)}\right) =& -2(g-1)[\psi(s) - \psi(z)] \\ &+ \frac{1}{2s-1}\frac{Z'(s)}{Z(s)} - \frac{1}{2z-1}\frac{Z'(z)}{Z(z)},\end{aligned}$$

where ψ is the Psi (or Digamma) function given in Section 1.2 and $Z(s)$ denotes the *Selberg Zeta function* on $\mathbf{M}$ defined by

$$Z(s) := \prod_{\{\tau\}_p}\prod_{n=0}^{\infty}\{1 - \exp[-(s+n)\, l(\tau)]\}.$$

(c) Prove that

$$\begin{aligned}\frac{1}{2s-1}\frac{Z'(s)}{Z(s)} =& \mathcal{B} + \frac{1}{s(s-1)} + (g-1)\,[\psi(s)+\gamma] \\ &+ \sum_{n=1}^{\infty}\left(\frac{1}{\lambda_n + s(s-1)} - \frac{1}{\lambda_n}\right),\end{aligned}$$

where γ is the Euler-Mascheroni constant and

$$\mathcal{B} := \lim_{z\to 1+}\left(\frac{1}{2z-1}\frac{Z'(z)}{Z(z)} - \frac{1}{z(z-1)}\right) = \frac{1}{2}\frac{Z''(1)}{Z'(1)}.$$

(d) Deduce the following Laurent series expansion about $s = 1$:

$$\frac{1}{2s-1}\frac{Z'(s)}{Z(s)} = \frac{1}{s-1} + (\mathcal{B}-1) + \sum_{n=1}^{\infty} a_n\,(s-1)^n,$$

$$\begin{aligned}a_n = (-1)^{n+1}\Bigg[&1 + 2(g-1)\,\zeta(n+1) \\ &+ \sum_{j=0}^{[n/2]}(-1)^{j+1}\binom{n-j}{j}\zeta_\Delta(n+1-j)\Bigg] \qquad (n \in \mathbb{N}),\end{aligned}$$

where $\zeta(s)$ is the Riemann Zeta function and

$$\zeta_\Delta(s) := \sum_{n=1}^{\infty} \frac{1}{\lambda_n^s}.$$

(e) Prove that the finite part of $\zeta_\Delta(s)$ at $s = 1$ is given by

$$\mathrm{FP}\,\zeta_\Delta(1) = 2(g-1)\,\gamma + \mathcal{B} \ \ (:= \gamma_\Delta)\,.$$

(f) Prove that

$$\zeta_\Delta'(0) = 2(g-1)\,\mathcal{C} - \log(Z'(1)),$$

where

$$\begin{aligned}\mathcal{C} :=& -\frac{\pi}{2}\int_0^\infty \frac{t^2+\frac{1}{4}}{\cosh^2(\pi\,t)}\left[1 - \log\left(t^2 + \frac{1}{4}\right)\right] dt \\ =& \frac{1}{4} - \frac{1}{2}\log(2\pi) - 2\,\zeta'(-1).\end{aligned}$$

(g) By considering the functional determinant of Δ on $\mathbf{M}$

$$D_\Delta(z) := \det{}'(-\Delta + z) \quad (z \in \mathbb{C}),$$

where the prime indicates that the zero mode has been omitted and (by Zeta regularization)

$$D_\Delta(0) := \exp\left[-\zeta_\Delta'(0)\right],$$

prove that

$$\begin{aligned}D_\Delta(z) =& D_\Delta(0)\,\exp\left(\gamma_\Delta\,z\right)\prod_{n=1}^{\infty}\left\{\left(1+\frac{z}{\lambda_n}\right)\exp\left(-\frac{z}{\lambda_n}\right)\right\} \\ =& Z'(1)\,\exp\left[\gamma_\Delta\,z - 2(g-1)\,\mathcal{C}\right]\prod_{n=1}^{\infty}\left\{\left(1+\frac{z}{\lambda_n}\right)\exp\left(-\frac{z}{\lambda_n}\right)\right\}.\end{aligned}$$

(h) Prove the following fundamental relation for the Selberg Zeta function:

$$Z(s) = Z'(1)\,s(s-1)\,V(s),$$

where, for convenience,

$$\begin{aligned}V(s) :=& \exp\left[\gamma_\Delta\,s(s-1)\right]\,\left\{(2\pi)^{1-s}\,\exp[s(s-1)]\,G(s)\,G(s+1)\right\}^{\frac{\mathcal{A}}{2\pi}} \\ &\cdot \prod_{n=1}^{\infty}\left\{\left[1+\frac{s(s-1)}{\lambda_n}\right]\exp\left[-\frac{s(s-1)}{\lambda_n}\right]\right\}\end{aligned}$$

and $G(s)$ denotes the Barnes G-function given in Section 1.3.

(i) Prove that

$$Z(s) = s(s-1)\,D_\Delta(s(s-1))\,\left[(2\pi)^{1-s}\,\exp[\mathcal{C} + s(s-1)]\,G(s)\,G(s+1)\right]^{2(g-1)}.$$

[*Cf.* Steiner (1987); see also Voros (1986)]

CHAPTER 6

MISCELLANEOUS RESULTS

Over one and a half centuries ago, von Staudt and Clausen proved independently (and published almost simultaneously) a useful theorem for the calculation of the higher-order Bernoulli numbers, which has recently been generalized and proved in many different ways in various investigations dealing with evaluations of the Bernoulli and Euler polynomials at rational arguments. Closed-form summations of several trigonometric series have also been of recent concern by various authors. In addition to presenting some of these developments in this chapter, we establish several integrals involving a certain periodic function and various other families of definite integrals involving logarithms of the Gamma function by introducing some not-too-familiar mathematical constants which stem from the use of the Euler-Maclaurin summation formula 1.3(68).

6.1 Bernoulli and Euler Polynomials at Rational Arguments

We begin by recalling a theorem involving the Bernoulli numbers discovered in 1840 (and proved independently and published almost simultaneously) by C. von Staudt (1798–1867) and T. Clausen (1801–1885).

Theorem 6.1 (von Staudt-Clausen Theorem). *Let* B_n *denote the Bernoulli numbers defined by* 1.6(2). *Then*

$$B_{2n} = I_n - \sum_{p-1|2n} \frac{1}{p} \quad (n \in \mathbb{N}), \tag{1}$$

where I_n *is an integer and the sum is taken over all primes* p *such that* $p-1$ *divides* $2n$.

As noted by Carathéodory (1958, p. 283), the von Staudt-Clausen theorem (1) is very useful for the calculation of higher-order Bernoulli numbers, since the value of I_n follows even from a preliminary estimate of B_{2n} that one might obtain in one way or another (see also Problem 6 of Chapter 1).

Bartz and Rutkowski (1993) gave a simple proof of a generalization of the von Staudt-Clausen theorem (1) to hold true for the Bernoulli polynomials, which also reduces to another interesting and inspiring result [Almkvist and Meurman (1991, p. 104, Theorem 2)]:

$$k^n \left[B_n\left(\tfrac{h}{k}\right) - B_n\right] \in \mathbb{Z} \quad (h,\, k \in \mathbb{N}), \tag{2}$$

Almkvist and Meurman (1991, p. 104 *et seq.*) proved (2) in an elementary (yet rather complicated) way. Subsequently, simpler proofs of the Almkvist-Meurman theorem (2) have been given by Sury [1993] and Clarke and Slavutskii (1997). An analogous result involving the Euler polynomials instead was proved by Fox (1999).

All the aforecited works are concerned with evaluations of the Bernoulli and Euler polynomials at rational arguments.

The remainder of this section is based largely upon the work of Srivastava (2000b), who gave several remarkably shorter proofs of each of the Cvijović-Klinowski summation formulas and also considered their various generalizations and analogues.

The Cvijović-Klinowski summation formulas

By employing a number of properties and characteristics of certain Dirichlet and trigonometric series, Cvijović and Klinowski (1995a) evaluated the Bernoulli polynomials $B_n(x)$ with $n \in \mathbb{N} \setminus \{1\}$ and the Euler polynomials $E_n(x)$ with $n \in \mathbb{N}$ for $0 \leqq x \leqq 1$ when x is a rational number. For the sake of ready reference, we recall here the *main* results of Cvijović and Klinowski (1995a) as Theorem 6.2 and Theorem 6.3 below:

Theorem 6.2. *In terms of the Hurwitz Zeta function* $\zeta(s,a)$ *defined by* 2.2(1), *the Bernoulli polynomials* $B_n(x)$ *at rational arguments are given by*

$$B_{2n-1}\left(\frac{p}{q}\right) = (-1)^n \frac{2(2n-1)!}{(2q\pi)^{2n-1}} \sum_{j=1}^{q} \zeta\left(2n-1, \frac{j}{q}\right) \sin\left(\frac{2jp\pi}{q}\right) \tag{3}$$

$$(n \in \mathbb{N}\backslash\{1\};\ p \in \mathbb{N}_0;\ q \in \mathbb{N};\ 0 \leq p \leq q)$$

and

$$B_{2n}\left(\frac{p}{q}\right) = (-1)^{n-1} \frac{2(2n)!}{(2q\pi)^{2n}} \sum_{j=1}^{q} \zeta\left(2n, \frac{j}{q}\right) \cos\left(\frac{2jp\pi}{q}\right) \tag{4}$$

$$(n \in \mathbb{N};\ p \in \mathbb{N}_0;\ q \in \mathbb{N};\ 0 \leq p \leq q).$$

Theorem 6.3. *In terms of the Hurwitz Zeta function* $\zeta(s,a)$ *defined by* 2.2(1), *the Euler polynomials* $E_n(x)$ *at rational arguments are given by*

$$E_{2n-1}\left(\frac{p}{q}\right) = (-1)^n \frac{4(2n-1)!}{(2q\pi)^{2n}} \sum_{j=1}^{q} \zeta\left(2n, \frac{2j-1}{q}\right) \cos\left(\frac{(2j-1)p\pi}{q}\right) \tag{5}$$

$$(n \in \mathbb{N};\ p \in \mathbb{N}_0;\ q \in \mathbb{N};\ 0 \leq p \leq q)$$

and

$$E_{2n}\left(\frac{p}{q}\right) = (-1)^n \frac{4(2n)!}{(2q\pi)^{2n+1}} \sum_{j=1}^{q} \zeta\left(2n+1, \frac{2j-1}{2q}\right) \sin\left(\frac{(2j-1)p\pi}{q}\right) \tag{6}$$

$$(n \in \mathbb{N};\ p \in \mathbb{N}_0;\ q \in \mathbb{N};\ 0 \leq p \leq q).$$

As pointed out by Cvijović and Klinowski (1995a, p. 1535), the formula (3) was derived earlier, in a completely different way, by Almkvist and Meurman (1991, p. 107, Proposition 10), who applied this formula to show *eventually* that

$$q^{2n-1}\, B_{2n-1}\left(\frac{p}{q}\right) \quad (n \in \mathbb{N} \setminus \{1\})$$

is an integer for *all* positive integers p and q.

Srivastava's shorter proofs of Theorem 6.2 and Theorem 6.3

First Proof of Theorem 6.2. We begin by recalling the following known result [*cf.*, *e.g.*, Magnus *et al.* (1966, p. 27)]:

$$B_n(x) = -\frac{2\cdot n!}{(2\pi)^n}\sum_{k=1}^{\infty}\frac{1}{k^n}\cos\left(2k\pi x - \frac{n\pi}{2}\right) \tag{7}$$
$$(n \in \mathbb{N}\setminus\{1\} \quad \text{and} \quad 0 \leqq x \leqq 1;\ n = 1 \quad \text{and} \quad 0 < x < 1),$$

which, in view of the definition in 2.2(1) *and* the elementary series identity:

$$\sum_{k=1}^{\infty} f(k) = \sum_{j=1}^{q}\sum_{k=0}^{\infty} f(qk+j) \quad (q \in \mathbb{N}), \tag{8}$$

immediately yields

$$B_n\left(\frac{p}{q}\right) = -\frac{2\cdot n!}{(2q\pi)^n}\sum_{j=1}^{q}\zeta\left(n, \frac{j}{q}\right)\cos\left(\frac{2jp\pi}{q} - \frac{n\pi}{2}\right) \tag{9}$$
$$(n \in \mathbb{N}\setminus\{1\};\ p \in \mathbb{N}_0;\ q \in \mathbb{N};\ 0 \leq p \leq q),$$

where we have also set

$$x = \frac{p}{q} \qquad (p \in \mathbb{N}_0;\ q \in \mathbb{N};\ 0 \leq p \leq q).$$

Upon replacing n in (9) by $2n-1$ and $2n$, respectively, we obtain the assertions (3) and (4) of Theorem 6.2.

Second Proof of Theorem 6.2. Upon setting $s = n$ $(n \in \mathbb{N}\setminus\{1\})$ in 2.2(8) and applying 2.1(17) (with $\ell = n-1$) to the left member of the resulting identity, we arrive once again at the formula (9) which unifies the assertions (3) and (4) of Theorem 6.2, the missing case $p = 0$ for the Bernoulli numbers B_n being easily verified *directly* by means of the well-known identities 2.3(18) and

$$\zeta(s) = q^{-s}\sum_{j=1}^{q}\zeta\left(s, \frac{j}{q}\right) \quad (q \in \mathbb{N}). \tag{10}$$

First Proof of Theorem 6.3. We now show that Theorem 6.3 is actually a simple consequence of Theorem 6.2. Indeed, by appealing to the known relationship

1.6(60), we find from (9) (with n replaced by $n+1$) that

$$\begin{aligned}E_n\left(\frac{p}{q}\right) =& \frac{2}{n+1}\left[-\frac{2(n+1)!}{(2q\pi)^{n+1}}\sum_{j=1}^{q}\zeta\left(n+1,\frac{j}{q}\right)\sin\left(\frac{2jp\pi}{q}-\frac{n\pi}{2}\right)\right.\\ &\left.+2^{n+1}\frac{2(n+1)!}{(4q\pi)^{n+1}}\sum_{j=1}^{2q}\zeta\left(n+1,\frac{j}{2q}\right)\sin\left(\frac{jp\pi}{q}-\frac{n\pi}{2}\right)\right]\\ =& \frac{4\cdot n!}{(2q\pi)^{n+1}}\left[\sum_{j=1}^{2q}\zeta\left(n+1,\frac{j}{2q}\right)\sin\left(\frac{jp\pi}{q}-\frac{n\pi}{2}\right)\right.\\ &\left.-\sum_{j=1}^{q}\zeta\left(n+1,\frac{j}{p}\right)\sin\left(\frac{2jp\pi}{q}-\frac{n\pi}{2}\right)\right]\end{aligned} \tag{11}$$

$$(n\in\mathbb{N};\ p\in\mathbb{N}_0;\ q\in\mathbb{N};\ 0\leq p\leq q)\,.$$

Obviously, since

$$\begin{aligned}&\sum_{j=1}^{2q}\zeta\left(n+1,\frac{j}{2q}\right)\sin\left(\frac{jp\pi}{q}-\frac{n\pi}{2}\right)\\ =&\sum_{j=1}^{q}\zeta\left(n+1,\frac{2j-1}{2q}\right)\sin\left(\frac{(2j-1)p\pi}{q}-\frac{n\pi}{2}\right)\\ &+\sum_{j=1}^{q}\zeta\left(n+1,\frac{j}{q}\right)\sin\left(\frac{2jp\pi}{q}-\frac{n\pi}{2}\right),\end{aligned} \tag{12}$$

upon separating the even and odd terms, (11) leads us immediately to the formula:

$$E_n\left(\frac{p}{q}\right) = \frac{4\cdot n!}{(2q\pi)^{n+1}}\sum_{j=1}^{q}\zeta\left(n+1,\frac{2j-1}{2q}\right)\sin\left(\frac{(2j-1)p\pi}{q}-\frac{n\pi}{2}\right) \tag{13}$$

$$(n\in\mathbb{N};\ p\in\mathbb{N}_0;\ q\in\mathbb{N};\ 0\leq p\leq q)\,,$$

which yields the assertions (5) and (6) of Theorem 6.3 when n is replaced by $2n-1$ and $2n$, respectively.

Second Proof of Theorem 6.3. Just as in our first proof of Theorem 6.2, the unification (13) of the assertions (5) and (6) of Theorem 6.3 can be proven *directly* by merely applying the series identity (8) in the following known result [*cf.*, *e.g.*, Magnus *et al.* (1966, p. 30); see also 1.6(56) and 1.6(57)]:

$$E_n(x) = \frac{4\cdot n!}{\pi^{n+1}}\sum_{k=1}^{\infty}\frac{1}{(2k-1)^{n+1}}\sin\left((2k-1)\pi x-\frac{n\pi}{2}\right) \tag{14}$$

$$(n\in\mathbb{N}\ \text{ and }\ 0\leq x\leq 1;\ n=0\ \text{ and }\ 0<x<1),$$

with, of course,

$$x=\frac{p}{q}\quad(p\in\mathbb{N};\ q\in\mathbb{N};\ 0\leq p\leq q)\,.$$

Formulas involving the Hurwitz-Lerch Zeta function

By virtue of 1.2(53), each of the formulas (9) and (13), and their even and odd integer versions given by Theorem 6.2 and Theorem 6.3, can easily be restated in terms of the Polygamma function.

For the general Hurwitz-Lerch Zeta function $\Phi(z,s,a)$ defined by 2.5(1), it is easily seen by using the series identity (8) that

$$\Phi(z,s,a) = q^{-s}\sum_{j=1}^{q}\Phi\left(z^q, s, \frac{a+j-1}{q}\right)z^{j-1}, \tag{15}$$

which, in the special case when

$$z = \exp\left(\frac{2p\pi i}{q}\right) \qquad (p\in\mathbb{Z};\ q\in\mathbb{N}),$$

yields the summation formula:

$$L\left(\frac{p}{q}, s, a\right) = q^{-s}\sum_{j=1}^{q}\zeta\left(s, \frac{a+j-1}{q}\right)\exp\left(\frac{2(j-1)p\pi i}{q}\right) \tag{16}$$

for the generalized Zeta function $L(x,s,a)$ defined by 2.5(11) in terms of the Hurwitz Zeta function $\zeta(s,a)$.

For $z=1$, (15) reduces at once to the identity [Hansen (1975, p. 360, Entry (54.13.1))]:

$$\zeta(s,a) = q^{-s}\sum_{j=1}^{q}\zeta\left(s, \frac{a+j-1}{q}\right), \tag{17}$$

which, for $a=1$, yields the well-known result (10). On the other hand, by setting $a=\frac{1}{2}$ in (15) and (16), we have

$$\sum_{n=1}^{\infty}\frac{z^n}{(2n-1)^s} = (2q)^{-s}\sum_{j=1}^{q}\Phi\left(z^q, s, \frac{2j-1}{2q}\right)z^{j-1} \tag{18}$$

and

$$\sum_{n=0}^{\infty}\frac{e^{(2n+1)p\pi i/q}}{(2n+1)^s} = (2q)^{-s}\sum_{j=1}^{q}\zeta\left(s, \frac{2j-1}{2q}\right)\exp\left(\frac{(2j-1)p\pi i}{q}\right), \tag{19}$$

respectively.

Lastly, in their special cases when $a=1$, (15) and (16) yield the following companions of the summation formulas (18) and (19), respectively:

$$\sum_{n=1}^{\infty}\frac{z^n}{n^s} =: \mathrm{Li}_s(z) = q^{-s}\sum_{j=1}^{q}\Phi\left(z^q, s, \frac{j}{q}\right)z^j \tag{20}$$

and

$$\sum_{n=1}^{\infty} \frac{e^{2np\pi i/q}}{n^s} =: L\left(\frac{p}{q}, s\right)$$

$$= q^{-s} \sum_{j=1}^{q} \zeta\left(s, \frac{j}{q}\right) \exp\left(\frac{2jp\pi i}{q}\right). \tag{21}$$

In particular, when $s = \nu$ ($\nu > 1$), by simply equating the real and imaginary parts in (19) and (21), we immediately obtain the following summation formulas involving trigonometric series:

$$\sum_{n=0}^{\infty} \frac{\cos\left[(2n+1)p\pi/q\right]}{(2n+1)^{\nu}} = (2q)^{-\nu} \sum_{j=1}^{q} \zeta\left(\nu, \frac{2j-1}{2q}\right) \cos\left(\frac{(2j-1)p\pi}{q}\right); \tag{22}$$

$$\sum_{n=0}^{\infty} \frac{\sin\left[(2n+1)p\pi/q\right]}{(2n+1)^{\nu}} = (2q)^{-\nu} \sum_{j=1}^{q} \zeta\left(\nu, \frac{2j-1}{2q}\right) \sin\left(\frac{(2j-1)p\pi}{q}\right); \tag{23}$$

$$\sum_{n=1}^{\infty} \frac{\cos(2np\pi/q)}{n^{\nu}} = q^{-\nu} \sum_{j=1}^{q} \zeta\left(\nu, \frac{j}{q}\right) \cos\left(\frac{2jp\pi}{q}\right); \tag{24}$$

$$\sum_{n=1}^{\infty} \frac{\sin(2np\pi/q)}{n^{\nu}} = q^{-\nu} \sum_{j=1}^{q} \zeta\left(\nu, \frac{j}{q}\right) \sin\left(\frac{2jp\pi}{q}\right). \tag{25}$$

The assertions made by the Lemma of Cvijović and Klinowski (1995a, p. 1530) are *precisely* the special summation formulas (19) and (21) *with* $s = \nu$ ($\nu > 1$). Formula (21) with $s = \nu$ ($\nu > 1$) is, in fact, also one of the three *main* results in another paper by Cvijović and Klinowski (1995b, p. 207, Eq. (7)); the other two *main* results of Cvijović and Klinowski (1995b, p. 207, Eqs. (8a) and (8b)) are essentially the same as (23) and (22), respectively, *with* p replaced by $2p$. Cvijović and Klinowski (1995b, p. 208, Eqs. (9a) and (9b)) also gave the special summation formulas (23) and (22), respectively, *with* $q = Q$.

As already pointed out by Srivastava (2000b, p. 83), each of the trigonometric sums (22) to (25), and indeed also various *further* special cases of many of the summation formulas considered here, can be found to be derived in the work of Cvijović and Klinowski (1995a).

An application of Lerch's functional equation 2.5(29)

In view of the summation formula (16), Lerch's functional equation 2.5(29) with

$$a = \frac{p}{q} \qquad (p \in \mathbb{Z};\ q \in \mathbb{N})$$

yields

(26)

$$L\left(x,1-s,\frac{p}{q}\right)=\frac{\Gamma(s)}{(2p\pi)^s}\left\{\sum_{j=1}^{q}\zeta\left(s,\frac{x+j-1}{q}\right)\exp\left[\left(\frac{s}{2}-\frac{2(x+j-1)p}{q}\right)\pi i\right]\right.$$
$$\left.+\sum_{j=1}^{q}\zeta\left(s,\frac{j-x}{q}\right)\exp\left[\left(-\frac{s}{2}+\frac{2(j-x)p}{q}\right)\pi i\right]\right\}.$$

In terms of a *mild* generalization $\mathcal{B}_n(a,\alpha)$ of the classical Bernoulli polynomials $B_n(x)$, which is defined by 2.5(41), by setting $s = n$ $(n \in \mathbb{N})$ in (26) and using 2.5(40), Srivastava (2000b, p. 84) obtained the following novel analogue of (9) for the generalized Bernoulli polynomials $\mathcal{B}_n(a,\alpha)$:

(27)

$$\mathcal{B}_n\left(\frac{p}{q},e^{2\pi ix}\right)=-\frac{n!}{(2q\pi)^n}\left\{\sum_{j=1}^{q}\zeta\left(n,\frac{x+j-1}{q}\right)\exp\left[\left(\frac{n}{2}-\frac{2(x+j-1)p}{q}\right)\pi i\right]\right.$$
$$\left.+\sum_{j=1}^{q}\zeta\left(n,\frac{j-x}{q}\right)\exp\left[\left(-\frac{n}{2}+\frac{2(j-x)p}{q}\right)\pi i\right]\right\},$$

which holds true whenever each side exists. Indeed, in its *special* case when $x \in \mathbb{Z}$, Srivastava's formula (27) can easily be shown to reduce to the known result (9).

6.2 Closed-Form Summation of Trigonometric Series

By Weierstrass's M-test, it is easily seen that each of the classical trigonometric series:

(1)
$$C_\nu(\alpha) := \sum_{k=0}^{\infty}\frac{\cos(2k+1)\alpha}{(2k+1)^\nu} \quad\text{and}\quad S_\nu(\alpha) := \sum_{k=0}^{\infty}\frac{\sin(2k+1)\alpha}{(2k+1)^\nu}$$

converges uniformly for all real values of α when $\nu > 1$. Furthermore, since

(2)
$$C_\nu(-\alpha) = C_\nu(\alpha) = -C_\nu(\pi-\alpha)$$

and

(3)
$$S_\nu(-\alpha) = -S_\nu(\alpha) = -S_\nu(\pi-\alpha),$$

it is sufficient to evaluate $C_\nu(\alpha)$ and $S_\nu(\alpha)$ only over the range $0 \leq \alpha \leq \frac{1}{2}\pi$.

The series $C_{2n+1}(\alpha)$ and $S_{2n}(\alpha)$ $(n \in \mathbb{N})$ converge extremely slowly, and the problem of their numerical evaluation was addressed by (among others) Dempsey *et al.* (1990), Boersma and Dempsey (1992), and Gautschi (1991). For the particular case when $n = 1$, Dempsey *et al.* (1990) developed a procedure based upon Plana's summation formula together with Romberg's method of integration, which significantly improves the convergence and accuracy when compared with those resulting from *direct* summation. Boersma and Dempsey (1992), on the other hand, transformed each of $C_{2n+1}(\alpha)$ and $S_{2n}(\alpha)$ into a rapidly convergent series which is well-suited for their computation (when, for example, $n = 1$).

Recently, Cvijović and Klinowski (1995b) considered the general series in (1) when $\nu > 1$ and α is a rational multiple of 2π. In terms of the Riemann Zeta function $\zeta(s)$ and the Hurwitz Zeta function $\zeta(s,a)$ defined by 2.3(1) and 2.2(1), respectively, they first showed that [Cvijović and Klinowski (1995b, p. 207, Eqs. (8a) and (8b))]

(4) $$C_\nu\left(\frac{2p\pi}{q}\right) = \frac{1}{q^\nu}\left\{\zeta(\nu)\left(1-\frac{1}{2^\nu}\right) + \sum_{j=1}^{q-1}\zeta\left(\nu,\frac{j}{q}\right)\left[\cos\left(\frac{2jp\pi}{q}\right) - \frac{1}{2^\nu}\cos\left(\frac{4jp\pi}{q}\right)\right]\right\}$$

and

(5) $$S_\nu\left(\frac{2p\pi}{q}\right) = \frac{1}{q^\nu}\sum_{j=1}^{q-1}\zeta\left(\nu,\frac{j}{q}\right)\left[\sin\left(\frac{2jp\pi}{q}\right) - \frac{1}{2^\nu}\sin\left(\frac{4jp\pi}{q}\right)\right]$$
$$(p,q\in\mathbb{N};\ \nu>1).$$

Following the work of Srivastava (1999b), we now show that the summation formulas (4) and (5), as well as their equivalent variants when α is a rational multiple of π [*cf.* Cvijović and Klinowski (1995b, p. 208, Eqs. (9a) and (9b))], would follow easily from some general results involving the family of Dirichlet series in 2.5(1), which is usually called Lerch's function or Lerch's transcendent.

For convenience, we recall here the summation formulas 6.1(24), 6.1(25), 6.1(22), and 6.1(23) in the following (slightly modified) forms:

(6) $$\sum_{k=1}^{\infty}\frac{\cos(2kp\pi/q)}{(2k)^\nu} = (2q)^{-\nu}\sum_{j=1}^{q}\zeta\left(\nu,\frac{j}{q}\right)\cos\left(\frac{2jp\pi}{q}\right),$$

(7) $$\sum_{k=1}^{\infty}\frac{\sin(2kp\pi/q)}{(2k)^\nu} = (2q)^{-\nu}\sum_{j=1}^{q}\zeta\left(\nu,\frac{j}{q}\right)\sin\left(\frac{2jp\pi}{q}\right),$$

(8) $$C_\nu\left(\frac{p\pi}{q}\right) := \sum_{k=0}^{\infty}\frac{\cos[(2k+1)p\pi/q]}{(2k+1)^\nu} = (2q)^{-\nu}\sum_{j=1}^{q}\zeta\left(\nu,\frac{2j-1}{2q}\right)\cos\left(\frac{(2j-1)p\pi}{q}\right),$$

and

(9) $$S_\nu\left(\frac{p\pi}{q}\right) := \sum_{k=0}^{\infty}\frac{\sin[(2k+1)p\pi/q]}{(2k+1)^\nu} = (2q)^{-\nu}\sum_{j=1}^{q}\zeta\left(\nu,\frac{2j-1}{2q}\right)\sin\left(\frac{(2j-1)p\pi}{q}\right),$$

respectively. Since [*cf.* Equation 6.1(17) with $q = 2$]

$$\zeta(s,a) = 2^{-s}\left[\zeta\left(s,\ \frac{1}{2}a\right) + \zeta\left(s,\ \frac{1}{2}a + \frac{1}{2}\right)\right], \tag{10}$$

the second member of the summation formula (4) can be rewritten in the form:

$$\begin{aligned}
q^{-\nu}\sum_{j=1}^{q}\zeta\left(\nu,\ \frac{j}{q}\right)&\left[\cos\left(\frac{2jp\pi}{q}\right) - 2^{-\nu}\cos\left(\frac{4jp\pi}{q}\right)\right] \\
&= (2q)^{-\nu}\left\{\sum_{j=1}^{q}\left[\zeta\left(\nu,\ \frac{j}{2q}\right) + \zeta\left(\nu,\ \frac{j+q}{2q}\right)\right]\cos\left(\frac{2jp\pi}{q}\right)\right. \\
&\qquad \left. - \sum_{j=1}^{q}\zeta\left(\nu,\ \frac{j}{q}\right)\cos\left(\frac{4jp\pi}{q}\right)\right\} \\
&= (2q)^{-\nu}\left\{\sum_{j=1}^{q}\zeta\left(\nu,\ \frac{2j-1}{2q}\right)\cos\left(\frac{2(2j-1)p\pi}{q}\right)\right. \\
&\qquad + \sum_{j=1}^{q}\zeta\left(\nu,\ \frac{2j}{2q}\right)\cos\left(\frac{4jp\pi}{q}\right) \\
&\qquad \left. - \sum_{j=1}^{q}\zeta\left(\nu,\ \frac{j}{q}\right)\cos\left(\frac{4jp\pi}{q}\right)\right\} \\
&= (2q)^{-\nu}\sum_{j=1}^{q}\zeta\left(\nu,\ \frac{2j-1}{2q}\right)\cos\left(\frac{2(2j-1)p\pi}{q}\right),
\end{aligned} \tag{11}$$

where we have also applied the trigonometric identity:

$$\cos\left(\frac{2jp\pi}{q}\right) = \cos\left(\frac{2(j+q)p\pi}{q}\right) \qquad (j = 1, \ldots, q;\ q \in \mathbb{N};\ p \in \mathbb{Z}).$$

Thus the summation formula (4) is reduced to its *equivalent* form:

$$C_\nu\left(\frac{2p\pi}{q}\right) = (2q)^{-\nu}\sum_{j=1}^{q}\zeta\left(\nu,\ \frac{2j-1}{2q}\right)\cos\left(\frac{2(2j-1)p\pi}{q}\right), \tag{12}$$

which is precisely (8) with p replaced rather trivially by $2p$.

Next, since the $j = q$ term in (5) is obviously nil, it is not difficult to reduce the summation formula (5) similarly to its *equivalent* form:

$$S_\nu\left(\frac{2p\pi}{q}\right) = (2q)^{-\nu}\sum_{j=1}^{q}\zeta\left(\nu,\ \frac{2j-1}{2q}\right)\sin\left(\frac{2(2j-1)p\pi}{q}\right), \tag{13}$$

which is precisely (9) with p replaced rather trivially by $2p$.

Each of the summation formulas (6) to (9) can indeed be proven *directly* by means of the series identity 6.1(8). The derivation of (4) and (5) by Cvijović and Klinowski (1995b) made use of the relationship:

$$\chi_\nu(z) = \mathrm{Li}_\nu(z) - 2^{-\nu}\,\mathrm{Li}_\nu(z^2), \tag{14}$$

where

$$\chi_\nu(z) := \sum_{k=0}^{\infty} \frac{z^{2k+1}}{(2k+1)^\nu} \quad \text{and} \quad \mathrm{Li}_\nu(z) := \sum_{k=1}^{\infty} \frac{z^k}{k^\nu} \tag{15}$$

$$(|z| \leq 1;\ \Re(\nu) > 1)$$

denote Legendre's Chi function and the Polylogarithm function of order ν [*cf.*, *e.g.*, Lewin (1981)]. In fact, as observed by Cvijović and Klinowski (1995b, p. 208), the (equivalent) summation formulas (8) and (9) are obtained when we make use of the relationship:

$$\chi_\nu(z) = \frac{1}{2}\left[\mathrm{Li}_\nu(z) - \mathrm{Li}_\nu(-z)\right] \tag{16}$$

instead of (14).

6.3 Integrals Associated with the Use of the Euler-Maclaurin Summation Formula

Zhang (1990) evaluated sums of certain series of the Riemann Zeta function using the Euler-Maclaurin summation formula 1.3(68), in which evaluations of some integrals involving the function

$$P(x) = x - [x] - \frac{1}{2} \tag{1}$$

play an important rôle, where $[x]$ denotes the greatest integer less than or equal to x. With or without the use of the closed-form evaluations of series involving the Zeta function given in Section 3.4, Choi *et al.* (1998) established several integrals involving the periodic function $P(x)$ in (1). Moreover, by introducing some not-too-familiar mathematical constants, they also considered several families of definite integrals involving logarithms of Gamma functions. In this section we describe the investigations by Choi et al. (1998).

Integrals involving the function $P(x)$

We first consider the integral:

$$\begin{aligned}
\int_2^\infty \frac{P(x)}{x}\,dx &= \sum_{k=2}^{\infty}\left\{1 - \left(k+\frac{1}{2}\right)\log\left(1+\frac{1}{k}\right)\right\} \\
&= \sum_{k=2}^{\infty}\left\{\sum_{n=2}^{\infty}\frac{(-1)^{n+1}}{(n+1)k^n} + \sum_{n=2}^{\infty}\frac{(-1)^n}{2n\cdot k^n}\right\} \\
&= \sum_{n=2}^{\infty}\frac{(-1)^{n+1}}{n+1}\sum_{k=2}^{\infty}\frac{1}{k^n} + \sum_{n=2}^{\infty}\frac{(-1)^n}{2n}\sum_{k=2}^{\infty}\frac{1}{k^n} \\
&= \sum_{n=2}^{\infty}\frac{(-1)^{n+1}}{n+1}\{\zeta(n)-1\} + \frac{1}{2}\sum_{n=2}^{\infty}\frac{(-1)^n}{n}\{\zeta(n)-1\},
\end{aligned}$$

which, upon using 3.4(475) and 3.4(150), yields

(2)
$$\int_2^\infty \frac{P(x)}{x}dx = \frac{1}{2}\log\pi + 2\log 2 - 2.$$

Noting that
$$\int_1^2 \frac{P(x)}{x}dx = 1 - \frac{3}{2}\log 2,$$

we have

(3)
$$\int_1^\infty \frac{P(x)}{x}dx = \frac{1}{2}\log(2\pi) - 1,$$

which is a known result recorded by (for example) Zhang (1990, p. 66, Eq. (7)) and Rainville (1960, p. 29, Eq. (5)).

Next we consider the integral:

$$\begin{aligned}\int_1^\infty \frac{P(x)}{x^2}dx &= \sum_{k=1}^\infty \int_k^{k+1} \frac{x-k-\frac{1}{2}}{x^2}dx \\ &= \sum_{k=1}^\infty \left\{ \sum_{n=2}^\infty \frac{(-1)^{n-1}}{n\cdot k^n} + \frac{1}{2}\left(\frac{1}{k} - \frac{1}{k+1}\right)\right\} \\ &= \sum_{n=2}^\infty \frac{(-1)^{n-1}}{n} \sum_{k=1}^\infty \frac{1}{k^n} + \frac{1}{2} \\ &= \sum_{n=2}^\infty \frac{(-1)^{n-1}}{n}\zeta(n) + \frac{1}{2},\end{aligned}$$

which, in view of 3.4(23), yields

(4)
$$\int_1^\infty \frac{P(x)}{x^2}dx = \frac{1}{2} - \gamma,$$

which is a known result recorded by (for example) Zhang (1990, p. 66, Eq. (5)) and Rainville (1960, p. 32, Example 14).

Thirdly, we find from 2.2(1) and 2.3(2) that

$$\begin{aligned}\int_1^\infty \frac{P(x)}{2x+1}dx &= \frac{1}{2}\int_1^\infty \frac{P(x)}{x+\frac{1}{2}}dx \\ &= \frac{1}{2}\sum_{k=1}^\infty \left\{ \sum_{n=2}^\infty \frac{(-1)^{n+1}}{n+1}\frac{1}{(k+\frac{1}{2})^n} + \frac{1}{2}\sum_{n=2}^\infty \frac{(-1)^n}{n}\frac{1}{(k+\frac{1}{2})^n}\right\} \\ &= \frac{1}{2}\sum_{n=2}^\infty \frac{(-1)^{n+1}}{n+1}\left\{\zeta(n,\frac{1}{2}) - 2^n\right\} + \frac{1}{4}\sum_{n=2}^\infty \frac{(-1)^n}{n}\left\{\zeta(n,\frac{1}{2}) - 2^n\right\} \\ &= \frac{1}{2}\sum_{n=2}^\infty \frac{(-1)^{n+1}}{n+1}\{\zeta(n)-1\}2^n + \frac{1}{2}\sum_{n=2}^\infty \frac{(-1)^n}{n+1}\zeta(n) \\ &\quad + \frac{1}{4}\sum_{n=2}^\infty \frac{(-1)^n}{n}\{\zeta(n)-1\}2^n + \frac{1}{4}\sum_{n=2}^\infty \frac{(-1)^{n+1}}{n}\zeta(n),\end{aligned}$$

from which, using 3.4(473), 3.4(457), 3.4(145), and 3.4(23), we obtain

$$\int_1^\infty \frac{P(x)}{2x+1}dx = -\frac{3}{4} + \frac{1}{4}\log 2 + \frac{1}{2}\log 3, \tag{5}$$

which is a known result recorded by (for example) Zhang (1990, p. 68, Eq. (11)).

Formula (5) corresponds to the special case $z = \frac{1}{2}$ of a result given earlier by Rainville (1960, p. 30, Theorem 12).

Now we consider the integral:

$$\begin{aligned}
\int_1^\infty P(x)\log\left(1+\frac{1}{x}\right)dx &= \sum_{k=1}^\infty \int_k^{k+1}\left(x-k-\frac{1}{2}\right)\log\left(1+\frac{1}{x}\right)dx \\
&= -\frac{1}{2}\sum_{k=1}^\infty \left[(k^2+3k+2)\log\left(1+\frac{1}{k+1}\right) - (k^2+k)\log\left(1+\frac{1}{k}\right) - 1\right] \\
&= -\frac{1}{12} - \frac{1}{2}\left[\sum_{n=2}^\infty \frac{(-1)^{n-1}}{n+2}\{\zeta(n)-1\} + \sum_{n=2}^\infty \frac{(-1)^n}{n+1}\{\zeta(n)-1\}\right. \\
&\qquad \left. + \sum_{n=2}^\infty \frac{(-1)^n}{n+2}\zeta(n) + \sum_{n=2}^\infty \frac{(-1)^{n-1}}{n+1}\zeta(n)\right],
\end{aligned}$$

from which, using 3.4(518), 3.4(475), 3.4(503), and 3.4(457), we obtain

$$\int_1^\infty P(x)\log\left(1+\frac{1}{x}\right)dx = \log 2 - \frac{3}{4}, \tag{6}$$

which is a known result recorded by (for example) Zhang (1990, p. 66, Eq. (6)).

For $n \in \mathbb{N} \setminus \{1, 2\}$, we express the following integral in terms of the Riemann Zeta function:

$$\begin{aligned}
\int_1^\infty \frac{P(x)}{x^n}dx &= \sum_{k=1}^\infty \int_k^{k+1} \frac{x-k-\frac{1}{2}}{x^n}dx \\
&= \sum_{k=1}^\infty \left[\frac{1}{(n-1)(n-2)}\left\{\frac{1}{k^{n-2}} - \frac{1}{(k+1)^{n-2}}\right\}\right. \\
&\qquad \left. - \frac{1}{2(n-1)}\left\{\frac{1}{k^{n-1}} + \frac{1}{(k+1)^{n-1}}\right\}\right] \\
&= \frac{n}{2(n-1)(n-2)} - \frac{1}{n-1}\zeta(n-1). \qquad (7)
\end{aligned}$$

Another family of definite integrals

The following definite integral formula is well known:

$$\int_0^{\pi/2} \log\sin t\, dt = -\frac{\pi}{2}\log 2, \tag{8}$$

which appears in elementary calculus textbook and mathematical formula handbooks [*cf.*, *e.g.*, Spiegel (1968, p. 99. Entry 15.102)]. We evaluate a family of definite integrals of the form:

$$\int_0^{\pi/p} \log \sin t \, dt,$$

where p is any real number greater than 1. Setting

$$f(x) = \left(x + \tfrac{1}{p}\right) \log\left(x + \tfrac{1}{p}\right)$$

in the Euler-Maclaurin summation formula 1.3(68) with $a = 1$, we encounter a class of mathematical constants A_p given by:

(9)
$$\log A_p = \lim_{n\to\infty}\left[\sum_{k=1}^{n}\left(k+\frac{1}{p}\right)\log\left(k+\frac{1}{p}\right) - \left\{\frac{n^2}{2} + \left(\frac{1}{2}+\frac{1}{p}\right)n + \frac{1}{2p^2} + \frac{1}{2p} + \frac{1}{12}\right\}\log\left(n+\frac{1}{p}\right) + \frac{n^2}{4} + \frac{n}{2p}\right],$$

where the constants A_p depend on p $(p > 0)$.

By setting $n = 10^5$ in (9), we obtain the following approximate values of the constants A_p :

$$A_{0.5} = 0.1179\ldots \qquad A_1 = 0.9990\ldots$$
$$A_{1.5} = 1.2928\ldots \qquad A_2 = 1.3680\ldots$$
$$A_{2.5} = 1.3862\ldots \qquad A_3 = 1.3882\ldots$$
$$A_{3.5} = 1.3840\ldots \qquad A_4 = 1.3782\ldots$$
$$A_{4.5} = 1.3727\ldots \qquad A_5 = 1.3666\ldots$$
$$A_{5.5} = 1.3602\ldots \qquad A_6 = 1.3568\ldots$$

Similarly, by setting

$$f(x) = \left(x - \tfrac{1}{p}\right) \log\left(x - \tfrac{1}{p}\right)$$

in the Euler-Maclaurin summation formula 1.3(68) with $a = 1$, we encounter another class of mathematical constants C_p:

(10)
$$\log C_p = \lim_{n\to\infty}\left[\sum_{k=1}^{n}\left(k-\frac{1}{p}\right)\log\left(k-\frac{1}{p}\right) - \left\{\frac{n^2}{2} + \left(\frac{1}{2}-\frac{1}{p}\right)n + \frac{1}{2p^2} - \frac{1}{2p} + \frac{1}{12}\right\}\log\left(n-\frac{1}{p}\right) + \frac{n^2}{4} - \frac{n}{2p}\right] \quad (p > 1),$$

where (obviously) $C_p = A_{-p}$.

By setting $n = 10^5$ in (10), we obtain the following approximate values of the constants C_p :

$$C_{1.5} = 0.8856\ldots \qquad C_2 = 0.9673\ldots$$
$$C_{2.5} = 1.0287\ldots \qquad C_3 = 1.0723\ldots$$
$$C_{3.5} = 1.1036\ldots \qquad C_4 = 1.1271\ldots$$
$$C_{4.5} = 1.1453\ldots \qquad C_5 = 1.1599\ldots$$
$$C_{5.5} = 1.1724\ldots \qquad C_6 = 1.1821\ldots.$$

It should also be remarked in passing that the Glaisher-Kinkelin constant A in 1.3(2) can be obtained by setting $f(x) = x \log x$ in the Euler-Maclaurin summation formula 1.3(68).

In order to evaluate the following integral:

$$I_p := -\int_0^{1/p} \log \Gamma(1+t)dt \quad (p > 0),$$

we first rewrite the Weierstrass canonical product form 1.1(2) of the Gamma function as follows:

$$\{\Gamma(z+1)\}^{-1} = e^{\gamma z} \prod_{k=1}^{\infty} \left\{ \left(1 + \frac{z}{k}\right) e^{-z/k} \right\}, \tag{11}$$

simply by using 1.1(9). By taking logarithms on both sides of (11) and integrating the resulting equation from 0 to $1/p$, we find that

$$\begin{aligned}
I_p &= \frac{\gamma}{2p^2} + \sum_{k=1}^{\infty} \int_0^{1/p} \left\{ \log\left(1 + \frac{t}{k}\right) - \frac{t}{k} \right\} dt \\
&= \frac{\gamma}{2p^2} + \sum_{k=1}^{\infty} \left\{ \left(k + \frac{1}{p}\right) \log\left(k + \frac{1}{p}\right) - k \log k - \frac{1}{p} - \frac{1}{p} \log k - \frac{1}{2p^2 k} \right\} \\
&= \frac{\gamma}{2p^2} + \lim_{n\to\infty} \left\{ \sum_{k=1}^{n} \left(k + \frac{1}{p}\right) \log\left(k + \frac{1}{p}\right) - \frac{n}{p} \right. \\
&\qquad \left. - \sum_{k=1}^{n} k \log k - \frac{1}{p} \sum_{k=1}^{n} \log k - \frac{1}{2p^2} \sum_{k=1}^{n} \frac{1}{k} \right\} \\
&= \log A_p - \log A - \frac{1}{2p} \log(2\pi) \\
&\quad + \lim_{n\to\infty} \left[\left\{ \frac{n^2}{2} + \left(\frac{1}{2} + \frac{1}{p}\right) n + \frac{1}{2p^2} + \frac{1}{2p} + \frac{1}{12} \right\} \log\left(n + \frac{1}{p}\right) \right. \\
&\qquad \left. - \left\{ \frac{n^2}{2} + \left(\frac{1}{2} + \frac{1}{p}\right) n + \frac{1}{2p^2} + \frac{1}{2p} + \frac{1}{12} \right\} \log n - \frac{n}{2p} \right],
\end{aligned}$$

where we have used 1.1(3), 1.3(2), 3.6(19), and (9). Now, if we make use of the Maclaurin series expansion of $\log(1+x)$, we obtain

$$\begin{aligned}
I_p =& \log A_p - \log A - \frac{1}{2p} \log(2\pi) \\
&+ \lim_{n\to\infty} \left[\left\{ \frac{n^2}{2} + \left(\frac{1}{2} + \frac{1}{p}\right) n + \frac{1}{2p^2} + \frac{1}{2p} + \frac{1}{12} \right\} \log\left(1 + \frac{1}{pn}\right) - \frac{n}{2p} \right] \\
=& \log A_p - \log A - \frac{1}{2p} \log(2\pi) + \lim_{n\to\infty} \left[\left\{ \frac{n}{2p} + \frac{1}{2p} + \frac{3}{4p^2} + O\left(\frac{1}{n}\right) \right\} - \frac{n}{2p} \right] \\
=& \log A_p - \log A - \frac{1}{2p} \log(2\pi) + \frac{1}{2p} + \frac{3}{4p^2}.
\end{aligned}$$

We, therefore, get

$$\int_0^{1/p} \log\Gamma(1+t)\,dt = \log A - \log A_p + \frac{1}{2p}\log(2\pi) - \frac{1}{2p} - \frac{3}{4p^2} \quad (p>0). \tag{12}$$

Similarly, we have

$$\int_0^{1/p} \log\Gamma(1-t)\,dt = \log C_p - \log A + \frac{1}{2p}\log(2\pi) - \frac{1}{2p} + \frac{3}{4p^2} \quad (p>1). \tag{13}$$

By setting $p=1$ in (12), we obtain

$$\int_0^{1} \log\Gamma(1+t)\,dt = \log A - \log A_1 + \frac{1}{2}\log(2\pi) - \frac{5}{4}, \tag{14}$$

which, in view of 1.1(9), corresponds to 1.2(39), since

$$\begin{aligned}\log A_1 - \log A &= \lim_{n\to\infty}\left\{-\left(\frac{n^2}{2}+\frac{n}{2}+\frac{1}{12}\right)\log\left(1+\frac{1}{n}\right)+\frac{n}{2}\right\}\\ &= \lim_{n\to\infty}\left[\left\{-\frac{n}{2}-\frac{1}{4}+O\left(\frac{1}{n}\right)\right\}+\frac{n}{2}\right] = -\frac{1}{4}.\end{aligned}$$

By applying the familiar relation [*cf.* Equations 1.1(12) and 1.1(9)]:

$$\Gamma(1+t)\Gamma(1-t) = \frac{\pi t}{\sin(\pi t)} \quad (t\notin\mathbb{Z}),$$

we obtain

$$\begin{aligned}&\int_0^{1/p} \log\Gamma(1+t)\,dt + \int_0^{1/p} \log\Gamma(1-t)\,dt\\ &\quad = \frac{1}{p}\log\left(\frac{\pi}{p}\right) - \frac{1}{p} - \frac{1}{\pi}\int_0^{\pi/p}\log\sin t\,dt \quad (p>1).\end{aligned} \tag{15}$$

Finally, by combining (12) and (13) with (15), we obtain our desired result:

$$\int_0^{\pi/p} \log\sin t\,dt = \pi\left\{\log A_p - \log C_p - \frac{1}{p}\log(2p)\right\} \quad (p>1). \tag{16}$$

By setting $p=2$ in (16), we get

$$\int_0^{\pi/2} \log\sin t\,dt = \pi(\log A_2 - \log C_2 - \log 2), \tag{17}$$

which agrees with (8), since

$$\begin{aligned}\log A_2 - \log C_2 =& \frac{1}{2}\log 2 + \lim_{n\to\infty}\left\{-\left(\frac{n^2}{2}-\frac{1}{24}\right)\log\left(n+\frac{1}{2}\right)\right.\\ &\left.+\left(\frac{n^2}{2}-\frac{1}{24}\right)\log\left(n-\frac{1}{2}\right)+\frac{n}{2}\right\}\\ =&\frac{1}{2}\log 2 - \lim_{n\to\infty}\left\{\left(\frac{n^2}{2}-\frac{1}{24}\right)\log\left(1+\frac{1}{n-\frac{1}{2}}\right)-\frac{n}{2}\right\}\\ =&\frac{1}{2}\log 2 - \lim_{n\to\infty}\left[\left\{\frac{n}{2}+O\left(\frac{1}{n}\right)\right\}-\frac{n}{2}\right] = \frac{1}{2}\log 2.\end{aligned}$$

In view of the fact that $\lim_{x\to 0} x\log x = 0$, by letting $p = 1$ in (10), we have

(18)
$$\log C_1 = \lim_{n\to\infty}\left[\sum_{k=2}^{n}(k-1)\log(k-1) - \left(\frac{n^2}{2}-\frac{n}{2}+\frac{1}{12}\right)\log(n-1)+\frac{n^2}{4}-\frac{n}{2}\right],$$

the approximate numerical value of C_1 being $0.9990\ldots$.

Finally, by setting $p = 1$ in (16) and using (18), we are led to the well-known result:

(19)
$$\int_0^{\pi}\log\sin t\,dt = \pi(\log A_1 - \log C_1 - \log 2) = -\pi\log 2,$$

since

$$\begin{aligned}\log A_1 - \log C_1 &= \lim_{n\to\infty}\left\{-n\log\left(1+\frac{1}{n}\right) - \left(\frac{n^2}{2}-\frac{n}{2}+\frac{1}{12}\right)\log\left(1+\frac{2}{n-1}\right)+n\right\}\\ &= \lim_{n\to\infty}\left[\left\{-1+O\left(\frac{1}{n}\right)\right\}+\left\{1-n+O\left(\frac{1}{n}\right)\right\}+n\right]=0,\end{aligned}$$

where the Maclaurin series expansion of $\log(1+x)$ is used for the second equality.

Problems

1. Show that

$$\int_0^{\infty}\frac{P(x)}{z+x}dx = \left(z-\frac{1}{2}\right)\log z - z + \frac{1}{2}\log(2\pi) - \log\Gamma(z),$$

where $P(x)$ is given as in (1) and $|\arg(z)| \leqq \pi - \epsilon$ $(0 < \epsilon < \pi)$.

[Rainville (1960, p. 30, Theorem 12)]

2. Show that the constant A_4 in (9) with $p = 4$ can be expressed in the form:

$$\log A_4 = \frac{5}{64} + \frac{1}{2}\log 2 - \frac{1}{8}\log A - \frac{G}{2\pi},$$

where A is the Glaisher-Kinkelin constant defined by 1.3(2) and G is the Catalan constant defined by 1.3(16).

[Choi and Srivastava (1999a, p. 100, Eq. (2.30))]

3. Evaluate the following integral:

$$\int_1^{\infty}P(x)\log\frac{2x-1}{2x+1}dx.$$

[Choi *et al.* (1998, pp. 108–109, Eq. (3.7)); Zhang (1990, p. 69, Eq. (14))]

4. Assume that $n \in \mathbb{N}$ and set $w = \exp(i\pi/n)$. Let $\zeta(s,a)$ and $\chi_\nu(z)$ be the Hurwitz Zeta and Legendre Chi functions defined by 2.2(1) and 6.2(15), respectively. Then show that

$$\zeta\left(\nu, \frac{2j-1}{2n}\right) = \frac{1}{n}\sum_{k=1}^{n} (2n)^\nu \chi_\nu\left(w^k\right) w^{-k(2j-1)} \quad (j = 1, \ldots, n)$$

and

$$\chi_\nu\left(w^k\right) = \frac{1}{(2n)^\nu}\sum_{j=1}^{n} \zeta\left(\nu, \frac{2j-1}{2n}\right) w^{k(2j-1)} \quad (j = 1, \ldots, n).$$

[Cvijović and Klinowski (1999, p. 1625, Eqs. (7a) and (7b))]

5. Prove that

$$\zeta\left(2n, \frac{2p-1}{2q}\right) + \zeta\left(2n, 1 - \frac{2p-1}{2q}\right)$$
$$= (-1)^n \pi^{2n} \frac{(2q)^{2n}}{(2n-1)!}\sum_{r=1}^{q} E_{2n-1}\left(\frac{r}{q}\right) \cos\left(\frac{(2p-1)r\pi}{q}\right)$$

and

$$\zeta\left(2n+1, \frac{2p-1}{2q}\right) - \zeta\left(2n+1, 1 - \frac{2p-1}{2q}\right)$$
$$= (-1)^n \pi^{2n+1} \frac{(2q)^{2n}}{(2n)!}\sum_{r=1}^{q} E_{2n}\left(\frac{r}{q}\right) \sin\left(\frac{(2p-1)r\pi}{q}\right) \quad (n \in \mathbb{N}).$$

[Cvijović and Klinowski (1999, p. 1628, Eqs. (15a) and (15b))]

6. Show that the mathematical constants A_p and C_p defined by 6.3(9) and 6.3(10) are related as follows:

$$A_p = \left[\frac{p}{\pi}\sin\left(\frac{\pi}{p}\right)\right]^{\frac{1}{p}} \frac{G\left(1+\frac{1}{p}\right)}{G\left(1-\frac{1}{p}\right)} C_p \quad (p > 1).$$

[Choi and Srivastava (1999a, p. 95, Eq. (2.2))]

7. Prove that, if $f(x) \in \mathbb{Q}[x]$ is integer-valued on $\mathbb{N}_0$, that is, if $f(k) \in \mathbb{Z}$ for all $k \in \mathbb{N}_0$, and p is prime, then $f(a) \in \mathbb{Z}_{(p)}$ for all $a \in \mathbb{Z}_{(p)}$, where $\mathbb{Z}_{(p)}$ denotes the ring of rational numbers whose denominator is not divisible by the prime p.

[Clarke and Slavutskii (1997, p. 22, Proposition 1)]

8. Show that, for $\alpha > -1$,

$$\int_0^1 \frac{\rho^\alpha}{\rho^2 - 2\rho\cos\theta + 1} d\rho = \frac{1}{2q}\sum_{k=1}^{q-1} \left(\frac{\sin(k\theta)}{\sin\theta}\right) \psi\left(1 + \frac{\alpha-k}{2q}\right),$$

where

$$\theta = \frac{p\pi}{q} \quad (p,\, q \in \mathbb{N};\; 0 < p < 2q;\; (p,q) = 1),$$

and $(p,q) = 1$ denotes the fact that p and q are relatively prime.

[de Doelder (1984, p. 327)]

9. Prove that

$$\log \frac{G\left(1 + \frac{p}{2q}\right)}{G\left(1 - \frac{p}{2q}\right)} = -\frac{p}{2q} \log \left(\frac{1}{\pi} \sin \left(\frac{p\pi}{2q} \right) \right) + \frac{1}{8\pi q^2} \sum_{r=1}^{q-1} \left[\psi' \left(1 - \frac{r}{2q} \right) - \psi' \left(\frac{r}{2q} \right) \right] \sin \left(\frac{pr\pi}{q} \right)$$

$$(p \in \mathbb{N};\; q \in \mathbb{N} \setminus \{1\};\; 1 \leqq p \leqq 2q - 1;\; (p,q) = 1),$$

where G denotes the Barnes G-function considered in Section 1.3 and (p,q) is the same notation as in Problem 8.

[Choi and Srivastava (1999a, p. 95, Eq. (2.2))]

10. For $\Phi(z,s,a)$ defined by 2.5(1) *and* the Srivastava polynomials $S_n^m(z)$ defined by [*cf.* Srivastava (1972, p. 6, Eq. (1))]

$$S_n^m(x) := \sum_{k=0}^{[n/m]} \frac{(-n)_{mk}}{k!} A_{0,k}\, z^k$$

$$(m \in \mathbb{N};\; n \in \mathbb{N}_0),$$

where, in the *general* case, $\{A_{n,k}\}$ is a suitably bounded *double* sequence of complex numbers, show that

$$\sum_{n=0}^{\infty} \frac{(\lambda)_n}{n!} \Phi(x, \lambda + n, a) S_n^m(y) t^n = \sum_{n=0}^{\infty} \frac{(\lambda)_{mn}}{n!} A_{0,n} \Phi(x, \lambda + mn, a - t) \left\{ y(-t)^m \right\}^k$$

$$(|x| \leqq 1;\; |t| < |a|;\; \lambda \neq 1).$$

[*Cf.* Raina and Srivastava (1995, p. 302); see also Problem 6 (Chapter 3)]

11. Prove the following *classical* result analogous to those in (for example) Problem 26(b) (Chapter 2) [or Equation 4.2(2)] *and* Problem 2 (Chapter 4):

$$\zeta(2) = 3 \sum_{k=1}^{\infty} \frac{1}{k^2 \binom{2k}{k}},$$

which was given by Euler in 1748 [and also by Catalan (1865)].

[*Cf.* Melzak (1973, p. 85) and Comtet (1974, p. 89)]

Bibliography

Abouzahra, M. and Lewin, L. (1985). *The polylogarithm in algebraic number fields*, J. Number Theory **21**, 214–244.

Abouzahra, M., Lewin, L., and Hongnian, X. (1987). *Polylogarithms in the field of omega (a root of a given cubic): Functional equations and ladders*, Aequationes Math. **33**, 23–45.

Abramowitz, M. and Stegun, I. A. (1964). *Handbook of Mathematical Functions with Formulas, Graphs, and Mathematical Tables*, Applied Mathematics Series **55**, National Bureau of Standards, Washington, D.C.; Reprinted by Dover Publications, New York, 1965.

Aczél, J. (1966). *Lectures on Functional Equations and Their Applications*, Academic Press, New York and London.

Adamchik, V. S. (1997). *On Stirling numbers and Euler sums*, J. Comput. Appl. Math. **79**, 119–130.

Adamchik, V. S. (1998). *Polygamma functions of negative order*, J. Comput. Appl. Math. **100**, 191–199.

Adamchik, V. S. and Srivastava, H. M. (1998). *Some series of the Zeta and related functions*, Analysis **18**, 131–144.

Ahlfors, L. V. (1979). *Complex Analysis,* Third Ed., McGraw-Hill Book Company, New York.

Airey, J. R. and Webb, H. A. (1918). *The practical importance of the confluent hypergeometric function*, Philos. Mag. **36**, 129–141.

Alexeiewsky, W. P. (1894). *Über eine Classe von Funktionen, die der Gammafunktion analog sind*, Leipzig Weidmannsche Buchhandlung **46**, 268–275.

Almkvist, G. and Granville, A. (1999). *Borwein and Bradley's Apéry-like formulae for* $\zeta(4n+3)$, Experimental Math. **8**, 197–203.

Almkvist, G. and Meurman, A. (1991). *Values of Bernoulli polynomials and Hurwitz's Zeta function at rational arguments*, C. R. Math. Rep. Acad. Sci. Canada **13**, 104–108.

Al-Saqabi, B. N., Kalla, S. L., and Srivastava, H. M. (1991). *A certain family of infinite series associated with Digamma functions*, J. Math. Anal. Appl. **159**, 361–372.

Alzer, H. (1993). *Some Gamma function inequalities*, Math. Comput. **60**, 337–346.

Alzer, H. (1997). *On some inequalities for the Gamma and Psi functions*, Math. Comput. **66**, 373–389.

Alzer, H. (1999). *Inequalities for the Gamma function*, Proc. Amer. Math. Soc. **128**, 141–147.

Alzer, H. and Wells, J. (1998). *Inequalities for the Polygamma functions*, SIAM J. Math. Anal. **29**, 1459–1466.

Amdeberhan, T. (1996). *Faster and faster convergent series for* $\zeta(3)$, Electronic J. Combin. **3**(1), 2 pp. (electronic).

Anastassiadis, M. J. (1952). *Fonctions semi-monotones et semi-convexes et solutions d'une équation fonctionnelle*, Bull. Sci. Math. (2) **76**, 148–160.

Anastassiadis, M. J. (1957a). *Sur les solutions logarithmiquement convexes ou concaves d'une équation fonctionnelle*, Bull. Sci. Math. (2) **81**, 78–87.

Anastassiadis, M. J. (1957b). *Une propriété de la fonction Gamma*, Bull. Sci. Math. (2) **81**, 116–118.

Anastassiadis, M. J. (1959). *Définitions fonctionnelles de la fonction* $B(x, y)$, Bull. Sci. Math. (2) **83**, 24–32.

Anastassiadis, M. J. (1964). *Définition des Fonctions Eulériennes par des Équations Fonctionnelles*, Gauthier-Villars, Paris.

Anderson, G. D., Barnard, R. W., Richards, K. C., Vamanamurthy, M. K., and Vuorinen, M. (1995). *Inequalities for zero-balanced hypergeometric functions*, Trans. Amer. Math. Soc. **347**, 1713–1723.

Anderson, G. D., Vamanamurthy, M. K., and Vuorinen, M. (1990). *Functional inequalities for complete elliptic integrals and their ratios*, SIAM J. Math. Anal. **21**, 536–549.

Anderson, G. D., Vamanamurthy, M. K., and Vuorinen, M. (1992). *Functional inequalities for hypergeometric functions and complete elliptic integrals.*, SIAM J. Math. Anal. **23**, 512–524.

Anderson, G. D. and Qiu, S.-L. (1997). *A monotoneity property of the Gamma function*, Proc. Amer. Math. Soc. **125**, 3355–3362.

Andrews, G. E. (1979). *An introduction to Ramanujan's "Lost" Notebook*, Amer. Math. Monthly **86**, 89–108.

Anglesio, J. (1996a). *Problems and Solutions*, Amer. Math. Monthly **103**, 427.

Anglesio, J. (1996b). *Problems and Solutions*, Amer. Math. Monthly **103**, 903.

Apéry, R. (1979). *Irrationalité de* $\zeta(2)$ *et* $\zeta(3)$, in *Journées Arithmétiques de Luminy* (Colloq. Internat. CNRS, Centre Univ. Luminy, Luminy, 1978), pp. 11–13, Astérisque **61**, Soc. Math. France, Paris.

Apostol, T. M. (1950). *Generalized Dedekind sums and transformation formulae of certain Lambert series*, Duke Math. J. **17**, 147–157.

Apostol, T. M. (1951). *On the Lerch Zeta-function*, Pacific J. Math. **1**, 161–167.

Apostol, T. M. (1951). *Remark on the Hurwitz Zeta function*, Proc. Amer. Math. Soc. **2**, 690–693.

Apostol, T. M. (1954). *Some series involving the Riemann Zeta function*, Proc. Amer. Math. Soc. **5**, 239–243.

Apostol, T. M. (1973). *Another elementary proof of Euler's formula for* $\zeta(2n)$, Amer. Math. Monthly **80**, 425–431.

Apostol, T. M. (1974). *Mathematical Analysis*, Second Ed., Addison-Wesley Publishing Company, Menlo Park (California), London, and Don Mills (Ontario).

Apostol, T. M. (1976a). *Introduction to Analytic Number Theory*, Springer-Verlag, New York, Heidelberg, and Berlin.

Apostol, T. M. (1976b). *Modular Functions and Dirichlet Series in Number Theory*, Springer-Verlag, New York, Heidelberg, and Berlin.

Apostol, T. M. and Vu, T. H. (1984). *Dirichlet series related to the Riemann Zeta function*, J. Number Theory **19**, 85–102.

Apostol, T. M. (1999). *An elementary view of Euler's summation formula*, Amer. Math. Monthly **106**, 409–418.

Arakawa, T. and Kaneko, M. (1999a). *Multiple Zeta values, Poly-Bernoulli numbers, and related Zeta functions*, Nagoya Math. J. **153**, 189–209.

Arakawa, T. and Kaneko, M. (1999b). *On Poly-Bernoulli numbers*, Comment. Math. Univ. St. Paul. **48**(2), 159–167.

Artin, E. (1964). *The Gamma Function*, Holt, Rinehart and Winston, New York, Chicago, San Francisco, Toronto, and London.

Askey, R. (1978). *The q-Gamma and q-Beta functions*, Applicable Anal. **8**, 125–141.

Askey, R. (1980). *Ramanujan's extensions of the Gamma and Beta functions*, Amer. Math. Monthly **87**, 346–359.

Aular de Durán, J., Kalla, S. L., and Srivastava, H. M. (1995). *Fractional calculus and the sums of certain families of infinite series*, J. Math. Anal. Appl. **190**, 738–754.

Ayoub, R. (1974). *Euler and the Zeta function*, Amer. Math. Monthly **81**, 1067–1086.

Bailey, D. H., Borwein, J. M., and Crandall, R. E. (1997). *On the Khintchine constant*, Math. Comput. **66**, 417–431.

Bailey, D. H., Borwein, J. M., and Girgenshon, R. (1994). *Experimental evaluation of Euler sums*, Experimental Math. **3**, 17–30.

Bailey, W. N. (1935). *Generalized Hypergeometric Series*, Cambridge Tracts in Mathematics and Mathematical Physics, No. **32**, Cambridge University Press, Cambridge, London, and New York; Reprinted by Stechert-Hafner Service Agency, New York and London, 1964.

Bailey, W. N. (1936). *Series of hypergeometric type which are infinite in both directions*, Quart. J. Math. Oxford Ser. **7**, 105–115.

Bailey, W. N. (1951). *On the simplification of some identities of the Rogers-Ramanujan type*, Proc. London Math. Soc. (3) **1**, 217–221.

Balakrishnan, U. (1988). *A series for $\zeta(s)$*, Proc. Edinburgh Math. Soc. (2) **31**, 205–210.

Barnes, E. R. and Kaufman, W. E. (1965). *The Euler-Mascheroni constant*, Amer. Math. Monthly **72**, 1023.

Barnes, E. W. (1899a). *The theory of the Gamma function*, Messenger Math. **29**, 64–128.

Barnes, E. W. (1899b). *The theory of the G-function*, Quart. J. Math. **31**, 264–314.

Barnes, E. W. (1900). *Genesis of the double Gamma function*, Proc. London Math. Soc. **31**, 358-381.

Barnes, E. W. (1901). *The theory of the double Gamma function*, Philos. Trans. Roy. Soc. London Ser. A **196**, 265-388.

Barnes, E. W. (1904). *On the theory of the multiple Gamma functions*, Trans. Cambridge Philos. Soc. **19**, 374–439.

Bartz, K. and Rutkowski, J. (1993). *On the von Staudt-Clausen theorem*, C. R. Math. Rep. Acad. Sci. Canada **15**, 46–48.

Bateman, P. T. and Grosswald, E. (1964). *On Epstein's Zeta function*, Acta Arith. **9**, 365–373.

Beckenbach, E. F. (1948). *Convex functions*, Bull. Amer. Math. Soc. **54**, 439–460.

Beckenbach, E. F. (1964). *Superadditivity inequalities*, Pacific J. Math. **14**, 421–438.

Beckmann, P. (1971). *A History of π (Pi)*, Second Ed., The Golem Press, Boulder, Colorado.

Bendersky, L. (1933). *Sur la fonction Gamma généralisée*, Acta Math. **61**, 263–322.

Berndt, B. C. (1972a). *Two new proofs of Lerch's functional equation*, Proc. Amer. Math. Soc. **32**, 403–408.

Berndt, B. C. (1972b). *On the Hurwitz Zeta-function*, Rocky Mountain J. Math. **2**, 151–157.

Berndt, B. C. (1973). *Generalized Dedekind Eta-functions and generalized Dedekind sums*, Trans. Amer. Math. Soc. **178**, 495–508.

Berndt, B. C. (1975). *Elementary evaluation of $\zeta(2n)$*, Math. Mag. **48**, 148–153.

Berndt, B. C. (1977). *Modular transformations and generalizations of several formulae of Ramanujan*, Rocky Mountain J. Math. **7**, 147–189.

Berndt, B. C. (1985). *Ramanujan's Notebooks* (Part I), Springer-Verlag, New York.

Berndt, B. C. (1989). *Ramanujan's Notebooks* (Part II), Springer-Verlag, New York.

Beukers, F. (1979). *A note on the irrationality of* $\zeta(2)$ *and* $\zeta(3)$, Bull. London Math. Soc. **11**, 268–272.

Beyer, W. A. and Waterman, M. S. (1974a). *Error analysis of a computation of Euler's constant*, Math. Comput. **28**, 599–604.

Beyer, W. A. and Waterman, M. S. (1974b). *Decimals and partial quotients of Euler's constant and* $\ln 2$, Math. Comput. **28**, 667.

Beyer, W. A. and Waterman, M. S. (1978). *Decimals and partial quotients of Euler's constant and* $\ln 2$, Math. Comput. **32**, 317–318.

Bigler, H. (1888). *Über Gammafunktionen mit beliebigem parameter*, J. Reine Angew. Math. **102**, 237–254.

Bleimann, G. and Stark, E. L. (1980). *Kernels of finite oscillations and convolution integrals*, Acta Math. Acad. Sci. Hungar. **35**, 419–429.

Blyth, C. R. and Pathak, P. K. (1986). *A note on easy proofs of Stirling's theorem*, Amer. Math. Monthly **93**, 376–379.

Bochner, S. (1958). *On Riemann's functional equation with multiple Gamma factors*, Ann. of Math. (2) **67**, 29–41.

Bochner, S. and Chandrasekharan, K. (1956). *On Riemann's functional equation*, Ann. of Math. (2) **63**, 336–360.

Boas, Jr., R. P. and Wrench, Jr., J. W. (1971). *Partial sums of the Harmonic series*, Amer. Math. Monthly **78**, 864–870.

Boersma, J. and Dempsey, J. P. (1992). *On the numerical evaluation of Legendre's Chi-function*, Math. Comput. **59**, 157–163.

Boole, G. (1960). *A Treatise on the Calculus of Finite Differences*, Second Ed., Macmillan Company, London, 1872 (Reprinted by Dover Publications, New York).

Borwein, D. and Borwein, J. M.(1995). *On an intriguing integral and some series related to* $\zeta(4)$, Proc. Amer. Math. Soc. **123**, 1191–1198.

Borwein, D., Borwein, J. M., and Girgensohn, R. (1995). *Explicit evaluation of Euler sums*, Proc. Edinburgh Math. Soc. (2) **38**, 277–294.

Borwein, J. (1985). *Some modular identities of Ramanujan useful in approximating* π, Proc. Amer. Math. Soc. **95**, 365–371.

Borwein, J. M. and Borwein, P. B. (1987). *Pi and the AGM*, John Wiley and Sons, New York, Chichester, Brisbane, Toronto, and Singapore.

Borwein, J. and Bradley, D. (1997). *Empirically determined Apéry-like formulae for* $\zeta(4n+3)$, Experimental Math. **6**, 181–194.

Borwein, J. M., Bradley, D. M., and Broadhurst, D. J. (1997). *Evaluations of k-fold Euler/Zagier sums: a Compendium of results for arbitrary k*, Electronic J. Combin. **4**(2), 21 pp. (electronic).

Borwein, J. M., Bradley, D. M., Broadhurst, D. J., and Lisoněk, P. (1998). *Combinatorial aspects of multiple Zeta values*, Electronic J. Combin. **5**, 12 pp. (electronic).

Borwein, J. M., Bradley, D. M., and Crandall, R. E. (2000). *Computational strategies for the Riemann Zeta function*, J. Comput. Appl. Math. **121**, 247–296.

Borwein, J. M. and Girgensohn, R. (1996). *Evaluation of triple Euler sums*, Electronic J. Combin. **3**(1), 27 pp. (electronic).

Borwein, J. M. and Zucker, I. J. (1992). *Fast evaluation of the Gamma function for small rational fractions using complete elliptic integrals of the first kind*, IMA J. Numer. Anal. **12**, 519–526.

Bourguet, L. (1882). *Note sur les intégrales Eulériennes*, Acta Math. **1**, 295–296.

Bourguet, L. (1882). *Sur quelques intégrales définies*, Acta Math. **1**, 363–367.

Bourguet, L. (1883). *Sur les intégrales Eulériennes et quelques autres fonctions uniformes*, Acta Math. **2**, 231–232.

Bourguet, L. (1883). *Sur la fonction Eulérienne*, Acta Math. **2**, 296–298.

Bowman, F. (1947). *Note on the integral* $\int_0^{\frac{1}{2}\pi} (\log \sin\theta)^n \, d\theta$, J. London Math. Soc. **22**, 172–173.

Boyd, A. V. (1960). *Gurland's inequality for the Gamma function*, Skand. Aktuarietidskr. **1960**, 134–135.

Boyd, A. V. (1967). *Note on a paper by Uppuluri*, Pacific J. Math. **22**, 9–10.

Bradley, D. M. (1999). *A class of series acceleration formulae for Catalan's constant*, Ramanujan J. **3**, 159–173.

Branson, T. P. and Ørsted, B. (1991). *Explicit functional determinants in four dimensions*, Proc. Amer. Math. Soc. **113**, 669–682.

Brent, R. P. (1977). *Computation of the regular continued fraction for Euler's constant*, Math. Comput. **31**, 771–777.

Brent, R. P. (1978). γ *and* e^γ *to* 20700D *and their regular continued fractions to* 20000 *partial quotients*, Math. Comput. **32**, 311.

Brent, R. P. and McMillan, E. M. (1980). *Some new algorithms for high-precision computation of Euler's constant*, Math. Comput. **34**, 305–312.

Briggs, W. E. (1955-1956). *Some constants associated with the Riemann Zeta-function*, Michigan Math. J. **3**, 117–121.

Briggs, W. E. (1989). *Zeta function identity*, Problem 1302, Math. Mag. **62**, 275–276.

Briggs, W. E., Chowla, S., Kempner, A. J., and Mientka, W. E. (1955). *On some infinite series*, Scripta Math. **21**, 28–30.

Briggs, W. E. and Chowla, S. (1955). *The power series coefficients of* $\zeta(s)$, Amer. Math. Monthly **62**, 323–325.

Broadhurst, D. J. and Kreimer, D. (1997). *Association of multiple Zeta values with positive knots via Feynman diagrams up to* 9 *loops*, Phys. Lett. B **393**, 403–412.

Bromwich, T. J. I'A. (1926). *An Introduction to the Theory of Infinite Series,* Second Ed., Macmillan and Company, London.

Bruckman, P. S. and Georghiou, C. (1982). *An Euler sum*, Fibonacci Quart. **20**, 186–187.

Bruckner, A. (1960). *Minimal superadditive extensions of superadditive functions*, Pacific J. Math. **10**, 1155–1162.

Bruckner, A. M. (1962). *Tests for the superadditivity of functions*, Proc. Amer. Math. Soc. **13**, 126–130.

Bruckner, A. M. and Ostrow, E. (1962). *Some function classes related to the class of convex functions*, Pacific J. Math. **12**, 1203–1215.

Brun, V. (1970). *Leibniz' formula for* π *deduced by a "mapping" of the circular disc*, Nordisk Mat. Tidskr. **18**, 73–81.

Buschman, R. G. and Srivastava, H. M. (1993). *Asymptotic behavior of some power series with* ζ*-functions in the coefficients*, Monatsh. Math. **115**, 291–298.

Bustoz, J. and Ismail, M. E.-H. (1986). *On Gamma function inequalities*, Math. Comput. **47**, 659–667.

Butzer, P. L. and Hauss, M. (1992a). *Riemann Zeta function: Rapidly converging series and integral representations,* Appl. Math. Lett. **5**(2), 83–88.

Butzer, P. L. and Hauss, M. (1992b). *Integral and rapidly converging series representations of the Dirichlet L-functions $L_1(s)$ and $L_{-4}(s)$*, Atti Sem. Mat. Fis. Univ. Modena **40**, 329–359.

Butzer, P. L. and Jongmans, F. (1991). *Eugène Catalan and the rise of Russian science*, Acad. Roy. Belg. Bull. Cl. Sci (6) **2**, 59–90.

Butzer, P. L., Markett, C., and Schmidt, M. (1991). *Stirling numbers, central factorial numbers, and representations of Riemann Zeta function*, Resultate Math. **19**, 257–274.

Butzer, P. L. and Schmidt, M. (1990). *Central factorial numbers and their role in finite difference calculus and approximation*, in *Approximation Theory* (Kecskemét, 1990), pp. 127-150, Colloq. Math. Soc. János Bolyai **58**, Elsevier (North-Holland), New York, London, and Amsterdam, 1991.

Butzer, P. L., Schmidt, M., and Stark, E. L. (1988). *Observations on the history of central B-splines*, Arch. History Exact Sci. **39**, 137–156.

Butzer, P. L., Schmidt, M., Stark, E. L., and Vogt, L. (1989). *Central factorial numbers; Their main properties and some applications*, Numer. Funct. Anal. Optim. **10**, 419–488.

Butzer, P. L. and Stark, E. L. (1969). *On a trigonometric convolution operator with kernel having two zeros of simple multiplicity*, Acta Math. Acad. Sci. Hungar. **20**, 451–461.

Callebaut, D. K. (1965). *Generalization of the Cauchy-Schwarz inequality*, J. Math. Anal. Appl. **12**, 491–494.

Campbell, R. (1966). *Les Intégrales Eulériennes et Leurs Applications*, Dunod, Paris.

Carathéodory, C.(1958). *Theory of Functions of a Complex Variable*, Vol. **1** (Second English Ed. translated by F. Steinhardt), Chelsea Publishing Company, New York.

Carlitz, L. (1948). *q-Bernoulli numbers and polynomials*, Duke Math. J. **15**, 987–1000.

Carlitz, L. (1953). *Some congruences for Bernoulli numbers of higher order*, Quart. J. Math. Oxford Ser. (2) **4**, 112–116.

Carlitz, L. (1954). *q-Bernoulli and Eulerian numbers*, Trans. Amer. Math. Soc. **76**, 332–350.

Cassou-Noguès, P. (1979). *Analogues p-adiques des fonctions Γ-multiples*, in *Journées Arithmétiques de Luminy* (Colloq. Internat. CNRS, Centre Univ. Luminy, Luminy, 1978), pp. 43–55, Astérisque **61**, Soc. Math. France, Paris.

Catalan, E. (1864). *Sur une transformation de série*, Mém. Soc. Roy. Sci. Liège Sér. 2 **12**, 206–215.

Catalan, E. (1865). *Mémoire sur la transformation des séries et sur quelques intégrales défines*, Mém. Acad. Roy. Belg. **33**, 1–50.

Catalan, E. (1883). *Recherches sur la constante G, et sur les inégrales Eulériennes*, Mém. Acad. Imp. Sci. St.-Pétersbourg Sér. 7 **31**(3), 1–51.

Cayley, A. (1893/1894). *The numerical value of* $\Pi i = \Gamma(1+i)$, Messenger Math. **23**, 36–38.

Chandrasekharan, K. and Narasimhan, R. (1960). *Sur l'ordre moyen de quelques fonctions arithmétiques*, C. R. Acad. Sci. Paris **251**, 1333–1335.

Chassan, J. B. (1955). *A statistical derivation of a pair of trigonometric inequalities*, Amer. Math. Monthly **62**, 353–356.

Chaudhuri, J. (1967). *Some special integrals*, Amer. Math. Monthly **74**, 545–548.

Chen, C.-P. (1993). *The Riemann hypothesis and Gamma conditions*, J. Math. Anal. Appl. **173**, 258–275.

Chen, K.-Y. and Srivastava, H.M. (2000). *Some infinite series and functional relations that arose in the context of fractional calculus*, J. Math. Anal. Appl. **252**, 377–388.

Chen, M.-P. and Srivastava, H. M. (1998). *Some families of series representations for the Riemann* $\zeta(3)$, Resultate Math. **33**, 179–197.

Choe, B. R. (1987). *An elementary proof of* $\Sigma_{n=1}^{\infty} 1/n^2 = \pi^2/6$, Amer. Math. Monthly **94**, 662–663.

Choi, J. (1992). *On a generalization of the Hurwitz Zeta function* $\zeta(s,a)$, Indian J. Pure Appl. Math. **23**, 83–91.

Choi, J. (1994a). *Determinant of Laplacian on* S^3, Math. Japon. **40**, 155–166.

Choi, J. (1994b). *A proof of Stirling's formula*, Comm. Korean Math. Soc. **9**, 853–855.

Choi, J. (1996a). *Explicit formulas for the Bernolli polynomials of order* n, Indian J. Pure Appl. Math. **27**, 667–674.

Choi, J. (1996b). *A duplication formula for the double Gamma function* Γ_2, Bull. Korean Math. Soc. **33**, 289–294.

Choi, J. and Nam, Y. M. (1995). *The first Eulerian integral*, Kyushu J. Math. **49**, 421–427.

Choi, J. and Nash, C. (1997). *Integral representation of the Kinkelin's constant* A, Math. Japon. **45**, 223–230.

Choi, J. and Quine, J. R. (1992). *E. W. Barnes' approach of the multiple Gamma functions*, J. Korean Math. Soc. **29**, 127-140.

Choi, J. and Rathie, A. K. (1997). *An evaluation of* $\zeta(2)$, Far East J. Math. Sci. **5**, 393–398.

Choi, J., Rathie, A. K., and Srivastava, H. M. (1999). *Some hypergeometric and other evaluations of* $\zeta(2)$ *and allied series*, Appl. Math. Comput. **104**, 101–108.

Choi, J. and Seo, T. Y. (1997a). *Evaluation of some infinite series*, Indian J. Pure Appl. Math. **28**, 791–796.

Choi, J. and Seo, T. Y. (1997b). *The double Gamma function*, East Asian Math. J. **13**, 159-174.

Choi, J. and Seo, T. Y. (1998). *Integral formulas for Euler's constant*, Comm. Korean Math. Soc. **13**, 683–689.

Choi, J. and Seo, T. Y. (1999). *Identities involving series of the Riemann Zeta function*, Indian J. Pure Appl. Math. **30**, 649–652.

Choi, J. and Srivastava, H. M. (1996). *A reducible case of double hypergeometric series involving the Riemann Zeta function*, Bull. Korean Math. Soc. **33**, 107–110.

Choi, J. and Srivastava, H. M. (1997). *Sums associated with the Zeta function*, J. Math. Anal. Appl. **206**, 103–120.

Choi, J. and Srivastava, H. M. (1999a). *Certain classes of series involving the Zeta function*, J. Math. Anal. Appl. **231**, 91–117.

Choi, J. and Srivastava, H. M. (1999b). *Certain classes of infinite series*, Monatsh. Math. **127**, 15–25.

Choi, J. and Srivastava, H. M. (1999c). *An application of the theory of the double Gamma function*, Kyushu J. Math. **53**, 209–222.

Choi, J. and Srivastava, H. M. (2000a). *Certain classes of series associated with the Zeta function and multiple Gamma functions*, J. Comput. Appl. Math. **118**, 87–109.

Choi, J. and Srivastava, H. M. (2000b). *A certain family of series associated with the Zeta and related functions*, Preprint.

Choi, J. and Srivastava, H. M. (2000c). *Evaluation of higher-order derivatives of the Gamma function*, Univ. Beograd. Publ. Elektrotehn. Fak. Ser. Mat. **11**.

Choi, J. and Srivastava, H. M. (2000d). *A certain class of series associated with the Zeta function*, Integral Transform. Spec. Funct. (to appear).

Choi, J., Srivastava, H. M., and Quine, J. R. (1995). *Some series involving the Zeta function*, Bull. Austral. Math. Soc. **51**, 383–393.

Choi, J., Srivastava, H. M., Seo, T. Y., and Rathie, A. K. (1999). *Some families of infinite series*, Soochow J. Math. **25**, 209–219.

Choi, J., Srivastava, H. M., and Zhang, N.-Y. (1998). *Integrals involving a function associated with the Euler-Maclaurin summation formula*, Appl. Math. Comput. **93**, 101–116.

Chowla, S. (1935). *Note on Dirichlet's L-functions*, Acta Arith. **1**, 113–114.

Christophe, L. M., Jr., Finn, M. V., and Crow, J. A. (1986). *Two series involving the Zeta function values*, Math. Mag. **59**, 176–178.

Chrystal, G. (1964). *Algebra: An Elementary Text-Book (Part II)*, Seventh Ed., Chelsea Publishing Company, New York.

Chu, J. T. (1962). *A modified Wallis product and some applications*, Amer. Math. Monthly **69**, 402–404.

Chudnovsky, G. V. (1979). *Padé approximations to the generalized hypergeometric functions*. I, J. Math. Pures Appl. **58**, 445–476.

Clarke, F. and Slavutskii, I. Sh. (1997). *The integrality of the values of Bernoulli polynomials and of generalized Bernoulli numbers*, Bull. London Math. Soc. **29**, 22–24.

Cohen, H. (1981). *Généralisation d'une construction de R. Apéry*, Bull. Soc. Math. France **109**, 269–281.

Cohen, H., Lewin, L., and Zagier, D. (1992). *A sixteenth-order polylogarithm ladder*, Experimental Math. **1**, 25–34.

Comtet, L. (1974). *Advanced Combinatorics: The Art of Finite and Infinite Expansions* (Translated from the French by J. W. Nienhuys), Reidel, Dordrecht and Boston.

Conway, J. B. (1978). *Functions of One Complex Variable*, Second Ed., Springer-Verlag, New York, Heidelberg, and Berlin.

Cooper, R. (1957). *On a duplication formula*, Math. Gaz. **41**, 217–218.

Copson, E. T. (1935). *Theory of Functions of a Complex Variable*, Clarendon (Oxford University) Press, Oxford, London, and New York.

Coxeter, H. S. M. (1935). *The functions of Schläfli and Lobatschefsky*, Quart. J. Math. Oxford Ser. **6**, 13–29.

Crandall, R. E. and Buhler, J. P. (1994). *On the evaluation of Euler sums*, Experimental Math. **3**, 275–285.

Cvijović, D. and Klinowski, J. (1995a). *New formulae for the Bernoulli and Euler polynomials at rational arguments*, Proc. Amer. Math. Soc. **123**, 1527–1535.

Cvijović, D. and Klinowski, J. (1995b). *Closed-form summation of some trigonometric series*, Math. Comput. **64**, 205–210.

Cvijović, D. and Klinowski, J. (1997). *New rapidly convergent series representations for* $\zeta(2n+1)$, Proc. Amer. Math. Soc. **125**, 1263–1271.

Cvijović, D. and Klinowski, J. (1999). *Values of the Legendre Chi and Hurwitz Zeta functions at rational arguments*, Math. Comput. **68**, 1623–1630.

Dabrowski, A. (1996). *A note on the values of the Riemann Zeta function at positive odd integers*, Nieuw Arch. Wisk. (4) **14**, 199–207.

Darling, D. A. (1997). *An integral giving Euler's constant*, Amer. Math. Monthly **104**, 881.

Daudé, H., Flajolet, P., and Vallée, B. (1997). *An average-case analysis of the Gaussian algorithm for lattice reduction*, Combin. Probab. Comput. **6**, 397–433.

Davis, H. T. (1935). *An extension to Polygamma functions of a theorem of Gauss*, Bull. Amer. Math. Soc. **41**, 243–247.

Davis, H. T. (1962). *The Summation of Series*, The Principia Press of Trinity University, San Antonio, Texas.

Day, W. A. (1970). *On monotonicity of the relaxation functions of viscoelastic materials*, Proc. Cambridge Philos. Soc. **67**, 503–508.

de Bruijn, N. G. (1937). *Integralen voor de ζ-functie van Riemann*, Mathematica (Zutphen) **B5**, 170–180.

de Doelder, P. J. (1991)). *On some series containing $\psi(x) - \psi(y)$ and $(\psi(x) - \psi(y))^2$ for certain values of x and y*, J. Comput. Appl. Math. **37**, 125–141.

De Lillo, N. J. (1982). *Advanced Calculus with Applications*, Macmillan Company, New York.

Dempsey, K. M., Dajin, L., and Dempsey, J. P. (1992). *Plana's summation formula for $\Sigma_{m=1,3,\dots}^{\infty}$ $m^{-2}\sin(m\alpha)$, $m^{-3}\cos(m\alpha)$, $m^{-2}A^m$, $m^{-3}A^m$*, Math. Comput. **55**, 693–703.

Denjoy, A. (1954). *Une démonstration de l'identité fondamentale de la fonction $\zeta(s)$ de Riemann*, J. Analyse Math. **3**, 197–206.

D'Hoker, E. and Phong, D. H. (1986a). *On determinant of Laplacians on Riemann surface*, Comm. Math. Phys. **104**, 537–545.

D'Hoker, E. and Phong, D. H. (1986b). *Multiloop amplitudes for the bosonic polyakov string*, Nuclear Phys. B **269**, 204–234.

D'Hoker, E. and Phong, D. H. (1989). *Functional determinants on Mandelstam diagrams*, Comm. Math. Phys. **124**, 629–645.

Diaconis, P. and Freeman, D. (1986). *An elementary proof of Stirling's formula*, Amer. Math. Monthly **93**, 123–125.

Dikiĭ, L. A. (1961). *Trace formulas for Strum-Liouville differential operators*, Amer. Math. Soc. Trans. **18**, 81–115.

Dilcher, K. (1996). *Sums of products of Bernoulli numbers*, J. Number Theory **60**, 23–41.

Dinghas, A. (1958/1959). *Zur throrie der Gammafunktion*, Math.-Phys. Semester. **6**, 245–252.

Dinghas, A. (1974). *Zur Charakterisierung der Riemannschen Zeta-Funktion durch Hadamardsche Funktionalgleichungen*, Rev. Roumaine Math. Pures Appl. **19**, 995–1007.

Dittrich, W. and Reuter, M. (1983). *Effective QCD-Lagrangian with ξ-function regularization*, Phys. Lett. B **128**, 321–326.

Dixon, A. L. and Ferrar, W. L. (1936). *A class of discontinuous integrals*, Quart. J. Math. Oxford Ser. **7**, 81–96.

Dufresnoy, J. et Pisot, Ch. (1963). *Sur la relation fonctionnelle $f(x+1) - f(x) = \varphi(x)$*, Bull. Soc. Math. Belg. **15**, 259–270.

Dutka, J. (1985). *On some Gamma function inequalities*, SIAM J. Math. Anal. **16**, 180–185.

Eagle, A. (1928/1929). *A simple theory of the Gamma function*, Math. Gaz. **14**, 118–127.

Edwards, H. M. (1974). *Riemann's Zeta Function*, Academic Press, New York and London.

Edwards, J. (1954). *A Treatise on the Integral Calculus with Applications: Examples and Problems*, Vols. 1 and 2, Chelsea Publishing Company, New York.

Efrat, I. (1988). *Determinants of Laplacians on surfaces of finite volume*, Comm. Math. Phys. **119**, 443–451; erratum, *ibid.* **138**(1991), 607.

Elebert, Á and Laforgia, A. (2000). *On some properties of the Gamma function*, Proc. Amer. Math. Soc. (Article electronically published on March 1), 7 pp. (electronic).

Eliezer, C. J. and Daykin, D. E. (1967). *Generalizations and applications of Cauchy-Schwarz inequalities*, Quart. J. Math. Oxford Ser. (2) **18**, 357–360.

Elizalde, E. (1985). *Derivative of the generalized Riemann Zeta function* $\zeta(z,q)$ *at* $z=-1$, J. Phys. A: Math. Gen. **18**, 1637–1640.

Elizalde, E. (1986). *An asymptotic expansion for the first derivative of the generalized Riemann Zeta function*, Math. Comput. **47**, 347–350.

Elizalde, E., Odintsov, S. D., Romeo, A., Bytsenko, A. A., and Zerbini, S. (1994). *Zeta Regularization Techniques with Applications*, World Scientific, Singapore, New Jersey, London, and Hong Kong.

Elizalde, E. and Romeo, A. (1990). *An integral involving the generalized Zeta function*, Internat. J. Math. and Math. Sci. **13**, 453–460.

Elizalde, E. and Soto, J. (1985). ζ*-regularized Lagrangians for massive quarks in constant background mean-fields*, Ann. Phys. **162**, 192–211.

Elsner, C. (1995). *On a sequence transformation with integral coefficients for Euler's constant*, Proc. Amer. Math. Soc. **123**, 1537–1541.

Elstrodt, J. (1985). *Note on the Selberg trace formula for the Picard group*, Abh. Math. Sem. Univ. (Hamburg) **55**, 207–209.

English, B. J. and Rousseau, G. (1997). *Bounds for certain harmonic sums*, J. Math. Anal. Appl. **206**, 428–441.

Erber, T. (1961). *The Gamma function inequalities of Gurland and Gautschi*, Skand. Aktuarietidskr. **44**, 27–28.

Erdélyi, A., Magnus, W., Oberhettinger, F., and Tricomi, F. G. (1953). *Higher Transcendental Functions*, Vol. **I**, McGraw-Hill Book Company, New York, Toronto, and London.

Erdélyi, A., Magnus, W., Oberhettinger, F., and Tricomi, F. G. (1955). *Higher Transcendental Functions*, Vol. **III**, McGraw-Hill Book Company, New York, Toronto, and London.

Estermann, T. (1947). *Elementary evaluation of* $\zeta(2k)$, J. London Math. Soc. **22**, 10–13.

Euler, L. (1736a). *Methodvs vniversalis serievm convergentivm svmmas*, Comment. Acad. Sci. Imp. Petropolit. **8**, 3–9.

Euler, L. (1736b). *Methodvs vniversalis series svmmandi*, Comment. Acad. Sci. Imp. Petropolit. **8**, 147–158.

Euler, L. (1775). *Meditationes circa singulare serierum genus*, Novi Comment. Acad. Sci. Imp. Petropolit. **20**, 140–186.

Ewell, J. A. (1990). *A new series representation for* $\zeta(3)$, Amer. Math. Monthly **97**, 219–220.

Ewell, J. A. (1991). *On values of the Riemann Zeta function at integral arguments*, Canad. Math. Bull. **34**, 60–66.

Ewell, J. A. (1995). *On the Zeta function values* $\zeta(2k+1)$, $k=1,2,\ldots$, Rocky Mountain J. Math. **25**, 1003–1012.

Feller, W. (1939). *Completely monotone functions and sequences*, Duke Math. J. **5**, 661–674.

Fine, N. J. (1951). *Note on the Hurwitz Zeta-function*, Proc. Amer. Math. Soc. **2**, 361–364.

Fink, A. M. (1982). *Kolmogorov-Landau inequalities for monotone functions*, J. Math. Anal. Appl. **90**, 251–258.

Fisher, R. A. (1924/1925). *Theory of statistical estimation*, Proc. Cambridge Philos. Soc. **22**, 700–725.

Flajolet, P. and Prodinger, H. (1999). *On Stirling numbers for complex arguments and Hankel contours*, SIAM J. Discrete Math. **12**, 155–159.

Flajolet, P. and Salvy, B. (1998). *Euler sums and contour integral representations*, Experimental Math. **7**, 15–35.

Forder, H. G. (1957). *Duplication formulae*, Math. Gaz. **41**, 215–217.

Forsyth, A. R. (1900). *Theory of Functions of a Complex Variable*, Cambridge University Press, Cambridge, London, and New York.

Foster, B. L. and Phelps, R. R. (1957). *Problem* 4712 proposed by J. V. Whittaker, Amer. Math. Monthly **63**(1956), 669; *ibid.* **64**, 677–678.

Fox, G. J. (1999). *Euler polynomials at rational numbers*, C. R. Math. Rep. Acad. Sci. Canada **21**, 87–90.

Frenzen, C. L. (1992). *Error bounds for the asymptotic expansion of the ratio of two Gamma functions with complex argument*, SIAM J. Math. Anal. **23**, 505–511.

Gandhi, J. M. (1970). *A conjectured representation of Genocchi numbers*, Amer. Math. Monthly **77**, 505–506.

Gautschi, W. (1959). *Some elementary inequalities relating to the Gamma and incomplete Gamma function*, J. Math. and Phys. **38**, 77–81.

Gautschi, W. (1974). *A harmonic mean inequality for the Gamma function*, SIAM J. Math. Anal. **5**, 278–281.

Gautschi, W. (1974). *Some mean value inequalities for the Gamma function*, SIAM J. Math. Anal. **5**, 282–292.

Gautschi, W. (1991). *On certain slowly convergent series occurring in plate contact problems*, Math. Comput. **57**, 325–338.

Georghiou, C. and Philippou, A. N. (1983). *Harmonic sums and the Zeta function*, Fibonacci Quart. **21**, 29 – 36.

Gerst, I. (1969). *Some series for Euler's constant*, Amer. Math. Monthly **76**, 273–275.

Giesy, D. P. (1972). *Still another elementary proof that $\Sigma\, 1/k^2 = \pi^2/6$*, Math. Mag. **45**, 148–149.

Glaisher, J. W. L. (1872a). *On the history of Euler's constant*, Messenger Math. **1**, 25–30.

Glaisher, J. W. L. (1872b). *On the integrals $\int_0^\infty \sin(x^n)\, dx$ and $\int_0^\infty \cos(x^n)\, dx$*, Messenger Math. **1**, 106–111.

Glaisher, J. W. L. (1877). *On the product $1^1\, 2^2 \cdots n^n$*, Messenger Math. **7**, 43–47.

Glaisher, J. W. L. (1893). *On products and series involving prime numbers only*, Quart. J. Math. **26**, 1–74.

Glaisher, J. W. L. (1894/1895a). *Note on a relation connecting constants analogous to Euler's constant*, Messenger Math. **24**, 24–27.

Glaisher, J. W. L. (1894/1895b). *Expressions for Gamma functions in terms of complete elliptic integrals*, Messenger Math. **24**, 27–47.

Glaisher, J. W. L. (1898). *The Bernoullian function*, Quart. J. Math. **29**, 1–168.

Glaisher, J. W. L. (1912/1913). *Summations of certain numerical series*, Messenger Math. **42**, 19–34.

Glaisher, J. W. L. (1914/1915). *Relations connecting quantities of the form $1+2^{-n}+3^{-n}+4^{-n}$&c.*, Messenger Math. **44**, 1–10.

Glasser, M. L. (1968). *Some integrals of the arctangent function*, Math. Comput. **22**, 445–447.

Gokhale, D. V. (1963). *On an inequality for Gamma function*, Skand. Aktuarietidskr. **46**, 213–215.

Gonek, S. M. (1984). *Mean values of the Riemann Zeta-function and its derivatives*, Invent. Math. **75**, 123–141.

Gosper, R. W., Jr. (1976). *A calculus of series rearrangements*, in *Algorithms and Complexity: New Directions and Recent Results* (J. F. Traub, Editor), pp. 121-151, Academic Press, New

York, London, and Toronto.

Gosper, R. W., Jr. (1978). *Decision procedure for indefinite hypergeometric summation*, Proc. Nat. Acad. Sci. U. S. A. **75**, 40–42.

Gosper, R. W., Jr. (1997). $\int_{n/4}^{m/6} \ln \Gamma(z)\, dz$, Fields Inst. Comm. **14**, 71–76.

Gould, H. W. (1972a). *Explicit formulas for Bernoulli numbers*, Amer. Math. Monthly **79**, 44–51.

Gould, H. W. (1972b). *Cominatorial Identities: A Standardized Set of Tables Listing* 500 *Binomial Coefficient Summations*, Morgantown, West Virginia.

Gould, H. W. (1978). *Euler's formula for nth differences of powers*, Amer. Math. Monthly **85**, 450–467.

Gradshteyn, I. S. and Ryzhik, I. M. (1980). *Tables of Integrals, Series, and Products* (Corrected and Enlarged Ed. prepared by A. Jeffrey), Academic Press, New York.

Graf, J. H. (1907). *Berechnung von* $\Gamma(a)\,\Gamma(a+\frac{1}{n})\,\Gamma(a+\frac{2}{n})\cdots\Gamma(a+\frac{n-1}{n}) : \Gamma(na)$, Arch. Math. Phys. **11**, 206–209.

Graham, R. L., Knuth, D. E., and Patashnik, O. (1989). *Concrete Mathematics*, Addison-Wesley Publishing Company, Reading, Massachusetts.

Gram, J.-P. (1895). *Note sur le calcul de la fonction* $\zeta(s)$ *de Riemann*, Overs. Over. D. K. D. Vidensk. Selsk. Forh. 303–308.

Greenberg, R., Marsh, D. C. B., and Danese, A. E. (1967). *A Zeta-function summation*, Amer. Math. Monthly **74**, 80–81.

Grosswald, E. (1970). *Die Werte der Riemannschen Zetafunktion an ungeraden Argumenatellen*, Nachr. Wiss. Göttingen Math.-Phys. Kl. II **1970**, 9–13.

Grosswald, E. (1972). *Comments on some formulae of Ramanujan*, Acta. Arith. **21**, 25–34.

Guinand, A. P. (1947). *Some formulae for the Riemann Zeta-function*, J. London Math. Soc. **22**, 14–18.

Gurland, J. (1956a). *On Wallis' formula*, Amer. Math. Monthly **63**, 643–645.

Gurland, J. (1956b). *An inequality satisfied by the Gamma function*, Skand. Aktuarietidskr. **39**, 171–172.

Hadamard, J. (1909). *Sur une propriété de la fonction* $\zeta(s)$, Bull. Soc. Math. France **39**, 59–60.

Hansen, E. R. (1975). *A Table of Series and Products*, Prentice-Hall, Englewood Cliffs, New Jersey.

Hardy, G. H. (1901/1902). *A new proof of Kummer's series for* $\log \Gamma(a)$, Messenger Math. **31**, 31–33.

Hardy, G. H. (1912). *Note on Dr. Vacca's series for* γ, Quart. J. Pure Appl. Math. **43**, 215–216.

Hardy, G. H. (1949). *Divergent Series*, Clarendon (Oxford University) Press, Oxford, London, and New York.

Haruki, H. (1986). *A new characterization of Euler's Gamma function by a functional equation*, Aequationes Math. **31**, 173–183.

Heilbronn, H. (1934). *On the class-number in imaginary quadratic fields*, Quart. J. Math. Oxford Ser. **5**, 150–160.

Heilbronn, H. (1936). *On real characters*, Acta Arith. **2**, 212–213.

Heilbronn, H. and Linfoot, E. H. (1934). *On the imaginary quadratic corpora of class-number one*, Quart. J. Math. Oxford Ser. **5**, 752–755.

Hejhal, D. A. (1975). *The Selberg trace formula for congruence subgroups*, Bull. Amer. Math. Soc. **81**, 752–755.

Hejhal, D. A. (1976). *The Selberg Trace Formula for* $\mathbf{PSL}(2,\mathbb{R})$, Vol. **1**, Lecture Notes in Mathematics **548**, Springer-Verlag, Berlin, Heidelberg, and New York.

Hejhal, D. A. (1983). *The Selberg Trace Formula for* $\mathbf{PSL}(2,\mathbb{R})$, Vol. **2**, Lecture Notes in Mathematics **1001**, Springer-Verlag, Berlin, Heidelberg, New York, and Tokyo.

Henrici, P. (1986). *Applied and Computational Complex Analysis*, John Wiley and Sons, New York, Chichester, Brisbane, Toronto, and Singapore.

Hervé, M. (1954). *Itération des transformations analytiques dans le bicercle-unité*, Ann. Sci. École Norm. Sup. (3) **71**, 1–28.

Hjortnaes, M. M. (1954). *Overføring av rekken* $\Sigma_{k=1}^{\infty}(1/k^3)$ *til et bestemt integral*, in *Proceedings of the Twelfth Scandanavian Mathematical Congress* (Lund; August 10–15, 1953), pp. 211–213, Scandinavian Mathematical Society, Lund.

Hoffman, M. E. (1992). *Multiple harmonic series*, Pacific J. Math. **152**, 275–290.

Hoffman, M. E. (1997). *The algebra of multiple harmonic series*, J. Algebra **194**, 477–495.

Hoffman, M. E. (1999). *Derivative polynomials, Euler polynomials, and associated integer sequences*, Electronic J. Combin. **6**, 13 pp. (electronic).

Hölder, O. (1886). *Über eine Transcendente Funktion*, Göttingen, Dieterichsche Verlags-Buchhandlung, 514–522.

Hölder, O. (1928). *Über eine von Abel untersuchte Transzendente und eine merkwürdige Funktionalbeziehung*, Berichte über die Verhandlungen der Sacchsischen Akademie der Wissenschaften zu Leipzig. Mathematisch-Physische Klasse **80**, 312–325.

Holme, F. (1970). *En enkel beregning av* $\Sigma_{k=1}^{\infty}\frac{1}{k^2}$, Nordisk Mat. Tidskr. **18**, 91–92.

Hovstad, R. M. (1972). *The series* $\Sigma_{k=1}^{\infty}\frac{1}{k^{2p}}$*, the area of the unit circle and Leibniz' formula*, Nordisk Mat. Tidskr. **20**, 92–98.

Humbert, P. (1921a). *Sur les polynômes hypergéométriques*, C.R. Acad. Sci. Paris **172**, 1282–1283.

Humbert, P. (1921b). *Formule de multiplication pour la fonction de Kummer*, $\Phi(\alpha, \gamma, x)$, C. R. Acad. Sci. Paris **173**, 217–219.

Hurwitz, A. (1882). *Einige eigenschaften der Dirichlet'schen funktionen* $F(s) = \Sigma\,(D/n)\cdot 1/n^s$, *die bei der bestimmung der classenanzahlen binärer quadratischer formen auftreten*, Zeitschr. Math. Phys. **27**, 86–101.

Hurwitz, A. (1899). *Über die entwickelungscoefficienten der lemniscatischen functionen*, Math. Ann. **51**, 196–226.

Hutchinson, J. I. (1925). *On the roots of the Riemann Zeta function*, Trans. Amer. Math. Soc. **27**, 49–60.

Hutchinson, J. I. (1929). *Properties of functions represented by the Dirichlet series* $\Sigma\,(a\nu + b)^{-s}$, *or by linear combinations of such series*, Trans. Amer. Math. Soc. **31**, 322–344.

Ingham, A. E. (1932). *The Distribution of Prime Numbers*, Cambridge University Press, Cambridge, London, and New York.

Ireland, K. and Rosen, M. (1982). *A Classical Introduction to Modern Number Theory*, Springer-Verlag, New York.

Ishibashi, M. (2000). *Multiple cotangent and generalized Eta functions*, Ramanujan J. **4**, 221–229.

Ismail, M. E.-H., Lorch, L., and Muldoon, M. (1986). *Completely monotonic functions associated with the Gamma function and its q-analogues*, J. Math. Anal. Appl. **116**, 1–9.

Ivić, A. (1985). *The Riemann Zeta-Function*, John Wiley and Sons, New York.

Jacobi, C. G. J. (1834). *De usu legitimo formulae summatoriae Maclaurinianae*, J. Reine Angew. Math. **12**, 263–272.

Janos, W. and Srivastava, H. M. (1999). *Solution to Problem* 2.98, Univ. Beograd. Publ. Elektrotehn. Fak. Ser. Mat. **10**, 121–122.

Jensen, J. L. W. V. (1887). *Sur la fonction $\zeta(s)$ de Riemann*, C.R. Acad. Sci. Paris **104**, 1156–1159.

Jensen, J. L. W. V. (1906). *Sur les fonctions convexes et les inégalités entre les valeurs moyennes*, Acta Math. **30**, 175–193.

John, F. (1939). *Special solutions of certain difference equations*, Acta Math. **71**, 175–189.

Johnson, W. W. (1905/1906). *Note on the numerical transcendents S_n and $s_n = S_n - 1$*, Bull. Amer. Math. Soc. **12**, 477–482.

Jolley, L. B. W. (1961). *Summation of Series*, Second Ed., Dover, New York.

Jordan, C. (1965). *Calculus of Finite Differences*, Third Ed., Chelsea Publishing Company, New York.

Jordan, P. F. (1973). *Infinite sums of Psi functions*, Bull. Amer. Math. Soc. **79**, 681–683.

Kac, M. (1966). *Can one hear the shape of a drum?*, Amer. Math. Monthly **73**, 1–23.

Kahane, J.-P. and Mandelbrojt, S. (1958). *L'équation fonctionnelle de Riemann et la formule sommatoire de Poisson*, Ann. Sci. École Norm. Sup. (3) **75**, 57–80.

Kairies, H.-H. (1969). *Zur axiomatischen charakterisierung der Gammafunktion*, J. Reine Angew. Math. **236**, 103–111.

Kairies, H.-H. (1970). *Über die logarithmische der Gammafunktion*, Math. Ann. **184**, 157–162.

Kairies, H.-H. and Muldoon, M. E. (1982). *Some characterizations of q-factorial functions*, Aequationes Math. **25**, 67–76.

Kalla, S. and Villalobos, A. (1991). *Transformadas de Laplace y Mellin en teoria de numeros*, Math. Notae **35**, 13–23.

Kaneko, M. (1995). *A recurrence formula for the Bernoulli numbers*, Proc. Japan Acad. Ser. A Math. Sci **71**, 192–193.

Kanemitsu, S. (1978). *On some sums involving Farey fractions*, Math. J. Okayama Univ. **20**, 101–113.

Kanemitsu, S., Kumagai, H., and Yoshimoto, M. (2000). *Sums involving the Hurwitz Zeta function*, Ramanujan J. **5**(2001).

Kanemitsu, S., Katsurada, M., and Yoshimoto, M. (2000). *On the Hurwitz-Lerch Zeta-function*, Aequationes Math. **59**, 1–19.

Karatsuba, A. A. and Voronin, S. M. (1992). *The Riemann Zeta-Function* (Translated from the Russian by Neal Koblitz), de Gruyter Expositions in Mathematics, Vol. **5**, Walter de Gruyter and Company, Berlin.

Karatsuba, E. A. (1995). *Fast computation of the Riemann Zeta function $\zeta(s)$ for integer values of s*, Problemy Peredachi Informatsii **31**(4), 69–80 (in Russian); English translation in Problems Inform. Transmission **31**, 353-362.

Karatsuba, E. A. (1996). *On the fast calculation of the Riemann Zeta function for integer argument*, Dokl. Akad. Nauk **349**, 463 (in Russian).

Karatsuba, E.A. (1998). *Fast computation of the values of the Hurwitz Zeta function and Dirichlet L-series*, Problemy Peredachi Informatsii **34**(4), 62–75 (in Russian); English translation in Problems Inform. Transmission **34**, 342-353.

Katayama, K. (1973). *On Ramanujan's formula for values of Riemann Zeta-function at positive odd integers*, Acta Arith. **22**, 149–155.

Katsurada, M. (1996). *Power series with the Riemann Zeta-function in the coefficients*, Proc. Japan Acad. Ser. A Math. Sci. **72**, 61–63.

Katsurada, M. (1999). *Rapidly convergent series representations for* $\zeta(2n+1)$ *and their* χ*-analogue*, Acta Arith. **40**, 79–89.

Kazarinoff, D. K. (1956). *On Wallis' formula*, Edinburgh Math. Notes **40**, 19–21.

Kečkić, J. D. and Vasić, P. M. (1971). *Some inequalities for the Gamma function*, Publ. Inst. Math. (Beograd) (N. S.) **11**(**25**), 107–114.

Kemp, W. (1973). *On Gamma function inequalities*, Skand. Aktuarietidskr. **56**, 65–69.

Kenter, F. K. (1999). *A matrix representation for Euler's constant,* γ, Amer. Math. Monthly **106**, 452–454.

Kershaw, D. (1983). *Some extensions of W. Gautschi's inequalties for the Gamma function*, Math. Comput. **41**, 607–611.

Khintchine, A. (1934). *Metrische Kettenbruchprobleme*, Compositio Math. **1**, 361–382.

Kimberling, C. H. (1973). *Two-dimensional complete monotonicity with diagonalization*, Amer. Math. Monthly **80**, 789–791.

Kimberling, C. H. (1974). *A probabilistic interpretation of complete monotonicity*, Aequationes Math. **10**, 152–164.

Khintchine, A. (1964). *Continued Fractions*, Third Ed., University of Chicago Press, Chicago.

Kinkelin, V. H. (1860). *Über eine mit der Gamma Funktion verwandte Transcendente und deren Anwendung auf die Integralrechnung*, J. Reine Angew. Math. **57**, 122–158.

Klambauer, G. (1975). *Mathematical Analysis*, Marcel Dekker, New York.

Klamkin, M. S. (1951). *A summation problem: Advanced Problem* 4431, Amer. Math. Monthly **58**, 195; *ibid.* **59**(1952), 471–472.

Klamkin, M. S. (1955). *Another summation*, Amer. Math. Monthly **62**, 129–130.

Klusch, D. (1987). *Asymptotic equalities for the Lipschitz-Lerch Zeta-function*, Arch. Math. (Basel) **49**, 38–43.

Klusch, D. (1989). *A hybrid version of a theorem of Atkinson*, Rev. Roumaine Math. Pures Appl. **34**, 721–728.

Klusch, D. (1992). *On the Taylor expansion of the Lerch Zeta-Function*, J. Math. Anal. Appl. **170**, 513–523.

Kluyver, J. C. (1927). *On certain series of Mr. Hardy*, Quart. J. Pure Appl. Math. **50**, 185–192.

Knopp, K. (1951). *Theory and Application of Infinite Series*, Second English Ed. (Translated from the Second German Ed. and Revised in accordance with the Fourth German Ed. by R. C. H. Young), Hafner Publishing Company, New York.

Knopp, M. I. (1970). *Modular Functions in Analytic Number Theory*, Markham Publishing Company, Chicago.

Knuth, D. E. (1962). *Euler's constant to* 1271 *places*, Math. Comput. **16**, 275–281.

Knuth, D. E. (1976). *Evaluation of Porter's constant*, Comput. Math. Appl. **2**, 137–139.

Knuth, D. E. (1992). *Two notes on notation*, Amer. Math. Monthly **99**, 403–422.

Koblitz, N. (1977). *p-Adic Numbers, p-Adic Analysis, and Zeta-Functions*, Graduate Texts in Mathematics, Vol. **58**, Springer-Verlag, New York, Heidelberg, and Berlin.

Koblitz, N. (1982). *On Carlitz's q-Bernoulli numbers*, J. Number Theory **14**, 332–339.

Koecher, M. (1979/1980). *Letter*, Math. Intelligencer **2**, 62–64 (in German).

Koecher, M. (1987). *Klassische Elementare Analysis*, Birkhäuser Verlag, Basel.

Kölbig, K. S. (1986). *Nielsen's generalized Polylogarithms*, SIAM J. Math. Anal. **17**, 1232–1258.

Koshliakov, N. S. (1928/1929). *Application of the theory of sum-formulae to the investigation of a class of one-valued analytical functions in the theory of numbers*, Messenger Math. **58**, 1–23.

Krull, W. (1948). *Bemerkungen zur differenzengleichung* $g(x+1) - g(x) = \varphi(x)$, Math. Nachr. **1**, 365–376.

Krull, W. (1949) . *Bemerkungen zur differenzengleichung* $g(x+1) - g(x) = \varphi(x)$. II, Math. Nachr. **2**, 251–262.

Kubota, T. (1973). *Elementary Theory of Eisenstein Series*, Kodansha Limited, Tokyo; John Wiley and Sons, New York, London, Sydney, and Toronto.

Kuczma, M. (1968). *Functional Equations in a Single Variable*, PWN-Polish Scientific Publishers, Warszawa.

Kumagai, H. (1999). *The determinant of the Laplacian on the n-sphere*, Acta Arith. **91**, 199-208.

Kummer, E. E. (1836). *Über die hypergeometrische Reihe*, J. Reine Angew. Math. **15**, 39–83 and 127–172.

Kummer, E. E. (1840). *Über die transcendenten, welche aus wiederholten integrationen rationaler formeln entstehen*, J. Reine Angew. Math. **21**, 74–90.

Kummer, E. E. (1847). *Beitrag zur theorie der funktion* $\Gamma(x) = \int_0^\infty e^{-v}\, v^{x-1}\, dv$, J. Reine Angew. Math. **35**, 1–4.

Kurokawa, N. (1991). *Multiple sine functions and Selberg Zeta functions*, Proc. Japan Acad. Ser. A Math. Sci. **67**, 61–64.

Kurokawa, N. (1992). *Gamma factors and Plancherel measures*, Proc. Japan Acad. Ser. A Math. Sci. **68**, 256–260.

Labelle, G. and Laforest, L. (1995). *Combinatorial variations on multidimensional quadtrees*, J. Combin. Theory Ser. A **69**, 1–16.

Laforgia, A. (1984). *Further inequalities for the Gamma function*, Math. Comput. **42**, 597–600.

Lammel, E. (1966). *Ein beweis, dass die Riemannsche Zetafunktion* $\zeta(s)$ *in* $|s-1| \leq 1$ *keine nullstelle besitzt*, Univ. Nac. Tucumán Rev. Ser. A **16**, 209–217.

Lan, Y. (1999). *A limit formula for* $\zeta(2k+1)$, J. Number Theory **78**, 271–286.

Landau, E. (1953). *Handbuch der Lehre von der Verteilung der Primzahlen*, Second Ed., Chelsea Publishing Company, New York.

Lange, L. J. (1999). *An elegant continued fraction for* π, Amer. Math. Monthly **106**, 456–458.

Lavoie, J. L., Grondin, F., Rathie, A. K., and Arora, K. (1994). *Generalizations of Dixon's theorem on the sum of a* ${}_3F_2$, Math. Comput. **62**, 267–276.

Lavoie, J. L., Grondin, F., and Rathie, A. K. (1996). *Generalizations of Whipple's theorem on the sum of a* ${}_3F_2$, J. Comput. Appl. Math. **72**, 293–300.

Lazarević, I. B. and Lupaş, A. (1974). *Functional equations for Wallis and Gamma functions*, Univ. Beograd. Publ. Elektrotehn. Fak. Ser. Mat. Fiz. No. **461–497**, 245–251.

Le, T. Q. T. and Murakami, J. (1995). *Kontsevich's integral for the Homfly polynomial and relations between values of multiple Zeta functions*, Topology Appl. **62**, 193–206.

Lee, J. (1999). *Integrals of Bernoulli polynomials and series of Zeta function*, Comm. Korean Math. Soc. **14**, 707–716.

Legendre, A. M. (1826). *Traité des Fonctions Elliptiques et des Intégrales Eulériennes*, Vol. **2**, Huzard-Courcier, Paris.

Lehmer, D. H. (1933). *Factorization of certain cyclotomic functions*, Ann. of Math. (2) **34**, 461–479.

Lehmer, D. H. (1939). *Note on an absolute constant of Khintchine*, Amer. Math. Monthly **46**, 148–152.

Lehmer, D. H. (1940). *On the maxima and minima of Bernoulli polynomials*, Amer. Math. Monthly **47**, 533–538.

Lehmer, D. H. (1975). *Euler constants for arithmetical progressions*, Acta Arith. **27**, 125–142.

Lehmer, D. H. (1985). *Interesting series involving the central binomial coefficient*, Amer. Math. Monthly **92**, 449–457.

Lehner, J. and Newman, M. (1969). *Sums involving Farey fractions*, Acta Arith. **15**, 181–187.

Lenstra, A. K., Lenstra, Jr., H. W., and Lovász, L. (1982). *Factoring polynomials with rational coefficients*, Math. Ann. **261**, 515–534.

Lerch, M. (1887). *Note sur la fonction* $\mathcal{R}(w,x,s) = \Sigma_{n=0}^{\infty}(e^{2\pi inx}/(n+w)^s)$, Acta Math. **11**, 19–24.

Lerch, M. (1903). *Démonstration élémentaire de la formule*, Enseign. Math. **5**, 450–453.

Leshchiner, D. (1981). *Some new identities for* $\zeta(k)$, J. Number Theory **13**, 355–362.

LeVeque, W. J. (1956). *Topics in Number Theory*, Vol. **II**, Addison-Wesley Publishing Company, Reading (Massachusetts) and London.

Lévy, P. (1936). *Sur le développement en fraction continue d'un nombre choisi au hasard*, Compositio Math. **3**, 286–303.

Lewin, L. (1958). *Dilogarithms and Associated Functions*, Macdonald, London.

Lewin, L. (1981). *Polylogarithms and Associated Functions*, Elsevier (North-Holland), New York, London, and Amsterdam.

Lewin, L. (1982). *The dilogarithm in algebraic fields*, J. Austral. Math. Soc. Ser. A **33**, 302–330.

Lewin, L. (1984). *The inner structure of the dilogarithm in algebraic fields*, J. Number Theory **19**, 345–373.

Lewin, L. (1986). *The order-independence of the polylogarithmic ladder structure-implications for a new category of functional equations*, Aequationes Math. **30**, 1–20.

Lewin, L. (Editor) (1991). *Structural Properties of Polylogarithms* (Mathematical Surveys and Monographs, Vol. **37**), American Mathematical Society, Providence, Rhode Island.

Lewin, L. and Rost, E. (1986). *Polylogarithmic functional equations: A new category of results developed with the help of computer algebra (MACSYMA)*, Aequationes Math. **31**, 223–242.

Liang, J. J. Y. and Todd, J. (1972). *The Stieltjes constants*, J. Res. Nat. Bur. Standards Sect. B Math. Sci. **76**, 161–178.

Lichtenbaum, P. (1931). *Über die* $\chi(s) = \Sigma_{k=0}^{\infty}(-1)^k(2k+1)^{-s}$, Math. Zeitschr. **33**, 641–647.

Lindelöf, E. (1905). *Le Calcul des Résidus*, Gauthier-Villars, Paris.

Lipschitz, R. (1857). *Untersuchung einer aus vier Elementen gebildeten Reihe*, J. Reine Angew. Math. **54**, 313–328.

Lipschitz, R. (1889). *Untersuchung der Eigenschaften einer Gattung von unendlichen Reihen*, J. Reine Angew. Math. **105**, 127–156.

Lorch, L. (1984). *Inequalities for ultraspherical polynomials and the Gamma function*, J. Approx. Theory **40**, 115–120.

Lorch, L., Muldoon, M. E., and Szegö, P. (1970). *Higher monotonicity properties of certain Sturm-Liouville functions*. III, Canad. J. Math. **22**, 1238–1265.

Loxton, J. H. (1984). *Special values of the Dilogarithm function*, Acta Arith. **43**, 155–166.

Lucht, L. G. (1990). *Mittelwertungleichungen für Lösungen gewisser differenzengleichungen*, Aequationes Math. **39**, 204–209.

Luke, Y. L. (1969). *The Special Functions and Their Approximations*, Vols. **I** and **II**, Academic Press, New York and London.

Magnus, W., Oberhettinger, F., and Soni, R. P. (1966). *Formulas and Theorems for the Special Functions of Mathematical Physics*, Third Enlarged Edition, Die Grundlehren der Mathematischen Wissenschaften in Einzeldarstellungen mit besonderer Berücksichtigung der Anwendungsgebiete, Bd. **52**, Springer-Verlag, New York.

Maligranda, L., Pečarić, J. E., and Persson, L. E. (1995). *Stolarsky's inequality with general weights*, Proc. Amer. Math. Soc. **123**, 2113–2118.

Malmstén, C. J. (1847). *Sur la formule* $hu_x = \Delta u_x - \frac{h}{2}\Delta u_x + \frac{B_1 h^2}{1\cdot 2}\cdot \Delta u_x'' - \frac{B_2 h^4}{1\cdots 4}\cdot \Delta u_x^{iv} + etc$, J. Reine Angew. Math. **35**, 55–82.

Mann, H. B. (1962). *An inequality suggested by the theory of statistical inference*, Illinois J. Math. **6**, 131–136.

Markett, C. (1994). *Triple sums and the Riemann Zeta function*, J. Number Theory **48**, 113–132.

Matsumoto, K. (1998). *Asymptotic series for double Zeta, double Gamma, and Hecke L-functions*, Math. Proc. Cambridge Philos. Soc. **123**, 385–405.

Matsumoto, K. (2001). *Corrigendum and Addemdum to "Asymptotic series for double Zeta, double Gamma, and Hecke L-functions"*, Math. Proc. Cambridge Philos. Soc. (to appear).

Matsuoka, Y. (1960). *Asymptotic formula for Vallée Poussin's singular integrals*, Sci. Rep. Kagoshima Univ. **9**, 25–34.

Matsuoka, Y. (1982). *On the values of a certain Dirichlet series at rational integers*, Tokyo J. Math. **5**, 399–403.

Mayer, A. E. (1939). *Konvexe lösung der funktionalgleichung*, Acta Math. **70**, 57–62.

McKean, H. P. (1972). *Selberg's trace formula as applied to a compact Riemann surface*, Comm. Pure Appl. Math. **25**, 225–246.

McKean, H. P. and Singer, I. M. (1967). *Curvature and the Eigenvalues of the Laplacian*, J. Differential Geom. **1**, 43–69.

Mellin, Hj. (1883). *Über die transcendente funktion* $Q(x) = \Gamma(x) - P(x)$, Acta Math. **2**, 261–295.

Mellin, Hj. (1883/1884). *Eine verallgemeinerung der gleichung* $\Gamma(1+x)\,\Gamma(1-x) = \pi x/\sin \pi x$, Acta Math. **3**, 102–104.

Mellin, Hj. (1886a). *Zur theorie der Gammafunktion*, Acta Math. **8**, 37–80.

Mellin, Hj. (1886b). *Om en ny klass af transcendenta funktioner, hvilka äro nära beslägtade med Gammafunktionen*, Acta Soc. Sci. Fenn. **14**, 355–385.

Melzak, Z. A. (1973). *Companion to Concrete Mathematics*, Vol. **I**: *Mathemtical Techniques and Various Applications*, John Wiley and Sons, New York, London, Sydney, and Toronto.

Melzak, Z. A. (1976). *Companion to Concrete Mathematics*, Vol. **II**: *Mathematical Ideas, Modeling and Applications*, John Wiley and Sons, New York, London, Sydney, and Toronto.

Menon, P. K. (1961). *Some series involving the Zeta function*, Math. Student **29**, 77–80.

Mikolás, M. (1950). *Farey series and their connection with the prime number problem.* I, Acta Sci. Math. (Szeged) **13**, 93–117.

Mikolás, M. (1957). *Integral formulae of arithmetical characteristics relating to the Zeta-function of Hurwitz*, Publ. Math. Debrecen **5**, 44–53.

Miller, J. and Adamchik, V. S. (1998). *Derivatives of the Hurwitz Zeta function for rational arguments*, J. Comput. Appl. Math. **100**, 201–206.

Minakshisundaram, S. (1949). *A generalization of Epstein Zeta functions*, Canad. J. Math. **1**, 320–327.

Minakshisundaram, S. and Pleijel, A. (1949). *Some properties of the eigenfunctions of the Laplace-operator on Riemann manifolds*, Canad. J. Math. **1**, 242–256.

Minc, H. (1963). *Upper bounds for permanents of* (0, 1)*-matrices*, Bull. Amer. Math. Soc. **69**, 789–791.

Minc, H. and Sathre, L. (1964/1965). *Some inequalities involving* $(r!)^{1/r}$, Proç. Edinburgh Math. Soc. (2) **14**, 41–46.

Mitchell, K. (1949). *Tables of the function* $\int_0^z \frac{-\log|1-y|}{y}\,dy$, *with an account of some properties of this and related functions*, Philos. Mag. **40**, 351–368.

Moak, D. S. (1980). *The q-Gamma function for* $q > 1$, Aequationes Math. **20**, 278–285.

Mordell, L. J. (1958a). *On the evaluation of some multiple series*, J. London Math. Soc. **33**, 368–371.

Mordell, L. J. (1958b). *On the evaluation of some multiple series*, J. London Math. Soc. **33**, 371–375.

Motohashi, Y. (1997). *Spectral Theory of the Riemann Zeta-Function*, Cambridge University Press, Cambridge, New York, and Melbourne.

Muldoon, M. E. (1972). *Some characterizations of the Gamma function involving the notion of complete monotonicity*, Aequationes Math. **8**, 212–215.

Muldoon, M. E. (1978). *Some monotonicity properties and characterizations of the Gamma function*, Aequationes Math. **18**, 54–63.

Nash, C. and O'Connor, D. (1993). *Ray-Singer torsion, topological field theories and the Riemann Zeta function at* $s = 3$, in *Low-Dimensional Topology and Quantum Field Theory* (Proceedings of a NATO Advanced Research Workshop held at the Isaac Newton Institute at Cambridge, U. K.; September 6–12, 1992) (H. Osborn, Editor), pp. 279–288, Plenum Press, New York and London.

Nash, C. and O'Connor, D. (1995). *Determinants of Laplacians, the Ray-Singer torsion on lens spaces and the Riemann Zeta function*, J. Math. Phys. **36**, 1462–1505.

Neville, E. H. (1951). *A trigonometrical inequality*, Proc. Cambridge Philos. Soc. **47**, 629–632.

Nielsen, N. (1923). *Traité Élémentaire des Nombres de Bernoulli*, Gauthier-Villars, Paris.

Nielsen, N. (1965). *Die Gammafunktion*, Chelsea Publishing Company, Bronx and New York.

Nörlund, N. E. (1920). *Mémoire sur les ploynômes de Bernoulli*, Acta Math. **43**, 121–196.

Odlyzko, A. M. and Schönhage, A. (1988). *Fast algorithms for multiple evaluations of the Riemann Zeta function*, Trans. Amer. Math. Soc. **309**, 797–809.

Ojha, P. K. and Singh, I. N. (1999). *A discussion on two methods for finding the value of the Riemann Zeta function* $\zeta(s)$ *when* $s = 2$, Math. Ed. (Siwan) **33**, 24–28.

Olkin, I. (1959). *An inequality satisfied by the Gamma function*, Skand. Aktuarietidskr. **42**, 37–39.

Olver, F. W. J. (1974). *Asymptotics and Special Functions*, Academic Press, New York.

Osgood, B., Phillips, R., and Sarnak, P. (1988). *Extremals of determinants of Laplacians*, J. Funct. Anal. **80**, 148–211.

Ostrowski, A. M. (1960). *On the zeros of Bernoulli polynomials of even order*, Enseign. Math. **6**, 27–47.

Ostrowski, A. (1969a). *Note on Poisson's treatment of the Euler-Maclaurin formula*, Comment. Math. Helv. **44**, 202–206.

Ostrowski, A. (1969b). *On the remainder term of the Euler-Maclaurin formula*, J. Reine Angew. Math. **239/240**, 268–286.

Papadimitriou, I. (1973). *A simple proof of the formula* $\Sigma_{k=1}^{\infty} k^{-2} = \pi^2/6$, Amer. Math. Monthly **80**, 424–425.

Patin, J. M. (1989). *A very short proof of Stirling's formula*, Amer. Math. Monthly **96**, 41–42.

Pitteway, M. L. V. and Castle, C. M. A. (1988). *On the subtractive version of Euclid's algorithm*, Bull. Inst. Math. Appl. **24**, 17–21.

Pochhammer, L. (1892). *Bemerkungen über das integral* $\Gamma(a)$, Math. Ann. **41**, 157–166.

Popov, B. S. (1998). *Expressions of Laguerre polynomials through Bernoulli polynomials*, Mat. Bilten **22**(**48**), 15–18.

Post, E. L. (1918/1919). *The generalized Gamma functions*, Ann. Math. (2) **20**, 202–217.

Powell, E. O. (1943). *An integral related to the radiation integrals*, Philos. Mag. **34**, 600–607.

Powell, E. O. (1952). *A table of the generalized Riemann Zeta function in a particular case*, Quart. J. Mech. Appl. Math. **5**, 116–123.

Prudnikov, A. P. (1996). *On the Euler problem of summation of harmonic series*, Integral Transform. Spec. Funct. Informat. Bull. **1**(1), 5–7. (In Russian).

Prym, F. E. (1877). *Zur theorie der Gammafunktion*, J. Reine Angew. Math. **82**, 165–172..

Quenouille, M. H. (1947-1950). *The evaluation of probabilities in a normal multivariate distribution, with special reference to the correlation ratio*, Proc. Edinburgh Math. Soc. (2) **8**, 95–100.

Quine, J. R. and Choi, J. (1996). *Zeta regularized products and functional determinants on spheres*, Rocky Mountain J. Math. **26**, 719–729.

Quine, J. R., Heydari, S. H., and Song, R. Y. (1993). *Zeta regularized products*, Trans. Amer. Math. Soc. **338**, 213–231.

Raina, R. K. and Nahar, T. S. (2000). *A note on a certain class of functions related to Hurwitz Zeta function and Lambert transform*, Tamkang J. Math. **31**, 49–56.

Raina, R. K. and Srivastava, H. M. (1995). *Certain results associated with the generalized Riemann Zeta functions*, Rev. Técn. Fac. Ingr. Univ. Zulia **18**, 301–304.

Rainville, E. D. (1960). *Special Functions*, Macmillan Company, New York; Reprinted by Chelsea Publishing Company, Bronx, New York, 1971.

Ramachandra, K. (1964). *Some applications of Kronecker's limit formulas*, Ann. of Math. (2) **80**, 104–148.

Ramanujan, S. (1916/1917). *A series for Euler's constant* γ, Messenger Math. **46**, 73–80.

Ramanujan, S. (1957). *Notebooks of Srinivasa Ramanujan*, Vols. **I** and **II**, Tata Institute of Fundamental Research, Bombay.

Ramaswami, V. (1934). *Notes on Riemann's* ζ*-function*, J. London Math. Soc. **9**, 165–169.

Rasch, G. (1931). *Notes on the Gamma-function*, Ann. Math. **32**, 591–599.

Richmond, B. and Merlini, D. (1997). *Srirling numbers for complex arguments*, SIAM J. Discrete Math. **10**, 73–82.

Richmond, B. and Szekeres, G. (1981). *Some formulas related to Dilogarithms, the Zeta function and the Andrews-Gordon identities*, J. Austral. Math. Soc. Ser. A **31**, 362–373.

Riemann, B. (1859). *Über die Anzahl der Primzahlen unter einer gegebenen Grösse*, Monatsber. Akad. Berlin, pp. 671–680.

Riordan, J. (1968). *Combinatorial Identities*, John Wiley and Sons, New York, London, and Sydney.

Riordan, J. and Stein, P. R. (1973). *Proof of a conjecture on Genocchi numbers*, Discrete Math. **5**, 381–388.

Robbins, H. (1955). *A remark on Stirling's formula*, Amer. Math. Monthly **62**, 26–29.

Rogers, L. J. (1907). *On function sum theorems connected with the series* $\Sigma_{n=1}^{\infty} \frac{x^n}{x^2}$, Proc. London Math. Soc. (2) **4**, 169–189.

Rosenbaum, R. A. (1950). *Sub-additive functions*, Duke Math. J. **17**, 227–247.

Rosser, J. B. (1950). *Real roots of real Dirichlet L-series*, J. Res. Nat. Bur. Standards **45**, 505–514.

Rosser, J. B. (1951). *Transformations to speed the convergence of series*, J. Res. Nat. Bur. Standards **46**, 56–64.

Rowe, C. H. (1931). *A proof of the asymptotic series for* $\log \Gamma(z)$ *and* $\log \Gamma(z+a)$, Ann. of Math. (2) **32**, 10–16.

Ruben, H. (1967). *Variance bounds and orthogonal expansions in Hilbert space with an application to inequalities for Gamma functions and* π, J. Reine Angew. Math. **225**, 147–153.

Ruijsenaars, S. N. M. (2000). *On Barnes' multiple Zeta and Gamma functions*, Adv. in Math. **156**, 107–132..

Rutledge, G. and Douglass, R. D. (1934). $\int_0^1 \frac{\log u}{u} \log^2(1+u)\,du$ *and related definite integrals*, Amer. Math. Monthly **41**, 29–36.

Sandham, H. F. (1949a). *Advanced Problem* 4353, Amer. Math. Monthly **56**, 414.

Sandham, H. F. (1949b). *A logarithmic transcendent*, J. London Math. Soc. **24**, 83–91.

Sarnak, P. (1987). *Determinants of Laplacians*, Comm. Math. Phys. **110**, 113–120.

Satoh, J. (1989). *q-Analogue of Riemann's* ζ*-function*, J. Number Theory **31**, 346–362.

Schmidt, H. (1958). *Eine bemerkung zum aufbau der lehre von der* Γ*-funktion*, Arch. Math. (Basel) **9**, 297–299.

Schwatt, I. J. (1915). *Note on the expansion of a function*, Philos. Mag. **29**, 65–67.

Schwatt, I. J. (1931). *The expansion of* $\tan^n x$ *by Maclaurin theorem*, Tôhoku Math. J. **33**, 150–152.

Selberg, A. (1956). *Harmonic analysis and discontinuous groups in weakly symmetric Riemannian spaces with applications to Dirichlet series*, J. Indian Math. Soc. (N. S.) **20**, 47–87.

Selberg, A. and Chowla, S. (1967). *On Epstein's Zeta-function*, J. Reine Angew. Math. **227**, 86–110.

Sellliah, J. B. (1976). *An inequality satisfied by the Gamma function*, Canad. Math. Bull. **19**, 85–87.

Seo, T. Y., Choi, J., Gang, J. S., and Ok, B. M. (1993). *Note on the analytic continuation of the multiple Hurwitz Zeta function* $\zeta_n(s, a)$, Kyungpook Math. J. **33**, 127–131.

Shafer, R. E. and Lossers, O. P. (1969). *Euler's constant*, Amer. Math. Monthly **76**, 1077–1079.

Shail, R. (2000). *A class of infinite sums and integrals*, Math. Comput. (Article electronically published on March 1), 11 pp. (electronic).

Shallit, J. D. and Zikan, K. (1986). *A theorem of Goldbach*, Amer. Math. Monthly **93**, 402–403.

Sharpe, H. J. (1883/1884). *On the Gamma functions of a complex*, Messenger Math. **13**, 111–112.

Shen, L.-C. (1995). *Remarks on some integrals and series involving the Stirling numbers and* $\zeta(n)$, Trans. Amer. Math. Soc. **347**, 1391–1399.

Shintani, T. (1980). *A proof of the classical Kronecker limit formula*, Tokyo J. Math. **3**, 191-199.

Singh, R. J. and Verma, D. P. (1983). *Some series involving Riemann Zeta function*, Yokohama Math. J. **31**, 1–4.

Sitaramachandrarao, R. (1987). *A formula of S. Ramanujan*, J. Number Theory **25**, 1–19.

Sitaramachandrarao, R. and Sivaramasarma, A. (1979). *Some identities involving the Riemann Zeta function*, Indian J. Pure Appl. Math. **10**, 602–607.

Sitaramachandrarao, R. and Sivaramasarma, A. (1980). *Two identities due to Ramanujan*, Indian J. Pure Appl. Math. **11**, 1139–1140.

Sitaramachandrarao, R. and Subbarao, M. V. (1984). *Transformation formulae for multiple series*, Pacific J. Math. **113**, 471–479.

Skau, I. S. og Selmer, E. S. (1971). *Noen anvendelser av finn holmes methode for beregning av* $\Sigma_{k=1}^{\infty}\frac{1}{k^2}$, Nordisk Mat. Tidskr. **19**, 120–124.

Slater, L. J. (1951/1952). *Further identities of the Rogers-Ramanujan type*, Proc. London Math. Soc. (2) **54**, 147–167.

Slater, L. J. (1966). *Generalized Hypergeometric Functions*, Cambridge University Press, Cambridge, London, and New York.

Slavić, D. V. (1975). *On inequalities for* $\Gamma(x+1)/\Gamma(x+1/2)$, Univ. Beograd. Publ. Elektrotehn. Fak. Ser. Mat. Fiz. No. **498–541**, 17–20.

Smart, J. R. (1973). *On the values of the Epstein zeta function*, Glasgow Math. J. **14**, 1–12.

Sonin, M. N. (1889). *Sur les termes complémentaires de la Formule sommatoire d'Euler et de celle de Stirling*, Ann. Sci. Ecole Normale Supâerieure **3**, 257–262.

Spanier, J. and Oldham, K. B. (1987). *An Atlas of Functions*, Hemisphere Publishing Corporation (Harper and Row Publishers), Washington, New York, and London; Springer-Verlag, Berlin, Heidelberg, New York, London, Paris, and Tokyo.

Spence, W. (1809). *An Essay on the Theory of the Various Orders of Logarithmic Transcendents; with an Inquiry into Their Applications to the Integral Calculus and the Summation of Series*, John Murray and Archibald Constable and Company, Edinburgh and London.

Spiegel, M. R. (1968). *Mathematical Handbook*, McGraw-Hill Book Company, New York.

Spiess, O. (1945). *Die Summe der reziproken Quadratzahlen*, in *Festschrift zum 60 Geburtstag von Prof. Dr. Andreas Speiser* (L. V. Ahlfors *et al.*, Editors), pp. 66-86, Füssli, Zürich.

Srivastava, H. M. (1972). *A contour integral involving Fox's H-function*, Indian J. Math. **14**, 1–6.

Srivastava, H. M. (1986). *Summation of a class of series involving the Riemann Zeta function*, Rev. Técn. Fac. Ingr. Univ. Zulia **9**, 79–82.

Srivastava, H. M. (1987). *Some infinite series associated with the Riemann Zeta function*, Yokohama Math. J. **35**, 47–50.

Srivastava, H. M. (1988a). *A unified presentation of certain classes of series of the Riemann Zeta function*, Riv. Mat. Univ. Parma (4) **14**, 1–23.

Srivastava, H. M. (1988b). *Sums of certain series of the Riemann Zeta function*, J. Math. Anal. Appl. **134**, 129–140.

Srivastava, H. M. (1990). *A class of infinite series involving binomial and factorial coefficients*, Internat. J. Math. Ed. Sci. Tech. **21**, 75–79.

Srivastava, H. M. (1992). *A simple algorithm for the evaluation of a class of generalized hypergeometric series*, Stud. Appl. Math. **86**, 79–86.

Srivastava, H. M. (1997). *Certain families of rapidly convergent series representations for $\zeta(2n+1)$*, Math. Sci. Res. Hot-Line **1**(6), 1–6 (Research Announcement).

Srivastava, H. M. (1998). *Further series representations for $\zeta(2n+1)$*, Appl. Math. Comput. **97**, 1–15.

Srivastava, H. M. (1999a). *Some rapidly converging series for $\zeta(2n+1)$*, Proc. Amer. Math. Soc. **127**, 385–396.

Srivastava, H. M. (1999b). *A note on the closed-form summation of some trigonometric series*, Kobe J. Math. **16**, 177–182.

Srivastava, H. M. (2000a). *Some simple algorithms for the evaluations and representations of the Riemann Zeta function at positive integer arguments*, J. Math. Anal. Appl. **246**, 331–351.

Srivastava, H. M. (2000b). *Some formulas for the Bernoulli and Euler polynomials at rational arguments*, Math. Proc. Cambridge Philos. Soc. **129**, 77–84.

Srivastava, H. M. (2000c). *Some families of rapidly convergent series representations for the Zeta functions*, Taiwanese J. Math. **4**, 569–598.

Srivastava, H. M., Glasser, M. L., and Adamchik, V. S. (2000). *Some definite integrals associated with the Riemann Zeta function*, Z. Anal. Anwendungen **19**, 831–846.

Srivastava, H. M. and Karlsson, P. W. (1985). *Multiple Gaussian Hypergeometric Series*, Halsted Press (Ellis Horwood Limited, Chichester), John Wiley and Sons, New York, Chichester, Brisbane, and Toronto.

Srivastava, H. M., Lavoie, J.-L., and Tremblay, R. (1983). *A class of addition theorems*, Canad. Math. Bull. **26**, 438–445.

Srivastava, H. M. and Manocha, H. L. (1984). *A Treatise on Generating Functions*, Halsted Press (Ellis Horwood Limited, Chichester), John Wiley and Sons, New York, Chichester, Brisbane, and Toronto.

Srivastava, H. M. and Miller, E. A. (1990). *A simple reducible case of double hypergeometric series involving Catalan's constant and Riemann's ζ-function*, Internat. J. Math. Ed. Sci. Tech. **21**, 375–377.

Srivastava, H. M. and Todorov, P. G. (1988). *An explicit formula for the generalized Bernoulli polynomials*, J. Math. Anal. Appl. **130**, 509–513.

Srivastava, H. M. and Tsumura, H. (2000a). *A certain class of rapidly convergent series representations for $\zeta(2n+1)$*, J. Comput. Appl. Math. **118**, 323–335.

Srivastava, H. M. and Tsumura, H. (2000b). *New rapidly convergent series representations for $\zeta(2n+1)$, $L(2n,\chi)$, and $L(2n+1,\chi)$*, Math. Sci. Res. Hot-Line **4**(7), 17–24 (Research Announcement).

Srivastava, H. M. and Tsumura, H. (2001). *Certain classes of rapidly convergent series representations for $L(2n,\chi)$ and $L(2n+1,\chi)$*, Acta Arith. (to appear).

Stark, E. L. (1969). *Another proof of the formula $\Sigma_{k=1}^{\infty} 1/k^2 = \pi^2/6$*, Amer. Math. Monthly **76**, 552–553.

Stark, E. L. (1974). *The series $\Sigma_{k=1}^{\infty} k^{-s}$, $s = 2, 3, 4, \ldots$, once more*, Math. Mag. **47**, 197–202.

Steffensen, J. F. (1950). *Interpolation*, Second Ed., Chelsea Publishing Company, New York.

Steiner, F. (1987). *On Selberg's Zeta function for compact Riemann surfaces*, Phys. Lett. B **188**, 447–454.

Stieltjes, T. J. (1878). *Een en ander over de integraal* $\int_0^1 l\,\Gamma(x+u)\,du$, Nieuw Arch. Wisk. **2**, 100–104.

Stieltjes, T. J. (1887). *Table des valeurs des sommes* $S_k = \Sigma_1^\infty n^{-k}$, Acta Math. **10**, 299–302.

Stoica, G. (1987). *O Relaţie de recurenţă ın studiul funcţiei Zeta a lui Riemann (A recurrence formula in the study of the Riemann Zeta function)*, Stud. Cerc. Mat. **39**, 261–264.

Stolarsky, K. B. (1991). *From Wythoff's Nim to Chebyshev's inequality*, Amer. Math. Monthly **98**, 889–900.

Subbarao, M. V. and Sitaramachandrarao, R. (1985). *On some infinite series of L. J. Mordell and their analogues*, Pacific J. Math. **119**, 245–255.

Sury, B. (1993). *The value of Bernoulli polynomials at rational arguments*, Bull. London Math. Soc. **25**, 327–329.

Suryanarayana, D. (1974). *Sums of the Riemann Zeta function*, Math. Student **42**, 141–143.

Sweeney, D. W. (1963). *On the computation of Euler's constant*, Math. Comput. **17**, 170–178.

Szekeres, G. (1953). *Some asymptotic formulae in the theory of partitions.* II, Quart. J. Math. Oxford Ser. (2) **4**, 96–111.

Szmidt, J. (1983). *The Selberg trace formula for the Picard group* $\mathbf{SL}(2,\mathbb{Z}[i])$, Acta Arith. **42**, 391–424.

Terras, A. (1976). *Some formulas for the Riemann Zeta function at odd integer argument resulting from Fourier expansions of the Epstein Zeta function*, Acta Arith. **29**, 181–189.

Terras, A. (1985). *Harmonic Analysis on Symmetric Spaces and Applications*, Vol. I, Springer-Verlag, New York.

Titchmarsh, E. C. (1927). *On an inequality satisfied by the Zeta-function of Riemann*, Proc. London Math. Soc. (2) **28**, 70–80.

Titchmarsh, E. C. (1951). *The Theory of the Riemann Zeta-Function*, Clarendon (Oxford University) Press, Oxford, London, and New York; Second Ed. (Revised by D. R. Heath-Brown), 1986.

Titchmarsh, E. C. (1964). *The Zeta-Function of Riemann*, Stechert-Hafner Service Agency, New York and London.

Todorov, P. G. (1985). *Une formule simple explicite des nombres de Bernoulli généralisés*, C. R. Acad. Sci. Paris Sér. I Math. **301**, 665–666.

Todorov, P. G. (1993). *Explicit and recurrence formulas for generalized Euler numbers*, Funct. Approx. **22**, 113–117.

Tornheim, L. (1950). *Harmonic double series*, Amer. J. Math. **72**, 303–314.

Trimble, S. Y., Wells, J., and Wright, F. T. (1989). *Superadditive functions and a statistical application*, SIAM. J. Math. Anal. **20**, 1255–1259.

Tsumura, H. (1987). *On a p-adic interpolation of the generalized Euler numbers and its applications*, Tokyo J. Math. **10**, 281–293.

Tsumura, H. (1990). *On the values of a q-analogue of the p-adic L-functions*, Mem. Fac. Sci. Kyushu Univ. Ser. A **44**, 49–60.

Tsumura, H. (1991). *A note on q-analogues of the Dirichlet series and q-Bernoulli numbers*, J. Number Theory **39**, 251–256.

Tsumura, H. (1994). *On evaluation of the Dirichlet series at positive integers by q-calculation*, J. Number Theory **48**, 383–391.

Underwood, R. S. (1928). *An expression for the summation* $\Sigma_{m=1}^{n} m^p$, Amer. Math. Monthly **35**, 424–428.

Uppuluri, V. R. A. (1966). *On a stronger version of Wallis' formula*, Pacific J. Math. **19**, 183–187.

Vacca, G. (1910). *A new series for the Eulerian constant* $\gamma = \cdot 577\ldots$, Quart. J. Pure Appl. Math. **41**, 363–364.

van der Poorten, A. (1978/1979). *Some wonderful formulae... footnotes to Apéry's proof of the irrationality* $\zeta(3)$, Séminaire Delange-Pisot-Poitou (Théorie des nombres) **29**, 7 pp.

van der Poorten, A. (1979). *Some wonderful formulas... an introduction to Polylogarithms*, Queen's papers in Pure Appl. Math. **54**, 269–286.

van der Poorten, A. (1979). *A proof that Euler missed... Apéry's proof of the irrationality of* $\zeta(3)$, Math. Intelligencer **1**, 195–203.

van Engen, H. (1938). *Concerning Gamma function expansions*, Tôhoku Math. J. **45**, 124–129.

Vardi, I. (1988a). *Integrals, an introduction to analytic number theory*, Amer. Math. Monthly **95**, 308–315.

Vardi, I. (1988b). *Determinants of Laplacians and multiple Gamma functions*, SIAM J. Math. Anal. **19**, 493–507.

Verma, D. P. (1961). *A note on Euler's constant*, Math. Student **29**, 140–141.

Verma, D. P. and Kaur, A. (1983). *Summation of some series involving Riemann Zeta function*, Indian J. Math. **25**, 181–184.

Vignéras, M. -F. (1979). *L'équation fonctionnelle de la fonction zêta de Selberg du groupe moudulaire* $PSL(2, Z)$, in *Journées Arithmétiques de Luminy* (Colloq. Internat. CNRS, Centre Univ. Luminy, Luminy, 1978), pp. 235–249, Astérisque **61**, Soc. Math. France, Paris.

Voros, A. (1986). *The Hadamard factorization of the Selberg Zeta function on a compact Riemann surface*, Phys. Lett. B **180**, 245–246.

Voros, A. (1987). *Special functions, spectral functions and the Selberg Zeta function*, Comm. Math. Phys. **110**, 439–465.

Watson, G. N. (1920). *An expansion related to Stirling's formula, derived by the method of steepest descents*, Quart. J. Pure Appl. Math. **48**, 1–18.

Watson, G. N. (1936). *Ramanujan's integrals and Gauss's sums*, Quart. J. Math. Oxford Ser. **7**, 175–183.

Watson, G. N. (1937a). *A note on Spence's logarithmic transcendant*, Quart. J. Math. Oxford Ser. **8**, 39–42.

Watson, G. N. (1937b). *A note on Lerch's functions*, Quart. J. Math. Oxford Ser. **8**, 43–47.

Wechsung, G. (1966). *Über Kummers funktionalgleichung für den Pentalogarithmus*, Jahresber. Deutsch. Math.–Verein. **68**, 140–143.

Wechsung, G. (1970). *Über die unmöglichkeit des vorkommens von funktionalgleichungen gewisser struktur für Ploylogarithmen*, Aequationes Math. **5**, 54–62.

Weisberger, W. I. (1987). *Conformal invariants for determinants of Laplacians on Riemann surfaces*, Comm. Math. Phys. **112**, 633–638.

Weisberger, W. I. (1987). *Normalization of the path integral measure and the coupling constants for bosonic strings*, Nuclear Phys. B **284**, 171–200.

Whitehouse, D. J. and Phillips, M. J. (1982). *Two-dimensional discrete properties of random surfaces*, Philos. Trans. Roy. Soc. London Ser. A **305**, 441–468.

Whittaker, E. T. and Watson, G. N. (1963). *A Course of Modern Analysis: An Introduction to the General Theory of Infinite Processes and of Analytic Functions; With an Account of the*

Principal Transcendental Functions, Fourth Ed., Cambridge University Press, Cambridge, London, and New York.

Wilf, H. S. (1997). *Advanced Problem* 10588, Amer. Math. Monthly **104**, 456.

Williams, G. T. (1953). *A new method of evaluating* $\zeta(2n)$, Amer. Math. Monthly **60**, 19–25.

Williams, K. P. (1912/1913). *The asymptotic form of the function* $\psi(x)$, Bull. Amer. Math. Soc. **19**, 472–479.

Williams, K. S. (1971). *On* $\Sigma_{n=1}^{\infty}(1/n^{2k})$, Math. Mag. **44**, 273–276.

Williams, K. S. and Zhang, N.-Y. (1993). *Special values of the Lerch Zeta function and the evaluation of certain integrals*, Proc. Amer. Math. Soc. **119**, 35–49.

Wilton, J. R. (1922/1923). *A proof of Burnside's formula for* $\log\Gamma(x+1)$ *and certain allied properties of Riemann's* ζ*-function*, Messenger Math. **52**, 90–93.

Wilton, J. R. (1927). *A note on the coefficients in the expansion of* $\zeta(s,x)$ *in power of* $s-1$, Quart. J. Pure Appl. Math. **50**, 329–332.

Wrench, Jr., J. W. and Shanks, D. (1966). *Questions concerning Khintchine's constant and the efficient computation of regular continued fractions*, Math. Comput. **20**, 444–448.

Wu, T.-C., Leu, S.-H, Tu, S.-T., and Srivastava, H. M. (1999). *A certain class of infinite sums associated with Digamma functions*, Appl. Math. Comput. **105**, 1–9.

Zerr, G. B. M. (1898). *Summation of series*, Amer. Math. Monthly **5**, 128–135.

Zhang, N.-Y. (1990). *Euler's constant and some sums related to the Riemann Zeta function*, Theory Practice Math. (Beijing) **4**, 62–70. (In Chinese).

Zhang, N.-Y. and Williams, K. S. (1993a). *Some infinite series involving the Riemann Zeta function*, in *Analysis, Geometry and Groups: A Riemann Legacy Volume, Parts* I *and* II (H. M. Srivastava and Th. M. Rassias, Editors), Part II, pp. 691–712, Hadronic Press, Palm Harbor, Florida.

Zhang, N.-Y. and Williams, K. S. (1993b). *Some series representations of* $\zeta(2n+1)$, Rocky Mountain J. Math. **23**, 1581–1592.

Zhang, N.-Y. and Williams, K. S. (1993c). *Application of the Hurwitz Zeta function to the evaluation of certain integrals*, Canad. Math. Bull. **36**, 373–384.

Zhang, N.-Y. and Williams, K. S. (1995). *Values of the Riemann Zeta function and integrals involving* $\log(2\sinh\frac{\theta}{2})$ *and* $\log(2\sin\frac{\theta}{2})$, Pacific J. Math. **168**, 271–289.

Zhao, J. (1999). *Analytic continuation of multiple Zeta functions*, Proc. Amer. Math. Soc. **128**, 1275–1283.

Zucker, I. J. (1985). *On the series* $\Sigma_{k=1}^{\infty}(k!)^2/(2k)!\,k^{-n}$ *and related sums*, J. Number Theory **20**, 92–102.

Zucker, I. J. and Robertson, M. M. (1976a). *Some properties of Dirichlet L-series*, J. Phys. A: Math. Gen. **9**, 1207–1214.

Zucker, I. J. and Robertson, M. M. (1976b). *A systematic approach to the evaluation of* $\Sigma_{(m,n\neq 0,0)}(am^2+bmn+cn^2)^{-s}$, J. Phys. A: Math. Gen. **9**, 1215–1225.

AUTHOR INDEX

SUBJECT INDEX

G

H

I

MIX
Papier aus verantwortungsvollen Quellen
Paper from responsible sources
FSC® C105338

If you have any concerns about our products,
you can contact us on
ProductSafety@springernature.com

In case Publisher is established outside the EU,
the EU authorized representative is:
Springer Nature Customer Service Center GmbH
Europaplatz 3, 69115 Heidelberg, Germany

Printed by Libri Plureos GmbH
in Hamburg, Germany